Practice of
High Performance
Liquid Chromatography

Applications, Equipment
and Quantitative Analysis

Editor: H. Engelhardt

With Contributions by
K. Aitzetmüller, J. Asshauer, P.R. Brown,
H. Colin, Th. Crispin, J.P. Crombeen, T. Daldrup,
H. Engelhardt, G. Guiochon, I. Halász,
A.P. Halfpenny, H. Hulpke, B.L. Karger, J.C. Kraak
H.-J. Kuss, M. Martin, P. Michalke, H.J. Möckel,
B.A. Persson, S. Szathmary, G. Tittel, M. Uihlein,
H. Ullner, H. Wagner, A. Wehrli, U. Werthmann

With 189 Figures and 79 Tables

Springer-Verlag
Berlin Heidelberg New York Tokyo

Editor:
Professor Dr. Heinz Engelhardt
Angewandte Physikalische Chemie
Universität des Saarlandes
D-6600 Saarbrücken/FRG

ISBN 3-540-12589-2 Springer-Verlag Berlin Heidelberg New York Tokyo
ISBN 0-387-12589-2 Springer-Verlag New York Heidelberg Berlin Tokyo

Library of Congress Cataloging-in-Publication Data
Main entry under title:
Practice of high performance liquid chromatography.
Includes index.
1. Liquid chromatography. I. Engelhardt, Heinz, 1936–   .
[DNLM: 1. Chromatography, High Pressure Liquid. QD 117.C5 P8954]
QD 117.C5 P68 1985   543'.0894   85-26083
ISBN 0-387-12589-2 (U.S.)

© by Springer-Verlag Berlin · Heidelberg 1986
Printed in Germany

Typesetting, Offsetprinting and Bookbinding: Universitätsdruckerei H. Stürtz AG Würzburg
2154/3020-543210

Preface

During its short 20 year history High Performance Liquid Chromatography (HPLC) has won itself a firm place amongst the instrumental methods of analysis. HPLC has caused a revolution in biological and pharmaceutical chemistry. Approximately two thirds of the publications on HPLC are concerned with problems from this area of life science. Biotechnology, where it is necessary to isolate substances from complicated mixtures, is likely to give further impetus to the dissemination of modern liquid chromatography in columns, particularly on the preparative scale.

This book presents, by means of examples, the application of HPLC to various fields, as well as fundamental discussions of chromatographic methods.

The quality of the analytical result is decisively dependent on the qualities of the equipment employed (by Colin, Guiochon, and Martin). Especially the demands are discussed that are placed on the components of the instrument including those for data acquisition and processing. The section on "quantitative analysis" (by Aßhauer, Ullner) covers besides the principles also the problems of ensuring the quality of the data in detail. The basic problems arising by enlarging the sample size to preparative dimensions and the requirements put on the aparatus are discussed in the section on "preparative applications" (by Wehrli).

Sample preparation by means of column switching (by Hulpke and Werthmann) or by off-line processing (by Uihlein) constitutes one of the most important operations in analysis in complex matrices, which can be employed not only to remove undesired and interfering accompanying substances but also to enrich trace components.

A further section on fundamentals deals with liquid-liquid partitioning (by Kraak and Crombeen). This method has fallen into unwarranted neglect, but, nevertheless, offers the possibility of carrying out the most various separations just by altering the coating of the stationary phase within the column. The separation can be varied and optimized additionally by means of secondary chemical equilibria (by Karger and Persson) which can be adjusted by altering the conditions.

In the application part there is a general discussion of the possibilities of HPLC in pharmaceutical (by Wagner and Tittel) and toxicological (by Daldrup, Michalke and Szathmary) analysis, for the determination of nucleic acids (by Brown and Halfpenny), and also of the specific application for the detection of psychotropic drugs and their metabolites in body fluids (by Kuss).

The problems occuring with components that do not possess a chromophore, yet which can still be successfully analysed, are illustrated by the example of the lipids (by Aitzetmüller). Various derivatization possibilities are available for the amino acids before or after separation. The advantages and disadvantages of the various methods are discussed as well as of the different systems for the separation of proteins (by Engelhardt). Apart from the determination of ions the separation of inorganic compounds has been neglected; it is, nevertheless, possible to analyse many inorganic compounds successfully by HPLC (by Möckel). The rational combination of the various chromatographic methods, including preparative applications, makes it possible to separate, fractionate and identify the non-volatile components from crude oil fractionation and coal liquification (by Halász and Crispin).

I would like to express my special thanks to the contributors to this book for their patience and willingness to prepare the review articles.

The help of my coworkers A. Groß, M. Mauß and G. Radünz in the preparation of the subject index is greatly appreciated.

Saarbrücken, November 1985   H. Engelhardt

List of Contributors

Dr. K. Aitzetmüller
Feldbahnstraße 64
2085 Quickborn/FRG

Dr. J. Asshauer
Hoechst AG
Analytisches Labor G 836
6230 Frankfurt/M.-Höchst/FRG

Prof. Dr. Ph. R. Brown
University of Rhode Island
Department of Chemistry
Kingston, RI 02881/USA

Dr. H. Colin
Ecole Polytechnique
Laboratoire de Chimie Analytique Physique
Route de Saclay
91120 Palaiseau/France

Dr. Th. Crispin
Bayer AG
PH-EP-AQ-L
5090 Leverkusen-Bayerwerk/FRG

Dr. J.P. Crombeen
Laboratory for Analytical Chemistry
University of Amsterdam
Nieuwe Achtergracht 166, 1014 WV
Amsterdam/The Netherlands

Priv.-Doz. Dr. T. Daldrup
Institut für Rechtsmedizin
Universität Düsseldorf
Moorenstr. 1
4000 Düsseldorf 1/FRG

Prof. Dr. H. Engelhardt
Universität des Saarlandes
Angewandte Physikalische Chemie
6600 Saarbrücken/FRG

Prof. G. Guiochon
Ecole Polytechnique
Laboratoire de Chimie Analytique Physique
Route de Saclay
91120 Palaiseau/France

Prof. Dr. I. Halász
Universität des Saarlandes
Angewandte Physikalische Chemie
6600 Saarbrücken/FRG

Dr. Anne P. Halfpenny
University of Rhode Island
Department of Chemistry
Kingston, RI 02881/USA

Dr. H. Hulpke
Bayer AG
Sparte Pflanzenschutz
Pflanzenschutzzentrum Monheim, Geb. 6100
5090 Leverkusen-Bayerwerk/FRG

Prof. Dr. B.L. Karger
Inst. of Chemical Analysis
Northeastern University
Boston, Massachusetts, 02115/USA

Dr. J.C. Kraak
Laboratory for Analytical Chemistry
University of Amsterdam
Nieuwe Achtergracht 166, 1014 WV
Amsterdam/The Netherlands

Dr. H.-J. Kuss
Universität München
Psychiatrische Klinik
Abteilung für Neurochemie
Nußbaumstr. 7
8000 München 2/FRG

Dr. M. Martin
Ecole Polytechnique
Laboratoire de Chimie Analytique Physique
Route de Saclay
91120 Palaiseau/France

Dr. P. Michalke
Institut für Rechtsmedizin
Universität Düsseldorf
Moorenstr. 1
4000 Düsseldorf 1/FRG

Dr. H.J. Möckel
Hahn-Meitner-Institut für Kernforschung-Berlin
Bereich Strahlenchemie
1000 Berlin 39/FRG

Dr. B.A. Persson
Department of Analytical Chemistry
AB Hässle
S-43183 Mölndal/Sweden

Dipl.-Pharm. S. Szathmary
Institut für Rechtsmedizin
Universität Düsseldorf
Moorenstr. 1
4000 Düsseldorf 1/FRG

Dr. G. Tittel
Laboratorium für Analysen und Trenntechnik
Dr. Tittel GmbH
Maria-Eich Str. 7
8032 Gräfelfing/München/FRG

Dr. M. Uihlein
Hoechst AG
Biochemie II D 528
6230 Frankfurt/M.-Höchst/FRG

Dr. H. Ullner
Hoechst AG
6230 Frankfurt/M.-Höchst/FRG

Prof. Dr. H. Wagner
Institut für Pharmazeutische Biologie
Universität München
Karlstr. 29
8000 München 2/FRG

Dr. A. Wehrli
Sandoz Ltd.
Pharmaceutical Division
Preclinical Research
4002 Basel/Switzerland

Dipl.-Ing. U. Werthmann
Bayer AG
Sparte Pflanzenschutz
Pflanzenschutzzentrum Monheim, Geb. 6100
5090 Leverkusen-Bayerwerk/FRG

Table of Contents

Liquid Chromatographic Equipment

Henri Colin
VAREX Corporation, 12221 Parklawn Drive, Rockville, MD, 20853/USA

Georges Guiochon
Chemistry Department, Georgetown University, Washington, DC, 20057/USA

Michel Martin
Laboratoire de Chimie Analytique Physique, École Polytechnique,
91120 Palaiseau/France

Introduction

The development of liquid chromatography has been quite rapid over the last
15 years and new generations of equipments have appeared every fifth year or so.
Although most problems are not really solved yet, they seem to be fairly well understood. Extremely efficient and fast intruments are available, but to achieve acceptably
good results the user must understand some of the basic principles of their design.
    The amount of literature published in the field of HPLC instrument alone is
enormous and could not be even properly scanned, so this chapter is not an exhaustive review of this field, but rather an outline limited to the basic principles and
the current solutions to the main instrumental problems with some glimpses to
some new developments which are going to materialize soon. For example, it was
considered as impossible to merely discuss the principles of most detectors described
in the literature or even to present an exhaustive list of them. The discussion of
the performances of current commercial equipments is also outside the scope of
this chapter and for a list of the instrument manufacturers the reader is referred
to the journals which publish yearly such listings[1, 2] and to the nearly exhaustive
review by McNair[3]. Also several books have been published in this field which
contain still useful material, evenso that kind of literature tends to age very rapidly
in the seventies[4–6]. Discussions on basic principles of equipment remain valid much
longer than descriptions of actual instruments or of their performances.
    In the recent past, it has been shown that LC analysis can be carried out very
rapidly, using short (3–5 cm long) columns packed with fine (3–5 μm) particles.
Very large amount of information can be generated in a short time. This has readily
necessitated the implementation of system automation. The problems associated
with the development of advanced instruments of that kind appear at two levels:
hardware (automatization) and software (data acquisition and processing). Thus
a new generation of instruments is appearing which will soon be of great importance
in laboratories, the intelligent instruments which can optimize automatically the
conditions of a separation after the analyst's requirements, analyze automatically
a large number of samples, each one in different conditions, or control a process
in a feed-back loop. The purpose of the last two sections is to give an overview
of the present state of automation in LC and venture some prospective considerations.

1 Pumps

1.1 Introduction

A pumping system is an essential part of a liquid chromatograph and one should
not forget that its operating characteristics will affect the result of an analysis,
even if it is hidden in a glittering case with shining microprocessor-controlled indi-
cators. It is therefore quite important to know and understand the principle on
which a given pumping unit is built.

Commercial instrumentation has been significantly modified in the last few years, partly in response to the recent evolution of liquid chromatography. Specifically, four trends impose new requirements on the pumping systems. Firstly, for the sake of high efficiency, columns are packed with small particles, 5 μm and now 3 μm in average size[1]. This implies that higher pressure drops are necessary to percolate the eluent at the required velocity. For example, the pressure drop per unit column length necessary to work at optimum velocity on columns packed with 3 μm particles, which is the pressure drop below which a column should not be operated, is typically 1 MPa/cm[2] and the corresponding hold-up time per unit length 10 s/cm. In these conditions, one might expect to get *ca.* 1 500 theoretical plates/cm. For 5 μm particles, these values become 0.2–0.25 MPa/cm, 17 s/cm and 1 000 theoretical plates/cm. Secondly, LC analyses performed on such columns tend to be faster than before[3], columns being used again at flow velocities markedly larger than the optimum, which reinforces the requirement for high pressure capability of pumps. Typical hydraulic conductivity ($\eta/k = Pt_0/L^2$) with 3 μm packed columns are at least, for most water-methanol eluent mixtures, about 10 MPa·s/cm², which means that the elution of an inert compound in 10 seconds on a 10 cm long column will require a 100 MPa pressure drop. Thirdly, the need for reduction of consumption of sometimes expensive chromatographic eluents as well as for a decrease in dilution, hence in detection limit of samples available only in very restricted amounts (clinical applications) or for direct coupling of the LC column to a mass spectrometer has led to the development of narrow-bore packed columns with internal diameter of 1 mm instead of 4 or 4.6 mm for conventional columns[4]. The flow-rate in such small bore columns is thus decreased by a factor of about 20 for similar performances. When the typical flow-rate in conventional columns is 1 ml/min, which corresponds to a hold-up time of 1 min on a 10 cm long column, it becomes 50 μl/min at a similar velocity in a 1 mm i.d. column. Fourthly, progress in the characterization and predictability of retention in reversed-phase chromatography has led to the common use of ternary and even quaternary solvent systems[5]. It is therefore desirable to be able to change quickly the eluent from one analysis to the next one and to adequately select any composition of, at least, three component mobile phases.

Accordingly, the desirable general characteristics of modern liquid chromatography pumping systems are indicated below and these pumps are classified according to their operation principle and degree of sophistication to meet these requirements.

1.2 Desirable Characteristics of Pumping Units

1.2.1 Convenience of Use

A major quality of a pump is its ease of operation which implies, for example, that the desired flow-rate should be easily and reproducibly selectable, and that the pump should be easy to prime, or better, self-priming. It should not be prone to stop running because of bubble formation in the pump head, or, alternatively, should be constructed in such a way as to avoid bubble formation. Such formation may come from either unsufficiently tightened connections in the pump inlet line (air suction) or release of gases (mainly oxygene) dissolved in the solvent or, even, cavitation of volatile solvents if the pressure in the pump chamber falls below the vapor pressure of the eluent while the pump head is being refilled. It may also be found convenient that a pumping system incorporates adjustable high and low pressure limits in order to avoid high pressure build-up with subsequent pump or column damaging and to detect possible eluent leakage with non pressure-regulated pumps, respectively. In some instances, solvent degassing capabilities will be welcome not only to reduce potential bubble-induced pump depriming, but also to avoid bubble formation at the column outlet leading to noisy detector baseline since the solubility of gases in liquids decreases linearly with decreasing pressure, according to Henry's law, or to reduce the interference of dissolved gases with the detection sensitivity (for example, fluorescence quenching by dissolved oxygen). Furthermore, a pump should be safe, solvent spillage onto electrical parts should not occur. Besides, all parts in contact with solvents should be chemically inert, which may become a severe requirement when some buffer systems are used.

1.2.2 Rapid Solvent Change, Small Hold-up Volume

As mentioned above, the pump should provide facilities for rapid solvent change, which implies that it should have a small hold-up volume and that this volume should be easily flushed and rinsed. Non-flushed Bourdon-type tubes, sometimes used for damping purposes and/or pressure measurement, must be avoided as replacement of one solvent by another in such tubes can be done only by diffusion, which may take considerable time and lead to unstable and undesirable eluent composition. A small hold-up volume is also required for low pressure gradient and recycle operation in order to minimize composition distortion and zone remixing, respectively. As a pump should also preferably work from an unlimited solvent reservoir, in order to limit attendance, automatic pump refilling will be appreciated.

1.2.3 Maximum Operating Pressure

The desirable maximum operating pressure of the pumping system depends on the analytical conditions. For routine separations of a small number of compounds requiring moderate efficiencies (about 10000 theoretical plates) in a relatively short time (5 or 10 minutes), pressure capabilities of 20 MPa are largely sufficient[6]. If, however, faster and/or highly efficient separations are desired, then, as seen above, pressure drops larger than 100 MPa may be required. One should note that some solvents must be used with caution at such high pressures to avoid piston breaking, as the increase of the melting temperature of most organic solvents with increasing pressure is about 0.15–0.4° C/MPa[7], while the melting point of ice I decreases with increasing pressure by about 0.1° C/MPa[8]. Besides, addition of small amounts of other components to a solvent decreases its melting temperature.

1.2.4 Flow-Rate Range

The range of selectable flow-rates that a pump should deliver will depend on the operating conditions, especially column diameter and speed of separation. The smallest flow rate will be selected when working at the velocity corresponding to the optimum of the plate height curve for the biggest particles to be used and the least diffusive solutes. Taking 0.33×10^{-5} cm^2/s for the diffusion coefficient of these solutes and 10 µm for the largest particle size currently used, the optimal velocity,

Table 1. Desired selectable flow-rate range of pumping systems as a function of column type

Column type	Small-bore	Conventional	Semi-preparative
Column internal diameter (mm)	1	4.6	10
Minimum flow-rate (ml/min)	0.003	0.07	0.3
Maximum flow-rate (ml/min)	0.6	14	60

Column porosity: 0.65, typical of columns used for reversed phase chromatography

below which one should not work as both resolution and analysis time deteriorate, becomes *ca.* 0.01 cm/s. The largest flow-rate will be used for fast separations, with eluent velocities reaching as much as 2 cm/s. The flow-rates corresponding to these extreme velocities are indicated in Table 1 for three column internal diameters, 1 mm for small-bore columns, 4.6 mm for conventional columns and 10 mm for semi-preparative columns. Preparative, i.e. production-aimed, columns are not considered here.

The range of available flow-rates should therefore be 200-fold for a given column type and larger if the pump has to be used with columns of different types. The values given in Table 1 are only indicative and can be extended either to lower flow-rates if larger particles are used, as might be the case in the search of very high efficiencies with long columns due to the limited available pressure[4, 9], or to larger flow-rates if elution times smaller than 0.5 s/cm are desired, provided that the maximal available pressure is not exceeded. Besides, when the internal diameter of the small bore column becomes smaller than 1 mm[10, 11], the desired flow-rates are decreased in proportion to the square of the column diameter. One of the road-block in the development of narrow bore columns and capillary columns has been the lack of suitable pumps, which obviously would have to be specially designed.

Furthermore, to give versatility to a pumping system, the number of selectable flow-rates, in cases where they are quantified, should be at least 50 or 100. Additionnally, in order to have a sufficient number of digitally selectable flow-rates in a restricted relative flow range, these flows should be distributed according to a geometrical progression rather than to an arithmetical one.

1.2.5 Flow-Rate Stability

It is highly desirable that the pumping system delivers a constant, stable flow. Three levels of flow instabilities can be considered and their effects on chromatographic performances have been investigated[12]. Flow noise, characterized by an average time period (or volume scale) which is small compared to the peak width, does in most cases negligibly or slightly affect peak retention times and areas, but increases the baseline noise of flow-sensitive detectors, like differential refractometers and, to a lesser extent, UV photometers. Flow wander, characterized by a time period about equal to the average peak base width, has a small effect on peak retention times but a considerable one on peak areas. Finally, flow drift, characterized by a time period that is of the order of peak retention times or larger affects both peak retention times and areas in the same direction[12].

Flow instabilities are created by the pump action and may be due to several causes. Some relate to incorrect operation of the pump and should be eliminated with care. For example, check valve malfunction due to build-up of bubbles or the presence of small solid particles in the solvent can be avoided by appropriate degassing of the eluent, as discussed above, and by filtering of the solvent. This can be done by using filter frits in the inlet check valve for particles remaining the eluent and in the outlet valve for particles arising from wear of the piston seals, respectively. Other instabilities are associated with the periodic refill of the pump head. During this refill and the time needed to sufficiently compress the solvent for opening the outlet valve, no flow is delivered from this head. This results in flow and pressure pulsations if no means are taken to compensate for the flow variations from this head. The resulting effect of these pulsations will depend *on their amplitude* (which is related to the number of pump heads, to the piston displacement profile, to the possible use of flow feedback loops, to the ratio of the volume of the pump head to the volume displaced by the piston, to the solvent compressibility and seal elasticity), *on the efficiency of their damping* (which is related to the ratio of the system volume to the stroke volume, to the column pressure drop, the solvent compressibility and the possible use of damping devices) *and on their time* (or volume) *period* compared to the peak retention time (or volume) and peak width.

If several heads are used, the pulsation frequency is a small multiple of the individual piston frequency, usually 2 for duplex and 6 for triplex pumps, as seen below. However, the volumetric period of pulsations is approximately equal to the solvent volume displaced by each piston. Therefore, the ratios of this piston volume to the average peak retention volume and width are of prime importance in determining the type of flow instability created by the pressure pulsations. Typical values of retention volumes and peak widths are indicated in Table 2 as a function of column type and capacity factor.

It appears from this table that for a typical piston displaced volume of 100 μl, the pressure pulsations, if not satisfactorily reduced will mostly contribute to flow noise with semi-preparative columns, flow wander with conventional columns and flow drift with small-bore columns. It is therefore highly important that the pumping system reduces significantly the pressure pulsations associated with the piston refill

Table 2. Typical retention volumes and peak widths as a function of column type and capacity factor, k' [a]

Column type		Small-bore	Conventional	Semi-preparative
Column internal diameter		1 mm	4.6 mm	10 mm
$k'=0$	Retention volume	130 μl	2.7 ml	13 ml
	Peak width	5 μl	0.11 ml	0.51 ml
$k'=2$	Retention volume	380 μl	8.1 ml	38 ml
	Peak width	15 μl	0.32 ml	1.5 ml
$k'=6$	Retention volume	890 μl	18.9 ml	89 ml
	Peak width	36 μl	0.76 ml	3.6 ml

[a]   Theoretical plate number: 10,000, column length: 25 cm. Column porosity 0.65, typical of columns used in reversed phase chromatography

stroke when working with conventional columns. This is still more important with small-bore columns.

1.2.6 Flow Precision and Accuracy

Flow precision is desired in order to obtain data on peak retention times and areas which are reproducible from one analysis to the next one. It depends on the extent of pressure pulsations, as discussed above, and on the proper operation of the pumping system. Besides elimination of problems arising from bubbles and small particles, it is affected by the correct functioning of mechanical and electronic parts on which the chromatographer has little or no action.

Flow accuracy is desired for physico-chemical measurements or studies of retention mechanisms and to facilitate exchange of reproducible retention data. Flow accuracy is limited by the different types of flow instabilities. However, when the flow is stable, its value measured at the column outlet may not be equal to the selected one, even if the pump is working properly.

Moreover, a variation in the pressure due, for example, to a column or mobile phase change, will induce a change in the measured flow-rate for a given selected flow. This is due, essentially, to the compressibility of the solvent and, to a lesser extent, to seal elasticity. Indeed, as mentioned above, at the beginning of its forward stroke, the piston (or the membrane in diaphragm reciprocating pumps, see below) must compress the liquid in the pump head up to the column inlet pressure before to open the pump outlet check valve and to deliver the solvent to the column. If all the liquid contained in the pump head were displaced by the piston, then the flow-rate measured at the column outlet would be equal to the piston flow. But this does not occur as, for mechanical reasons, when the forward motion of the piston ends, some amount of solvent remains in the cylinder. So part of the piston motion is used to compress a mass of liquid which is not displaced and, consequently, the flow-rate measured, during one piston stroke, at column outlet is less than the volume displaced by the piston. It can easily be established that the relative flow-rate loss, $\Delta F/F$, is given by:

$$\Delta F/F = ((V_H/V_P) - 1) \chi \Delta P \tag{1}$$

where V_H is the volume of the head, V_P the volume swept by the piston, χ the solvent compressibility, and ΔP the column pressure drop, assuming the column outlet is at atmospheric pressure. The ratio V_H/V_P is, therefore, very critical in determining the relative error in flow-rate. Unfortunately, it is rarely indicated in pump specifications. However, a value as high as 5 for V_H/V_P is not untypical. In such a case, with a typical solvent compressibility of 10^{-3} MPa^{-1} and a pressure drop of 25 MPa, the loss in flow-rate reaches 10%. For this reason, some pumping systems incorporate a solvent compressibility correction.

Two kinds of flow rate fluctuations should be distinguished because of their very different effects: long term fluctuations affect essentially the reproducibility of retention data, while short-term fluctuations control the reproducibility of peak area measurements, essential for quantitative analysis[19]. As discussed in Section 3.1.1, the peak area given by an LC detector is almost always proportional to the reverse of the eluent flow-rate and consequently the reproducibility of relative peak areas, i.e. relative concentrations of components measured in quantitative analysis is of the order of the relative stability of the flow rate over a time equal to the peak width (4σ).

Halasz and Vogtel[19] have shown that an easy and rapid check of the short term flow rate fluctuations of a chromatographic equipment can be made by carrying out a few quantitative analysis of a simple mixture (4–5 components eluted over a k'-range of 1 to 3–6, the relative concentration being such that all peaks have similar heights). They have found that for modern good quality pumps this stability can be as good as *ca.* 0.5%. Such a result would be very hard to improve on and is not easy to equal: careful maintenance of the pumps is necessary. The test should be repeated periodically as if there is a problem with a pump the fluctuations usually increase suddenly by an order of magnitude or more.

1.2.7 Properties of the Solvents

It is important to keep in mind some of the properties of the solvents which are going to be pumped.

Table 3 gives the compressibility, the viscosity and the pressure dependence of the viscosity for a number of solvents used in LC[15].

The compressibility $\left(\chi = -\dfrac{1}{v}\dfrac{dv}{dp} \right)$ decreases with increasing pressure, so the values given in the table cannot be used for linear extrapolation for over a few hundreds atmosphere. The compressibility of liquids is of the order of 1×10^{-4} bar^{-1} which means that the volume of a liquid decreases by *ca.* 1% when the pressure increases by 100 atm. The compressibility increases markedly with increasing temperature. It usually becomes large above or around the boiling point under atmospheric pressure (compare nC_5, nC_6 and nC_7 at 25° C) and of course becomes infinite at the critical point.

Table 3

	Temperature (°C)	Compressibility $\chi \times 10^4$ (bar^{-1})	Temperature (°C)	Viscosity η_0 (cP)	$\dfrac{1}{\eta_0}\dfrac{d\eta}{dP} \times 10^3$ (bar^{-1})
n-pentane	25	3.14	30	0.220	1.06
n-hexane	25	1.61	30	0.296	1.15
n-heptane	25	1.42	30	0.355	1.09
n-octane	25	1.20	30	0.483	1.12
Benzene	25	0.96	30	0.566	1.22
Chloroform	25	0.97	30	0.519	0.625
Carbon tetrachl.	25	1.07	30	0.845	1.25
Diethylether	20	1.87	30	0.212	1.11
Acetone	25	1.24	30	0.285	0.684
Ethylacetate			30	0.390	0.810
Methanol	20	1.23	30	0.520	0.470
Ethanol	20	1.11	30	1.003	0.585
Water	25	0.46	30	0.80	0.053
Acetonitrile			25	0.345	

The viscosity of solvents increases steadily with increasing pressure, by about 1% when the pressure increases by 10 atm, except for water. The viscosity decreases rapidly with increasing temperature, by about 1–2% per °C.

There is no general rule to derive the compressibility, the viscosity or its pressure dependence for a mixture of solvents.

1.3 Classification of Pumping Systems

Commercially available pumping systems have been described in details some years ago[13]. At this time, the pumps were classified in four groups, namely, pneumatic pumps, syringe-type pumps, reciprocating pumps and hydraulic amplifier pumps. In addition, different feedback systems used for motor speed, flow or pressure corrections, were described. The evolution of pumping systems in the past five years has been important and today, more than 90% of these systems are of the mechanical reciprocating type. Pneumatic pumps are now mostly used for preparative purposes (e.g., Jobin-Yvon Modulprep system) or column packing (Haskel pumps). Hydraulic amplifier pumps are not anymore manufactured. In the following, actual pumping systems are classified in four categories: non reciprocating pumps, one-headed, two-headed (duplex) and three-headed (triplex) mechanical reciprocating pumps. In Table 4 are indicated, for some commercially available pumping systems, the type of pump, the maximum working pressure, the selectable flow-rate range, the minimum flow increment when the flow-rate is digitally selectable (Note: in all such systems but one, the selectable flow-rates are distributed according to an arithmetic progression), and the volume displaced by the piston at full stroke, when these informations are available. It should be noted that some of these systems are not available as an individual module.

1.3.1 Non-Reciprocating Pumps

Pneumatic pumps use a source of a pressurized gas to drive the liquid mobile phase to the column either directly in a holding coil or a stainless steel cylinder (Gow Mac 80–500 system, with 300 ml solvent capacity), or by means of a flexible or moving separator to avoid substantial dissolution of the gas in the upper part of the liquid. These systems are cheap. However, they are substantially solvent-consuming, solvent change is not rapid and the maximum available pressure is usually limited to less than 10 MPa for safety reasons. Other advantages and disadvantages, as well as somewhat detailed descriptions of these systems and derived pneumatic amplifier pumps, still sometimes used, have been previously discussed[13]. Pneumatic amplifiers offer the simplest and cheapest way to achieve extremely high pressures (Haskel pump, above 500 MPa). Pulsations are large, however, and may have destructive mechanical effects through metal fatigue.

The most widely used non-reciprocating pumps are of the syringe-type. In these pumps, the liquid enclosed in a cylinder is displaced towards the column by a piston moving at a constant speed, giving a constant flow-rate when steady-state conditions are reached. The capacity of the cylinder is generally quite larger than the volume required to perform an analysis. Besides possible piston speed variations due to the use of a stepping motor, the major problem of these pumps in isocratic

Table 4. Characteristics of LC pumping systems

Manufacturer and/or distributor	Model	Pumping system								Volume displaced at full stroke (μl)	Maximum pressure (MPa)	Flow-rate range (ml/min)[h]	Flow-rate increment (ml/min)[i]
		Non-reciprocating[a]	Reciprocating										
			Pump type[b]	Flow regulation[c]	1 head[d]	2 parallel heads[d]	2 heads in series	3 heads[d]	Electronic correction[e]				
ACS (Applied Chromatography Systems)	351		P	L	XS				XC	7	42 10[f]	0.01–9.99 0–40[f]	0.01
	400		P	L	XS					7	42	0.1–10 0–50[f] 0.05–2[g]	cont.
Beckman	110 B		P	F	XR				XC	140	42 21	0.1–9.9 0.01–9.9[p] 0.28–27.9[f]	0.1 0.01
	114 M		P	F	XR				XC, P	90	42	0.001–10 0–29[f]	0.001
Bio-Rad	1330		P	F		XT				100	42	0.1–9.9	0.1
Biotronik	IC-5000		P	F		XT					50	0.01–9.99	0.01
Bischoff	2200		P	F			X		XC	45[k]	55 55	0.1–5 0.02–1[f] 0.5–20[f]	0.1 0.01 0.1
Brownlee Labs	MPLC Micropump	XS								(10 ml)	50	0.001–2	0.001
Bruker-Merck-IBM	LC 21 or 31		P	F				XS	XC, P	90	49	0.1–20.0	0.1

Manufacturer	Model										
Büchi	B-681	P	L and/ or F	X				260	4	3–29	cont.
Cecil	CE 2010	M	L	XS				70	28	0–10 / 0–5^g	cont.
Dionex	Series 2000i	P	F	XT		XC, P		100	13	0.1–9.9	0.1
Eldex	A-60-S	P	L	X				50	17.5	0.1–3	cont.
	A-30-S	P	L	X				50	35	0.05–1.5	cont.
	B-94	P	L	X				80	35	0.16–7.5	cont.
	B-100-S	P	L	X				80	35	0.27–8.0	cont.
	AA-72-S	P	L		X			50	35	0.12–7.2	cont.
	AA-94	P	L		X			50	42	0.16–9.4	cont.
	AA-100-S	P	L		X			50	35	0.17–10.0	cont.
	BBB-4	P	L				X	320	42	0.1–80	cont.
	E	P	L	X				42	7	0.2–5.0	cont.
EM Science	MACS 100	P	F	XR		XC		40	42	0.005–5	0.005
ERC	BCMA-JN1	XP						(1.7 l)	15	–	cont.
Gilson	Model 302	P	F	XR		XC		40	42	0.005–5 / 0.0005–5^p	0.005 / 0.0005
								80^f	21	0.01–10^f / 0.001–10^p	0.01 / 0.001
	Model 303	P	F	XR		XC		40	60	0.005–5 / 0.0005–5^p	0.005 / 0.0005
								80^f	60	0.01–10^{f,j} / 0.001–10^p	0.01 / 0.001
Gira	MP 831.G	P	F			XC, P	X	500^k	45	0.01–5	0.01
Gow-Mac	80–500 system	XP						(300 ml)	7	–	cont.
Gynkotek	Model 300C	P	F				X	80^k	40	0.001–5	0.01

Table 4. (continued)

Manufacturer and/or distributor	Model	Non-reciprocating [a]	Pump type [b]	Flow regulation [c]	1 head [d]	2 parallel heads [d]	2 heads in series	3 heads [d]	Electronic correction [e]	Volume displaced at full stroke (μl)	Maximum pressure (MPa)	Flow-rate range (ml/min) [h]	Flow-rate increment (ml/min) [i]
Hewlett-Packard	1090 (DR5)		M	L	XS					8.3	44	0.001–5	0.001
Hitachi-Merck	Model 655-A12		P	F			X		XC, P	100^k	40 / 10^f	0.01–9.9 / 0.3–30^f	0.001 / 0.1
HPLC Technology	RR/066		P	F	XS					7	42	0.1–9.9	cont.
	MP 085		P	F	XR					50	42	0.5–5.0	cont.
ISCO	LC 5000	XS								(500 ml)	21	0.025–6.7	
	Model 2300		P	F	XR				XC		42	0.1–9.9	0.1
	μLC-500 Micropump	XS								(50 ml)	70	0.0002–0.6	cont.
Jasco	BIP-1		P	F		XT			XC	80	50	0.01–9.9	0.01 (from 0.01–0.99) / 0.1 (from 0.1–9.9)
	Familic-100N	XS								(0.5 ml)	10	0.001–0.029	0.001
	Familic-300S		P	F				XT	XC	80	50	0.01–9.9	0.01 (from 0.01–0.99) / 0.1 (from 0.1–9.9)
	Twincle		P	F		X					42	0.1–9.9	0.1
Knauer	Type 64		P	F			X		XC	20^k / 8^k / 96^k	40 / 40 / 10	0.1–9.9 / 0.02–2^f / 0.5–50^f	0.1 / 0.02 / 0.5

Manufacturer	Model										
Kontron	Model 414		P	F	XR		XC	142	40	0.1–9.9	0.1
										0.01–3.3^f	
	Model 420		P	F		X			60	0.01–2	
										0.05–10^f	
										0.1–20^f	
Kratos	Spectroflow 400		P	F		X		80^k	40	0.01–5	0.01
Labotron	LDP 20	X2S						(13.5 ml)	5	0.005–0.27	cont.
	LPD 21	X2S						(13.5 ml)	5	0.05–2.7	cont.
Latek	P 400		P	F		XT			40	0.1–10	cont.
LDC/Milton Roy	ConstaMetric III		P	F		XT		100	42	0.1–9.99	cont.
										0.03–3.33^f	cont.
	ConstaMetric 3000		P	F		XT			42	0.10–9.99	0.01
	MiniMetric		P	F	XR			100	42	0.05–5.0	cont.
	MicroMetric	XS	P	F				(7 ml)	70	0.001–0.450	0.001
	MiniPump VS		P	F	XR			50	42	0.5–5.0	cont.
Lewa	FLMk		M	L	XS			71	70	0–2.6	cont.
										0–5.1^g	cont.
										0–10.6^g	cont.
								196	31	0–7.1^f	cont.
										0–14.1^g	cont.
										0–29.5^g	cont.
LKB	2150		P	F		XT	XC, P	36.5	35	0.01–5	0.01
Major Scientific Instruments	Model 301		P	F		X			10	2–99	0.1
	Model 310		P	F		X			70	0.2–9.99	0.01
Milton-Roy	Minipump		P	L	XS			100	25	0–3.3	0.1
									25	0–6.6^g	cont.
										0–10^g	cont.
Orlita	DMP AE 10.4		M	L	XS			125	50	0–12.5	cont.

Table 4. (continued)

Manufacturer and/or distributor	Model	Non-reciprocating[a]	Pump type[b]	Flow regulation[c]	1 head[d]	2 parallel heads[d]	2 heads in series	3 heads[d]	Electronic correction[e]	Volume displaced at full stroke (µl)	Maximum pressure (MPa)	Flow-rate range (ml/min)[h]	Flow-rate increment (ml/min)[i]
Perkin-Elmer	Series 4		P	F	XR				XC	100	42	0.01–9.9	0.01 (from 0.01–0.99) 0.1 (from 0.1–9.9)
	Series 10		P	F	XR				XC	100	42	0.1–9.9	0.1
	Series 100		P	F	XS					100	42	0.5–5.0	cont.
	Series 400		P	F	XR				XC	100	42	0.01–10.0	0.01
Pharmacia	P-500	X2S								(10 ml)	4	0.017–8.32	0.017
Pye-Unicam/ Philips	PU 4015		P	F		XT			XC	80	42	0.01–10	0.01
	PU 4003		P	F		XT			XC, P	80	42	0.02–10	0.01
Shimadzu	LC-4A		P	F		XT			XC	100	50[m]	0.01–20	0.01
	LC-5A		P	F	XR				XC	100	50	0.001–9.9	0.001 (from 0.001–0.099) 0.1 (from 0.1–9.9)
	LC-6A		P	F	XR				XC	100	50[n]	0.1–9.9 0.01–9.9[p]	0.1 0.01
Spectra-Physics	SP 8700		P	F			X		XC, P	225[k]	42	0.1–10	0.01
Scientific Systems (SSI)	Model 200		P	F	XR				XC		42 84[f] 14[f]	0.05–9.99 0.05–5.0[f] 0.15–15[f]	
	Model 300		P	F	XR				XC		42 14[f]	0.5–5.0 1.5–15	

Tokyo Rikakikai	Eyela PLC-10		P	F	X			32	20	0.1–2.0 / 0.1
Touzart et Matignon	EC 93		P	F	XR	XC	300	40	0.1–30	0.1
Tracor	951		P	F	XT	XC, P	100	70 / 17.5	0.1–9.99 / 0.04–40[f]	0.01 / 0.04
	955		P	F	XT	XC	100	42	0.01–9.99	0.01
Varian	Series 5000		P	F	XR	XC	90	42	0.01–15.0	0.01
	2010		P	F	X	XC		52	0.01–9.9	0.01
Waters/Millipore	QA-I	XS								–
	501		P	F	X		(45 ml) 100	14 42	0.5, 1, 2 or 4 / 0.1–9.9 / 0.025–10[p]	0.1 / 0.01[n]
	510		P	F	XT	XC	100 50	42	0.1–9.9 / 0.05–4.95[f] / 0.225–22.5[f]	0.1 / 0.05 / 0.225
	590		P	F	XT	XC	100 50 225	49	0.001–10 / 0.0005–5[f] / 0.00225 to 22.5[f]	0.001 / 0.0005 / 0.00225

[a] Non reciprocating pumps: P = Pneumatic, S = Syringe, 2S = Double Syringe
[b] Pump type: P = piston type reciprocating pump, M = membrane type reciprocating pump
[c] Flow regulation of reciprocating pump: F = by change of the piston motion frequency, L = by change of the piston displacement length
[d] R = Rapid refill reciprocating pump, S = basically sinusoidal piston flow, T = basically trapezoidal piston flow
[e] Electronic correction: C = of compressibility induced flow loss, P = of flow pulsations
[f] By change of the pump head
[g] By change of the drive motor or of the motor speed
[h] With some pumps, the larger flow rates cannot be used at the maximum pressure
[i] cont. = continously adjustable flow rate
[j] Other pump heads are available for semi-preparative and preparative work
[k] Volume displaced at full stroke by the working piston
[m] Maximum pressure: 50 MPa up to 10 ml/min, 20 MPa from 10 to 20 ml/min
[n] Maximum pressure: 50 MPa up to 5 ml/min, 25 MPa from 5 to 10 ml/min
[p] With an external microprocessorized controller

conditions comes from the sometimes long time required to reach the steady-state pressure in the cylinder because of the eluent compressibility[14]. Indeed, the time required to achieve 95% of the steady-state pressure depends on the volume of the cylinder V_0, the piston flow-rate F_0, the liquid compressibility χ, and the steady-state pressure ΔP_0 and may be approximated by:

$$t_{0.95} = 3 \, \frac{V_0}{F_0} \, \chi \, \Delta P_0 \tag{2}$$

If ΔP_0 reaches 20 MPa and χ equals typically 10^{-3} MPa^{-1}, then this transitory time becomes 6% of the time needed for the piston to completely flush the cylinder, if the piston moves during the pressurization of the reservoir at the same speed as during the analysis. In such conditions, the transitory time during which no injection should be made, can be as long as 15 or 30 minutes. Miniaturization does not change anything to these conclusions, if V_0 and F_0 are reduced in the same ratio (cf. Eq. (2)).

Nine syringe-type pumps are listed in Table 4. The Isco (Instrumentation Speci-alities Co.) LC 5000 pump is a conventional pump adapted to low flow-rates with a large capacity ($V_0 = 500$ ml). The integrated pumping system of the routine Waters QA-I Quality Analyzer has a 45 ml reservoir which is supposed to be sufficiently large for performing an analysis with a capacity factor up to 15 (see Table 2). The cylinder is refilled after each analysis and whatever the selected flow-rate, sample injection is made after a 3–4 minutes pressurization period. The Jasco (Japan Spec-troscopic Co.) Model Familic-100N is a syringe pump of a 500 µl capacity designed for use with narrow bore columns at flow-rates less than 29 µl/min. The Pharmacia P-500 pump and Labotron LDP pumps are double-syringe pumps in which one piston is delivering the solvent to the column while the other one is refilled from the reservoir. The materials in contact with the liquid (borosilicate glass, titanium and fluoroplastics) allow the use of most solvents, including chlorinated hydrocar-bons, detergents and oxidizing agents. However, the operating pressure is limited to 4–5 MPa. Therefore, the duration of the flow and pressure spike occurring when solvent delivery is switched from one syringe to the other does not usually exceed 1 or 2% of the time elapsed between two consecutive switchings. The Brownlee Labs MPLC Micropump and some other syringe pumps listed in Table 4 are especially designed for high-pressure work with small bore columns.

The main advantage of syringe pumps lies in a potentially quite stable flow delivery since no check valve has to be actuated during an analysis.

1.3.2 Single Head Mechanical Reciprocating Pumps

The general description of reciprocating pumps has been given with details some years ago[13]. Basically, the movement of a plunger piston causes suction of the eluent during the backward stroke and delivery of the pressurized solvent during the forward stroke. Generally, the pump head is equipped with ball check-valves to insure the positive direction of the flow.

In the past, the main distinction between reciprocating pumps and syringe pumps was based on the higher piston frequency and the smaller volume of solvent delivered during each piston cycle by the former. Today, this distinction is rather vague since, as can be seen from Table 4, the volume of solvent delivered during one cycle by some syringe pumps designed for low flow-rates may be almost as small as the one delivered by some reciprocating pumps. Nevertheless, the distinction can still be based on the fact mechanical reciprocating pumps specifically designed for LC usually have devices for minimizing the flow and pressure pulsations associated with the movement of the piston as well as the flow variations during one piston forward stroke.

In the past few years, a large number of single head mechanical reciprocating pumps became commercially available, as these pumps are probably the most convenient to use, they are relatively easy to flush and rinse, they are cheaper than multiple head pumps at similar levels of sophistication and they are more reliable as most maintenance problems occur in connection with check-valves.

1.3.2.1 Single Head with Basically Sinusoidal Piston Displacement

The simplest single-head pump is probably the Milton-Roy Minipump, used for industrial applications for more than 25 years and still used for routine and simple LC applications. The piston has a sinusoidal movement, consequently the flow profile at low pressures is also sinusoidal during the forward stroke and there is no solvent delivery during half of the piston cycle while the pump head is refilled with solvent. Therefore there are considerable flow and pressure pulsations which must be dampened for satisfactory operation, by appropriate devices, preferably of the flow-through type. The Lewa FLMk pump is a similar industrial pump with sinusoidal piston movement which differs, however, from the previous one, by the fact that a flexible metallic membrane separates the piston chamber filled with an hydraulic oil from the pump head containing the eluent. This configuration reduces considerably the wearing of the seal and leakage at this seal and facilitates solvent change-over. However, the V_H/V_P ratio for membrane reciprocating pumps is usually quite higher than for piston pumps, which results in larger compressibility-induced flow-rate losses. In both pumps, the flow-rate is selected by adjusting the length of the piston course with a constant piston displacement frequency. In the Orlita DMP AE 10.4 membrane pump, as previously described[13] a variably positioned counter-piston limits the effective course, at constant length and frequency, of the working piston, thus the flow profile is a part of a half sinusoid.

The ACS (Applied Chromatography Systems) pump is unique in that the constant piston frequency is very high (23 strokes per second). Therefore, the pulsations have a period (0.04 sec) which is quite smaller than the response time of most chromatographic detectors and are not detected. The record of the piston force, which is related to the pressure in the pump head, by means of a strain gauge transducer, in connection with the knowledge of the piston position, allows the measure of the flow actually delivered at the pump head pressure by the piston, and its eventual correction through a servo-motor modifying the piston course length.

In the recently developed DR5 pump of the Hewlett Packard 1090 liquid chromatograph, three dual-syringe metering pumps continuously deliver a solvent component at low pressure to a 9 µl mixing chamber. A single-head diaphragm high pressure reciprocating pump delivers the ternary solvent mixture to a pulse damper. The piston of this pump moves at high frequency (10 Hz) at constant displacement length. However, the piston course length which effectively displaces the flexible membrane is controlled by the combined flow-rates of the three low-pressure syringe pumps. Since the membrane motion displaces nearly all the solvent contained in the pump head, that is V_H/V_P in equation 1 is very close to 1, there is negligible flow-rate loss associated with solvent compressibility.

1.3.2.2 Single-Head Pumps with Rapid Refill of the Pump Head

The more recent generation of single head reciprocating pumps uses a specially contoured cam to insure a rapid refill of the pump head as well as a nearly constant flow during the delivery stroke. In addition, they allow a variation of the cam rotation speed, during the rotation cycle, to further reduce the refill time and, in some instances, the time required to compress the liquid in the pump head up to the column inlet pressure in order to open the pump outlet check valve. All these pumps are of the piston type with constant piston displacement length. Some pumps are equipped with a flow-through pulse damper. In most of these pumps, the pressure, measured by a pressure transducer between the pump outlet and the column inlet or estimated from the measurement of the motor load or torque, at the end of the forward stroke is used as a reference for the determination of the time during which the cam moves at high rotation speed. In addition, it serves to correct the delivery cam rotation speed for possible flow losses due to compressibility. The solvent compressibility is either estimated for a typical solvent (Touzart et Matignon EC 93 pump) or may be adjusted by a screw potentiometer (Beckman 110B, Isco Model 2300, Kontron 414 and Perkin-Elmer series 4, 10 and 400, SSI pumps). The Gilson pumps have no compressibility correction but allow digital selection of the refill time as well as piston displacement at high cam rotation speed for solvent compression.

The recently introduced Beckman Model 114M is unique in that it includes a pressure transducer in the pump head, allowing the compression time to be precisely determined as well as measurement of the actual solvent compressibility thus performing automatic flow compressibility correction using appropriate electronic circuitry and greatly minimizing the pulsations even if gas bubbles are present.

Although the Perkin-Elmer Series 4 pump contains two pistons in series, it should be considered as a single head pump of the rapid refill type, as only the second piston serves to pressurize the liquid and deliver it to the column. The first piston moves in a low pressure chamber in opposition with the second one, slowly refilling this low volume chamber and rapidly discharging its content into the high pressure side. This virtually eliminates the problem of bubble formation as well as facilitates low pressure gradient operation. For similar reasons, in the Varian Model 5000 pump, the pump inlet check valve is replaced by an inlet needle valve mechanically linked to the piston. A flow controller providing a constant backpressure is used after the pump to compensate for compressibility differences among solvents.

1.3.3 Double Head Mechanical Reciprocating Pumps

1.3.3.1 Reciprocating Pumps with two Parallel Heads

Some of the previously described single-head pumps with simple sinusoidal piston movement are available in duplex or multiplex configuration, the same motor powering the two heads which are 180° out-of-phase. As a result, the average flow-rate is doubled if the piston course length in both heads is the same, however the amplitude of the pulsations remains unchanged. Usually, such duplex pumps can feed two separate chromatographic columns.

To reduce, and possibly, eliminate this pulsation problem, parallel dual head pumps with special piston displacement profile were introduced about 12 years ago. Basically, the profile of the flow delivered by one head, at least at low pressure, has the shape of a symmetrical trapeze. The second head is refilling while the first is in its constant flow phase, and starts delivering the solvent when the flow rate of the first head is linearly decreasing. As a result, the combined flow from the two heads is, at least in principle, constant. Such flow profiles are obtained using non-circular drive gear and crank arms or one or two specially designed (generally computer-calculated) cams driving the two 180° out-of-phase pistons. All these pumps are of the piston type, with constant piston displacement length at a frequency adjusted according to the selected flow-rate.

However, while these pumps give a definite improvement over duplex sinusoidal pumps, a pump head does not deliver solvent as soon as its piston is moving forwards, because of the solvent compressibility. This mismatch between the two head flows gives some residual, more or less important, pulsations as well as a total flow loss, in agreement with Equation 1. The several dual parallel head pumps with trapezoidal flow profile differ in level of sophistication to reduce or eliminate this problem.

For some pumps (Bio-Rad Model 1330, Jasco BIP-1, Pye-Unicam 4015, Shimadzu LC4A, Tracor 955), an electronic compressibility correction is done from the measure of pressure (or motor load for the Tracor 955 pump) made with a pressure transducer placed just downstream of or right inside the tee connection where the two head effluents are mixed, using a solvent compressibility value adjusted by a screw potentiometer or entered into a keyboard controller. This flow correction, however, does not reduce the amplitude of the residual pulsations and some manufacturers include damping coils which may be short-circuited in some instances.

In the Dionex Series 2000, LKB 2150, Micromeritics 750, Pye-Unicam 4003 and Tracor 951 pumps, a double feedback scheme is used in order to both reduce the flow pulsations and perform automatically the flow compressibility correction. Basically, when only one piston is delivering the eluent, during the constant flow part of the trapezoidal profile, a tachometer is used to insure a constant rotation speed of the cam(s) and the pressure value given by a fast response pressure transducer is memorized. At the end of this constant flow period, when the two pistons are both delivering solvent and their flow profiles overlap, the constant flow feedback mode is switched to a constant pressure mode using the reference pressure value previously memorized. Then, the rotation speed of the cam is modified to provide a constant pressure. As in isocratic chromatography the flow-rate is proportional to the pressure drop, this insures that the flow delivered to the column during this overlapping period is identical to the flow delivered by one piston during the nonoverlapping period. By this way, the flow-rate at the column inlet pressure is equal to the selected flow-rate, but the flow measured at the column outlet will depend on the pressure drop and on the solvent, because of the decompression of the liquid in the column.

1.3.3.2 Reciprocating Pumps with two Heads in Series

With two head in series pumps, only two check valves are used (instead of four for pumps with two parallel heads) at the inlet and outlet of the upstream head. The second head piston displaces a volume which is, depending on the model, one-half, one-third or one-fourth of the volume displaced by the upstream head piston (also called working or principal piston). The two pistons are driven by either specially designed or circular cams which are out-of-phase. During the working piston forward stroke, the downstream piston is moving backwards and its chamber is refilled at high pressure in such a way that the flow-rate delivered to the column is constant, at least in principle. When the working piston chamber is refilling at low pressure, the downstream piston (sometimes called relieving piston) is delivering solvent to the column. This greatly helps reducing the pressure pulse associated with the working piston backward stroke. However, because the solvent in the upstream chamber must be compressed before to be delivered, there remains a residual pulsation which may be smoothed by a pulse damper. In the Knauer Type 64 and Waters M-45 pump, using the indication of a manometer and the compressibility value manually adjusted through a screw potentiometer, the cam rotation speed is modified to perform this correction. The Spectra-Physics SP 8700 pump uses a double feedback system, basically similar to the one described in the previous section. The reference pressure, memorized during a constant flow period of the pump, serves to adjust the cam rotation speed, in a constant pressure mode, during the piston change-over period. If the eluent composition is continuously changing, then a complex algorithm is used to calculate the correction to be brought to the cam rotation speed to make-up for possible deviations from the selected flow during the few past piston cycles.

1.3.4 Triple Head Mechanical Reciprocating Pumps

When three heads with identical sinusoidal piston movement are combined together with a 120° phase shift between each, the resulting total flow output is periodic. The average flow is 95.5% of the maximum flow delivered by one head and the pulsation is reduced to 14.0% of this average flow. Moreover, the period of the pulsations is 1/6 the period of the pistons and may become, at relatively large flow-rates, smaller than the response time of the pressure sensors (pressure transducer or detector), which gives apparent pulsations of amplitude smaller than the actual one, or even apparently pulseless records. The volume of solvent eluted during one pulsation period is one half of the volume displaced by one piston stroke.

While simple head sinusoidal piston movement pumps designed for industrial purposes may be assembled in a triplex configuration, some manufacturers have specially designed three head reciprocating pumps of the piston type, for LC applications. In this case, the flow-rate is adjusted by variations of the cam rotation speed, at full piston displacement. The spatial arrangement of the three heads varies with the model: triangular vertical assembly with parallel horizontal pistons for the Jasco Familic-300S pump, star configuration in horizontal plan for the Bruker-Merck-IBM pump. The three pistons of the Jasco Familic-300S pump are driven by one cam with a three dimensional geometry rotating around an horizontal axis. The individual flow profiles of each head of this pump are of an isosceles trapezoidal shape, the time of the constant flow period of one head being the same as the time of each linear flow period. Therefore, in absence of compressibility effect, there is no pulsation with this pump.

The flow loss due to solvent compressibility is electronically compensated using the actual pressure indication of a pressure transducer and the compressibility value manually adjusted by means of a screw potentiometer in the Jasco Familic-300S pump. In the Bruker-Merck-IBM pump, an electronic double feedback control, based on pressure measurement, is used to simultaneously correct the short term pressure fluctuations due to the piston displacement profile and insure a long term constant flow when the column flow resistance changes, for example, in gradient elution.

1.3.5 Description of Available Pumps

A large number of pumping systems used in liquid chromatography, for which informations were available to us, are described in Table 4. We thank the French LC manufacturers and representatives for their kindness to furnish information on their products. Some of these products are described in the text. We emphasize that this does not imply any endorsement from us. We simply wanted to describe the pumping principles, but one should not forget that a theoretically very good pumping system may be found unsatisfactory if the pump is poorly manufactured or maintained, or is misused.

1.4 Electroosmosis

Electroosmosis has been suggested repeatedly as a principle for pumping solvents through LC columns[16]. This is applicable only in the case of solvents with a relatively large conductivity such as those used in reversed phase LC. Under the influence of an applied electrical field the liquid moves at a linear velocity proportional to the field and function of the electrical properties of the solution. The great difference with conventional pumping is that the liquid movement is a piston flow, the velocity profile being nearly rectangular and not parabolic as in viscous flow. This could afford larger column efficiencies than conventional flow, especially using wide bore capillary columns, although results are still not fully conclusive.

Nevertheless electroosmotic pumping may become a useful technique in the future, in spite of the dangers associated with the handling of large voltages.

1.5 Gradient Elution Systems

When analyzing complex samples with components of interest covering a broad retention range, it is necessary to modify the elution conditions during the analysis to insure a sufficient resolution of the first eluted components and maintain the analysis time within reasonable limits. This is usually done by changing the eluent composition, increasing the eluotropic strength with time. With the recent trend to use three and even four eluent components, this is not a trivial experimental matter.

Gradient elution systems can be classified in two groups, low pressure and high pressure systems[17]. In the first group, the eluent components are mixed in proportions varying with time at low pressure and the mixture is pumped in order to be delivered at high pressure to the column. In the second group, components or mixtures of fixed composition are each pumped by separate pumps and, then, mixed at high pressure in a ratio varying with time.

With low pressure gradient systems, a single pump is used. Three and, sometimes four, solvents are delivered to a mixing chamber via either one or several time proportioning solenoid valves or several low pressure, precision metering pumps. Then, the mixture is taken up by the high pressure pump. Static or dynamic mixers are used to blend the solvents. In order to limit the delay between the time of formation of a specific composition and the time it enters the column, as well as limit the distortion of the gradient by diffusion, the system must have as small as possible a volume. Because the solvent mixing may not be sufficient before entering the pump, it is essential that the solenoid valve controller is synchronized with the piston displacement profile. In cases where this profile is not linear, the solvent mixture pumped may have a completely different composition from the one given by solenoid valves operated on a time basis[18]. In addition, when one eluent component has to be delivered during a time not much larger than the switching time of the valve, the accuracy of the gradient composition may be impaired.

With high pressure gradient systems, two or three pumps are used, each one pumping a different component of the eluent at a rate which depends on the eluent composition. The sum of the flow-rates delivered by each pump is kept constant during the run. Problems arise if the pumps do not have a device for automatic correction of pressure pulsations and of flow losses due to solvent compressibility because the column inlet pressure is constantly changing during a run. This is especially true when one of the pumps is operated at a low flow rate. During the compression part of its forward stroke, this pump is not delivering the nominal amount of solvent component due to compressibility losses. Moreover, this phenomenon is not reproducible due to pressure fluctuations. Even with pumps equipped with such correction devices, the flow-rate entering the column may vary because of the fact that associating solvents, blended in a high pressure static or dynamic mixing chamber, have a non-zero excess mixing volume.

Therefore, while the concept of mixing solvents in proportions varying during the analysis looks simple, it is not trivial to experimentally implement it and to obtain that the composition of the mixture entering the column conforms to the desired profile. Admittedly, however, these problems can be solved without excessive difficulties in the 10–90% composition range. Considerable progress, due to the use of microprocessorized controllers, allows today satisfactory results in the entire composition range with some gradient systems.

A simple test of the flow-rate stability of a gradient system has been suggested by Halasz and Vogtel[19]. A simple mixture is analyzed by isocratic chromatography first using one pump, then using the gradient system, but each pump pumping the same solvent. The flow-rate fluctuations are easy to notice. Further tests of the flow-rate stability of a real concentration gradient are then easier to make and account for.

2 Sample Introduction Systems

The amount of sample which is injected into the column, after the final preparation step, is determined by the fact that the components of interest have to be detected with adequate accuracy after separation on the column. This amount depends on the sensitivity of the detector for these components and on the extent of dilution undergone in the column. Therefore, the volume to be injected is determined by the concentration of these components in the sample.

The finite size of the sample volume contributes to the final width of the solute zone, at the detector level. It is usual to consider that the injection process is independent of the column migration process. Accordingly, the contribution to the total variance of the zone from these processes are additive and the injection contribution cannot be smaller than $V_{inj}^2/12$ if V_{inj} is the volume injected[1]. This corresponds to the variance of the rectangular plug. It is, obviously, desirable for this contribution to be small compared to the column one. The trend, mentioned in the previous section, towards a wider use of narrow bore columns, imposes therefore that the injected volume is small. For example, for a 1 mm i.d., 15 cm long column packed with 5 μm particles and operated in optimum conditions, a plug injection of 1 μl of sample will contribute to 18% of the variance generated for an unretained solute by the column. This contribution being proportional to $1/(1+k')^2$ will be only 2% of the column variance at $k'=2$.

The practical situation is worse as any sample introduction device distorts the sample plug, increasing the injection variance, sometimes far above the $V_{inj}^2/12$ value[2-4]. The complete description of the injection process and of its contribution to peak broadening is very complex[5] and has not yet been successfully undertaken. It depends on the type of the injection system used, on the connection pattern between the injector and the column, on the injected volume as well as on the injection time. It is however possible to experimentally determine the injection contribution to band broadening[2-4] and estimate the maximum injection volume acceptable without impairing significantly the resolution[6].

Accordingly, a good injection system should have the smallest possible contribution to peak broadening. It should be convenient to use, able to operate at high pressures, chemically inert with the eluent and the sample and reproducible. In some instances, the injection should be possible at the high temperature required for sample solubility, as for the steric exclusion chromatographic analysis of polyethylene samples. Besides, automation of the sampling system should be possible.

Sample introduction systems are usually classified into two groups: syringe injectors and sampling valves[7]. This classification may be somewhat misleading as most recent valves make use of a syringe not only to fill the valve loop but also to precisely determine the volume of sample to be injected. Nevertheless, because of major differences between these two types of injectors, we retain this classification in the following.

2.1 Syringe Injectors

With this device, the needle of the syringe, which contains the desired amount of sample to be injected, is introduced onto the top of the pressurized column through a self-sealing elastomeric membrane (septum) and its content is discharged in the eluent stream. This is the simplest form of sample introduction. However, it has some inherent drawbacks. Depending on the eluent used, the rubber septum may swell and loose much of its elasticity. To alleviate this problem, some septa have a PTFE coating on one face. In addition, after repeated injections, particles of septum may be torn out and deposited on the column top, partially plugging this column and sometimes causing sample loss or peak distortion by adsorption on these particles. Use of a guide may help the needle to puncture the septum at the same position at each injection and reduces septum particle breakage.

The maximum pressure at which a syringe can be used for on-line injections is limited either by leakage between the piston and the syringe barrel, or by the force the operator can exert on his finger to counteract the uplift force of the pressurized liquid on the piston and push it down. Practically, with most commercial high pressure syringes, one can inject microliter quantities up to 10–20 MPa, depending on the operator. For larger sample amounts, with increased piston diameter, this maximum pressure diminishes.

In cases where on-line high pressure injection cannot be made, one may stop the eluent flow which causes the column inlet to be depressurized, insert the syringe needle into the septum, inject the sample, withdraw the needle and turn on the pump. Practically, it is found that, when properly carried out, on-line syringe injection and stop-flow injection have similar levels of reproducibility and contribution to peak broadening[8].

With these injectors the reproducibility of retention times is found as good as 0.2–0.5%. However, the reproducibility of plate numbers is in the range 10–15% for on-line injection and 3–5% for stop-flow injection, while the reproducibility of peak areas are about 5–15% and 3–5% for on-line and stop-flow injection, respectively[8]. However, these reproducibility ranges depend greatly on the operator skill. In some instances, fast injections have proved better than slow injections while in other cases, they may give badly tailing or even doublet peaks. It has also been found that injection through a porous PTFE plug deposited on the top of the packing gives a lower contribution to peak broadening than injection onto the top of this plug[9]. These types of syringe injection can hardly be automated.

2.2 Sample Valve Injectors

Because they are convenient to use, allow sample injection at pressures up to 50 MPa and give results nearly independent of the operator skill, sampling valves are today the most used injection devices. Moreover, they are easily automated as seen in a subsequent section. While several types of valves are used in modern LC, they all have the same basic characteristics. The sample is loaded at nearly atmospheric pressure by means of a syringe in a non-flushed flow line (loop or cavity), while the eluent flows directly from the pumping system to the column. Actuating the valve switches the flow line connections, allowing the eluent to flush the sample containing line into the chromatographic column. However, it has been found that the temporary interruption of flow through the column during the valve switching may reduce the column lifetime[10]. Injection valves can be subdivided into fixed-volume valves and variable-volume valves.

The first LC sampling valves were of the former type. The simple devices are six-port rotary valves. The sample storage line is loaded by means of a load syringe. Excess sample is used to insure that the storage line is completely filled. This line may be either an external loop with adequate volume which may be replaced when injection volume is to be changed or an internal cavity of fixed volume grooved at the surface of the rotor or drilled into the rotor. These devices deliver highly reproducible sample amounts, sometimes with less than 0.2% error. However, thorough flushing of the sample loop is required before the next load to eliminate the sample trapped.

More recently, variable-volume valve injectors have been introduced. They contain a needle port, which can be closed or sealed at high pressure, to insert the syringe. The desired sample amount is injected or sucked into the loop which becomes partially filled with sample. Then solvent line switching causes the sample to be injected into the column. It is essential with these devices that the sample side of the sampling loop comes first into the column rather than the solvent side to avoid excessive dilution and band broadening. In some models a flow restrictor is inserted in the direct line between the pump and the column. The opening of two-port valves allows the solvent to preferably flow through the sampling loop and introduce the sample in the column. In another model a three-way valve is used to switch the flow line through the sample loop and simultaneously open the ball check-valve located at the other end of the sample loop. In most recent models, variable-volume sample injection can be done by means of four or six-port rotating valves. In the latter case, special arrangement of the needle port as well as of the rotor-stator interface may suppress the need of sample loop rinsing before the next injection, except for trace analysis.

The reproducibility of variable-volume sample injectors is dependent on the precision of the sample delivery syringe; as seen above, it is generally better than for syringe injectors. On the other hand the contribution of sampling valves to band broadening is usually larger than that of syringe injectors[8]. In spite of this fact, the convenience of use of the valves as well as their automation capability make them a very useful device in liquid chromatographs.

3 Detectors

Almost since modern liquid chromatography was incipient, when HPLC stood for high pressure liquid chromatography, but columns were still several meters long and packed with 80 µm particles, detectors were already considered as the weak part of the chromatograph. More than 15 years later, in spite of tremendous developments they still are. Many possible developments in the field of instrumental chroma-tography are still hampered or even just made impossible because of the lack of detector having suitable characteristics: one of the best examples is the open tubular column[1, 2]. Nevertheless, there have been so many publications in this field that it is impossible to give a reasonably complete list of them and do more than present a flavor of the present state of the art.

The first section deals with the general properties of detectors, then the principles of the most important detectors, the refraction index detector and the UV absorption detector are described and the main problems arising with their design and use are discussed. Finally a catalog of various detectors is given. Among them some are useful for specific applications, others are of historical interest and a few show promises because of unusual properties. This catalog certainly reflects the prejudices of the authors.

3.1 General Properties of Detectors

There has been a large number of reviews published in this field[3–5] and many lists of important properties and of some lesser ones have been given. These proper-ties can be summarized as follows:
- the *sensitivity* relates the quantitative response or detector signal to the amount of sample introduced; together with the *noise* it determines the *detection limit*.
- the *selectivity* relates the sensitivity to the chemical nature of the sample. It is a property of the detector *and* of two compounds (or possibly two classes of compounds). It is the ratio of the detector sensitivities for these compounds.
- the *linearity* indicates whether the detector response depends linearly on the sample size and if yes in which range. The *dynamic linear range* is the range of sample size for which a signal is detected and is a linear function of this size.
- the contribution of the detector to band broadening is the amount of variance contributed to solute zones when they migrate through the detector. It should be small compared to the variance originating in the column.
- the ease of operation and reliability of the detector are less easy to quantitize but they are very important practical factors.

Ideally, each property of a detector, and especially the range of linearity, the response factor and the detection limits should be determined directly, by pumping solutions of known concentrations through the detector cell. The influence of the various parameters can then be studied easily.

Determinations made by injecting known amounts of solutes in a chromatograph are easier to carry out, but they depend on both the detector *and the column* proper-ties and although easier to carry out, are more difficult to account for. The influence of column chracteristics should be studied and substracted before general conclusions are drawn.

3.1.1 Detector Response

There are two kinds of detectors. Those which respond to changes of the solute concentration in the eluent and those which respond to changes of the solute mass flow rate into the detector cell[6, 7].

Detectors of the first group are concentration sensitive detectors. Their signal is proportional to the concentration of solute in the eluent and independent of the solute mass-flow rate, i.e. of the flow-velocity of the eluent in the detector cell. They are generally non-destructive like the differential refractometers and the spectrophotometers. When the solvent flow is stopped, the signal remains constant [6], with the following restrictions: (i) the base line and the response factor sometimes depend on the flow-rate; (ii) the concentration of solute in the detector cell changes slowly because of molecular diffusion. Within these limits, the peak area is inversely proportional to the flow rate and the peak height remains approximatively constant if the flow rate is close to the optimum (minimum plate height). The response decreases if a scavenger flow is added to the column eluent to sweep faster the detector cell.

Detectors of the second group are sensitive to the mass-flow rate of solute to the detector, i.e. the signal is proportional to the product of solute concentration by mobile phase flow-rate. If the flow-rate is zero the signal is zero whatever the concentration. They are usually destructive. When the eluent flow is stopped, the signal decreases to zero very rapidly, quasi-exponentially, with a time constant equal to the detector response time. The peak area is independent of the flow rate (assuming the response factor does not depend on the flow rate) and the peak height increases constantly with increasing flow rate. Electrochemical detectors and the mass spectrometer belong to that group.

Most LC detectors belong to the first kind. The peak area measured with all detectors is the integral of the signal versus time, while the sample size is, in the case of a concentration sensitive detector, the integral of the concentration versus volume of eluent flowing through the detector cell. Accordingly the peak area is proportional to the sample size, the response factor and the reverse of the flow-rate [6, 7, 87]. Accordingly it is important to stabilize the flow-rate so that its short-term fluctuations, over a time equal to the peak width, be smaller than the reproducibility the analyst wants to achieve (cf. Sect. 1.2.6). In the case of detectors of the second group flow-rate fluctuations have lesser effect if they do not affect the response factor.

3.1.2 Detector Sensitivity. Detection Limit

They depend on the signal noise and the response factor at low concentration. It is best expressed by the detection limit, which is the amount of material necessary to obtain a signal equal to a given number of times, usually 2 or 3, the base line noise.

It is important to distinguish between the detector sensitivity which is the concentration (for detectors of group 1) or the mass flow rate (for detectors of group 2) giving a signal equal to twice the noise and which depends only on the detector (and the solute) and the detection limit of a given analytical procedure which also depends on the nature and performance of the column used. Chromatography is a dilution process and the extent of that dilution, hence the maximum concentration of the solute band in the detector cell, depends on the analytical conditions, i.e. analysis time and column efficiency.

For example, if the maximum sample volume is introduced in the column (i.e. the sample volume which increases the column plate height by 10%) the solute is approximately 5 times more dilute in the column effluent than in the original sample[8]. But it is not always possible to inject the sample without diluting it previously in an adequate solvent as the stationary phase might be overloaded, especially for the main components of the mixture. Furthermore, the requirement regarding the maximum sample volume applies to the less retained components in the mixture. Accordingly, a dilution factor of 100 is not unusual, and a detector sensitivity of 1 ppm results in a detection limit of 100 ppm in the actual sample[5]. The uncautious analyst may be disappointed. The maximum sample volume permitting a plate height increase of no more than 10%, for example, is proportional to $1 + k'$[8,9].

The noise has a critical influence on the sensitivity and most of the progresses made by instrument companies have been in the reduction of that noise. This is the only possible way to decrease the detection limits: the use of faster, more efficient columns results in less dilution during the chromatographic separation, but these columns accept proportionally smaller sample volumes, so the dilution ratio for the same overload factor (i.e. plate height increase due to finite sample volume) remains constant. The effect of noise can be minimized by correcting for base line drift (low frequency noise) and by signal integration (high frequency noise). Noise also affects quantitative results because the algorithms for signal detection are very sensitive to noise and often result in peak area measurements for trace compounds being less reproducible than peak height measurements. The most effective but most sophisticated way of reducing the noise is to use the Fourier transform and replace the unwanted components by straight lines before restoring a clean signal (cf. Part 5.2). This requires the use of a computer. Anyway, the noise whose frequency is close to that of the peak (i.e. between *ca.* $5/\sigma$ and $1/5\,\sigma$) cannot be corrected.

The origin of noise is to be found in the electronics associated to the detector and in the fluctuations of the physical parameters in the detector environment: light source intensity for UV photometers, temperature and pressure changes for the refraction detector cell ... For non-selective detectors such fluctuations change the background, i.e. result in base-line noise. The contribution of pumping fluctuations or cycles to the base-line noise occurs through this mechanisms (cf. Part 1.2.5). For selective detectors, when the response for the eluent is negligible, only the response factor is fluctuating. This results in a noise which increases with the signal, does not affect the detection limit, which may be very low, but may reduce slightly the reproducibility.

Finally it must be emphasized that detection limits should be determined from response factors measured with very small concentrations of solute, of the order of 10–50 times the detection limit. Extrapolation from measurements made at large concentrations are often fallacious.

3.1.3 Detector Linearity. Dynamic Linear Range

The response of an ideal detector is proportional to either the concentration or the mass flow rate of solute. Such an ideal, linear behavior is rare. At large concentrations, of course, a detector can exhibit deviation from linearity, usually with a saturation effect. More dangerous, and often overlooked is the fact that the response of detectors may change markedly but progressively with concentration over a large concentration range.

The current practice of plotting the detector response versus the sample size in double logarithmic coordinates is misleading as such a plot dampens fluctuations. A plot of the ratio of detector signal to sample size versus logarithm of sample size is much more instructive: a large range of sample size can be explored on the graph while fluctuations will still appear clearly. Inhomogeneity of photocells, non-linear behaviour of electronics ... will explain some unexpected curves.

The dynamic linear range is the ratio of the sample concentration or sample size for which the response deviates by more than 5% (or sometimes 10%) from a linear behavior to the detection limit (in the same unit). A good detector should have a dynamic linear range exceeding 10^3. Some detectors reach 2×10^4. Sometimes the concentration range in which the detector behaves linearly can be changed by one or two orders of magnitude by adjusting one parameter (i.e. the wavelength of a UV photometer).

Deviations from linear behavior can result from the physical principle used, from the equipment design, from the parts used to make it or from the column itself. Beer's law is not linear but logarithmic; a correction can be calculated and the detector is "linearized" (cf. 3.3); the refraction index is not a linear function of the solute concentration at large concentration and the correction through molar refraction is not usually calculated. In refractometry, one of the conventional design uses a prism which deviates a light beam which originally lights equally two photocells; when the solute concentration increases the rotating light beam lights one cell more than the other until this last one gets no light at all: the detector is saturated. Lack of homogeneity in glass windows, lenses, photocells, light source and electronic circuits can also result in lack of linearity of an instrument.

It must be emphasized that many optical detectors and especially photometric detectors, operate on laws which are valid only at infinite or nearly infinite dilution. This is, for example, the case of Beer's law[87] which is valid in a range of absorbance which does not exceed usually 1 (i.e. reduction of light intensity on the photocell by a factor 10). This is further discussed in Sect. 3.3.1.

Finally, when the column is overloaded the peak height does not increase linearly with sample size but peak area does, at least if the end of the peak can be properly detected. Also at low sample size a significant part of some components may be quasi-irreversibly adsorbed on the packing material.

3.1.4 Detector Selectivity

Most detectors have response factors which vary widely from one component to another. The selectivity characterizes this variation in response with the chemical nature of the compounds studied. It is difficult to give numbers, however and there is no general agreement in the literature on a quantitative definition. The selectivity for one compound by respect to another one could be the ratio of their response factors. Ratios exceeding 1×10^4 are not rare. Unfortunately, in a given group or family of compounds the response factor is rarely constant so the ratio of response factors for two groups is a kind of average number which can hardly be used for predicting the response factor of other compounds.

It would be useful to have non-selective detectors with response factor identical for all compounds and selective ones for which the response factor could be adjusted to be very large for a group of compounds (and the same for all of them) and negligible for the others. Such ideal detectors do not exist.

Non-selective detectors respond also to the mobile phase and are sensitive to changes in its density. Their sensitivity is accordingly limited[9]. Properties like refraction index, dielectric constant and density vary from one compound to the other, so the response factor may vary in a range of about 1 to 100. It is considered to be impossible, even with these detectors, to derive an approximative figure of the composition of a mixture without calibration. However, synovec and Yeung have shown recently that it is possible to calibrate detectors for unknown samples if they can be analyzed using several solvent systems[88].

The response of selective detectors also depend on one or several adjustable parameters which permits both very sensitive and selective detection (wavelength in UV adsorption, potential difference for electrochemical detectors, wavelengths of incident and observed radiation for fluorescent detector ...).

The mass spectrometer is both highly selective in the selected ion monitoring mode and (relatively) non-selective in total ion current.

3.1.5 Contribution to Band Broadening

As soon as the solute band leaves the column, separation mechanisms become inexistant, but band broadening through axial diffusion and resistance to radial mass-transfer goes on. Long lines of empty tubes (because of the Poiseuille flow profile and the slowness of diffusion in liquids), important dead pockets (for the last reason) and turbulences at places where the cross-section of the tubing available to the mobile phase changes abruptly, are the most important sources of band broadening and should be banished[3-5, 10-14]. The response time of the detector is also important to consider.

The peak variance σ_t^2, as recorded, results of three contributions:

$$\sigma_t^2 = \sigma_i^2 + \sigma_c^2 + \sigma_d^2$$

arising respectively in the sampling system (σ_i^2 cf. Part 2), in the column, (σ_c^2) and in the detector, σ_d^2. This last contribution depends on the volume of the detector cell and connecting tubings, on their shape and on the response time of the detector. The variance contribution of short tubes is small if they are well designed[15] and will thus be neglected here. The variance contribution of the detector can be measured directly using a very fast sampling system and a narrow, short capillary tube in place of the column[16]. It depends markedly on the flow-rate, being controlled at low flow rate by mass transfer processes and at high flow-rates by the response time. The influence of these factors has been studied in details[17-19].

The time constant, τ, of a detector describes the speed at which it responds to an instantaneous change of concentration of solute in the detector cell. Although an instantaneous change cannot really take place it is not difficult to find chromatographic conditions in which the peak width at column outlet is small compared to the time constants of many detectors. This time constant is usually defined assuming that the detector, the signal transducer, the amplifier and the recorder behave as a first order system and that the signal tends exponentially towards the steady state value. Accordingly, the deviation between the "true" value and the signal is still 5% after a time equal to 3 τ and 1% after 4.5 τ.

Although there is no band remixing if the detector time constant is too large, the result is the same, some valuable information is lost. It is generally admitted that the time constant of the detector should be between 10 and 20 times smaller than the peak width; otherwise the peak is broadened, becomes smaller and tails. Too small a time constant can be detrimental in terms of detection limits, because the bandwidth of the signal is larger and less noise is filtered out.

An order of magnitude of the requirements can be obtained by assuming that σ_d^2 should be no larger than $\sigma_c^2/10$. Obviously the column diameter becomes significant and specifications will be 20 times more drastic for 1 mm i.d., narrow bore packed columns than for the conventional 4.5 mm i.d. columns. For open tubular columns, they are still beyond our capabilities[1, 2]. For a conventional column, the most drastic situation is achieved with the very recent 3 cm long columns packed with 3 μm particles[20]. The column variance for the non retained peak is 2.2×10^{-5} cm^6 ($\sigma_c^2 = (\varepsilon_t \pi R_c^2)^2$ L H, with R_c the column radius, L the column length, H = 2 dp, the plate height, dp the particle average diameter and ε_t the total porosity). In order to observe a detector contribution 10 times smaller, and assuming total mixing in the detector cell (and neglecting other contributions) we need a cell volume smaller than 2 μl, which is obtained now with several commercial instruments. Similarly, the elution time of the non retained peak will be of the order of 10 s, with an efficiency of 5000 plates, and a time standard deviation of 0.14 s, requiring a response time below 50 milliseconds.

Of course those are the most drastic specifications. Many analyses are slower, use longer columns and do not achieve a much better efficiency, so the requirements are such that a number of analyses can still be performed with 10 μl cells and 0.2 s response times. Those equipments are obsolete, however, and especially in routine analyses with automatic equipments where large gains in productivity can be achieved by a significant reduction in analysis time, they will contribute markedly to band broadening and reduce the apparent column performance. They are still quite convenient for exploratory work and many more conventional applications.

3.1.6 Practicality

This cannot be put in equations nor modelized but it is an important quality of detectors. It applies to the principles but also to each design which may exhibit its plusses and its flaws.

The sensitivity to ambient parameters is one of the many facets of this quality. Some differential refractometers take hours to stabilize. Others give a steady base line in half an hour or so. This depends on the design and on the quality of the ambient parameter controls.

Detectors like all the parts of the chromatograph must be compatible with all solvents, at least all the conventional ones in LC. Only corrosion resistant material should be used. Some detectors, because of their principle cannot be used with a limited number of solvents: (hydrocarbons, chlorinated hydrocarbons) for the electrochemical detectors, most buffers for the mass spectrometer. Such exclusions should extend to a limited number of solvents for a general purpose detector.

The cell design must avoid eddies and dead zones which contribute markedly to band broadening, it must minimize the influence of flow-rate changes on the base line and signal. The cell must be easy to disassemble, wash, and reassemble; change in solvent should be easy and rapid and they should not trap gas bubbles or small silica particles. They must be resilient and withstand pressure at least up to a few atmospheres without expanding which would make them respond to flow rate pulsations.

The detector signal should be available in digital form as well as analog and the time constant should be adjustable easily.

3.1.7 Problems Specific to Microbore Columns

Recently the use of packed columns with an inner diameter much narrower than that of conventional columns (4–4.5 mm) has been advocated for a number of reasons, some more valid than others. Narrow bore columns need a flow-rate of solvent in proportion to the square of their diameter, which permits appreciable savings with conventional columns or the use of exotic solvents. Similarly they need a smaller sample and when the sample size available is small, less dilution and thus lower detection limits are possible. These columns are not more efficient than conventional ones, however and suffer from a few drawbacks which result essentially from the very tight specifications placed on instruments[1].

The sample size, the connection volumes and the detector cell volumes should be reduced in proportion to the square of the column diameter if comparable performances are to be achieved and this is not an easy task. For example the detector cell volume should be less than 1 μl. This is difficult to achieve with a UV detector without some reduction in the optical path length, hence an increase in the detection limits which is opposite to the rationale in using microbore columns. The specifications can be relaxed somewhat by chosing columns longer than the conventional ones and operating then at a higher flow velocity to a comparable analysis time. But then the inlet pressure is large, the dilution is increased which again is not satisfactory.

Considerable effort is devoted by manufacturers in the design of small detector cells, but it is not sure yet this is a fruitful trend.

3.1.8 Conclusions

In spite of the drastic appearance of all the requirements discussed above, thousands of detectors work properly everyday. It should not be forgotten that in a complex instrument like a liquid chromatograph, the performances of the whole are no better than those of the weakest element. The detector should not be much less good that the other units of a LC, as it is sometimes, but on the other hand good analytical results cannot be achieved without proper use of the instrument and especially proper care in the handling of the mobile phase, keeping the flow rate reasonably constant as well as the composition of the eluent[21]. Gradient elution analysis have special requirements[3–5] which makes them impossible to carry out with detector which have a non-negligible response for the solvents used.

Detectors of several different principles are available in a large number of models. Many of them offer interesting compromises between contradictory requirements, to the point that it may be difficult to choose. Non technical considerations like price, quality of service available locally ... are also important. The quality of detectors available, their performance and reliability have improved considerably over the last 10 years, to the point that few technical problems remain unsolved. The most important are:

- there is no selective, sensitive detector available for some important classes of compounds: sugars, fats and their derivatives, and to a lesser degree some underivatized amino acids and peptides.
- there is a lack of reliable detectors for trace analysis. Those available work only with a limited number of compound groups.
- there is no detector commercially available with an equivalent cell volume (cf. 3.1.5) below 0.1 µl, which would be necessary for open tubular columns[2].

Work is in progress in all those areas, however, and some remarkable developments should be made in the forthcoming years.

At present a large fraction of analyses made by LC are carried out using a UV absorption detector. Other detectors which are used to an important extent are the differential refraction detectors, the fluorescence detectors, the reaction detectors and the electrochemical detectors. A large number of other detectors are used for some very specific analysis, are used to a very little extend by highly specialized scientists or have been suggested and abandonned. The mass-spectrometer could become an important detector for LC, but is still the object of intense investigation.

Detectors based on so-called hyphenated techniques, i.e. using mass spectrometry, infra-red spectroscopy and nuclear magnetic resonance in connection with LC are discussed in a separate chapter, as well as reaction detectors. The more conventional detectors are discussed now, briefly.

3.2 The Differential Refraction Detector and Related Detectors

The related detectors discussed at the end of this section are using changes in the dielectric constant or density of the mobile phase during elution of the sample.

3.2.1 The Differential Refraction Detector[3-5, 22-25]

This detector measures the variation of the refraction index of the column effluent. As the refraction index of a liquid changes by about 4×10^{-4} for a temperature change of $1°$ C and 4×10^{-5} per pressure change of 1 atm, it is necessary to control very carefully these environmental parameters. Making a differential measurement by respect to pure eluent permits a greater stability by using a symmetrical design which minimizes the effect of temperature fluctuations: it is easier to achieve a temperature difference of less than 1×10^{-4} °C between the reference and measurement cells than to reduce the fluctuations of the absolute temperature of these cells below 1×10^{-2} °C. Nevertheless, it is extremely difficult to achieve a base line noise smaller than $2-5 \times 10^{-8}$ index unit with a pure solvent and 1×10^{-7} with a solvent mixture, although with extremely careful controls a commercially available equipment is claimed to achieve 1×10^{-9}. The detector cannot be used in gradient elution.

Although there are some exceptions (e.g. CS_2), the refraction index of organic compounds is between 1.30 and 1.50. The difference between the refraction index of two compounds can be as large as 0.17 (benzene in methanol) or as low as 0.003 (diethylether in acetonitrile) but will usually be of the order of 0.10. This explains that the detector must always be calibrated. Furthermore, with a good RI detector, the minimum detectable signal being around 1×10^{-7}, the detector sensitivity will not exceed 1 ppm and the detection limit is usually in tens of ppm (cf. Sect. 2.1.2).

The refraction index does not vary linearly with the concentration of solute. The RI detector is not linear above 2–5% which is not too serious a limitation in LC, but it is not possible to use the difference in refraction indices of the solvent and solute to derive response factors. The additivity of molar refractions can be used but it is not a very popular approach[22, 23].

Finally the detector is sensitive to flow rate fluctuations or rather to pressure pulsations as the detector cell has to be maintained under slight back-pressure to prevent the formation of bubbles and permit the exit of the solvent stream through the outlet tubing. As the capacity of the LC system is small the column dampens the pump pulsations only slightly. With a back pressure of 0.5 atm and a pressure pulsation as low as 1%, a noise contribution of 2×10^{-7} RI is observed. The back pressure should be kept minimal[22].

There are several methods to measure the refraction index: the deviation of a light beam by a prism, the critical angle of reflection, the Christiansen effect and interferometry[4].

The most popular method is still the deviation of a parallel light beam by a set of two prismatic cells one containing the solvent, the other the eluent, arranged so that their deviations are in opposite directions. When the eluent is the pure solvent no net deviation occurs and the beam lights equally two photocells. A change in the composition of the eluent generates a proportional change in its refraction index and a proportional rotation of the beam. The difference in currents delivered by the two cells is a measure of that composition.

The amount of light energy reflected at the interface between a glass and a solution is a function of the difference in refraction indices of those two media. Whereas the first method gives a constant response factor, with this method the response factor depends on the refraction index itself.

These two methods have permitted the design of very simple and reliable instruments which have comparable performances (sensitivity around ca. 1×10^{-7} RI unit, cell volume ca. 10 µl, time constant ca. 30 ms). This is demonstrated by the small number of papers in this field in the last 20 years. It does not seem possible to exceed these performances and especially to achieve lower detection limits[22]. Accordingly, some interest has been focused to the use of interferometric measurements, using a laser beam, a Perot-Fabry interferometer and two light beams going through the reference and the measure cell, respectively[24, 25]. Interferences between the two beams result in a phase shift proportional to the RI difference. The response is sinusoidal and the detector requires a microcomputer for proper data handling. Extremely low detection limits are claimed (in the 1×10^{-9} to 1×10^{-8} range).

These performances will be useful only if the detector is properly integrated in a very carefully controlled chromatographic system. Whereas a 10 ppm sensitivity is rather easily achieved, further improvement requires higher and higher control of cell temperature and pressure and of mobile phase composition. The entire chromatograph, including the mobile phase tank must be temperature controlled within 1×10^{-2} °C. A syringe pump, with the complex ancillary devices permitting delivery of a really constant flow will be used to maintain the utmost stable cell pressure required. Since the mobile phase is often a solvent mixture it should be kept at constant temperature continuously agitated and degassed with a slow helium stream. Fluctuations in column temperature result in fluctuations of the eluent composition as one solvent is adsorbed preferably to the other one and this excess depends on temperature. The cost of such a system is large.

3.2.2 Dielectric Constant [26-29]

There have been a very large number of papers published on this detector principle, a few of which are referred to here, as most of this work has not had great practical results. The dielectric constant at high frequency and the refraction index are related, but a dielectric constant detector has been very slow to develop [29]. The design of a cell small enough to satisfy the LC requirements but large enough to permit sensitive detection is difficult.

Although a detector of this type is available now, it does not seem the method will become highly popular as the results are only as good as those obtained with a current RI detector: the detection limits obtained in reversed phase LC are around 100 ppm [29]. It would be difficult to improve them and the stability problem will be the same as for the RI detector.

3.2.3 Density [30]

Problems similar in nature but still worse in importance are encountered when differential density measurements are attempted [30]. This principle has been abandoned.

3.3 UV Absorption Detector [3-5, 31-38]

The UV photometer is the workhorse detector of LC. It has been used since the early beginning and continues to be the favorite. Although in the early times single wavelength detectors were the only model in existence, spectrophotometers are now widely used. There are two possibilities with these instruments, either their monochromator can be set on a given wavelength for the entire analysis in which case there is little difference with a single wavelength detector, except for the additional flexibility, or the entire spectrum can be scanned rapidly enough so a number of spectra are available for each band width. In this last case new performances are available and special methods of data handling must be used.

An excellent review has been recently published by Abbott and Tusa [31].

3.3.1 UV Photometry

There has been an extremely large number of publications involving descriptions of UV or UV-visible photometers. In most cases the equipment described is not original or the modifications are minor and the work really deals with applications of this detector.

The principle is extremely simple. A light source delivers a monochromatic parallel light beam which goes through a cell swept by the column eluent, and falls on a photocell. A signal proportional to the amount of light received is measured and recorded. The light source of the simple UV photometers is often a mercury lamp permitting work at 254 nm (e.g. 280 nm or 365 nm using proper interference filters to isolate the desired mercury line). The need for lower wavelength lines has led to the development of miniaturized zinc (214 nm) and cadmium (229 nm) gas discharge lamps. A very large fraction of organic compounds absorb at 214 nm, so the UV photometer with a zinc lamp is almost a universal detector, although its response factor can vary by orders of magnitude from one compound to another one.

To increase detection selectivity, variable wavelength detectors are used. They incorporate a deuterium lamp and sometimes a tungsten lamp to extend the wavelength range to the visible domain. Lamp stability has been greatly increased in the last few years through careful control of their operating parameters[31]. A monochromator permits the selection of the desired wavelength. The noise level has been reduced to the level of ca. 5×10^{-5} AU, while flow sensitivity was also reduced. In current cell design, noise, flow rate and temperature sensitivity originate in the part of the light beam which strikes the cell wall[31]. The cell volume is now of the order of 1 µl, while detectors with a cell volume as small as 0.1 µl have been described[2]. The response time is of the order of 50 ms. There is no difficulty in finding and operating a UV photometer allowing maximum performance of any narrow bore or microbore packed column[31], or even of wide-bore open tubular columns (down to ca. 40 µm i.d.[2]).

The light detector is a silicon photodiode. The recent generation diodes have good response, down to below 200 nm. Advances in diode technology, as well as in gas discharge lamps make conceivable the design of detectors with a noise equivalent to ca. 1×10^{-6} a.u.[31].

Even a variable wavelength detector, however, is not very flexible. The variation in response factor with wavelength for a given compound, or from compound to compound at a constant wavelength is very large. To optimize signal-to-noise ratio, time programmable detectors are used. The wavelength and the absorbance range change by program during the analysis, which is only possible for routine analysis. This technique has been made possible by the recent developments in electronics (fast photodiodes), optics (low stray-light, inexpensive gratings) and microcomputers[31-33]. Better performances can be obtained with spectrophotometers which record the spectrum of the column effluent during the analysis.

3.3.2 UV Spectrophotometers

Although this terminology is used in the commercial literature to describe variable or programmable wavelength detectors, (cf. section above) we consider here only those detectors which record the UV spectrum of the eluent during the analysis, at a large enough frequency (several tenths of Hz to several Hz). From these spectra UV-chromatograms on any desired wavelength can be reconstructed and evaluated.

Such records appear as three dimensional surfaces, where the absorbance is a function of both wavelength and time. For maximum use of the information contained they should be treated as such, using double integrals and not as the combination of two grids (UV spectra and UV-chromatograms) [34].

The first attempts made use of a silica prism placed after the UV cell and which dispersed the polychromatic light on the entrance slit of a Vidicon camera, thus permitting the record of the intensity of transmitted light of the entire spectrum as a function of time [35].

More recently, Milano et al. [36] have shown the potential advantages of using a diode array to record the dispersed light and illustrated some of the more sophisticated use of the detector, like using the derivative spectrum ($dA/d\lambda$ where A is the absorbance, as a function of λ) and considering the variation of $dA/d\lambda$ versus time, for a trace compound, at a wavelength for which the absorbance of a major, interfering compound is maximum, $dA/d\lambda$ being zero for this last component the trace compound can be detected at a low concentration. Similarly several interfering compounds can be quantitized simultaneously if their UV spectra are known. As the ratio of absorbance at any two wavelength is a characteristic of the spectrum and independent of the concentration, consideration of the variation of this ratio with time permits an easy check of the purity of an eluted band. Also comparison of a group of such ratios with those measured on the spectrum of an authentic compound is a test of identity which is much more informative when it is negative than when it is positive, as the relative lack of information in an UV spectrum prevents to use it for positive identification [31, 36-38].

These advantages have been fully recognized and the progress in the diode array and microcomputers have made these detectors highly popular at meetings as topics for communications, as exhibition features as well as, hopefully, selling items [31]. Progress in data handling are still required.

3.3.3 Quantitative Analysis with a UV Detector

Because the UV detectors, photometers and spectrophotometers, are the most important group of detectors in LC, it is useful to point out some of the common pitfalls in their use in quantitative analysis which are not always recognized by analysts [87].

One of the important limitations of the precision in quantitative analysis results from the short term fluctuations of the eluent flow-rate, since the UV-detectors are concentration sensitive (cf. Sects. 1.2.6 and 3.1.1). Experimental methods to measure this contribution in isochratic and in gradient elution chromatography have been described [87].

A second problem results from the limited range of validity of Beer's law, strictly valid only at infinite dilution [87]. The concentration range where the law is valid is further limited by the combination effect of the finite bandwidth of the monochromator (2 to 10 nm, depending on the equipment) and the variation of the extinction coefficient with the wavelength [87]. The detector sensitivity increases with bandwidth, because the signal increases while the noise and drift result from the photocells. The signal is thus the sum of the absorption on all wavelengths contained in the bandwidth and because of the logarithmic nature of Beer's law is not proportional to the absorption on the average wavelength. This deviation is small for compounds for which absorption is maximum at this wavelength. It is larger for compounds for which this wavelength is on the side of an absorption band[87].

It may be difficult to change wavelength during an analysis, so analysts should be careful to check the validity of Beer's law for compounds for which they want quantitative analysis and if need be to determine calibration curves or adapt the detector parameters by reducing the monochromator bandwidth, when signal to noise ratio is not a problem, or changing sequentially the wavelength or turning to a diode array spectrophotometer.

3.4 Other Optical Detectors

Whereas ionization detectors are most powerful in gas chromatography, the higher density of the liquid phase, the tighter requirements regarding the detector cell volume and the more difficult task of distinguishing between a mobile phase and a solute molecule make optical detectors the most popular general principle in LC.

This section does not discuss infra-red absorption as it could have, as this topic is dealt with in the chapter on hyphenated techniques. The simple IR photometer or spectrophotometer has never become popular as not sensitive enough, and hardly compatible with many solvents. The advance of Fast Fourier Transform IR spectroscopy has made the technique more sensitive and able to supply much more information, an entire IR spectrum.

Various techniques of fluorescence are rapidly becoming popular. Phosphorescence, polarimetry, photoacoustic detection, inductively coupled plasma, the laser thermal lens are either used for some specific applications or actively studied as detection principles. Some of these techniques will become quite useful in the near future.

3.4.1 Fluorescence Detector

When molecules absorb high frequency electromagnetic radiations (UV or near UV visible) they are excited to higher electronic states from which they return to the fundamental state by one of several possible channels, one of which at least being the emission of one or several photons, with a frequency always lower than that of the absorbed photon. When this emission follows very shortly the absorption, the process is called fluorescence. The fluorescence spectrum is a function of the excitation wavelength. The diagram of intensity of the emission light versus wavelengths of the excitation and emission lights is a three dimensional surface characteristics of the molecule. This principle is the basis of several detectors which have become very popular in the recent past. For discussions on the principle of the method the reader is referred to the relevant literature on fluorescence spectroscopy.

In principle, a UV light beam is focussed in the detector cell, swept by the column eluent. Light emitted in the perpendicular direction is collected on the photocell and its intensity is a measurement of the concentration of eluate in the effluent stream[39]. It is obviously easy to combine such a fluorescence detector to a UV photometer[40].

In simple fluorometers, the excitation light is the emission line of a mercury gas discharge lamp and the entire emission spectrum is collected and its total intensity measured[39,41]. In spectrofluorometers it is possible to adjust the wavelength of the excitation light. In the most sophisticated instruments the wavelength of the emission light can also be adjusted, thus enhancing both sensitivity and selectivity. Then a few compounds are detectable, but at extremely low concentration levels, with any given setting. For this reason programmable instruments as well as instruments using a diode array to permit the record of the entire emission spectrum as a function of time are being developed.

In comparison to other detection methods, fluorescence offers greater detection sensitivity, lower detection limits, and in the same time less sensitivity to fluctuations of flow rate, temperature and pressure, since it measures the intensity of emitted light. On the other hand it is highly selective and a large number of compounds just cannot be detected this way. Its dynamic linear range varies widely from compound to compound due to the peculiarity of fluorescence mechanism, the possibility of quenching and self-absorption or absorption by other solutes or impurities. It must always be checked, never assumed. Highly pure solvents, free from fluorescing impurities are required.

Interaction during collisions between the excited species and the molecules of some compounds may result in desactivation of the excited species without the emission of a photon. This quenching effect of some compounds like non-fluorescing polynuclear hydrocarbons can be used to detect them with the fluorescence detector, if the mobile phase is spiked with 5 ppm aniline by recording the decrease in the fluorescence emitted by aniline which accompanies the elution of these hydrocarbons[42].

The development of lasers has permitted a decrease of about ten times in the detection limits versus conventional light sources [31,43] in spite of the enormously larger energy available[44]; some further advances are probable. The use of two-photon excited fluorescence[44], controlled by different quantum mechanical selection rules, and the use of temporal resolution, using a pulsed laser[45] which permits discrimination against fluorescence by solvent impurities or cell walls, to enhance detection sensitivity of long fluorescence lifetime species, offer challenging opportunities[31].

Combination of several of these techniques applied simultaneously can generate a deluge of information whose handling will delight computer experts and puzzle analytical chemists for years[46]. Provided their work is properly coordinated, fabulous results can be expected.

3.4.2 Phosphorescence[47]

Most organic compounds do not fluoresce. They decay from their excited singlet state to ground level through a triplet-state whose lifetime is much longer and is usually deactivated by collision before it can emit a photon.

An exception to this rule is biacetyle whose strong phosphorescence is observed in thoroughly deoxygenated solutions at room temperature. Sensitized room temperature phosphorescence is based on direct collisional energy transfer from an eluate triplet state to the biacetyle triplet state. Biacetyle is added to the mobile phase. Sensitive detection of chlorobiphenyles and chloronaphthalenes has been achieved this way[47].

3.4.3 Inductively Coupled Plasma [48, 49]

The column effluent is nebulized and fed into an argon plasma formed at atmospheric pressure and sustained by inductive coupling to a high frequency magnetic field [48, 49]. The eluate vapor is dissociated into free atoms, which are excited and ionized in the plasma. The emission light of the plasma is collected, analyzed with a grating and the intensity of wavelengths corresponding to selected elements is recorded.

This technique is widely used in elemental analysis as its response is free of matrix effects and offers sensitivity at the ppb level. It is compatible with LC effluents, although the plasma lacks stability with the hydrocarbons used in normal phase LC. This detector is not suitable to analysis of non-metallic elements (C, H, O, N, P) but offers excellent sensitivity and selectivity for metals, and will be especially useful in enzyme analysis [48, 49].

3.4.4 Polarimetry [50]

The rotation of the polarization plane of a planar polarized light beam can be measured to 1×10^{-3} degree. This permits the detection of optical isomers with strong optical activity. Most compounds which have one or several asymmetrical carbon atoms have a small or even negligible activity, but some have a large one, especially close to the main UV absorption band. This detector can be suitable for detection of biochemical compounds, and has the advantage of being insensitive to large changes in eluent composition (gradient elution). Its detection limit is in the 1–100 ppm range, unfortunately, which drastically reduces its attractiveness.

3.4.5 Photoacoustic Detection [51, 52]

Non-fluorescing molecules which absorb a photon are raised to an excited electronic state and subsequently return to ground state following a deactivating collision. The net result of this phenomenon is a strong local temperature raise, resulting in a pressure jump which can be detected by a sensitive microphone. The use of a pulsed laser, affording a high energy light source with high frequency modulation, permits a sensitive detection, down to 10–100 ppb depending on the compounds used and experimental setting [51, 52].

This detector is potentially more sensitive than a UV absorption detector [31], but the light source should have a short enough wavelength, so its light is absorbed by most solutes, but of course not by the solvent. Suitable lasers are complex and expensive.

3.4.6 Laser Thermal Lens [53]

A laser beam passing through a solution is slightly absorbed. The heat generated being larger in the center of the beam, and the refraction index of liquids being strongly temperature dependent, the radial temperature gradient results in the solution behaving as a lens. The effect on the laser beam can be detected and recorded [53] by measuring the overflow of the laser beam beyond a limiting aperture.

This detector is attractive, owing to its simplicity, but not yet sensitive enough.

3.5 Electrochemical Detectors

Although according to a recent survey[54] the optical detectors (based on absorbance and fluorescence) are still the workhorses of HPLC detection (more than 71% of the applications in 1980–1981), the electrochemical (EC) methods are slowly emerging as a powerful tool for trace analysis. The popularity of the currently available EC detectors stems from their high sensitivity (picogram and femtogram range) and selectivity of detection: without the use of secondary reactions, the number of compounds which can undergo electrolysis at a preselected potential is limited. Furthermore, EC detectors are highly suitable for operation with microbore columns, which makes them very attractive in view of the recent trends towards miniaturization of LC equipment[55, 56].

Contrary to the situation with other LC detectors, an EC detector is a transducer which permits a direct conversion of chemical information into electrical current. Furthermore, the EC reaction takes place at the electrode surface which makes it possible to reduce the cell volume to a thin layer of minimal thickness. Thus, cell volumes of less than a microliter can be easily obtained[55–59]. Recently, an EC detector with a cell volume of less than 1 nL was developed and used in conjunction with capillary columns[60].

The overall detection sensitivity of these detectors depends on the mass transfer and electron transfer processes, which are complex functions of the solute concentration gradient in the cell, solvent flow rate and viscosity, temperature, cell geometry and reaction thermodynamics and kinetics[61].

Flow-through EC cells can be classified according to their geometries into three basic categories: channel type, tubular and wall-jet (for detailed description of each type, see Ref.[61] and those contained therein). Depending on the extent of electrolysis, EC detectors can be further subdivided into amperometric and coulometric[61]. In the former, only approximately 5% of the total number of electroactive molecules will undergo the reaction, while in the latter, the conversion efficiency is 100%. However, contrary to the expectations, coulometric detectors afford lower detection sensitivity since an increase in electrode area is necessarily accompanied by a concurrent increase of background current and noise, and thus lower S/N ratios[62]. In addition, larger cell volumes necessary for accommodation of electrodes and lower flow rates can degrade the column performance.

The majority of EC detectors operate on a 3-electrode potentiostatic system which assures the constancy of the working potential against potential drift and ohmic drop. The positioning of the reference and auxiliary electrodes with respect to the working electrode is important since it may affect the noise and response time[61]. Two types of electrode materials enjoy a widespread popularity for the analyses of easily oxidizable and reducible substances: carbon paste and glassy carbon. While the former gives a lower background current, it cannot be used with mobile phases containing more than 20–30% of organic modifier. Conversely, glassy carbon electrodes are more rugged and compatible with organic solvents. Their useful potential range is wide: $+1.2$ to -0.80 V. Mercury and amalgamated gold are employed for reductions.

Since EC detectors are very efficient antennas capable of picking up 50 or 60-H_z noise, the use of adequate shielding (Faraday cage) is mandatory. Furthermore, pump oscillations and temperature variations may cause problems and should be dampened or eliminated, as well as certain mobile phase contaminants (oxygen, trace metals, etc. ...).

In addition to the single working electrode arrangements, multiple electrodes designs have also been used in order to further increase the specificity of EC detection and to expand its use[63]. When the two working electrodes are connected in parallel and maintained at different potentials, the generated currents can be recorded simultaneously and their ratios used for solute characterization. With the serial arrangement, however, the upstream electrode serves as a post-column reactor for the successive reaction at the second electrode. Thus reaction specificity can be greatly enhanced for reversible or quasi-reversible reactions. The upstream electrode can also be replaced with a coulometric cell if complete elimination of the interferences is desired[64].

In addition to the currently used controlled potential methods, several derivative methods, such as differential pulse, reverse pulse and square wave voltammetry have been proposed and used in order to enhance the selectivity and sensitivity[65]. However, their potential has not yet been fully exploited.

Because of the high sensitivity of EC detection, its tandem operation with HPLC, particularly in the reversed-phase mode, is becoming an extremely useful tool for the analysis of trace compounds in complex matrices of biological, environmental and pharmaceutical origins[66]. Non-electroactive compounds can be analyzed using the concept of post-column chemical reactions[67]. This type of detector is discussed in a separate chapter.

The polarographic detectors have been completely replaced by amperometric detectors, due to the problems encountered in the design of a microdrop mercury electrode and the limited anodic potential range in continuous operation, which precludes its use in the electrooxidation mode[4].

3.6 Other Detectors

Sophisticated spectrometric techniques used to scan the effluent of an LC column and derive mainly qualitative information regarding the structure of unknown components of complex mixtures are discussed in a separate chapter which deals with LC – fast Fourier transform infra-red spectroscopy, with LC – mass spectrometry and with LC – nuclear magnetic resonance.

Similarly, reaction detectors which would have deserved a section in this chapter are dealt with in detail in a separate chapter.

Discussed here are detectors which are still or have become of marginal interest. Some of them may become of significant or great importance in the future if they hold the promises of their first designers or if miniaturization of LC columns imposes the development of specific detectors. Due to their still uncertain future they cannot be discussed in detail, however.

3.6.1 Transport Detectors

The principle of these detectors consists in eliminating the volatile solvent and carrying the non volatile solute in a separate area where it is detected and quantitized.

Originally developed by James, Ravenhill and Scott[68] who used a moving wire, and by Haahti and Nikkari[69] who used a gold chain, this detector has been further improved by Scott and Lawrence[70]. Although it has been quite popular in the early seventies it has fallen out of favor because of lack of sensitivity, linearity and reproducibility, these last problems being due to mechanical difficulties. The eluent stream coats a wire[68] or chain[69] which passes through it, but unfortunately can take only a small fraction, depending on the surface tension of the solution. The film is carried through a low temperature oven where the solvent vaporizes, leaving the non-volatile solute on the wire or chain, and then a high temperature oven where it is pyrolyzed[68, 69] or oxidized into CO_2[70].

The pyrolysis products are swept to a flame ionization detector[56, 69], as well as the methane obtained by reduction of the carbon dioxide[70]. The sensitivity never exceeded a few µg/ml of solvent which is hardly as good as a good refraction index detector[4].

A different principle has been suggested by Charlesworth[71] and used by different authors[72, 73]. The column effluent is nebulized in a warm gas stream which carries the solution droplets through a drift tube, where the solvent is vaporized, and then through a parallel light beam. The droplets or solid particles of non-volatile solutes scatter light. The scattered light is collected and used as a measure of solute concentration. This detector is extremely simple and reliable. Its sensitivity is at present comparable to that of a good RI detector for sugars and fats[72, 73] and it is remarkably insensitive to changes in the mobile phase composition[71 – 73]. The sensitivity could probably be improved by one order of magnitude[73]. The dynamic linear range of the detector is still controversial[71, 73]. This detector could become competitive with the refraction index detector for the analysis of carbohydrates, fats and related compounds which do not absorb in the conventional UV range.

3.6.2 Spray Impact Detector[74]

The eluent stream is forced through a narrow tube to form a jet impinging at high speed on an electrode and forming a spray. In this process electric charges are separated by a still obscure process, and a current flows through the electrode, the spray droplets being positively and the electrode negatively charged. The current intensity depends strongly on the concentration of organic compounds in the mobile phase. With pure water, detection limits between 2 ppm and 1 ppb in the eluent reaching the detector have been reported. With water/organic solvent mixtures the detection limits are higher; while they are still in the ppb range for polar, ionized species like nitrophenol, they are in the 10–100 ppm range for hydrocarbons[74].

The detector has some attractive features like the small dead volume, the fast response and potentially very low detection limits for ionized compounds. It has been totally ignored for the last 10 years and some further work is necessary to understand better its mechanism and improve its performances.

3.6.3 Radioactivity Detectors

Since the beginning of column chromatography, biochemists using radio-labelled compounds have attempted to monitor their elution using different kinds of counters. The main source of problem is the design of a suitable compromise between the requirements of HPLC (i.e. low dead volume, fast response) and those of radioactive countings, which are just the opposite, to achieve good counting efficiency and low detection limits[4, 75]. These detectors are necessary for people working with radio labelled compounds. Probably they should design a LC using wider bore columns than usual to allow the use of rather large volume counting cells, although their use would, admittedly, require large sample size.

3.6.4 Electrical Conductivity Detector

The measurement of the resistance of an ionic solution, using an alternative voltage to eliminate spurious effects due to electrode polarization can be used as a detection principle, for ionic species. One of the potential advantages of such a detector is the very small dead volume[76]. When used with buffer solutions detection limit is in the low ppm range[4]. With non ionic mobile phases of low electrical conductivity the detection limit can be reduced by several orders of magnitude.

The combination of the conductivity detector with ion-exchange chromatographic columns has led to a technique called "ion chromatography", which has known a rapid development since his first presentation by Small et al.[77]. In spite of the excellent selectivity of ion-exchange resins, the analysis of non UV-absorbing ions has been hampered essentially because of the lack of a suitable universal detector. In order to suppress the overwhelming conductivity of the background eluent electrolyte, Small et al. introduce, between the ion-exchange separation column and the conductivity cell, an ion-exchange suppressor column with opposite charge groups, *i.e.* an anion-exchange suppressor column for cation separation and vice-versa. The eluent is thus neutralized and the conductivity of the eluate ions can be conveniently measured. The suppressor column increases the analysis time and the band broadening and may give some undesired effects (ion-exclusion, reaction with some ions).

It has to be replaced or regenerated after a few analyses are performed but this operation can be done automatically in modern instruments. Alternatively, an ion-exchange hollow fiber may be used for eluent neutralisation[78]. The technique has been applied to organic as well as inorganic ions.

The fast and recent development of ion analysis with a conductivity detector has led some manufacturers, who, for patent reasons, cannot include a suppressor column in their instruments, to develop conductivity meters with electronic background suppression. This technique, however, cannot be applied to strongly conducting eluents and gives higher detection limits i.e. poorer sensitivity than the previous one.

3.6.5 Gas Chromatography Detectors

These detectors are sensitive and reliable. Some of them like the flame ionization detector are non-selective. Others like the electron capture detector, the flame photometric detector and the thermo-ionization detector are selective for some groups of compounds. Various techniques of vaporization or direct injection of the LC effluent in these detectors have been attempted.

The electron capture detector has been modified[79, 80] to serve as an LC detector. The effluent is nebulized in a stream of warm gas and vaporized. A fraction of this gas stream passes through the detector which is insensitive to the eluent (no water), and detects the solutes which have a strong electron affinity and are volatile at 300° C, maximum temperature at which the detector can be operated. Polar solvents like methanol or THF can be used, but they should be carefully deoxygenated[81]. This detector is very sensitive for some compounds, but its dynamic linear range is narrow and its use is restricted to volatile compounds. These drawbacks probably explain why it has slowly fallen into near oblivion.

The flame ionization detector and its related parents the *flame photometric detector* (selective for sulphur and phosphorus) and the *thermoionic detector* (selective for phosphorus and nitrogen) have been suggested by McGuffin and Novotny[82] for use with very narrow bore packed columns and open tubular columns. The flow rate of mobile phase through these columns is very small (less than 1 μl/min) and when nebulized is quite compatible with proper operation of these detectors. Water and acetonitrile give very small background current even with the flame ionization detector. Non-volatile solutes are carried to the detector flame as very small droplets and burn as well as volatile ones. Reasonable performances are obtained, especially if one takes into account the very small dead volume offered by these detectors, which allows their use with narrow capillary columns.

It is too early to predict whether these detectors will remain of academic interest or undergo a second career in LC. Certainly the high level of technological development they have achieved through their popularity in GC makes them easy to afford.

3.6.6 Polymer Specific Detectors

The objective of the analysis of polymers, especially those of synthetic origin, is to supply a molecular weight distribution (MWD) of the sample of known chemical structure. Classically, this is done by converting the elution curve obtained with a concentration sensitive detector to a MWD curve with the help of a calibration curve relating the elution volume to the molecular weight. Because the construction of such calibration curves is sometimes difficult or impossible, it has been found useful to combine the concentration sensitive detector with a molecular weight sensitive detector. Among the several classical static techniques of polymer characterization, two have been amenable to the flow-through requirements of modern LC (low cell volume, fast response). The first one is a viscosimetric detector based on pressure drop measurement in a small capillary tube[83, 84]. This requires, however, the previous knowledge of the constants in the Mark-Houwink equation relating the intrinsic viscosity to the molecular weight for each particular polymer type. The second one is a light scattering detector[85]. The use of the laser technology allows measurements of the scattered light to be made at low angle (less than 5°). This considerably simplifies the relationship between this scattered light and the solution properties. The measurement of the excess Rayleight factor basically provides a continuous measure of the product of the mass concentration by the molecular weight of the polymer solution flowing through the cell[86]. The combination with a concentration detector (refractive index, UV, IR or fluorescence detector) allows easy extraction of concentration and molecular weight, which gives the desired MWD curve.

4 Automatization of LC Chromatographs and Instrument Control

The automation or mechanization of a liquid chromatograph concerns the automatization of the various subunits in the chromatographic instrument such as a sample pretreatment unit (extraction, derivatization ...), a sample injector, a column switching unit, a post-column derivatization unit and a fraction collector and the cooperative operation of these different subunits. The term "instrument control" comprises all the electronic hardware which allows proper operation of the overall system. This includes the control of the eluent delivery system (flow feed-back, gradient elution ...) and the dispatching center for all command signals dedicated to autosampler, data acquisition, fraction collector, and the various external events (flags). Several interesting reviews have been published dealing with these problems[1-3].

4.1 Mechanization

4.1.1 Sample Pretreatment Module

The preparation of samples prior to chromatography is most often a preliminary unavoidable step. A survey of the literature seems to indicate that analysts had not focussed much effort in the past on the automation of this operation which is, however, generally time consuming. Automated sample preparation has been recently reviewed[4]. It will be made much easier by the recent appearance of robots dedicated to or trained for analytical chemistry.

The principal steps of sample pretreatment depend on the nature of the material to be analyzed (liquids or solids)[4]. Solid samples first require dissolution in or homogeneization with a proper solvent in order to extract compounds of interest. Critical parameters of this operation are the nature of the solvent and the crushing conditions (speed, duration and temperature). The extraction is generally followed by removal of solids using either filtration, dialysis or two-phase extraction.

The next steps are common for liquid and solid samples. They consist of solvent extraction and/or solvent exchange, and finally derivatization and concentration. Several techniques can be used for these purposes. Pre-column derivatization is less critical than post-column derivatization in terms of equipment since, in the former case, there is no problem of band broadening, remixing and sample dilution. Once the compounds of interest have been extracted, concentrated and derivatized, they are transferred to the sampling valve. In well designed systems, each sample is prepared automatically while the preceding sample is on the column, so the automatization of the preparation process increases the sample throughput rate[1].

Various examples of automated sample pretreatment published in the literature clearly demonstrate the advantages of the technique both in terms of sample throughput rate and of precision[5-9] since there is no way a chemist can repeat as reproducibly his movements and all the elementary times of the various steps of the sample preparation as an automatic system. For instance, the average day to day coefficient of variation for the chromatographic analysis of common tricyclic antidepressant drugs and their metabolites in serum or plasma was found to be in the range 5–6%, with an automated sample pretreatment including sample aspiration, adjustment of sample pH, extraction of the drugs as bases, back extraction of the ionized bases and injection onto a reversed phase column[9]. In these conditions, the carry-over was estimated to be 2–3%, the major source of sample loss being adsorption of the drugs onto the walls of the polymeric pump tubes used to transport the sample through the system, and which tend to behave as very low efficiency open tubular chromatographic columns[9].

A particular case of sample pretreatment is trace analysis. This involves precon-centration using specially designed precolumns, followed by sample desorption and transfer to the analytical column. The equipment associated with this operation consists mainly in multiport switching valves. The automation is quite easy and is made directly from the control unit of the module or from the instrument control station. The operation and optimization of sample preconcentration, including pre-column design is now well documented for a variety of problems (see Ref.[10] and References cited therein).

4.1.2 Automatic Sampling System

The purpose of an autosampler is to transfer a given volume of a sample from a vial to the column, with the maximum flexibility possible, and minimum loss of sample and cross-contamination.

Commercially available autosamplers have different degrees of sophistication. In the simplest case, the samples are injected in the order in which they are loaded on the carrousel and they are all analyzed, with the same operating parameters (injected volume, number of repetitions, time between consecutive injections). More advanced autosamplers allow greater versatility to control independently the injec-tion parameters for each sample.

There are basically two principles of operation used for transfering the sample from the vial to the injection valve (direct syringe injection through a rubber septum, although conventional in gas chromatography, has been abandoned in LC). The common procedure is a pneumatic transfer using a double needle. The needle is introduced into the sample vial through a septum. A flow of nitrogen coming from the first, shorter needle, pressurizes the vial and pushes the liquid sample up the second, longer needle, which dips in the sample. This needle is connected to the sampling valve by a piece of capillary tubing. Depending on the nitrogen pressure and on the time the vial is pressurized, a variable volume of sample is transferred to the sampling loop. This procedure has the advantage of extreme simplicity and long term dependable operation, but it presents several drawbacks. First, large vol-umes of sample are generally required in order to purge all the circuitry (typically, several hundreds of microliters for an injection of a few microliters). Second, it is difficult to control the volume of sample injected since, for a given nitrogen pressure and time of pressurization, the volume actually transferred to the loop depends on the sample viscosity which varies with its composition and temperature, and on the permeability of the sampling system circuitry which can change slowly with corrosion by samples or deposits. In fact, in most cases, reproducible injections can only be achieved when the loop of the sampling valve is completely filled with sample, which is not the best condition for good plug injection and maximum column performance.

The second principle of operation of an autosampler consists in transferring the sample from the vial to the injection valve using a syringe. It is in fact the automation of the manual procedure. Although this technique necessitates a more sophisticated mechanical equipment, it has the advantage of a much more efficient processing of the sample. The dead volume is limited to the volume of the syringe needle, and the amount of sample injected can be precisely controlled. For the injection of a few µl sample, no more than 10–15 µl total sample volume is necessary. Sophisticated and totally versatile autosamplers are based on this design.

The complete automation of an autosampler, that is the possibility to chromatograph each sample under different conditions is possible only if the system is microprocessor controlled, which in the same time permits easy introduction and change of the instructions. Furthermore, it is necessary that the autosampler be remote controlled and can itself control other components of the chromatographic chain, for example initiate an integrator or a data system, order the preparation of the analytical report of the sample, control a fraction collector, place the chromatograph on stand-by when the set of samples is analyzed.

A procedure for the evaluation of autosamplers has been recently published [11].

4.1.3 Column Switching Unit

The advantages of multidimensional systems have been demonstrated many times [12–15]. In the simplest case, this technique involves a change in the solvent composition (continuous or step gradient elution) and in the most sophisticated one, a stationary phase (column) change. The situation is somewhat similar to that of preconcentration. The hardware is mainly composed of multiport switching valves (and their pneumatic pilots) and can be easily automated. An interesting utilisation of column switching is the boxcar chromatography introduced by Snyder et al. [16].

4.1.4 Post-column Derivatization Unit

This technique is extremely useful and widely used in chromatography for both improvement (decrease) of detection limits and specificity of detection. Depending on the type of derivatization reaction, different steps are necessary. They include addition of suitable reagents, change of pH, delay loops, effluent storage, liquid-liquid extractions, heating, phase separation ... Sophisticated equipments, most often self-controlled are available and there is not much to say about their automation which is rather straightforward. There is an abundant literature on the instrumental aspect of post-column derivatization. Very detailed reviews have been published on the subject [17–20].

4.1.5 Fraction Collector

Fraction collection is associated with preparative chromatography (micro, mini, medium and large scales). Although there is not a large choice of equipment on the market at present it must be noted, however, that very often people need to collect the bands they have separated. Since a preparative application requires usually several identical consecutive injections, it is extremely interesting to automate this step.

Fraction collectors operate on one of two principles. The collection of the column effluent in a given vial can be made either on a time basis or on the basis of a signal generated by the detector (threshold or slope change). Both procedures have advantages and drawbacks, but it seems, however, that due to the high level of flow rate (and thus retention time) constancy now attainable with modern pumping systems, a time-based collection is to be preferred.

Critical points that have to be considered in the choice of a fraction collector include the number of collection vials and their volume, the dead volume of the circuitry (band broadening and zone remixing) and the response time (time required to move from one collecter to another one). The design of the system can be such that either the collection vials are mobile (and are moved close to the outlet of the column during collection) or the "head" of collection is mobile and is positioned above the corresponding vial. A "Christmas tree" valve system is equivalent. The first approach has the advantage of offering a very low dead volume, but places constraints on the collection vials and the response time. The opposite is true in the second case. It seems that the general trend is to choose the second approach which is more versatile, but introduces a greater dead volume.

It must be noted that the design of a fraction collector is much less critical than that of an autosampler. A simple collector can be easily home-made using a multiport switching-valve controlled by the instrument control station (external events, see next section) or by hand. The characteristics of various fraction collectors have been discussed in the literature[21-23]. An interesting application of sample post-column collection and manipulation is given in Ref.[24].

4.2 Instrument Control

There is no doubt that the trend in the design of future chromatographs will be the systematic use of microprocessors. Ideally, all the parts of a chromatograph should be self and remote-controlled using a standard procedure of instrument dialog (IEEE 488 for instance). This would yield the expected degree of instrument compatibility which is unfortunately almost inexistent at present.

It is necessary to distinguish between "integrated" instruments and modular ones. Integrated instruments use the same electronic unit to control the pumping system (including the gradient mode) and, to a more or less high level, the rest of the equipment (sample injection, detection, valve switching and other external events). In modular instruments, each part of the chromatograph has its own command unit and is independent of the rest of the equipment. Most often, a separate control unit is available which synchronizes the operation of the different parts of the chromatograph. This modular conception is by far more versatile than the integrated one, but often also more expensive. Furthermore, lack of standardization of communication procedures sometimes makes the assembly of units of various origins difficult. Many individual control units have been developed by analysts in order to automate a complete chain. These units are in general rather specific since they have been designed for particular operations.

It is impossible to discuss the automation of a pumping system or a detector without entering into the details of the design of these equipments. These points are also alluded to in other parts of this book but it is not the aim of this chapter to provide an in-depth discussion of these rather specialized aspects of modern instrument design. The basic principle in purchasing units from different manufacturers and assembling them into an automated chromatograph remains the old "caveat emptor".

The role of the instrument control unit is to achieve certain specific tasks for the complete automation of the chromatograph. This includes control of the pumps (solvent line selection, flow-rate adjustment ...) and of the detectors, particularly with sophisticated spectrophotometers (wavelength and sensitivity changes, scanning ...). Most often, the instrument control station can generate logic signals to command other parts of the equipment (external events). For instance, the autosampler previously programmed, can be started with an impulse generated by the instrument control unit after it has checked that the chromatograph is operating properly. Upon receiving coded pulses from the autosampler, the control unit can perform various tasks: valve switching, column switching or back flush, edition of analytical report, or put the equipment on stand-by. The situation is the same for the fraction collector. External events can also be used to control switching valves and other relays *i.e.* for column switching.

Many publications dealing with the complete automation of a given separation have appeared in the literature. Most of them report only on the use of an autosampler, an integrator (or a data system, see next section) and sometimes a fraction collector. More original papers are devoted to preparative chromatography applications, including sample prefractionation and sophisticated fraction collection in the case of overlapping peaks. In almost all cases, the associated electronic hardware is based on a microprocessor (see Refs. [24–28] for instance).

5 Data Acquisition, Processing and Handling

Although microprocessors appeared on the market in the early 70's, it is only since 1975 that the cost of these chips has become sufficiently low and their performances sufficiently high for routine applications in instrumentation. Microprocessors are particularly well suited for process control and iterative tasks. The concomitant development of memory chips and microsystems has brought a very elegant solution to the problems of data acquisition and processing.

It is clear that the complete automation of a chromatograph is only fully realized when the signal of the detector is automatically acquired and processed. This can be done in two different ways. The simplest solution is to use an integrator. Elaborated equipments are now available which permit on-line processing of the signal and compute retention times and peak areas with sophisticated calculation procedures for base-line drift corrections, integration of overlapping bands etc. Almost all top of the line chromatographs include now an integrator or a data system. Although integrators offer very interesting possibilities (including the control of external events), they nevertheless suffer from two different limitations. First, the calculations are mostly limited to the derivation of retention times (at peak maximum) and peak areas. It is impossible to use such techniques as fast Fourier transforms (FFT), deconvolution or correlation or even to calculate peak moments. Second, because the signal is processed "on the fly", it is impossible to replay the data with different integration conditions or to store the data for further manipulations.

The other way to process the chromatographic signals is to acquire and store them on a given support (cassettes, floppy disks ...) and then to proceed to the necessary calculations. The ideal solution is in fact a data system with both capabilities.

The following discussion will focus on the problem of data acquisition hardware (interfacing and transfer), processing (filtering and the associated techniques of signal to noise ratio improvement) and handling (peak recognition, base-line correction, deconvolution ...).

5.1 Data Acquisition

The design of interfaces for data acquisition and the relevant hardware problems have been discussed several times in the Refs.[1−7].

5.1.1 Analog to Digital Conversion

The output of a detector gives an analogic signal, most often a voltage. Acquisition and storage of the signal require its conversion into a digital form using an analog-to-digital converter (A/DC). This conversion must be made in such a way that a minimal amount of the information contained in the signal is lost. The accuracy and speed of the A/DC should be adjusted to those of the detector. The maximum frequency of data acquisition should be such that about 3 data points are taken during a time equal to the detector response time. It is important, for signal to noise ratio considerations, that the signal be integrated during this period, $i.e.$ the digital number acquired must be proportional to $\int_t^{t+\tau} y dt$, where y is the detector signal and $1/\tau$ the data acquisition frequency. If the integration time is smaller, the measurements are too noisy and information is lost.

The accuracy of the conversion also depends on the size of the number used. The outlet of the converter is a binary number. Each position of the binary word is a bit, and the number of bits determines the precision of the conversion. For instance, an 8-bit byte is equivalent to a decimal value between 0 and 255, so the smallest change in signal that can be quantified with this byte is 1/255 ($ca.$ 0.4%) which is not precise enough. In most cases, it is necessary to perform a scaling operation before the A/D conversion in order to ensure that the voltage is suitable for conversion. This is accomplished by an operational amplifier or, more conveniently, by an autoranging amplifier which automatically adjusts its amplification factor to maintain the outlet voltage within certain limits. Although more complex and expensive than a simple amplifier, the autoranging amplifier has the major advantage of allowing the conversion of a wide dynamic range signal with the same degree of resolution for small signals and for large ones.

There are four types of A/D converters: sample and hold successive approximation converters (SHC), single slope converters (SSC), dual slope converters (DSC) and voltage-to-frequency converters (VFC). In the SHC, the signal is hold and compared to a variable voltage until both signals are equal. The procedure is fast, well suited to multiplexing but very sensitive to noise (because the signal is integrated for a short time) and expensive. In the DSC, the signal is used to load a capacitor during a fixed period of time. Then the capacitor is allowed to discharge in such a way that the time of discharge is proportional to the signal. This converter is very accurate, easy to operate and not sensitive to noise (since it is basically an integration process) and the signal can be integrated during a large fraction of the time. It has the drawback of being slow, requiring at least 10–100 ms per conversion, which is merely adequate for fast LC and is much too slow for any application involving the use of fast Fourier transform. The principle of the SSC is similar except that the capacitor is charged to its maximum value. The time needed is inversely proportional to the signal. This converter is also simple to use and not sensitive to noise, but it is less accurate than the DSC. In the VFC, the signal is used to charge the capacitor of a RC oscillator, and the integrator is allowed to charge and discharge continuously, generating an output voltage whose frequency is proportional to the input voltage. This converter is cheap but slow and requires more complex computer programming.

It is difficult to recommend the use of a specific converter or even converter principle since the best choice depends on the rest of the equipment used and on the computer facilities available. It seems that VFC and SMC are the most commonly employed systems.

It must be noted that integrating converters (SSC and DSC) have a wider dynamic range but can work only at rather low sampling rates (maximum about 200 Hz). Successive approximation converters are capable of sampling rates as high as 100 Hz. Integrating converters have, however, a larger resolution than successive approximation converters, and they are frequently 16-bit devices whereas SHC are rather 12- or 8-bit ones. The problem of sampling rate is, in general, the most critical one (for closely reproducing the signal and for such applications as fast Fourier transform, FFT) which explains probably why 8-bit devices are becoming popular, also because of their low cost and in spite of their limited precision. The choice is thus depending on the applications the analyst has in mind.

5.1.2 Data Transmission

After the signal has been digitalized, it has to be transferred to the computer and stored. There are two types of transmission: serial and parallel. The format of the characters transferred is most often the familiar ASCII coding chart.

In the serial transmission, the bits of the characters are transferred one at a time. Since, in most cases, an A/D converter has a parallel output, this output must be first converted to a serial form. This is usually done with a shift register. There are two types of circuits for transmitting bits between the interface and the computer: 200 mA loop and RS 232C. 20 mA loop transmission is based on a current flow in a closed loop between the sender and the receiver. An instantaneous 20 mA current represents a "1" bit, whereas no current represents a "0" bit. In the RS 232C standard, the "1" state is represented by any voltage more negative than -3 V and the "0" state by any voltage more positive than $+3$ V.

The main advantage of 20 mA transmission is that it permits long distances between the devices (up to 300 m). It is also relatively insensitive to noise. There is, however, no possibility for modem control and handshaking which is the part of the dialogue between the "talker" and the "listener" which makes sure that the data can be send on the bus and/or has been correctly received. RS 232C is more sensitive to noise, especially at high baud rates and is therefore limited to short distances (15 m maximum). It has, however, modem and handshaking possibilities and other control lines which makes it by far more popular than the 20 mA loop. This latter type of transmission remains useful over long distances, however. The maximum speed of transfer is about 3,000 bauds.

The second mode of transmission is the parallel mode. It is characterized by a simultaneous transmission of the 8 bits of an octet on the computer data bus, using a series of dual inputs "and" gates. The procedure of transmission of an octet obeys strict rules of information exchange between the talker which sends the signal and the listener which receives it. The most popular standard of transmission between an instrument and a computer is the IEEE 488 standard. It fully defines the 16 lines bus (physical cable) between the computer and the interface. Eight of the lines are for parallel transmission of the ASCII character, three are for handshaking and the last fives for general bus management. Because each instrument has its own address on the bus, several peripherals (up to 15) can be connected to the same data bus. This mode of transmission is very versatile and simple to use, but it is more sensitive to noise than serial transmission. The maximum distance between the computer and the instrument is about 15 m. The speed of transfer is very fast, up to several hundred thousands bauds.

To illustrate further the comparison between the performances of the two transmission modes, let us mention that printers are usually connected to microcomputers through RS 232C serial transmission, while magnetic disks are connected through IEEE 488.

There are also two other approaches to getting information from an experiment into a computer: analog I/O and digital or parallel I/O. These approaches are, however, more difficult to use since they require the user to do much more in setting up an interface for real time data acquisition.

Digital I/O (input/output) was the first type of computer interface developed for analytical instruments and it is still often used. There are wide variations in the number of bits transmitted in parallel over data lines. Most digital I/O interfaces consist of one or more 16-bits or 32-bits modules. The number of bits is a multiple of 4 because data are binary coded decimals (BCD). 4 bits are thus necessary to code a figure. The standardization of how digital data are presented to the interface is now quite well achieved since most systems are TTL (transistor transistor logic) compatible.

5.1.3 Frequency of Acquisition

Data acquisition is a discontinuous process wich can be characterized by its rate. The frequency of acquisition (FA) has an upper limit which depends on the hardware of the interface, the type of transmission mode chosen and the computer memory capacity. It is clear, however, that FA must be adapted to the signal to be sampled, in order to avoid loss of information (when FA is too small), and also noise and too large an amount of data to be stored (when FA is too large).

According to the Nyquist-Shannon frequency rule, the sampling rate must be larger than twice the highest frequency component of the input (for FFT). Thus,

FA must be conveniently chosen with respect to the signal itself and the type of calculations to be made. It must also be indicated that FA should be a multiple of the power supply frequency to ensure that the induced noise is minimal. Moreover, it is generally admitted that, in order to describe an elution profile with enough accuracy, a minimum of 5 data points per standard deviation is necessary (about 20 points per peak)[8, 9]. Better results are obtained with a higher density of data points.

5.2 Data Manipulation

The purpose of this section is not to describe the methods of calculation and data processing. It is rather to summarize briefly the different possibilities of data manipulations. This section is divided into three parts: data processing (filtering, signal-to-noise improvement), data handling (peak recognition, base line correction, integration ...) and data interpretation. This last part is rather a prospective investigation of what could be done in the future with intelligent chromatographs, than an account of what is currently done, as most data interpretation carried out in chromatography is presently done by the analyst himself.

5.2.1 Data Processing

An important advantage of computer data acquisition and manipulation is the improvement of signal-to-noise ratio. Although the simplest solution is to use an analog filter (RC circuit), it is also the worst one (unless the detector response is indeed too fast for the chromatographic column, which is a rare situation indeed) since it modifies the form of the signal through an exponential convolution and can give erroneous estimates of the analysis time and the peak area. Numerical techniques are more preferable, but it must be emphasized that filtering which is a convolution operation leads to both smoothing of noise, which is the desired result, and smoothing of waveforms, which itself results into analytical errors. Some form of compromise, maximizing the first results and minimizing the second ones must thus be found out.

Numerical filtering can be carried out either in the time domain or in the frequency domain[10]. Filtering in the time domain includes various techniques such as least-squares smoothing and correlations. A first procedure of filtering uses a rectangular convolution function. This involves taking a certain number of data points, multiplying each data by the appropriate coefficient, summing up and dividing by the number of points. This yields one averaged data point. Then, the first data point is dropped, a new one is added to the set and the whole process is repeated. Another possibility is curve smoothing by fitting an nth degree polynomial. The parameters of the polynomial are adjusted to minimize the sum of the squares of the residuals. Details regarding this procedure are given in Ref.[11]. Fast digital filters are based on this concept[12].

The use of correlation is based on the fact that signals are predictable whereas random noise is not[10]. This approach uses a chromatographic technique much more complex than usual, since a large number of injections must be carried out, 2^n to reduce the noise by a factor $2^{n/2}$. These cross-correlation techniques make use of a pseudorandom binary sequence to modulate the injection function, which is considered as a synthetic version of noise. The cross-correlation integral between the input and output functions produces a time average value close to zero at all times, except where all the signals are exactly in phase. The signal enhancement thus obtained can be remarkable, provided the system behaves linearly[13]. The procedure can even be improved (to handle base line drifts) if the cross-correlation is made on the first order derivative of the signal[14]. There is no application of this method to trace analysis by LC which is known to us yet.

Working in the frequency domain (Fourier transformation) offers unique advantages as illustrated by infra-red spectroscopy. Noise frequencies are easily discriminated from signal frequencies and signal enhancement methods become simple multiplication (convolution) or division (deconvolution) operations. The principle of the Fourier transformation is to consider a signal waveform as made up of simple sine waveforms of varying frequencies and phase relationships, and to convert the time domain signal to its frequency counterpart. Fast Fourier transform (FFT) is a special algorithm which permits the Fourier transform of a signal to be carried out in real time, either by strictly electronic means or with computer software. The most important feature of Fourier analysis is that convolution in the time domain is equivalent to multiplication of the respective Fourier transforms in the frequency domain. Thus, filtering a signal just consists in multiplying the Fourier transforms of the signal and the filter, and taking the inverse transformation to restore the filtered signal. Another characteristic of the Fourier analysis is that the number of Fourier frequencies necessary to describe a peak is inversely related to its time domain width. A direct consequence is that Fourier techniques may offer a better approach to data reduction for computer storage[15]. The data filtering in the frequency domain can thus consist in cancelling the undesired part of the Fourier transform and interpolating it to replace the suppressed part. This often needs several trials before giving optimum results and is best achieved by interactive process involving the analyst and the computer monitor. Spikes, spurious signals can easily be eliminated by this method[14]. Another method for signal-to-noise enhancement consists in accumulating and summing chromatograms obtained from similar injections. This technique is time consuming and only seldom used.

5.2.2 Data Handling

After the data have been acquired and first processed, the next step concerns peak detection and integration. There are different methods for peak recognition, the most frequently used being the combination of threshold level sensing and slope sensing[16]. The difficulties encountered in the design of the proper sequence of control of a conventional electronic integrator have been clearly discussed[17]. A more powerful method which requires, however, more calculations is to use an adapted filter. This technique uses the convolution product of the signal by the second derivative of the filter. It gives particularly interesting results when there is a base-line drift[14, 18]. The use of the pseudo-derivative of the signal, PDS[19], provides a good criterion for peak detection, because its variations for each peak are proportional to the peak height. Peak detection is more sensitive because the PDS noise is smaller than the derivative noise and the amplitude of variation of PDS is independent of the retention time, but only related to peak size. The detection of retained peaks is thus improved.

The algorithm for peak integration taking into account base line drifts, peak overlapping, etc. ... are described in the brochure of any electronic integrator. The reader is referred to them. Several publications describing programs for processing chromatographic data are available in the literature (see Refs. [20–22] for instance). It must be stressed, however, that most of these programs have been designed by electronic engineers with no serious understanding of chromatography and its problems. Extrapolation of solvent band profile, interpolation of base line, the various methods of allocation of band area in the case of bands which are not completely resolved are based on simplistic models of peak profile. Their use is extremely dangerous for the analyst and can result in major errors in quantitative analysis, especially for those models which do not show clearly after which principle they are operating for this critical step of band area allocation. The use of computer data acquisition and handling is to be preferred if the analyst can devote some time to instruct the computer to perform the treatment he does wish. The equipment and provisions for interactive dialogue is thus a necessary feature of the instrument.

Various other types of calculation can be made on stored data. For instance, the evaluation of the moments is very important for profile characterization [23]. It must be recalled that the mere calculation of the column efficiency requires the measurement of the first two moments, the calculation based on peak width at half height being most often a very optimistic evaluation of column efficiency [24]. The situation is the same for the evaluation of peak symmetry (or skew coefficient). Another interesting possibility offered by stored data is the adjustment of physico-chemical models and the calculation of the best values of the parameters of the model. An abundant literature is available on that subject (see Ref. [25] for instance). This possibility is particularly interesting for the determination of various physico-chemical parameters [26].

Resolution enhancement and quantification of overlapping peaks is also possible to some extent. Deconvolution methods can be used [27] as well as techniques based on profile normalization which give remarkably accurate results [28]. This method permits, by comparing peak profiles, to determine whether a peak in a complex chromatogram does correspond to a pure compound or not, providing the main component of the peak is known and available in pure form. This is a very sensitive test. Other types of applications based on stored data have been described recently [29].

Another advantage of acquiring and storing data is the possibility of documentation (data retrieval, data base management, security storage) and improved data presentation (logarithmic time scale, "zooming" on one part of the chromatogram ...).

5.2.3 Towards Intelligent Systems

Three conditions must be verified for a system to be called "intelligent": mechanical automation, storage of data and evaluation of the chromatograms and adjustment of the operating conditions (feed-back) on the basis of the analyst requirements and the results of the previous analyses. The last point to be discussed deals with the evaluation of the data in chromatographic terms (estimation of retention times which is associated to peak identification, calculation of the resolution, comparison between chromatograms ...) and the algorithms used for the ajustment of operating conditions. A first approach towards that goal has been published by Glajch et al. [30].

One of the major roadblock is the lack of a simple method to solve the problem of peak identification in successive analyses made on the same mixture in different conditions. This does not mean "absolute" identification of the solutes as chemicals, but merely determining how the retention time of a given solute change with experimental conditions. This is difficult since changing such conditions is often associated with inversions in the elution order and/or simultaneous elution of several compounds. An interesting possibility is based on the evaluation of peak area. This has been discussed in recent publications [31, 32]. It is sometimes possible to use the conventional thermodynamic relationship accounting for the variation of retention with solvent composition. Another approach recently introduced is to use either a UV scanning detector or a diode-array UV spectrometer. Both permits identification of the peaks on the basis of their spectra [33]. This method seems to be very promising.

Another difficult problem is the comparison of various chromatograms by a computer in order to find out which is the best one. This problem is closely related to the one facing the analyst who has to give a clear definition of his requirements, easy to implement in each case. Working with a computer obliges him to a deeper reflection about his work. The best chromatogram is not necessarily that one which has the maximum number of peaks or the maximum overall resolution as the analyst may be interested only in a few of the components of the mixture. Various approaches have been described to evaluate a chromatogram in terms of mathematical parameters [30, 34-36] but none of them yet is general enough to deal with all the possible situations that can be encountered in optimization problems.

It must be pointed out that there are many possibilities of optimization for a given problem. Schematically, the different steps of the optimization process are as follows. First, the analyst must define exactly what he expects: maximum overall resolution in a given time or without a time limit, "isolation" of one or several solutes in a chromatogram (i.e. only such and such solutes should be well enough resolved from the other components), maximum peak height for some compounds ... Second, some preliminary experiments must be carried out with selected mobile phases, the composition of which is calculated by the computer from a single previous injection [37], and the retention time and column plate number must be measured for each compound of interest. Then the computer can start its search for optimum conditions within the range of conditions allowed. Not only the solvent conditions, but also the type of column and the temperature are important parameters that have to be taken into account for the complete optimization of a separation.

The final step, which may be the most critical one, is to estimate the retention times (or capacity ratios) as a function of the solvent composition, temperature ... This can be done using empirical models or mathematical algorithms [30, 37]. Then, once the optimum conditions have been determined, the separation is carried out and the experimental chromatogram is compared to the calculated one. If there are significant differences, the parameters of the model or of the algorithm are adjusted, and the process is repeated.

It is clear that the development of the software associated to such an optimization is complex. There is no doubt however that such programs will be developed in the future. All the changes of the experimental parameters will be done automatically, the sole role of the analyst being then to define the type of optimization he wants ... and to load the autosampler with the samples.

6 Literature Cited

Part 0: Introduction

1. Analytical Chemistry. Laboratory guide. Issue Nr. 5 (April)
2. Journal of Chromatographic Science. International chromatography Guide, February Issue
3. McNair, H. M.: J. Chromatogr. Sci. 20, 537 (1982)
4. Instrumentation for High performance Liquid Chromatography. J. F. K. Huber Ed., Journal of chromatography Library n° 13, Elsevier, Amsterdam 1978
5. Detectors for Liquid Chromatography, J. Kennedy, Wiley, New York 1975
6. Liquid Chromatography Detectors, R. P. W. Scott, Journal of Chromatographic Library n° 11, Elsevier, Amsterdam 1977

Part 1: Pumps

1. Halasz, I., Schmidt, H. and Vogtel, P.: J. Chromatogr. 126, 19 (1976)
2. Note: 1 MPa $= 10^6$ N·m^{-2} = 10 bars = 9.87 atm = 145 psi = 10.2 kg·cm^{-2} = 10^7 dyne·cm^{-2}
3. Di Cesare, J. L., Dong, M. W. and Atwood J. G.: J. Chromatogr. 217, 369 (1981)
4. Scott, R. P. W. and Kucera, P.: J. Chromatogr. 125, 251 (1976)
5. Glajch, J. J., Kirkland, J. J., Squire, K. M. and Minor, J. M.: J. Chromatogr. 199, 57 (1980)
6. Martin, M, Eon, C. and Guiochon, G.: J. Chromatogr. 99, 357 (1974)
7. Brigdman, P. W.: The Physics of High Pressure, G. Bell and Sons ed., 1931, chap. 7
8. Techniques de l'Ingénieur, Paris, 1955, K680-5
9. Guiochon, G.: J. Chromatogr. 185, 1 (1979)
10. Ishii, D., Asai, K., Hibi, K., Jonokuchi, T. and Nagaya, M.: J. Chromatogr. 144, 157 (1977)
11. Yang, F. J.: J. Chromatogr. 236, 265 (1982)
12. Schrenker, H.: Int. Lab., Jul.–Aug. 1978, 67
13. Martin, M. and Guiochon, G.: In: Instrumentation for high-performance liquid chromatography, J. F. K. Huber ed., Elsevier, Amsterdam 1978, chap. 2
14. Martin, M., Blu, G., Eon, C. and Guiochon, G.: J. Chromatogr. 112, 399 (1975)
15. Martin, M., Blu, G. and Guiochon, G.: J. Chromatogr. Sci. 11, 692 (1973)
16. Tsuda, T., Nomura, K. and Nakagawa, G.: J. Chromatogr. 248, 241 (1982)
17. Martin, M. and Guiochon, G.: In: Instrumentation for high-performance liquid chromatography, J. F. K. Huber ed., Elsevier, Amsterdam, 1978, chap. 3
18. Laurent, C., Billiet, H. A. H., Vandam, H. C. and De Galan, L.: J. Chromatogr. 218, 83 (1981)
19. Halasz, I. and Vogtel, P.: J. Chromatogr. 142, 241 (1977)
20. Halasz, I.: Angew. Chem. (Int. Ed.) 21, 50 (1982)

Part 2: Sample Introduction Systems

1. Sternberg, J. C.: Advances in Chromatography, J. C. Giddings and R. A. Keller ed., Vol. 2, Marcel Dekker, New York. (1966) 205
2. Huber, J. F. K., Hulsman, J. A. R. J. and Meijers, C. A. M.: J. Chromatogr. 62, 79 (1971)
3. Karger, B. L., Martin, M. and Guiochon, G.: Anal. Chem. 46 (1974) 1640
4. Kirkland, J. J., Yau, W. W., Stoklosa, H. J. and Dilks, C. H. Jr.: J. Chrom. Sci. 15, 303 (1977)
5. Colin, H., Martin, M. and Guiochon, G.: J. Chromatogr. 185, 79 (1979)
6. Martin, M., Eon, C., Guiochon, G.: J. Chromatogr. 108, 229 (1975)

7. Kraak, J. C.: In "Instrumentation for High-Performance liquid Chromatography", J. F. K. Huber ed., Elsevier, Amsterdam, 1978, chap. 4
8. Colin, H., Diez-Masa, J. C., Martin, M., Jaulmes, A. and Guiochon, G.: Communication presented at the IVth International Symposium on Column Liquid Chromatography, Boston, May 1979
9. Webber, T. J. N. and McKerrel, E. H.: J. Chromatogr. 122, 243 (1976)
10. Di Cesare, J. L., Dong, M. W., Gant, J. R.: Chromatographia 15, 595 (1982)

Part 3: Detectors

1. Guiochon, G.: In "Microcolumn HPLC". P. Kucera Ed., Elsevier, Amsterdam, 1983, chap. 1
2. Kucera, P., Guiochon, G.: J. Chromatogr. 283, 1 (1984)
3. Kennedy, J.: "Detectors for liquid chromatography", Wiley, New York 1975
4. Scott, R. P. W.: "Liquid chromatography detectors", Elsevier, Amsterdam 1977
5. Poppe, H.: In "Instrumentation for high performance liquid chromatography". J. F. K. Huber Ed., Elsevier, Amsterdam 1978, chap. 7
6. Halasz, I.: Anal. Chem. 36, 1428 (1964)
7. Guiochon, G.: J. Chromatogr. 14, 378 (1964)
8. Guiochon, G., Colin, H.: Proc. of 2nd Congress on Environmental Analysis, Barcelona 1981. Pergamon, London 1982, p. 569
9. Guiochon, G.: Chromatographia 5, 571 (1972)
10. Guiochon, G.: In: High Performance Liquid Chromatography, C. Horvath Ed., Academic Press, New York 1980, p. 1
11. Martin, M., Eon, C., Guiochon, G.: J. Chromatogr. 108, 229 (1975)
12. Colin, H., Martin, M., Guiochon, G.: J. Chromatogr. 185, 79 (1979)
13. Guiochon, G.: J. Chromatogr. 185, 3 (1979)
14. Kirkland, J. J.: Anal. Chem. 40, 391 (1968)
15. Atwood, J. G., Golay, M. J. E.: J. Chromatogr. 218, 97 (1981)
16. Atwood, J. G., Goldstein, J.: Perkin Elmer analytical study 123, Norwalk 1982
17. Schmauch, L. J.: Anal. Chem. 31, 225 (1959)
18. Sternberg, J.: "Advances in chromatography", J. C. Giddings, R. A. Keller Eds, M. Dekker, New York 2, 205 (1966)
19. McWilliam, I. G., Bolton, H. C.: Anal. Chem. 41, 1755 (1969)
20. DiCesare, J. L., Dong, M. W., Atwood, J. G.: J. Chromatogr. 217, 369 (1981)
21. Bakalyar, S. R., Henry, R. A.: J. Chromatogr. 126, 327 (1976)

Differential refractometers (see also 3–5, 9)

22. Colin, H., Jaulmes, A., Guiochon, G., Corno, J., Simon, J.: J. Chromatogr. Sci. 17, 485 (1979)
23. Kempe, J.: Chem. Tech. 33, 375 (1981)
24. Hazebroek, H. F.: J. Phys. E (Sci. Instr.) 5, 180 (1972)
25. Woodruff, S. D., Yeung, E. S.: Anal. Chem. 54, 2124 (1982)

Dielectric constant (see also 3–5)

26. Conlon, R. D.: Anal. Chem. 41, 107 A (1969)
27. Haderka, S.: J. Chromatogr. 91, 167 (1974)
28. Vespalec, R.: J. Chromatogr. 108, 243 (1975)
29. Mowery, R. A.: J. Chromatogr. Sci. 20, 551 (1982)

Density

30. Fornstedt, N., Porath, J.: J. Chromatogr. 42, 376 (1969)

Photometers and spectrophotometers (see also 3–5)

31. Abbott, S. R., Tusa, J.: J. Liquid Chromatogr. 6 (S-1), 77 (1983)
32. Burce, G., Klotter, K.: Amer. Lab. March 1982, p. 74

33. Readman, J. W., Brown, L., Rhead, M. M.: Analyst 106, 122 (1981)
34. Edwards, T. R.: Anal. Chem. 54, 1519 (1982)
35. Bylina, A., Sybilska, D., Grabowski, Z. R., Koszewski, J.: J. Chromatogr. 83, 357 (1973)
36. Milano, M. J., Lam, S., Grushka, E.: J. Chromatogr. 125, 315 (1976)
37. Dessy, R. E., Reynolds, W. D., Nunn, W. G., Moler, F. G.: Clin. Chem. 22, 1472 (1976)
38. Dessy, R. E., Reynolds, W. D., Nunn, W. G., Titus, C. A., Moler, F. G.: J. Chromatogr. 126, 347 (1976)

Fluorescence (see also 3–5)

39. Cassidy, R. M., Frei, R. W.: J. Chromatogr. 72, 293 (1972)
40. Steichen, J. C.: J. Chromatogr. 104, 39 (1975)
41. Asmus, P. A., Jorgenson, J. W., Novotny, M.: J. Chromatogr. 126, 317 (1976)
42. Su, S. Y., Jurgensen, A., Bolton, D., Winefordner, J. D.: Anal. Letters 14 (AI), 1 (1981)
43. Diebold, G. J., Zare, R. N.: Science 196, 1439 (1977)
44. Yeung, E. S., Siepaniak, M. J.: Anal. Chem. 52, 1465A (1980)
45. Richardson, J. H., Larson, K. M., Haugen, G. R., Johnson, D. C., Clarkson, J. E.: Anal. Chim. Acta 116, 407 (1980)
46. Fell, A. F., Scott, H. P., Gill, R., Moffat, A. C.: Chromatographia 16, 69 (1982)

Other optical detectors

47. Donkerbroek, J. J., Van Eikema Hommes, N. J. R., Gooijer, C., Velthorst, N. H., Frei, R. W.: Chromatographia 15, 219 (1982)
48. Gast, C. H., Kraak, J. C., Poppe, H., Maessen, F. J. M. J.: J. Chromatogr. 185, 549 (1979)
49. Morita, M., Vehiro, T., Fuwa, K.: Anal. Chem. 52, 351 (1980)
50. Westwood, S. A., Games, D. E., Sheen, L.: J. Chromatogr. 204, 103 (1981)
51. Voigtman, E., Jurgensen, A., Winefordner, J. D.: Anal. Chem. 53, 1921 (1981)
52. Oda, S., Sawada, T.: Anal. Chem. 53, 471 (1981)
53. Leach, R. A., Harris, J. M.: J. Chromatogr. 218, 15 (1981)

Electrochemical detectors

54. Borman, S. A.: Anal. Chem. 54, 327A (1982)
55. Slais, K. and Kourilova, D.: J. Chromatogr. 258, 57 (1983)
56. Goto, M., Koyanagi, Y. and Ishii, D.: J. Chromatogr. 208, 261 (1981)
57. Goto, M., Sakurai, E. and Ishii, D.: J. Chromatogr. 238, 357 (1982)
58. Hirata, Y., Lin, P. T., Novotny, M. and Wightman, R. M.: J. Chromatogr. 181, 287 (1980)
59. Matysik, J., Soczewinski, E., Zminkowska-Halliop, E. and Przegalinski, M.: Chem. Anal. (Warsaw) 26, 463 (1981)
60. Slais, K. and Krejci, M.: J. Chromatogr. 235, 21 (1982)
61. Weber, S. G. and Purdy, W. C.: Ind. Eng. Chem. Prod. Res. Dev. 20, 593 (1981)
62. Krstulovic, A. and Colin, H.: Analusis 11, 111 (1983)
63. Shoup, R. E.: Current separations, p. 53, Bioanalytical Systems, 1982
64. Schieffer, G. W.: Anal. Chem. 52, 1994 (1980)
65. Bratin, K. and Kissinger, P. T.: J. Liq. Chromatogr. 4, 321 (1981)
66. Krstulovic, A. M., Colin, H. and Guiochon, G.: "Advances in chromatography", M. Dekker, New York, 24, 83 (1984)
67. Kissinger, P. T., Bratin, K., Davis, G. C. and Pachla, L.: J. Chromatogr. Sci. 17, 137 (1979)

Other detectors

68. James, A. T., Ravenhill, J. R., Scott, R. P. W.: Chem. Ind. 746 (1964)
69. Haahti, E. O. A., Nikkari, T.: Acta Chem. Scand. 17, 2565 (1963)
70. Scott, R. P. W., Lawrence, J. F.: J. Chromatogr. Sci 8, 65 (1970)
71. Charlesworth, J. M.: Anal. Chem. 50, 1414 (1978)

72. Macrae, R., Trugo, L. C., Dick, J.: Chromatographia 15, 476 (1982)
73. Stolyhwo, A., Colin, H. and Guiochon, G.: J. Chromatogr. 265, 1 (1983)
74. Mowery, R. A., Juvet, R. S.: J. Chromatogr. Sci. 12, 687 (1974)
75. Van Urk-Schoen, A. M., Huber, J. F. K.: Analytica Chimica Acta 52, 519 (1970)
76. Scott, R. P. W., Blackburn, D. W. J., Wilkins, T.: J. Gas Chromatogr. 5, 183 (1967)
77. Small, H., Stevens, T. S. and Bauman, W. C.: Anal. Chem. 47, 1801 (1975)
78. Stevens, T. S., Davis, J. C. and Small, H.: Anal. Chem. 53, 1488 (1981)
79. Nota, G., Palombari, R.: J. Chromatogr. 62, 153 (1971)
80. Willmott, F. W., Dolphin, R. J.: J. Chromatogr. Sci. 12, 695 (1974)
81. Chamberlain, A. T., Marlow, J. S.: J. Chromatogr. Sci 15, 29 (1977)
82. McGuffin, V. L. and Novotny, M.: J. Chromatogr. 218, 179 (1981)
83. Ouano, A. C.: J. Polym. Sci, A-1 10, 2169 (1972)
84. Lesec, J. and Quivoron, C.: Analusis 4, 399 (1976)
85. Kaye, W., Havlik, A. J. and McDaniel, J. B.: Polym. Lett. 9, 695 (1971)
86. Martin, M.: Chromatographia 15, 426 (1982)
87. Halasz, I. and Vogtel, P.: J. Chromatogr. 142, 241 (1977)
88. Synovec, R. E. and Yeung, E. S.: Anal. Chem. 55, 1599 (1983)

Part 4: Automation

1. Meakin, G. and Allington, R.: Am. Lab. 65 (Aug. 1980)
2. Erni, F., Krummen, K. and Pellet, A.: Chromatographia 12, 399 (1979)
3. Schrenker, H.: Int. Lab. 67 (July 1978)
4. Coverly, S. C.: Anal. Proc. 18, 491 (1981)
5. Roth, W., Beschke, K., Jauch, R., Zimmer, A. and Koss, F. W.: J. Chromatogr. 222, 13 (1981)
6. Gfeller, J. C., Huen, J. M. and Thevenin, J. P.: J. Chromatogr. 166, 133 (1978)
7. Gfeller, J. G., Huen, J. M. and Thevenin, J. P.: Chromatographia 12, 368 (1979)
8. Dolan, J. W., Van der Wal, Sj., Bannister, S. J. and Snyder, L. R.: Clin. Chem. 26, 871 (1980)
9. Bannister, S. J., van der Wal, Sj., Dolan, J. W. and Snyder, L. R.: Clin. Chem. 27, 849 (1981)
10. Werkhoven-Goewie, C. E., Brinkman, U. A. Th., Frei, R. W. and Colin, H.: J. Liq. Chrom. in press
11. Winkelbauer, P.: Am. Lab. 44 (May 1982)
12. Freeman, D. H.: Anal. Chem. 53, 2 (1981)
13. Colmsjoe, A. L. and MacDonald, J. C.: Chromatographia 13, 350 (1980)
14. Vestergaard, P.: J. Chromatogr. 111, 69 (1979)
15. Hulpke, H. and Verthmann, U.: Chromatographia 12, 390 (1979)
16. Snyder, L. R., Dolan, J. W. and van der Wal, Sj.: J. Chromatogr. 203, 3 (1981)
17. Frei, R. W.: J. Chromatogr. 165, 75 (1979)
18. Frei, R. W.: Chromatographia 15, 161 (1982)
19. Schmedt, G.: Angew. Chem. Int. Ed. Engl. 18, 180 (1979)
20. Krull, I. S., Lankmayr, E. P.: Am. Lab. 18 (May 1982)
21. Smith, A. I., McDermott, J. R., Biggins, J. A. and Boakes, R. J.: J. Chromatogr. 236, 489 (1982)
22. Garpe, L., Lundin, H. and Sjodahl, J.: Int. Lab. 62 (May 1982)
23. Bhown, A. S., Mole, J. E., Hollaway, W. L., Bennett, J. C.: J. Chromatogr. 156, 35 (1978)
24. Vestergaard, P., Bachman, A., Piti, T. and Kohn M.: J. Chromatogr. 111, 75 (1975)
25. Berger, D., Gilliard, B.: J. Chromatogr. 210, 33 (1981)
26. Vanderslice, J. T., Brown, J. F., Beecher, G. R., Maire, C. E. and Brownlee, S. G.: J. Chromatogr. 216, 338 (1981)
27. Radke, M., Willsch, H. and Welte, D. H.: Anal. Chem. 52, 406 (1980)
28. Bristow, P. A.: J. Chromatogr. 122, 277 (1976)

Part 5: Data Acquisition and Processing

1. Ashworth, H. A. and Augustine, R. L.: Rev. Sci. Instrum. 52, 105 (1981)
2. Lyne, P. M. and Scott, K. F.: J. Chrom. Sci. 19, 547 (1981)
3. Smith, S. L. and Wilson. C. E.: Anal. Chem. 54, 1439 (1982)
4. Reese, C. E.: J. Chrom. Sci. 18, 201 (1980)
5. Woerlee, E. F. G. and Mol, J. C.: J. Chrom. Sci. 18, 258 (1980)
6. Matthews, H. G.: Int. Lab. 60 (June 1982)
7. Liscouski, J. G.: Anal. Chem. 54, 849 A (1982)
8. Cram, S. P., Chesler, S.: Anal. Chem. 43, 1922 (1971)
9. Goedert, M., Guiochon, G.: Chromatographia 6, 76 (1973)
10. Annino, R.: In "Advances in Chromatography", J. C. Giddings, E. Grushka, J. Cazes, P. R. Brown Eds. M. Dekker, New York, vol. 15, p. 33 (1977)
11. Savitsky, A. and Golay, M. J. E.: Anal. Chem. 36, 1627 (1964)
12. Edwards, T. R. and Knight, R. D.: Instrument and control systems, 73 (Sept. 1974)
13. Annino, R., Gonnord, M. F. and Guiochon, G.: Anal. Chem. 51, 379 (1979)
14. Excoffier, J. L.: Thèse de Docteur Ingénieur, Université Pierre et Marie Curie, Paris 1982
15. Maldacker, T. A., Davis, J. E. and Rogers, L. B.: Anal. Chem. 46, 637 (1974)
16. Reese, C. E.: J. Chrom. Sci. 18, 249 (1980)
17. Baumann, F., Brown, A. C. and Mitchell, M. B.: J. Chrom. Sci. 8, 20 (1970)
18. van Rijswick, M. H. J.: Chromatographia 7, 491 (1974)
19. Excoffier, J. L. and Guiochon, G.: Chromatographia 15, 543 (1982)
20. Tarroux, P. and Rabilloud, T.: J. Chromatogr. 248, 249 (1982)
21. Bacon, G. D.: J. Chromatogr. 172, 57 (1979)
22. Weiman, B.: Chromatographia 7, 472 (1974)
23. Grushka, E.: J. Phys. Chem. 76, 2586 (1972)
24. Kirkland, J. J., Yau, W. W., Stoklosa, H. J. and Dilks, C. H. Jr.: J. Chrom. Sci. 16, 303 (1977)
25. Dondi, F., Betti, A., Blo, G. and Bighi, C.: Anal. Chem. 53, 496 (1981)
26. Haarhof, P. C., Vanderlinde, H.: Anal. Chem. 38, 573 (1966)
27. Dallura, N. J. and Juvet, R. S.: J. Chromatogr. 239, 439 (1982)
28. Rix, H., Colin, H. and Guiochon G: To be published
29. Lyne, P. M. and Scott, K. F.: J. Chrom. Sci. 19, 599 (1981)
30. Glajch, J. L., Kirkland, J. J., Squire, K. M. and Minor, J. M.: J. Chromatogr. 199, 57 (1980)
31. Issaq, H. J. and McNitt, K. L.: J. Liq. Chrom. 5, 1771 (1982)
32. Bounine, J. P., Colin, H. and Guiochon, G.: J. Chromatogr. 298, 1 (1984)
33. Drouen, A. C. J. H., Haddad, P., Bartha, A., Schoenmakers, P. J., Billiet, H. A. H. and de Galan, L.: VIIth International Symposium on Column Liquid Chromatography May 3–6, 1983, Baden-Baden. J. Chromatogr. 298, 1 (1984)
34. Morgan, S. L. and Deming, S. N.: Sep. Purif. Methods 5, 333 (1976)
35. Watson, M. W. and Carr, P. W.: Anal. Chem. 51, 1835 (1979)
36. Drouen, A. C. J. H., Billiet, H. A. H., Schoenmakers, P. J. and de Galan, L.: Chromatographia 16, 48 (1982)
37. Colin, H., Krstulovic, A., Guiochon, G. and Bounine, J. P.: Chromatographia 17, 209 (1983)

Quantitative Analysis in HPLC

Jürgen Asshauer
Hoechst AG, Werk Knapsack, 5030 Hürth/FRG

Helmut Ullner
Hoechst AG, 6230 Frankfurt/M 80/FRG

1 General Points of View and Definitions

1.1 Purposes and Limitations of Quantitative Analysis

The general object of analytics is to furnish information about the qualitative and quantitative composition of mixtures of substances. A quantitative statement on a lot of material (product) should describe the *true* value of a property of this material as well as possible. Both the statement and the product may be traded and transmitted to a customer. In most cases this involves a flow of countervalues in the form of goods or currency or information.

It must be possible to verify the statements on the whole of the material, because the customer must have the possibilities to check them. However, if the results of this check are exactly the same as those of the original analysis this is mere chance because an analysis always yields values which owing to the process by which information is obtained contain systematic and random errors and consequently deviate from the true value. As the results of quantitative statements are of economical, societal, and political importance the critical examination and interpretation of results is a most important step and a challenge to responsible analysts[1]. The significance of analytical results is reflected by the following questions which are often raised after reporting:

"What is the accuracy of these results? Do the results allow a decision?"

These questions clearly show that error diagnosis and description are among the most important tasks when quantitative analyses are planned and carried out. This implicates also the demand to publish the results in a form which enables the receiver too, to check and evaluate them critically.

Any method that is chosen for quantitative analysis is a complex process which might be compared with an information chain. The following Table 1 shows the most important parts of an analytical process in outlines.

Table 1 shows that quantitative analysis by any method can be achieved only in close connection with object and target. Separation and proper measurement

Table 1. Schematic representation of an analytical process

Object to be analysed	Whole of sample (e.g. container with chemical product)
Knowledge	Previous knowledge (quantity, state, origin, and concentration)
Subsample	Preparation of specimen by sampling methods
Analytical methods	Grinding, decomposition, and clean up
	Separation (e.g. HPLC)
	Measurement (e.g. photometric detection)
Target	Calculation
	Error diagnosis and critical evaluation
	Statement (analytical answer)
	Report and documentation

(illustrated by the example of HPLC) is but a small part of this chain of information accumulation.

Its influence on the errors that is to say precision and accuracy of the results cannot be isolated from the rest of the process (sampling, evaluation, and interpretation).

A qualified an experienced analyst must in any case be able to estimate the critical parts of the analytical process including all of the above mentioned steps. Every analyst has to answer to the question:

When and in which part of the process is it of decisive importance to aim at the highest possible precision?

To answer this it is necessary to know the opportunity and the order of magnitude of possible errors within the analytical process. The errors must be brought into line with the requirements resulting from the target. On the basis of these requirements analytical problems may be divided roughly into two groups which depend on the level of content.

A. Range of contents from 100% to 0.1%

Analyses of this kind are carried out, for example, for the following reasons:
Quality assurance of products.
Mass balance of producer and/or consumer.
Determination of prices and rates (taxes, customs duties).
Discussion of complaints.

The typical requirements with regard to accuracy and precision are within the range of $+/-1\%$ (relative).

B. Range of contents from 1 g/kg to the detection limit

Quantitative analyses are made, for example, for the following reasons:
Determination of trace compounds in materials.
Control of limiting values in marketed products and environmental samples.
Studies of relations between dose and effect of chemicals.
Determination of residues of interesting compounds (kinetics).
Determination of detection limits.

The typical requirements with regard to accuracy and precision are within the range of $+/-10\%$ (relative) or more. They depend to a high degree on the level of contents and the special problem.

The division into two groups and the beginning of trace analysis at quantities of less than 0.1% are more or less arbitrary and due to reasons of history. As will be shown below it would be more reasonable to choose a continuous way of representation.

The examples that are given serve to illustrate the large number of problems existing in the field of economy and society, which quantitative analytics have to supply with contributions and matter for discussion. Hence it follows directly that the analyst has to deal not only with the method of determination (for example HPLC) but also with all the problems of analysis ranging from sampling to publication. It should be taken into consideration that any passing on of analytical results is a form of publication.

2 Methodical Prerequisites

2.1 Responsibility of the Analyst

After the proper optimization of the chromatographic system it enables the analyst to separate and quantitative evaluate complex sample mixtures in amounts of micrograms or nanograms that is to say in amounts that can no longer be directly manipulated.

In most cases chromatographic separation is actually no longer carried out in order to yield pure substances (preparative chromatography) but to produce a measuring signal from which the interesting information with regard to qualitative and quantitative composition of the sample to be analyzed can be gathered.

The person in charge of the analysis has to carry out the following main tasks:
1) Elaborating the the steps of sample preparation (sampling, grinding, sieving, subsampling, and preconcentration).
2) Choosing and optimizing the separation system.
3) Thoroughly testing the elected separation system on its limiting factors: (see Sect. 4).
   Loadability of the column with sample
   Linearity of the detection system
   Estimation of detection and/or determination limits
4) Setting the method of evaluation and the necessary parameters especially the setting of the baseline definition parameters for the electronic integrator.
5) Recognizing sources of troubles and errors eliminating them with the help of statistical checkings and controls.
6) Finally judging the results of analysis with respect to the plausibility of the analytical statement.

Point 1) relates to general considerations regarding analysis and shall not be stressed further in this chapter. Point 2) has already been dealt with in detail in other chapters. The contents of points 3) to 6) will be discussed in the following paragraphs.

2.2 Principle of Quantitative Chromatographic Analysis

After a mixture has been separated into its constituents the latter can be determined quantitatively in an appropriate way. For reasons of history the usual methods of evaluation have been developed in gas chromatography. The basic principles of the separation processes are the same but the aggregation of the mobile phases is different. For this reason the principles of detection of the sample compounds in the mobile phase differ considerably from one another. In gas chromatography the detectors (HWD, FID) have a response that can easily be standardized for various substances. The value that is indicated for example by a Flame Ionisation Detector is approximately the same per carbon unit with the most hydrocarbons. Necessary correction factors are close to 1.0. For this reason the composition of a product can fairly well be estimated quantitatively by gas chromatography even if no calibration is made (100%-method).

Most of the detectors used in liquid chromatography are based on the principle of light absorption at different wavelengths. The photometric response of these detectors cannot be estimated at all because the extinction coefficients of sample substances may vary within many orders of magnitude. Even for purposes of rough estimation one needs a standard substance for each particular peak.

This difference is of decisive importance in practical operation because in HPLC the popular method of estimating components in a mixture according to the 100%-method (see 2.3.3) *cannot be used.* Liquid chromatographic analysis is generally based on the comparison of the chromatogram of the sample with the chromatogram of reference substances which for the before mentioned reason are indispensable in HPLC.

Liquid chromatographic analysis is almost exclusively carried out according to elution technique. The mobile phase is streaming continuously and the sample or reference solutions are introduced into the moving mobile phase discontinuously. Under ideal circumstances the components are completely separated from one another and detected by the detection system. The corresponding chromatogram shows peaks which are clearly separated by regions of the zero signal from the pure mobile phase (Fig. 1 b). These regions serve to fix the baseline for the following quantitative evaluation and calculation.

All methods of evaluation imply that the measured signal (peak area or peak height) is proportional to the quantity or concentration of the substance under study. As already mentioned in previous chapters practically all current liquid chromatographic detectors respond to the concentration of the measured compound in the mobile phase and it is the operators duty to guarantee that detector and recording devices are working in the linear range of their response curves.

After the chromatogram has been obtained the following two steps are carried out.

1) The base-line for the interesting peaks is defined and the signal value (peak area or peak height) for the peaks is determined.
2) The quantitative information is evaluated by means of calibration curves.

Where the peaks are incompletely resolved the base-line can be fixed on the basis of experience gained when the method was developed. The small peak shown

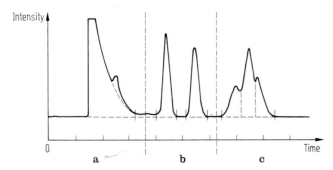

Fig. 1 a–c. Typical elution pattern as example for different possibilities to create the baseline to quantitate peaks. **a** Baseline of a peak on a tail. **b** The ideal situation of complete separation. **c** Processing by perpendicular division through the minimum

in Fig. 1a is processed as peak on the tail of a large peak; in Fig. 1c the peaks are processed by perpendicular division through the minimum (valley) between the peaks. These simple processes furnish approximate solutions only where attempts are made to compensate or minimize deviations from the "true" base-line under the peaks.

The use of electronic integrators or computers allows certain methods of evaluation to be fixed and repeated. These instruments often give the possibility to check the base-line to calculate the peak signals by reconstructing them on a screen or on the plotted chromatogram. Uncertainties in the course of the base-line often cause a considerable misinterpretation of the quantitative information. For this reason a *visual check of the results by means of the chromatogram should never be dispensed with.*

Because of the serious effects which base-line errors have on quantitative results, the measurements of not completely resolved peaks always yield a higher variation (cf. Sect. 5).

The answer to the question whether measurement of peak height or peak area is more precise depends to a large extent on the apparatus. The various influences are discussed in more detail in Sect. 4. The evaluation of peak areas is among other things more susceptible to baseline errors, so that preference is given to peak height evaluation in trace analysis where the signal to noise ratio is low. Peak height measurement on the other hand is strongly influenced by relatively small shifts of the retention volumes.

2.3 Methods of Evaluation

2.3.1 Absolute Calibration (External Standard Method)

According to the absolute calibration method the factor between injected sample amount and injected (known) standard amount is calculated and represented mathematically or graphically (calibration function). Very pure reference substances are required. For each component of a mixture a particular analytical function has to be established (see Fig. 2). Within the range of the analysis these curves should have a constant slope (linear) and the intercept should be zero or very near to zero. With such a graph or the corresponding mathematical expression one is able to determine the amount of the compound out of the analytical signal (peak area or height) in the sample provided that the injected amount of sample is constant or exactly known. If the peaks of standard and compound in the sample have similar dimensions it is often suffcient to perform a one point calibration (see formulas given below).

This method requires strict control of the analytical technic and absolute constancy of the conditions (adjustment of apparatus, conditions of separation, flow, and injected amounts). The method is very suitable, however, if all these requirements are met.

Instead of the actual injected amounts constant injection volumes may be used for calibration and analysis. This simplifies all subsequent calculations, which can now be based on the concentrations known from the sample or standard solution preparation. It is indicated for these purposes to use loop injectors or automatic

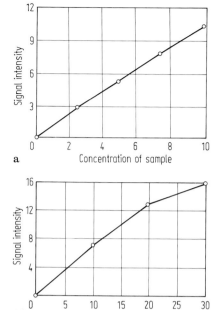

Fig. 2a and b. Schematic representation of calibration functions for absolute calibration. a Linear calibration curve; b calibration curve with non-linear parts (usual situation)

injection devices. Provided that the measurements are made in the linear region of the calibration curve (see Fig. 2) simple relations are obtained for result calculation.

Method of calculation in case of absolute calibration

a) Calibration

$$f_i = \frac{A_{i,E}}{c_{i,E}}$$

where

f_i = calibration factor of component i
$A_{i,E}$ = peak area or peak height of component i in the calibration solution (standard solution) (mm^2, mVsec, or mm, mV)
$c_{i,E}$ = concentration of component i in the calibration solution (mg/ml or mmol/ml)

b) Analysis

$$c_i = \frac{A_i}{f_i} = \frac{A_i * c_{i,E}}{A_{i,E}}$$

where

A_i = peak area or height of component i in the sample solution (dimensions as above)
c_i = concentration of component i in the *sample solution*

To convert the concentrations into the contents in the sample, the following equations are used:

$$x_i = \frac{c_i}{c_s} = \frac{A_i}{f_i\, c_s} = \frac{A_i * c_{i,E}}{A_{i,E} * c_s}$$

where

c_s = concentration of the sample in the measuring solution (weighed portion/volume)

x_i = content of component i in the *sample*

These formulas are only applicable if the injection volumes are constant. If they are not conversion has to start from the absolut masses.

$$m = c * V_{inj.}$$

2.3.2 Internal Standardization

Variations that are due to the equipment used, in particular variations of the injected volumes (quantities), can be eliminated to a large extent by the addition of a so-called "internal standard".

The method was originally developed in gas-chromatography to compensate for errors by variation of the injected amounts. Moreover it eliminates variations of the mobile phase flow from run to run (not within one run). It should be borne in mind, however, that the use of an internal standard means that *two* peaks have to be measured if *one* result is to be obtained. According to the error propagation law a consequence of this method is that the accuracy of the analysis is reduced by a factor of $\sqrt{2} = 1.4$. Moreover the fact that a second peak is present in the chromatogram has the consequence of an enhanced risk that overlapping of peaks may lead to unknown errors in the result.

On account of the development of more precise technics for the sample introduction, the "internal standard method" has lost importance in HPLC compared to the external standard procedure. It offers advantages, however, in HPLC with post column reaction if the standard is subjected to the same chemical reactions. A good example is the analysis of amino acids. If the internal standard substance is very similar to the compounds to be analysed it may be used also for compensation of recoveries in preconcentration or clean up processes. This procedure is often used in trace analysis whereas absolute calibration methods (see Sect. 2.3.1) are now preferred for the analysis of compounds at the percent level.

It is absolutely necessary that the substance used as internal standard complies with the following equirements:
1) It must not be present already in the sample (also in the future);
2) It must be completely separated from other components of the sample;
3) It should be eluted near the peak to be evaluated;
4) Its peak size should be similar to that of the peak to be evaluated;
5) It should not chemically react with other compounds of the sample;
6) It should be available very pure and have a good stability in storage.

It is often difficult to find substances meeting these requirements. Besides it may be necessary to use more than one internal standard substance in the course of one complex analysis.

Calculation methods for internal standardization

a) Calibration

$$F_{i/St} = \frac{A_{i,E} * c_{St,E}}{c_{i,E} * A_{St,E}} = \frac{f_i}{f_{St}}$$

where

$F_{i/St}$ = Relative calibration factor of component i relative to standard substance St (without dimensions)

$f_{i(St)}$ = Calibration factor of component i or the standard (see formula 1)

$A_{i,E}$ = Peak area or height of component i in the calibration solution (mm², mVsec oder mm, mV)

$A_{St,E}$ = Peak area or height of the standard substance in the calibration solution (mm², mVsec, or mm, mV)

$c_{i,E}$ or $c_{St,E}$ = Concentration of component i or the standard in the calibration solution (mg/ml oder mmol/ml)

b) Analysis

$$c_i = \frac{1}{F_{i/St}} * \frac{A_i * c_{St}}{A_{St}}$$

where

A_i = Peak area or height of component i in the sample solution (dimensions as above)

A_{St} = Peak area or height of the added standard substance in the sample solution (dimensions as above)

c_i = Unknown concentration of component i in the *sample solution*

c_{St} = Known concentration of the added standard in the *sample solution*

If conversion to the content in the sample is desired, then:

$$x_i = \frac{c_i}{c_s} = \frac{1}{F_{i/St}} * \frac{A_i * c_{St}}{A_{St} * c_s} * 100$$

where

c_s = Concentration of the sample in the solution to be measured (weight portion of sample/volume)

x_i = Unknown concentration of component i in the *sample* in percent

Instead of the concentrations the masses m can be used in all calculations.

2.3.3 Area Normalization Method (100%-method)

According to this method the signals (areas or heights) are added and the sum is assumed to represent 100%. The individual components of the sample are then calculated as portions. This method was originally developed and applied in gas chromatography as the response of most GC-detectors is very similar for various components. A result out of the sum of the areas thus has always yielded a reasonable estimate of concentrations with a low expenditure of work. A prerequisite is the complete elution and detection of all constituents of the sample. A good example for application of this method is the gas chromatographic analysis of petroleum distillates. In liquid chromatography, however, this procedure is not possible in the majority of cases because the substance related response of photometric detectors is very different and unpredictable. In UV-detection for example the molar extinction coefficients of different compounds may vary by some orders of magnitude at the designated measuring wavelength. When using refractive index detection within similar groups of substances (e.g. exclusion chromatography of polymers) the differences in response are small enough to use the 100% method. Nevertheless one must keep in mind that also in these cases the results contain an error due to the condition of equal response for all compounds.

Before the 100% method is used the following conditions must be fulfilled:
1) all components of the mixture are completely eluted;
2) all components of the mixture are covered by the detector in its linear range;
3) all components have the same detector response.

Calculation method for area normalisation (100% method).

a) With Substance Specific Correction Factors

$$x_i = \frac{f_i * A_i}{\sum (f_i * A_i)} * 100$$

where

x_i = content of the unknown component i in the sample solution in percent
f_i = substance specific correction factor, determined according to the absolute calibration method (see 2.3.1)
A_i = peak area or height of component i in the sample solution (mm², mVsec or mm, mV)

b) Simple Area Normalization Method

$$x_i = \frac{A_i}{\sum A_i} * 100$$

Here all substance specific correction factors must have the same value as indicated in the prerequisites.

2.3.4 Addition Method

In the trace analysis the small peaks are often influenced by other compounds of the sample which are not of interest. These influences occur in the course of the analytical run as well as in the preconcentration and clean up procedures (matrix influences). They may cause retention time shifts, peak broadening, overlapping of peaks, or bad recoveries. In order to obtain correct and reliable results one should carry out calibration by adding known amounts of the substance under study to a analytical sample of the same composition but without the interesting compound (blank sample). However, often a blank sample will not be available. In such cases the addition method is used. Various but known amounts of component i are added to a definite amount of sample solution and the mixture is subjected to chromatographic separation. The peak areas or heights are evaluated and plotted as a function of the added concentration of the component i (see Fig. 3). The desired content can then be calculated from the signal from the pure sample (with no additional standard). Calibration is carried out according to the absolute calibration method (2.3.1) on the basis of the added amounts of component i. If a plot of signal versus concentration or amount of added substance i has been obtained (as in Fig. 4) the desired value can be measured from the intercept of the analytical curve with the concentrations axis. Preparation of the sample and evaluation procedure is considerably facilitated by the use of equal portions of sample for the addition experiments (equal concentrations of unknown component i) and by pipetting the added amounts of component i out of a stock solution.

Relative Addition Method

If a peak k is clearly separated but which is not of interest for the analytical problem fulfills the criteria for an internal standard (see 2.3.2) it may be used as reference peak. The original sample (run 1) and a sample to which a certain amount of component i has been added (run 2) are chromatographed subsequently under the same conditions. The following calculations can then be used to evaluate the desired value.

$$x_j = \frac{m_{i,z} * A_{i,1} * A_{k,2}}{m_{s,2}(A_{i,2} * A_{k,1} - A_{i,1} \ 20 * A_{k,2})} * 100$$

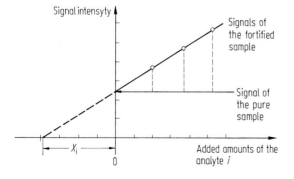

Fig. 3. Schematic representation of the calibration function using the addition method. The unknown amount X_i can be calculated out of the signals of the pure sample (containing X_i) and the fortified samples (containing X_i plus added amounts of analyte i). For calculation refer to 2.3.4

where

x_i = unknown content of component i in the sample solution in percent

$m_{s,2}$ = weighed amount of sample in run 2 (run 1 is not needed due to normalization). Other authors use run 1 for calculation. This leads to different formulas but identical results

$m_{i,z}$ = added amount of component i in run 2

$A_{i,1}$ = peak area or height of component i in run 1

$A_{i,2}$ = peak area or height of component i in run 2 (including the signal from the addition of $m_{i,z}$)

$A_{k,1}$ and $A_{k,2}$ = peak area or height of a undisturbed reference peak in run 1 and run 2

$m_{s,1}$ = is not required in this formula because of the normalization by the signals of the reference peak. The sample weights should be very similar. If the weighed portions m are identical and all other conditions are constant the signals of the reference peaks wil become identical and the formula changes to that of absolute addition method

b) Absolute addition method

$$x_i = \frac{m_{i,z} * A_{i,1}}{m_s * (A_{i,2} - A_{i,1})} * 100$$

The meaning of the symbols is the same as above.

The relative addition technique is a combination of the external and internal standard methods. Errors of flow and injection are compensated for by the use of a reference peak.

The evaluation methods described in this section are often incorporated as resident part of the software of electronic integrators or computers for chromatography and may then contain small deviations or additional possibilities which are not covered here. Reference is therefore made to the manuals of the used systems.

3 Sources and Representation of Errors

As has been stated at the beginning quantitative analysis is largely dependent on the recognition, evaluation, and description of the errors inherent to the results. These errors may be divided into two groups:

a) Influences which cause the results of an analytical procedure to shift in one direction are called systematic and the resulting errors are *systematic errors.*

b) Influences which cause the results of an analytical procedure to scatter around a mean value result in *random errors.*

This is illustrated by Fig. 4. In these examples method A yields results which are systematically deviating from the true value by an amount a and show a variation around the mean value (low precision, low accuracy). Method B yields values which vary considerably (large random error) around the true value but show no systematic deviation (low precision, high accuracy of the mean).

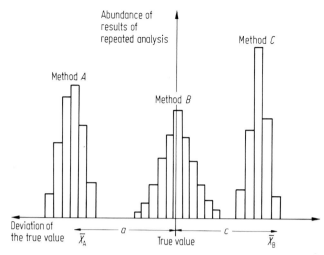

Fig. 4. Schematic representation of systematic and random errors. Method A and C yield results which are systematically too low or too high, while method B gives results which are near to the true value. The random deviations of every method result in a variation around the corresponding mean values. The random distribution is the broadest with method B

Method C yields results which are systematically too high and which have a very small random error (high precision, low accuracy).

In spite of their superior precision the results obtained by methods A and C are less correct and may therefore be useless.

It is not possible to make a clear and complete distinction between the two types of errors in each particular case as, for example, slight systematic deviations may be hidden under a broad dispersion of random errors. If the true value of a property (value) is not known it is difficult to recognize systematic errors. The following brief description of the two types of errors is intended to facilitate the understanding and the strategy for hunting errors.

3.1 Random Errors and International Standards for Their Measures

The random errors which contribute to the dispersion around a mean value can be recognized and described by means of statistical processes and characteristic values. The conditions to apply statistical methods are only fulfilled by random errors. The characteristic values thus determined are called precision data because they describe the precision (dispersion) of a process. Internationally accepted standards and processes are available to determine precision data.

Precision is defined as the extent to which results correspond to each other which are obtained when a definite process (analytical method) is repeatedly applied to identical samples (DIN 55350 part 13). Hence it follows that the influence of systematic errors is not taken into consideration when an analytical process is characterized by its precision data because systematic deviations would not be detected in repeated analysis.

The determination of precision data is standardized by international agreement: ISO-DIS 5725 – Precision of test methods

"Determination of Repeatability and Reproducibility by Inter-laboratory Tests"

The nomenclature used is based on the following standard: ISO 3534

"Statistics – Vocabulary and Symbols"

The most important parameters in the description of the precision of a test method are "repeatability r" and "reproducibility R". They are defined as follows (corresponding to ISO-DIS 5725):
   Repeatability r is the value below which the absolute difference between two individual test results obtained according to the same process on identical test material under the same conditions (same sample, same method, same operator, same apparatus, same laboratory, short space of time) can be expected to occur with a given probability. Unless otherwise stated this probability is 95%.
   Reproducibility R is the value below which the absolute difference between two individual test results obtained according to the same process on identical material but under different conditions (same sample, same method, different operator, different apparatus, different laboratories, and/or different time) can be expected to occur with a given probability. Unless otherwise stated this probability is 95%.
   These values are calculated from the variance σ_W^2 within the laboratory and the variance σ_L^2 between different laboratories.

repeatibility $r_{95} = 2\sqrt{2\,\sigma_W^2} = 2,83 \cdot \sigma_W$

reproducibility $R_{95} = 2\sqrt{2(\sigma_W^2 + \sigma_L^2)} = 2,83\sqrt{\sigma_W^2 + \sigma_L^2}$ *

When a batch of product is sold and tested in two different laboratories (vendor and customer) the situation is often as follows:
   Each laboratory created a mean value out of two individual determinations: $\bar{x}1$ and $\bar{x}2$ with $n = 2$.
   These mean values differ from one another. It has to be checked then whether a complaint is justified. For this purpose the so called critical difference is calculated from the precision data according to ISO-DIS 5725

The critical difference $D = \sqrt{R^2 - \dfrac{r^2}{2}}$ (for $n = 2$)

results from the data that have previously been determined by collaborative tests. The question whether there is a significant difference between the two mean values

---

* The sum of variances $(\sigma_W^2 + \sigma_L^2)$ can be approximated by the variance σ_R^2 obtained by collaborative tests. This method is used for example by the AOAC (Association of Official Analytical Chemists, USA)
   For the application of these precision data to practical business reference is made to DIN 51848 part 1.

can be decided by means of this critical difference D.

a) $|\bar{x}_1 - \bar{x}_2| \leq D$ no difference
b) $|\bar{x}_1 - \bar{x}_2| > D$ significant difference

Mention should once again be made of the fact that the above precision data merely describe the random dispersion of the results and not their systematic deviations. Consequently the critical difference between two measured values relates to the inherent dispersion of the method. Besides, the results may deviate in any way from the true value (see fig. 5). In the next chapter the main sources of such systematic deviations shall be described.

3.2 Systematic Errors.
General Sources of Deviations in Quantitative Analysis

It is difficult to discuss the group of systematic errors as a whole. Their sources and consequences are so different that general rules of dealing with them and describing them cannot be set up. On account of the serious effects which systematic errors have on the results in quantitative analysis it is, on the other hand, essential to find and eliminate them. The most important effects are often the same in various laboratories and trials and will be discussed now.

Between the concentration x of a substance and the measured signal y there is a mathematical correlation which is called analytical function. This function is obtained by calibration with known concentrations $x_1 - xn$. The resulting signals $y_1 - yn$ are plotted against the known concentrations $x_1 - xn$ (Fig. 5). A systematic error changes this function by shifting or biasing and therefore leads to misfindings.

In general there are two methods to test an analysis on systematic deviating results. The first is to use certified reference materials on which an analytical method that has just been developed can be tested (for example, material available from Community Bureau of References – BCR, Rue de la Loi 200, B-1049 Brussels; "Certified Reference Materials for Environmental Analysis"; or from EPA or NBS, USA).

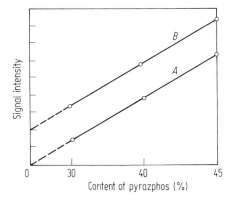

Fig. 5. Shift of the calibration function by an impurity HPLC with UV-detection. *A* Calibration curve for the fungicide Pyrazophos out of pure standard solutions. *B* Corresponding calibration curve out of technical solutions. The chromatograms are shown in Fig. 6

The second possibility to test for systematic deviations resides in the use of different or modified analytical methods. Which test method is preferably used in a particular case may largely depend on the problem concerned and on the resources available at a laboratory.

As an example the calibration curve obtained with pure solutions of a fungicide (HCPL with UV detection) and that obtained with commercial solutions of the same product are shown in Fig. 5. The commercial solutions contain an additional contaminant which stems from an auxiliary agent and is falsifying the elution band of the component to be measured. The chromatograms regarding these solutions are shown in Fig. 6. In principle it is possible to correct an influence of the kind mentioned above by calibrating with commercial products as is shown in Fig. 5 B. It is better however, to modify the separating properties of the system, so that the influence of the error is eliminated as has been done in the case shown in Fig. 6 d.

The sources of systematic errors can be divided into several groups in a way that is favourable in practical operation. The division makes it easier to determine the sources and to take them into account. It corresponds approximately to the division of the whole of the analytical complex represented in Table 1. The object of the summary given below is to remind of systematic errors as are daily experienced in large analytical laboratories when analytical operations are carried out.

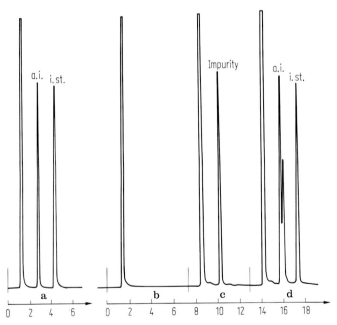

Fig. 6 a–d. Influence of an unrecognized impurity on quantitative results (refer to Fig. 5). **a** Test mix of a fungicide with an internal standard. **b** Test formulation without active ingredient containing methyl-naphtaline. **c** Test formulation with another batch of methyl-naphtaline. **d** Formulation with active ingredient after slight changes in mobile phase system. a.i. = active ingredient; i.st. = internal standard compound

*Groups of Errors Affecting the Correctness of Results Obtained
in Quantitative Analytics Outside HPLC*

a) Inhomogeneities in Sample Material

The inhomogeneity of sample material depends on the sample quality, crystallinity, grain size, and weighed portion. Errors which are due to inhomogeneities occur in analytical methods of any kind and in all levels of content. Every experienced analyst has met with cases of this kind. When solid samples are melted residual inhomogneities may, for example, remain in the viscous meld. Inhomogeneities in bulk goods are obvious. Even gases may be not uniform. With regard to the problems of sampling an international standard has been drwan up (ISO/DIS 6063, "Chemical products for industrial use – Sampling in order to determine a mean value ..."). Besides there are numerous monographs and other publications relating to this subject[2].

b) Losses Occurring in Processing Operations

In most cases losses of substance occur when analytical processes comprising several operations are carried out. It is advisable to check the rate of recovery in each series of analyses. The varying recoveries often limit repeatability and reproducibility of a method. Recovery rates of at least 70% are actually regarded in general as still admissible in the analysis of trace elements. The recovery is decreasing with decreasing concentration of analyt which is due to losses by processing, adsorption and chemical changes.

Table 2. General procedure in quantitative analytics

1. Definition of the problem (looking for the right question)
   Definition of the requirements with regard to accuracy
   Preparation of a plan

2. Selection or development of the analytical method
   Check for the accuracy of the mean (systematic errors)
   a) with reference samples
   b) by independent methods

3. Determination of the characteristics of precision (random errors)
   a) estimation in the light of experience
   b) determination of the within-laboratory repeatability r
   c) determination of the reproducibility R between different laboratories (collaborative test)

4. Determination of the sources of errors still present and adaptation to the requirements defined in step 1
   a) systematic errors
   b) random errors

5. New determination of the accuracy and precision data of the process (repeat steps 2 and 3), if required

6. Execution and representation of analyses
   Automatization, if necessary

c) Incorrect Characterization of the Calibration Standard

Like every relative method quantitation by HPLC is based on standard substances. No result can be more correct and more accurate than the certificate of the standard is correct and accurate.

The subject is dealt with in greater detail in a guideline published by WHO[3].

d) Contamination of Samples

The contamination of samples or processing products yields measured values which are too high (recoveries of more than 100%) or it may compensate a low recovery. The danger of contamination may occur at each point of the analysis. It can be avoided by careful planning, organization, and control only.

The pattern (scheme) of an analytical problem shown in table 1 can now be converted into an operating scheme in which both the detection and representation of random errors and the verification of the correctness of the results are taken into account. Table 2 represents this general mode of proceeding in a concise way.

4 Error Sources in HPLC and their Influences on Quantitation

Sources of errors that are not directly connected with the method of HPLC have already been dealt with in Sect. 3.2. Now all influences that are closely connected with HPLC will be discussed. These influences will be grouped according to the individual parts of the apparatus because they are often directly caused by special attributes or by malfunction of these parts.

4.1 Pump and Control System

The pump with its control system produces the required flow of the mobile phase for HPLC. By the flow resistance of the separation column and other parts of the apparatus results a pressure drop which can be easily measured and which amounts up to 400 bars in modern instruments. The mass flow of the mobile phase should be as constant as possible. This general statement has to be specified with regard to different ranges of frequency as these different ranges have different effects on the analysis.

a) Short-time variations of flow (Wavelength of 0.5 to 5 seconds)

Most of the flow variations are due to the construction and design of the flow delivery system because they have their origin in the piston stroke and its damping. With high amplification of the detector signal these effects are visible as signal noise with nearly all instruments. Differential refractive index detectors show up with the strongest effects. Most manufacturers try to smooth this signal noise by installing fixed or adjustable time constants (filters) in the detector amplifier. The consequences of these electronic filtering devices will be discussed under "detector".

b) Variations in the range of the signal transmission frequency

Variations having a wavelength of 2 to 100 seconds correspond to the peak width of the eluted bands (signal transmission frequency). Particular attention has to be paid to them. They may be caused by gas bubbles or microcontaminations in the pumping system. Flow variations in this range have strong effects on the measurement of the peak areas.

Current HPLC detectors measure the concentration of the sample in the detector cell. The measured actual concentrations are plotted on the paper of the recorder or in the memory of an integrator as a function of time and not as a function of flow (area = signal × time). Consequently the areas obtained in the time chromatogram vary when the relation between flow and time (the flow rate of the mobile phase) varies. The height of the peaks (concentration at the maximum) is not affected if the flow changes are not too enormous. Figure 7 demonstrates these effects.

Flow rate changes during the analysis are serious errors which should not occur. HPLC-runs with variable composition of the mobile phase (gradient-elution) are connected with unavoidable changes in flow rate because of the changes in viscosity and consequently pressure drop. It is obvious that under such conditions quantitative analyses must be less accurate.

The relative precision of the peak area and height measurement is shown by way of an example in table 3. The corresponding chromatogram is shown in Fig. 8. The example demonstrates good repeatability and it can be stated from these values that the flow constancy was better than 0.5%.

A record of the pressure drop may be a good indicator of the corresponding variations of flow and prevent misinterpretations.

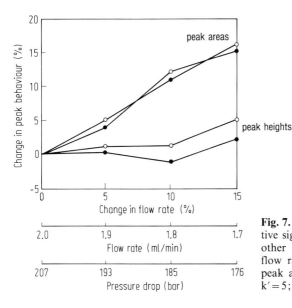

Fig. 7. Change of the measured quantitative signal as a function of flow rate. All other conditions constant. Changes in flow rate cause proportional changes in peak areas. ● Results from a peak with $k' = 5$; ○ Results from a peak with $k' = 8$

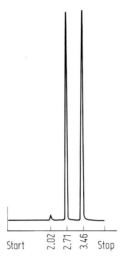

Start   2.02 2.71 3.46 Stop

Fig. 8. Chromatogram corresponding to Table 3. The peak with retention time 3.46 min is used as internal standard. Conditions: Column 25 cm length; ID 4.2 mm; Hypersil® 5; mob. phase 70% (vol) n-hexane, 30% (vol) CH_2Cl_2; detection UV 240 nm; evualation by plotting integrator. Component A: Triazophos (insecticide); Component B: Monolinuron (herbicide)

c) Slow variations of flow (drift)

The long-time constancy of flow influences the retention times as well as the peak areas and heights.

A drift in the retention times (retention volumes remain constant) determines the necessary size of the time window which serves to identify the peaks or attribute the corresponding correction factors. This is of great importance with regard to automatization by means of integrators or computers, controlled automatic sampling devices, and possibly fraction collectors because in most cases the individual components of the apparatus are time-controlled only.

A drift in flow rate affects the measured peak areas for the same reasons as mentioned in the previous section (cf. also Fig. 7). The change in pressure drop is not very suitable to monitor the long-time constancy of the flow because it is influenced by other effects as for example temperature, contamination of the separation column, and changes in the composition of the mobile phase. These effects do not only influence the peak areas and retention times but do also change the retention volumes and relative retention volumes (partition coefficients). It is quite possible that such changes occur during the course of several hours of analysis. As examples may be mentioned: the slow evaporation of volatiles out of the mobile phase mixture (e.g. due to degassing with Helium) or the precision of mixture and control in the case of gradient elution. Anyhow, the requirements on quantitative precision and accuracy need not be so high using gradient elution in comparison to isocratic elution analyses. To ensure qualitative and quantitative efficiency it is indicated to check the constancy of retention times of standard substances continuously. Each deviation signals a variation of the optimum conditions and a possible source of errors in quantitative operation. The retention time of a guide peak allows for correction of area measurements, if necessary.

A repeatability of retention times of less than 0.5% should be required.

4.2 Injection of Sample

Quantitative evaluation of chromatographic peaks depends to a large extent on the quality of the sample introduction. The sample injection process influences the absolute accuracy of the sample volume and the peak broadening effects. Both parameters effect the quantitative results.

a) Constancy and Correctness of the Injection Volume

Primarily, absolute calibration methods (see Sect. 2.3.1) with separate certified standards require that the injected sample volume is constant and can be repeated within certain limits. This requirement is known from all other analytical methods with external standardisation. If the injection-volume is inconstant one can use the internal standard method to compensate these variations by means of a reference peak (see Sect. 2.3.2).

For demonstration of the problem Table 3 shows values of repeated injections of a solution containing two components A and B. Figure 8 is representing the corresponding chromatogram. Absolute counts for area and height of the individual peaks are listed as well as the ratios for peaks A and B which can be assigned as the internal standard factors for a method where A is to be determined and B is added as internal standard. As can be seen from the last line of the table the coefficients of variation (VK) of the values for area or height that have directly been measured are not significantly improved by the formation of the ratios. This is a favourable case where it would be possible to work without internal standard. Modern sampling systems operate so precisely that in most cases an internal standard can be dispensed with (cf. Sect. 2.3.2 "Internal standardisation").

There are sampling devices of various designs for manual and automatic operation. They have already been described in another chapter of this book (cf. to

Table 3. Precision of replicate measurements of height or area of peaks in absolute and relative (internal standard) units. Sampling device: 20 µl loop injector (Rheodyne), manual; Detection: UV 240 nm, evaluation by calculating integrator

Run No.	Comp. A Height	Comp. B Height	Ratio A/B	Comp. A Area	Comp. B Area	Ratio A/B
1	920	928	0.9914	99092	127742	0.7757
2	920	929	0.9903	99256	128520	0.7723
3	921	930	0.9903	99396	128081	0.7760
4	925	934	0.9904	99763	128236	0.7779
5	929	935	0.9936	99575	128016	0.7778
6	923	938	0.9840	99648	127959	0.7787
7	930	936	0.9936	99996	128130	0.7804
8	927	937	0.9893	99443	128458	0.7741
9	926	932	0.9936	99729	128167	0.7781
10	924	929	0.9946	99309	127881	0.7766
Mean	924.5	932.8	0.9911	99521	128119	0.7768
SD	3.6	3.7	0.0031	272	242	0.0024
VK	0.39%	0.40%	0.31%	0.27%	0.19%	0.30%

SD = standard deviation, VK = coefficient of variation (relative standard deviation)

"Apparatus"). The user should not only rely on the specifications given by the manufacturer with regard to these parts of the apparatus as the process of sample introduction depends on many different parameters, for example the viscosity of the sample solution, the time required for filling the loop, and the rinsing volume. If an apparatus is used routinely it is altogether possible that it wears out by abrasion and consequently the results get worse slowly.

The quality of sample introduction into the chromatographic system should be checked from time to time and represented as it is shown in Table 3.

b) Influence on Peak Broadening

In addition to the effects on the absolute sample amount there are effects on peak shape and width produced by the sample introduction. The effects depend on the amount of sample and the quality of the sample solution itself.

In principle the influence of the sample amount on the peak shape is well known. Depending on the loadability of the separation system the width of the peaks begin to increase when a certain limiting value of the sample amount is exceeded and consequently the peak heights stop to increase proportionally. At the same time also the retention times start changing. Conditions of this kind are of course not desirable in quantitative evaluation and should prevail in special cases only in which the operation is carefully observed and controlled.

One origin of asymmetry and widening of peaks which is often observed in the analysis of complex samples resides in the composition of the sample solution. Particular in cases where polar mobile phases are used the solvent of the sample does not correspond exactly to the mobile phase. This may lead to solubility problems and non equilibrium conditions at the head of the separation column when the clear sample is introduced. Those problems result in severe effects on the peak shape. Some examples are given in Figs. 9 and 10.

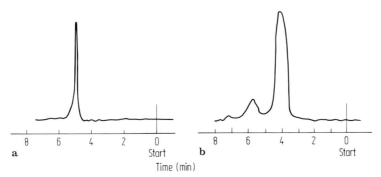

Fig. 9a and b. Influence of the composition of the sample solution on peak broadening. Trace analysis on an aromatic diamine with detection by fluorescence on an unpolar stationary phase. Mobile phase 50% (vol) methanol $+ 50\%$ (vol) $H_2O\,(P_H = 7)$. **a** 10 ng standard dissolved in mobile phase. **b** 100 ng standard dissolved in mobile phase but at $P_H = 2$, resulting in a broad peak with shorter retention time because of non equilibrium

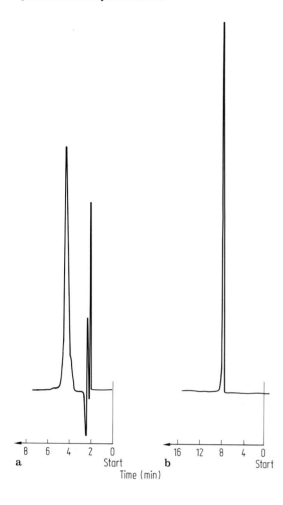

Fig. 10a and b. Chromatogram of a research pesticide on a reversed phase system; stationary phase SI-octadecylsilane; mobile phase. 70% methanol + 30% water; sample **a** 200 mg in 100 ml methanol; sample **b** 200 mg in 70 ml methanol and 30 ml water

As a basic rule it must be observed that sample solvent and mobile phase be as similar (identical) as possible.

4.3 Separation System, Mobile and Stationary Phase

The separation system is the central part in which the desired separation processes take place. This separation system, too, has influences on quantitative analysis which have to be observed. They can be divided into several groups:

a) Effects on the Sample Amount

If the sample amount decreases at any point of the total chromatographic system quantitative performance will be poor. Changes in the sample amount cannot easily be checked. The separation system contributes to such changes by three possible causes:

1) Leakages in the separation system are trivial errors and should be easy to recognize and to eliminate.

2) Very strong adsorption (high adsorption enthalpies) of certain portions of the sample by the stationary phase may occur. These adsorbed portions are delivered back very slowly and are no longer eluted as peaks but as fluctuations of the baseline. If the effect is reproducible it can be standardized. But in most cases the adsorbed portion is depending on the total sample amount, so that a standardisation is possible within narrow limits only. Similar effects can also be observed with new separation columns that have not been sufficiently conditioned.

The phenomenon that many calibration curves (especially in trace analysis) are not linear in the lower range can be interpreted by this influence easily.

3) The mobile phase often comprises more than one component which may moreover be contaminated by impurities. During analysis the samples come into close contact with this mixture for a few minutes only. However as has been required under Sect. 4.2 the samples are preferably dissolved in the mobile phase and may be allowed to stand for several hours before being analysed, which may be the case in particular with automatic sampling systems. In these complex mixtures it may happen that individual components of the sample react slowly with one another or with other constituents of the solvent and thereby change their chemical attributes or create new elution bands (ghost peaks). Such phenomena have been observed with different classes of samples e.g. ureas, oximes, aromatic amines, and all kinds of esters (beware of hydrolysis).

Effects of this kind can only be avoided if the analytical processes are subjected to *careful quality control* (see Sect. 5.2).

b) Effects on Width and Position of Peaks

Some effects on the peak shape have already been dealt with in the preceding paragraph which was concerned with the sample introduction system. If quantitative analyses are to be carried out the width and the position of the peaks must be optimized and kept constant. Separation systems are also subject to ageing. Alterations by ageing are most often due to the deposition of constituents of the sample which are not eluted. In the course of time the conditions of separation slowly deteriorate. Particularly in cases where the system works automatically this slow deterioration is not noticed. An example of the changes occuring in a separation system as a function of the number of analysed samples is shown in Fig. 11.

c) Interferences by Matrix Effects

A prerequisite for quantitation of an eluted peak is that only the interesting component is detected and measured.

Effects of the sample matrix are phenomena which are brought about by other components of the sample. Often not all of the constituents of a sample or sample solution are known. For this reason matrix effects must always be reckoned with. They can be avoided only by separation of the disturbing components.

Matrix effects may cause changes of the calibration function for the analyte in the sample in comparison to that of the pure standard solution. Analytical methods which incorporate separation processes should be less sensitive against interfer-

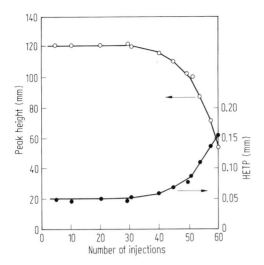

Fig. 11. Demonstration of the ageing of a separation system. Plasma samples were injected and analyzed on chloramphenicol (range 1–10 µg/ml sample). The peak areas remained constant. From: Wiese et al. (Ref. 4)

ences by matrix effects than unspecific determinations like photometrie. Nevertheless impurities often cause serious errors.

Figure 6 may serve as an example. The analysis of a fungicidal separation was carried out routinely by HPLC with internal standard. The values regarding the active ingredient (a.i.) suddenly rose by a few percent. The reason was that the supplier had modified an auxiliary agent used in the preparation of the material. This agent contained an additional impurity the peak of which eluted together with the active ingredient and increased the corresponding areas. Figure 6 represents the chromatograms and Fig. 5 the corresponding calibration curves. After a slight modification of the separation system this could clearly be seen (Fig. 6d) and the effect on the calibration curve be eliminated.

As a principle it can be stated that the probability of wrong measurement in a complex sample is the higher the less specific detection is. This is illustrated by Fig. 12. A distillation residue was expected to contain traces of benzidine as a result of the manufacturing process. The chromatogram of this residue (Fig. 12a) shows a large number of sharp peaks. The hatched one was attributed to benzidine by means of retention time comparison. The reference chromatogram is shown in Fig. 12b. The peak in the original chromatogram was fortified by addition of benzidine; the evaluation by the standard addition method (Sect. 2.3) resulted in 140–160 mg benzidine/kg sample (ppm). As in the light of other knowledge this value seemed to be too high the identity of the peak was examined by registering the complete UV-spectra at the peak maximum by means of a diode-array-detector. It was shown that the peak contained mainly an unknown compound and benzidine could not be identified. This separation system does not enable to give an answer on the question of benzidine content in this sample.

This was a typical example for false positive results. Similar effects are obtained by very late eluting peaks from preceding analyses. A case of this kind is shown in Fig. 13 which also demonstrates the danger incorporated with automatization. Repeated analytical cycles of an extract yielded varying quantitative results. This was due to late eluting compounds from the preceding run which often were inte-

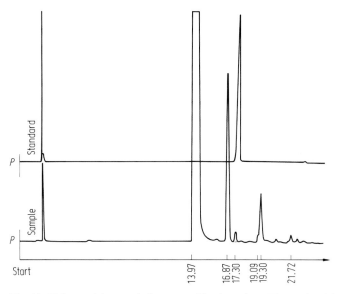

Fig. 12. False positive result by unspecific detection. The peak with retention time 17.3 was interpreted and quantitated as being Benzidine. The UV-spectrum taken with a diode-array detector proved the absence of Benzidine afterwards

grated together with the main component. The peaks could well be seen when amplification was increased. The effects were in the order of one to two percent.

General Requirements to Avoid Interferences

Separation of a sample must be as complete as possible to avoid overlapping by another component. Consequently quantitative results obtained under insufficient resolution are subject to frequent errors. Furthermore maximum resolution is required even to recognize the danger of overlapping peaks because otherwise false positive results as in Fig. 12 are possible. For this purpose chromatographic resolution is required.

The distinction of two elution peaks 1 and 2 necessitates a certain minimum of resolution R(1, 2). This resolution R of two peaks depends on the separation system and can be described by the following formula:

$$R(1, 2) = \frac{\sqrt{N}}{4} \frac{\alpha - 1}{\alpha} \frac{k'}{1 + k'} \quad \text{for } k'_1 \cong k'_2$$

The variables of the equation contain the retention k', the relative retention $\alpha(1, 2)$, and the theoretical plate number N of the separation system. If we assume a given apparatus with separation column then N is fixed. Now the values for retention and relative retention can be calculated for a given minimum resolution R(1, 2). Figure 14 shows an example illustrating an analysis of alkyl-benzenes on a commercially available ODS-separation column. The most important parameters are calculated from this chromatogram and indicated in form of a list. It can clearly

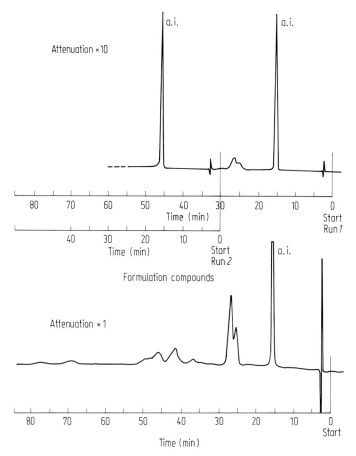

Fig. 13. Influence of late eluting compounds on automatically repeated analysis runs. The small peaks at 45 min. yield some percent deviation of the active ingredient measurement in the next cycle

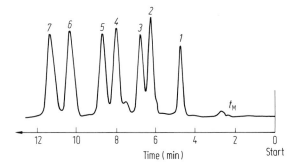

Peak Nr.	k'	α	R
1	1.00		
2	1.60		
		1.14	0.99
3	1.83		
4	2.30		
		1.14	1.09
5	2.63		
6	3.30		
		1.14	1.30
7	3.75		

Fig. 14. Separation power calculated from the chromatogram. Chromatogram of alkylbenzenes on a reversed phase system L = 30 cm, N ~ 3,000, dp ~ 10 μm

Table 4. Set of calculated retention parameters for a separation system with given efficiency (N = 3000) and required R (A, B)

Given	a) R = 1 (no baseline between peaks) Necessary		c) R = 0.5 (to detect a shoulder) Necessary	
k′	α	k′	α	k′
Compound A	Compound B		Compound B	
0.2	1.78	0.11	1.28	0.16
0.5	1.28	0.39	1.12	0.45
1.0	1.17	0.85	1.08	0.93
2.0	1.12	1.79	1.06	1.89
3.0	1.11	2.70	1.05	2.86
5.0	1.10	4.55	1.05	4.76
10	1.09	9.17	1.04	9.62

be seen that quantitative evaluation is impossible if the resolution R is below a value of 1.0. In cases were the resolution is below 0.5 it is not even possible to detect a shoulder as sign for interferences. A theoretical calculation based on the values of the chromatogram if Fig. 14 shows the importance of absolute retention k′ for the separation (cf. also Ref. [5]).

In other words:

To resolve two peaks at k′ = 0.2 with a relative retention of 1.1 and a resolution of 0.5 (to detect a shoulder) N = 17,000 theoretical plates are required.

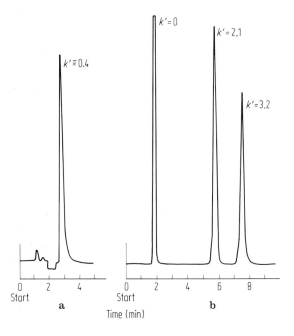

Fig. 15. a Elution pattern where quantitative evaluation is hazardous. b Elution pattern where interferences with α = 1.06 could be detected

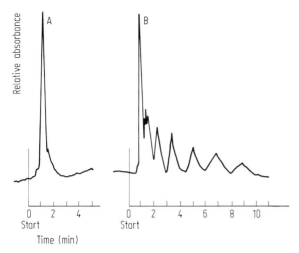

Fig. 16. Typical example for hazardous quantitative evaluation of peaks with k' < 1. **A** Chromatogram of pure LAS (Linear Alkylbenzene Sulfonate). **B** Chromatogram of LAS (A) in the presence of polyoxyethylene nonylphenylether (n = 9)

Figure 15 represents the required separation power in the form of a chromatogram and Fig. 16 shows that these errors of much too low absolute retention k' often occur in practice and even are shown in the corresponding literature. In the paper on which Fig. 16 is based the standards added are always found in amounts of more than 100% which is due to the low k'-values and demonstrates errors by interferences.

From the examples given in this section some rules can be derived:
1) In quantitative analysis the separating power (theoretical plate number) should be as high as possible.
2) Quantitative evaluation of peaks having a k'-value of less than 1 is not reliable.
3) In routine analysis the constancy of the matrix (sample) must be checked again and again under modified conditions of separation.

It is not reasonable to optimize an analytical method with regard to minimal analysis time and minimal necessary resolution because the composition of future samples to be analysed as a matter of routine is not fixed. The danger of possible errors is shown by the example given in Fig. 12.

4.4 Detector

The detector converts a measurable physical property of the sample component in the mobile phase into an analytical signal. This signal is recorded as a function of time and the resulting figure is called the chromatogram. In liquid chromatography the analytical signal depends on the concentration of the measured component in the flow cell of the detector. This dependance is mathematically described by the analytical function. It is containing both the properties of the detector alone and the effects produced by other components of the system. Particulars are given in 2.1. The validity of the analytical function and thereby the operating range of the detector is limited by several influences.

4.4.1 Linear Range of the Calibration Function

The analytical function can be linearized for certain ranges of validity by suitable mathematical processes. In the case of photometric detectors, which are the most famous in HPLC, the logarithm of the light intensities is proportional to the concentration of the sample in the cell according to the law of Lambert and Beer.

$$\ln I_0/I = e\,d\,c \quad \text{for} \quad \lambda = \text{const.}$$

The symbols mean:

I_0 = light intensity in the absence of the sample ($c = 0$)
I = light intensity with the sample in the measuring cell
e = specific molar coefficient of extinction at the measuring wavelength (strongly depending on the individual sample)
d = length of the lightpath in the sample (cell thickness)
c = concentration of the measured compound
λ = wavelength

Modern detectors for HPLC linearize the signal already in the detector amplifier.

The linear range is limited by the signal equivalent noise of the system at the lower concentration side and by a bend of the analytical function at the high concentration side. This effect is due to a too small residual light current I when the concentration of the sample in the cell exceeds certain limits. Figure 2 showed such a unlinear analytical function. For this reason one should not exceed an extinction of about 1 AU. Above that value maximal 10% of the light energy will be available and the law of Lambert and Beer is no longer applicable.

4.4.2 Time Constant for Noise Reduction

The detector puts out a more or less constant signal (base line) if the system is operated only with the mobile phase. The signal always exhibits fluctuations which may have a large variety of origins (cf. 4.1).

Short-time fluctuations within the range of 1 to 10 hertz are called noise. For the purpose of noise suppression (to improve the detection limit) most detectors are provided with an electronic smoothing device which can be described as time constant. This time constant is either built in firmly or adjustable and has a size of about 0.5 to 8 sec.

A time constant of 1 sec, however falsifies every peak with a peakwidth (at half the height) of 10 sec or less. Errors in peak areas and heights may be considerably. In Fig. 17 one peak with a k'-value of 1.5 and a absolute retention time of 3.5 min was recorded repeatedly with different detector time constants, the other

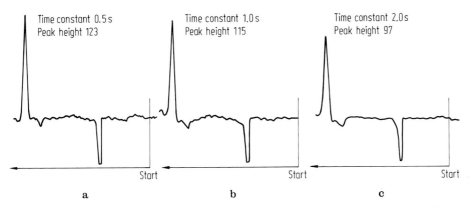

Fig. 17 a–c. Influence of noise suppression by a time constant in the detector. The peak heights are decreasing, the peak areas are increasing, the base-line is smoothed

conditions being identical. As a result of the smoothing effect which can be seen at the base-line the peak height is reduced by 7% by a time constant of 1 sec. The corresponding peak width at half height is increased by more than 10%. In this case the area of the peak increases. This cannot, however, be predicted generally. With a time constant of 2 sec the effect is becoming greater. As a general rule it may be concluded that the time constant of the detector or the amplifier should be at least 20 times smaller than the width of the peak to be measured at half its height (in time units).

4.4.3 Correctness of Wave-length and Band-width

The law of Lambert and Beer which is the basic equation for all photometric detectors can be applied only if the wavelength is constant. Deviations from the original wavelength may occur every day where the spectrometer is readjusted and will affect the extinction coefficients if they vary very much in the vicinity of the original wavelength (for example in the case of a sharp absorption maximum and steep flanks of the light absorption curve where small deviations result in strong signal changes).

On account of the increased light intensities an enlargement of the spectral band width of the detector results in an improvement of the signal to noise ratio, but, on the other hand, the calibration curve starts to bend because measurement is no longer carried out with one wavelength only and the law of Lambert and Beer is no longer applicable.

The effects of non-linear detectors have been described in detail by McDowell et al.[6]. The resulting errors cannot be repaired if only the areas or heights are stored in the integrator.

4.4.4 Temperature Effects

As mentioned above photometric detectors measure the concentration of the sample in the measuring cell (see Part 4.4.1). This concentration is a function of the tempera-

Table 5. Values of average absorbance change per degree centigrade for 6 solvents at specific wavelength [7]

Solvent	Wavelength	Absorbance	Average temperature coefficient
	nm	(AU)	AU/°C
Methanol	205	1.1	0.011
	210	0.62	0.003
	215	0.43	0.0025
	220	0.33	0.001
	230	0.21	0.0004
	245	0.10	0.0005
	260	0.08	0.0005
Water	205	–	0.0008
	215	–	0.0006
n-Hexane	200	0.95	0.0009
	210	0.43	0.0006
Iso-propanol	210	0.6	0.007
	215	0.3	0.0011
	230	0.12	0.0007
n-Heptane	225	0.99	0.0015
	240	0.41	0.0010
Acetonitrile	190	1.78	0.0017
	230	0.12	0.0013

ture because the density of the liquid mobile phase is a function of temperature. Consequently, variations of the temperature prevailing in the cell will cause variations of the signal supplied by the detector. The quantitative connection has been investigated and published by Campbell et al.[7]. Table 5 has been taken from this paper. The temperature coefficients of the absorption are listed for common solvents as examples for other organics at various wavelengths.

From the table the conclusion can be drawn that it is necessary to operate at a temperature that is as constant as possible, and to eliminate the slow temperature changes of up to 10 degrees that occur often in the laboratory from morning to evening.

4.5 Evaluation and Processing of Data

The detector delivers an intensity output signal which is plotted against time as chromatogram for example by a recorder. The intensity signals (peak height) as well as the elution peaks (peak area) contain quantitative information.

The data obtained can be evaluated in various ways:

4.5.1 Evaluation by Hand

Peak heights can be measured by hand relatively easily and rapidly. This method is still applied rather often in special cases. One example is trace analysis where the base lines often contain all kinds of noise and drift, here evaluation of peak heights by hand often yield results that are more correct than those of automatic integrators.

Measuring peak areas by hand is more sophisticated and for this reason is hardly in use today. If necessary however, the product of peak height and peak width at half the height may be used as equivalent of the peak areas. A fairly good measure of the peak areas is the weight of the peaks cut out of the recorder paper, for experience has shown that the strip chart recorder paper is very uniform.

4.5.2 Immediate Electronic Integration

The use of electronic integrators has been adopted for years. They process the analog signal output of the detector and provide areas or heights of peaks and the corresponding retention times. The principle shall be described shortly although there are numerous basic methods.

In the first step the analog signal of the detector is digitized (A/D-conversion). For this purpose the intensity (voltage) of the signal is determined at reasonable time intervals. The necessary time intervals are depending on the peak widths. The dynamic range of an integrator or A/D-converter indicates the range for the possible values of the intensities without changing the attenuation. In chromatography actually a typical input voltage range of 1 microvolt to 1 volt is required. That is in other words a dynamic range of $1:1,000,000$. This corresponds to a digital resolution of 20 bits. As A/D-converters with 20 bits resolution are seldom and expensive commercial integrators have for example 14 bits converters together with automatic range switching or have chosen the way over voltage frequency transmission and frequency counting.

The dynamic range defines the maximum accuracy of the system. Table 6 shows this accuracy of the data processing apparatus as a function of digital resolution of the converting system. The digital resolution of the data processor (computer) does not afford the accuracy but only the speed of operation.

The effective accuracy of measurement is of course determined by the whole analytical system and in most cases it is not better than about $1:10,000$ (0.1 mV in 1 V).

The individual values of the signal in their digital form are averaged for purposes of smoothing and then added in an appropriate way. The sums thus obtained are proportional to the areas under the elution curve if it is possible to subtract all portions that contain the areas between the electrical zero and the chromatographical zero (base line). The chief problems of integration reside in determining the accurate base line, recognizing the beginning and end of a peak, and recognizing the peak maxima. These problems are solved by calculating the differences between various numbers of successive intensity values (the first derivative of the curve). The differences on the undisturbed base line yield a measure for the noise which should not be intergrated and is determined by a slope sensitivity test before the analytic run. This test yields a threshold value for the peak to be recognized. Differences exceeding this threshold value indicate the beginning of a peak if they are positive and the flank of a peak if they are negative.

The time intervals for measuring the intensities by A/D-conversion (sampling rates) depend on the width of the peaks because at least 5 and preferably 10 pairs

Table 6. Maximum obtainable accuracy as a function of digital resolution of an Analog/Digital-converter

Number of bits (digital resolution)	Maximum accuracy	($= +/-1$ bit)
1	0.5	$= 50\%$
2	0.25	
3	0.125	
4	0.063	$= 6.3\%$
5	0.0031	$= 3\%$
6	0.0015	
7	7.8×10^{-3}	
8	3.9×10^{-3}	$= 0.4\%$
9	1.9×10^{-3}	
10	0.98×10^{-3}	
11	0.49×10^{-3}	
12	0.24×10^{-3}	$= 0.024\% = 240$ ppm
13	0.12×10^{-3}	
14	0.06×10^{-3}	$= 0.006\% = 60$ ppm
15	0.03×10^{-3}	
16	1.5×10^{-5}	$= 15$ ppm

According to Eichelberger and Günzler [8]

of time/intensity values (raw data) are needed to calculate the envelope of the peak and thereby the area with sufficient accuracy. The algorithms of calculation are different in commercial integrators.

Threshold values, sampling rates, and many other parameters for integration can be influenced by the operator and must be adapted to the problem. Great attention has to be paid to the preparation of this set of parameters. The manuals published by the manufacturers of integrators are extensive and often not easily understandable to chemists and laboratory personal. But if the manual instructions are not taken into consideration considerable misprogramming may occur. The influence on the results cannot be back-estimated because the raw data are not stored but immediately converted to areas or heights and retention times. Only these reduced data are sometimes stored for reporting and can be provided with correction factors and/or peak names.

If an analytical method has been well developed (complete separation of peaks) and the set of parameters for the integrator has been adapted to the problem the integrator yields correct and precise results at a high speed and reliability.

Today integrators are often connected with electronic plotters, so that a laboratory recorder can be dispensed with. In some times these integrators can even plot the electronically chosen base line into the chromatogram. This characteristic is very desirable because it allows to recognize possible errors very soon.

4.5.3 Integration after Storage of Raw Data

The decrease in the prices for laboratory computers, the increase of their performance, and the availability of large memory capacities have led to systems which can store the raw data of a chromatogram and enable peak detection errors and base line definition errors to be eliminated afterwords. This is preferably be done by representation of the raw chromatogram on a display unit. The analyst in charge is then able to check wether the base line is positioned correctly, if spikes have been detected, or wether the envelope curve used for calculation complies with the ideas of the operator.

Doubtlessly this requires much time and consequently is not a process that can be applied to each routine analysis run. However, it permits a critical examinaion of difficult and complex cases or the preparation of sets of parameters with optimal adaption to the problem for routine integration.

As the prices for personal computers with sufficient storage and working speed have already gone down considerably and will probably continue to fall it is to be expected that systems with raw data storage and suitable software will become commercially available to an increasing extent.

5 Practical Experiences

5.1 Published Results from Collaborative Work

As has been described in Sect. 3.1 random errors can be described by the quantities repeatability r and reproducibility R.

The determination of such precision data for analytical methods requires relatively much expenditure if it is intended to ascertain the reproducibility R by means of a collaborative laboratory comparison test. Often this expenditure is worth while only in the case of rather important commercial products. As example some results from collaborative work on plant protection chemicals in which different methods of analysis were applied are indicated in Table 7.

Out of the typical data $R = 2.5$ and $r = 1.2$ a critical difference D can be derived for two mean values obtained at different laboratories by two repeated determinations:

$$D_{krit} = 2.4\%$$

(For explanation refer to Sect. 3.1).

The above data clearly show that the reproducibility R between different laboratories is much worse than the repeatability r within one laboratory. This is due to the systematic errors which are difficult to eliminate or even calibrate when written methods or reports are used by different laboratories. The application of random theory and mathematics is normally not allowed to describe systematic errors. But as these errors are virtually independent from one another the values give useful information on the significance of errors between laboratories.

All the samples to be analysed were plant protection chemicals. The tests have been carried out from 1979 to 1982. Despite their variation the results show that the repeatability r within the laboratories was found always within the range of 0.6 to 1.5%. This means that the results of repeated measurements have a standard

Table 7. Published results from collaborative tests

Substance	Method	Range of content (%)	r_{95} (%)	R_{95} (%)	Ref.
Hostathion	HPLC, int.St.	60	1.2	2.4	[9]
		40	1.2	1.2	
Carbendazim	HPLC, ext.St.	100	1.0	2.0	[10 b]
		60	0.6	2.9	
		50	0.7	2.1	
Carbendazim	UV-Spectr.	100	1.5	2.9	[10 b]
		60	1.1	2.9	
		50	0.9	1.5	
Endosulfan	GC, int.St.	100	1.3	2.5	[10 a]
		50	0.7	1.3	
		30	0.7	0.9	
2,4,5-T	HPLC, int.St.	42		2.0	[11]
		20		1.3	
		22		1.4	
MCPA-Salz (1980)	HPLC, int.St.	41		2.1	[11]
		42		2.0	
		46		2.8	
MCPA-Salz (1978)	HPLC, int.St.	40		3.5	[12]
		40		4.6	
		50		4.5	

deviation of 0.2 to 0.5%. The reproducibility R between different laboratories is always twice as high and yields values from 0.9 to 4.6%, which corresponds to standard deviation of 0.3 to 1.6%.

In previous years, too, attempts were made to measure the quantitative precision of HPLC. Probably the most extensive and best documentated test was a trial performed by the ASTM (American Society for Testing and Materials) in 1979/1980. 78 laboratories took part in this collaborative work[13]. As in our opinion the results elucidate some very important facts this trial will be briefly described and discussed. The participants had to use a HPLC-apparatus with a reversed-phase column and isocratic elution. The samples contained 4 and 6 components, respectively, and had been weight and diluted as ready to analyse solutions before they were dispatched. The sample concentrations were chosen as to represent all components in the chromatogram without changing the attenuation. The sample chromatograms which were supplied together with the samples are shown in Fig. 18. Each laboratory had to calculate the average of three analytic runs.

The most important statistic characteristics of the test are represented in Table 8.

The evaluation of the data led to the following statements:

a) The quantitative results became much worse when the separation was incomplete (overlapping at the peak tail) (cf. to the compounds benzene and methyl benzoate in Table 8 B). Hence it follows that base-line-separation is a prerequisite for quantitative work.

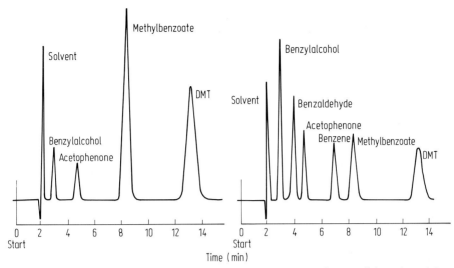

Fig. 18. Chromatograms of the samples dispatched by ASTM for a collaborative trial on quantitative precision (Ref.[13]). Stationary phase: silica-octadecylsilane. Mobile phase: methanol + water adapted to achieve the above chromatograms. Detection: any

Table 8. Results from a collaborative test on HPLC with 78 laboratories in USA under the guidance of ASTM in 1979/80 [13]

Component	Theoretical value (mg/ml)	Found mean value (mg/ml)	Reproducibility R_{95}[a]	
			(mg/ml)	(%)
A. Sample LC-79-1 containing 4 components (see Fig. 18) results based on debugged data (without outliers)				
Benzyl alcohol	0.420	0.421	0.045	11
Acetophenone	0.00797	0.0086	0.0011	13
Methyl benzoate	0.823	0.810	0.071	9
Dimethyl terephthalate	0.0806	0.0816	0.0097	12
B. Sample LC-79-2 containing 6 components (Fig. 18) results based on debugged data (without outliers)				
Benzyl alcohol	1.53	1.54	0.255	16
Benzaldehyde	0.0198	0.0197	0.0017	9
Acetophenone	0.0252	0.0258	0.0025	10
Benzene	1.04	1.04	0.246	24
Methyl benzoate	0.400	0.397	0.074	18
Dimethyl terephthalate	0.0540	0.0543	0.0059	11

[a] Reproducibility $R_{95} = 2.83 \times$ standard deviation of the trial (on the 95% confidence level)

  b) The debugged data have a very good accuracy of the mean. The found mean values are in good agreement with the theoretical values.
  c) the evaluation of either peak heights or peak areas did not result in different mean values. The dispersion of the values obtained by peak height evaluation were somewhat smaller.

d) The individual separation column, mobile phase, and detectors that were free to choose by the individual laboratory had no significant influence on the quality of the results if separation was good enough.

e) The best reproducibilities R(95) were about 9% corresponding to standard deviations of about 3%. These figures are quite high taking into account that sampling could not be a source of errors.

The represented results will perhaps be disappointing to many ambitious analysts but also the data shown in Table 7 indicate that a reproducibility R(95) of 2 to 4% is altogether normal.

Sometimes an analyst is concerned so intensively with "his" method that he overestimates its capabilities and efficiency outside the own laboratory. It is therefore advisable to consider the results measured in collaborative tests carried out according a large variety of different analytical methods. Figure 19 represents the results of many trials carried out by AOAC (Association of Official Analytical Chemists, USA) with fatty acid methyl esters by gas chromatography[14]. It can clearly be seen that the precision data depend on the level of content of the sample. The relative standard deviations (CV, coefficient of variation) in collaborative tests are the higher the lower the measured content, while the absolute reproducibilities R(95) still amount to 6% at a level of content of 100% and are thus equivalent to the above-mentioned HPLC results. Because of the importance of relative errors it is better, however, to take into consideration the relative standard deviations of the collaborative trials.

A similar dependence on the content level can be seen when in addition the results of collaborative tests with other analytical methods are taken into consideration. A review was published by Horwitz[15, 16] who compiled data from more than 150 independent collaborative trials guided by AOAC. These tests related to very different fields of samples and were carried out using different analytical technics (chromatography, atomic absorption spectrometry, spectrophotometry, electrochemistry, and bioassays). When the resulting data are plotted a curve is obtained which represents the quantitative quality of analyses on samples with different con-

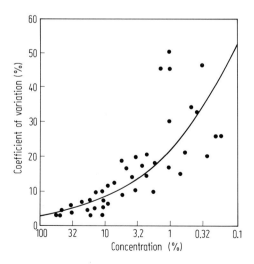

Fig. 19. The interlaboratory coefficient of variation for the gas chromatographic determination of methyl esters of fatty acids according to Firestone (Ref.[14])

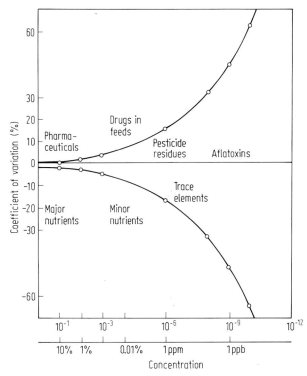

Fig. 20. The general curve relating interlaboratory coefficients of variation (expressed as powers of two on the right) with concentration (expresses as powers of 10) along the horizontal center axis according to Horwitz (Ref.[15, 16])

tent levels. The curve is shown in Fig. 20. Out of this curve one may estimate the achievable quantitative precision between different laboratories for a given level of content. The standard deviation that can be expected within one laboratory will in Horwitz's opinion be about half as high.

Other collaborative tests have proven that the precision data found in one laboratory (repeatability r) are better by a factor of roughly 2 relating to the precision data of collaborative tests (reproducibility R). This seems to indicate a sort of rule which may be worded as follows:

Repeatability r (within one laboratory) is always better than reproducibility R (out of collaborative work). Often the factor 2 is between the two values.

$$R = 2r$$

This principle is applicable irrespectively of matrix, analyte, and method used. One can only indulge in speculations about this behaviour. The systematic errors which occur in different laboratories and which in comparison with each other cannot be kept as constant as in one laboratory play an important part. Besides it is possible that the inhomogeneity of the divided sample has a distinct influence when a collaborative test is carried out.

From this rule conclusions can be drawn to the achievable precision data of an analytical method which is not yet tested collaboratively.

5.2 Quality Assurance of Analytical Results

In the Sects. 3 and 4 the various sources of errors that may have influence on the analytical result have been pointed out and discussed. In Sect. 5 practical results of laboratory comparison tests are described. The question arises now how to attain or improve such a quality of precision and accuracy. In addition to fastidious attention that is paid to the various sources of errors and a constant critical judgement of the obtained results there are further measures to ensure the quality of analysis.

They may be divided into three groups which will now briefly be described and discussed.

a) General Quality of the Laboratory

The subject of this paragraph does not seem to be closely connected to HPLC but nevertheless it is an important part of everydays analytical work. For this reason we will breafly describe our opinion on the requirements that an analytical laboratory has to meet.

Trained personnel and suitable rooms are fundamental prerequisites for quantitative analytical work. Layout and equipment of the rooms should take into account the problems of safety and contamination.

Planning and documentation of the analytical work also have a great influence on the performance of the laboratory (cf. Ref[17] "Data Aquisition and Data Quality Evaluation in Environmental Chemistry").

Standard operating procedures have to be prepared for all routine and development work that is carried out because it must be possible to repeat the same procedure lateron, if necessary. Short methods in standardized format have been proven to be very useful. An example out of our laboratory is given in Table 9. Short methods of this kind offer enough information to trained personnel and enable methods to be improved and modified with increasing experience without the need for complete reports in every new case.

Apparatus and its condition should comply with certain minimum requirements. They include regular maintenance and examination based on the maintenance manual or instructions and the detailed description of all modifications incorporated in the apparatus. They also include the regular calibration of the detectors and auxiliary instruments like balances and diluters. Documents regarding these calibrations should of course also be established and filed for a reasonable time. It is suitable to have a file on each apparatus which contains all minutes of maintenance and calibration and service instructions as well.

Reagents, reference standards, and standard solutions should be clearly labelled and stored suitably. Labelling should include: identity, purity, data of manufacture, and expiry date, if possible. HPLC technics always relate to standard substances. *No result can be more correct than the used standard substance and its description.*

Table 9. Example for a HPLC short method

Compound:	Na-fluorescein
Sample:	Waste-water
Concentration of Analyte:	0.1–100 µg/l
Method:	HPLC with external standard evaluation of peak heights

Conditions

1. Separation column	12.5 cm length, 0.46 cm I.D., SS
2. Stationary phase	Hypersil® ODS, dp 5 µm
3. Mobile phase	60% (vol) methanol; 40% (vol) water, ph = 1 (H_2SO_4); flow 1.2 ml/min Retention time of analyte: 2.6 min
4. Detection wavelength	Fluorescence (Shimadzu RF 530); Ex 440 nm; Em 519 nm Range 2×10 mV
5. Standard solutions in mobile phase	10 mg Na-fluorescein in 100 ml (standard I); 10 ml standard I in 1000 ml (standard II); 1 ml standard II in 1000 ml (standard III); injection volume 20 µl = 20 ng
6. Sample treatment	Dilute 25 ml of sample with 25 ml of mobile phase; if peak height is more then twice that of standard III, dilute once more according to estimated concentrations in the sample

b) Statistical Data

There are only few cases where an analyst will rely on a single result obtained with one sample because it is quite obvious that weighing or diluting errors or evaluation mistakes sometimes cause outliers. It is state of the art therefore to analyse duplicate samples and, if the results differ more than expected, to analyse a third portion. In HPLC the final sample solution is most often not limited in volume for one injection and there is no reason not to carry out two runs of analysis with this solution. The time consumption for duplicate runs is often short in comparison to the preparation time. Duplicate or even triplicate runs are facilitated if automatic injection systems are used. The data thus obtained would allow for extensive checking of the results.

Table 10 shows an example and the significance of the output of the checks. Average values 1 and 2 together with the corresponding standard deviations are tabulated and show with the help of a test value if the sample solutions differ significantly from one another. The basic principle is that the standard deviations of runs out of one sample solution are compared with the standard deviations of runs out of different sample solutions from one sample. The quotient out of the two standard deviations is used as test value which indicates the quality of the results. If the test value defined as in Table 10 is smaller than 1 the runs carried out with the same sample solution differ too much. This may indicate malfunction of the injection system. If the test value is higher than 10 (or whatever your threshold value may be) (see runs 12–15 in Table 10 as example) it indicates serious errors outside the HPLC system (e.g. sampling or dilution). So this test value indicates errors and enables the coarse localisation.

Table 10. Example for a group of HPLC-runs for one sample and the testing of the results

Raw Data				Results					
Run No.	Sample	Conc. mg/100 ml	Area	Result (%)	Average 1 (%) n = 2	SD 1	Average 2 (%) n = 4	SD 2	SD 2/SD 1 [a]
1	Stand.- solution	1006.5	15.96	40.00					
2	1 a	693.08	21.03	76.54					
3	1 a	693.08	21.06	76.65	76.60	0.08			
4	1 b	637.02	19.31	76.46					
5	1 b	637.02	19.31	76.46	76.46	0.00	76.52	0.09	2.3
6	Stand.- solution	1006.5	15.99	40.00					
7	2 a	992.78	15.73	38.89					
8	2 a	992.78	15.74	39.92	39.91	0.02			
9	2 b	1016.9	16.16	40.01					
10	2 b	1016.9	16.18	40.06	40.04	0.04	39.97	0.08	2.7
11	Stand.- solution	1006.5	15.94	40.00					
12	3 a	669.27	15.64	59.02					
13	3 a	669.27	15.62	58.95	58.99	0.05			
14	3 b	656.46	15.77	60.68					
15	3 b	656.46	15.78	60.71	60.70	0.02	59.84	0.99	25
16	Stand.- solution	1006.5	15.95	40.00					

SD = standard deviation; Average 1 = average from 2 runs out of 1 sample solution; Average 2 = average from 4 runs out of 2 sample solutions.

[a] Test value, increases sharply if the sample solutions differ significantly.
   (Calculating a standard deviation out of 2 values is of course not correct from the mathematical side. But in this case it is very suitable for checking. These numbers must not be compared to the precision data represented in the Tables 7 and 8)

The test values shown in Table 10 indicate that the weighed portions 3a and 3b differ significantly from one another. At that time the analyst in charge still has the opportunity to act, if necessary, and to establish subsequent values or to eliminate the outliers. If programmable calculating integrators are used together with automatic injection equipment these tests can be performed automatically by software control and will not waste time and resources.

c) Quality Control Charts

The data obtained from Table 10 do not allow to make a statement on the accuracy of the measurements. Besides, in the analysis of many fields like environmental monitoring or body fluids the available sample amounts do not allow more than one run on one sample. In these cases test values as indicated in Table 10 cannot be obtained but the need for quality ensurance is even greater.

ANALYTISCHE BIOCHEMIE

HOECHST AG

QUALITATSREGELKARTE   VOM 09.07.80   BIS 21.07.80
PRAPARAT:LASIX   EINWAAGE NR.: 328.   VOM 08.07.80
MATRIX:SERUM   METHODE:HPLC   MESSPLATZ NR.: 3.

	KWI		KWII		KWIII	
SOLLWERT	500(NG/ML)		100(NG/ML)		10(NG/ML)	
2 S	446. -	554.	77. - 123.		7. - 13	
3 S	419. -	581.	65.5 - 134.5		5.5 - 14.5	

NR.	DATUM	IST	DATUM	IST	DATUM	IST
1	09.07	534	09.07	102	09.07	9.9
2	09.07	483	09.07	81	09.07	8.2
3	10.07	576	10.07	119	09.07	10.0
4	10.07	488	10.07	121	10.07	9.9
5	16.07	426	16.07	134	10.07	13.6
6	16.07	502	16.07	110	16.07	11.7
7	17.07	513	17.07	96	16.07	14.1
8	17.07	503	17.07	102	17.07	10.5
9	20.07	533	20.07	111	17.07	13.4
10	20.07	489	20.07	114	20.07	13.0
11	21.07	496	21.07	105	20.07	11.0
12	21.07	523	21.07	83	21.07	13.8
13	21.07	483				

MW		503.8		106.5		11.6
S.D.		35.2		15.3		1.9

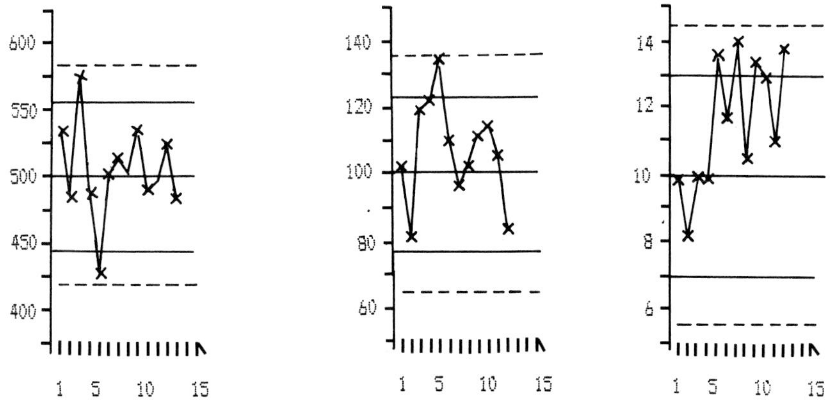

Fig. 21. Quality control chart edited by data system according to Uihlein et al. (Ref.[19])

The only solution to that problem is the use of homogeneous and stable reference samples which have a matrix similar to the samples to be analysed[18]. These reference samples may be Standard Reference Materials (SRM) which have been tested collaboratively and are commercially available, or they may be in-house samples that have been characterized and reported as good as possible.

Reference samples are introduced into the analytical process at regular intervals under code numbers that are unknown to the personnel. The representation of the corresponding results as a function of time clearly shows the precision of the analytical process that can be attained. If the absolute content of the reference material is known with sufficient accuracy the results also show the degree of correctness that can be reached. Systematic errors can be recognized as trends from the course of the curves or values.

A good example is represented in Fig. 21[19]. An automized system including an efficient data processing unit can prepare control charts without much additional expenditure, provided the corresponding software has been written and tested. We hope that the manufacturers of chromatographic data processing units will soon offer comfortable software for control and data management tasks as described above.

6 References

1. Currie, L.A.: "Scientific Uncertainty And Societal Decisions: The Challenge To The Analytical Chemist". Analytical Letters, 13(Al), 1–31 (1980)
2. Kratochvil, B., Wallace, D., Taylor, J.K.: "Sampling for Chemical Analysis." Anal. Chem. 56, 113R–129R (1984)
3. "General Guideline for the Establishment, Maintenance, and Distribution of Chemical Reference Substances." WHO, Technical Report Series Nr. 681 (1982)
4. Wiese, Martin, Hermansson: Chromatographia 15, 737–742 (1982)
5. Glajch, J.L., Kirkland, J.J.: Anal. Chem. Vol. 55, 319A–326A (1983)
6. McDowell, L.M., Barber, W.E., Carr, P.W.: Anal. Chem. 53, 1373–1376 (1981)
7. Campbell, J.E., Hewins, M., Lynch, R.J., Shrewsbury D.D.: Chromatographia 16, 162–165 (1982)
8. Eichelberger, W., Günzler, H.: „Analytiker Taschenbuch Bd 1", Ed. H. Kienitz et al. Springer Verlag, Heidelberg-New York, p. 43–62 (1980)
9. Aßhauer, J.: Chromatographia 15, 71–74 (1982)
10. Schmidt, P., Aßhauer J.: In: Pesticides, CIPAC Methods and Proceedings
    a) Series 2, 1980, Heffers Printers London, p 167
    b) Series 3, 1981, Heffers Printers London, p 144
11. Grorud, R.B., Stevens, T.S.: J. Assoc. Off. Anal. Chem., Vol. 63, 873–878 (1980)
12. Stevens, T.S., Grorud, R.B.: J. Assoc. Off. Anal. Chem, Vol. 62, 738–741 (1979)
13. "An Evaluation of Quantitative Precision in HPLC." American Society for Testing and Materials (ASTM). J. Chromatogr. Sc. Vol. 19, 338–348 (1981)
14. Firestone, D., Horwitz, W.: J. Assoc. Off. Anal. Chem., Vol. 62, 709–721 (1979)
15. Horwitz, W.: Anal. Chem., Vol. 54, 67A (1982)
16. Horwitz, W., Kamps, L.R., Boyer, K.W.: J. Assoc. Off. Anal. Chem., Vol. 63, 1344–1354 (1980)
17. "Guidelines for Data Acquisition and Data Quality Evaluation in Environmental Chemistry." ACS-Subcommittee on Environmental Improvement et al. Anal. Chem. 52, 2242–2249 (1980)
18. "Quality Assurance of Chemical Measurements." J. K. Taylor in Anal. Chem. 53, 1588A–1596A (1981)
19. Uihlein, M., Ostermann, H., Hajdu, P.: Microchimica Acta (Wien), (I) 159–169 (1981)

Preparative Application of HPLC

A. Wehrli
Sandoz Ltd., Pharmaceutical Division, Preclinical Research, 4002 Basel/Switzerland

1 Introduction

Chromatography is primarily a separation technique. When applied to the collection of pure material, it is called preparative chromatography. The aim is to extract as much material in as pure a state as possible.

For this purpose, two chromatographic variants are used:

a) batch or conventional liquid chromatography (Fig. 1a).
b) continuous liquid chromatography (Fig. 1b).

In batch chromatography one phase moves (the mobile phase) and the other remains stationary.

In continuous chromatography the conventional stationary phase is moved countercurrently in a "moving bed" against the mobile phase. By adjusting the flow-rate of the bed and the mobile phase, one can enrich and separate a component or even a group of components in one direction or the other.

This type of chromatography, where the apparatus used is normally highly specialized, has recently been reviewed[1, 2]. Therefore, and because continuous liquid

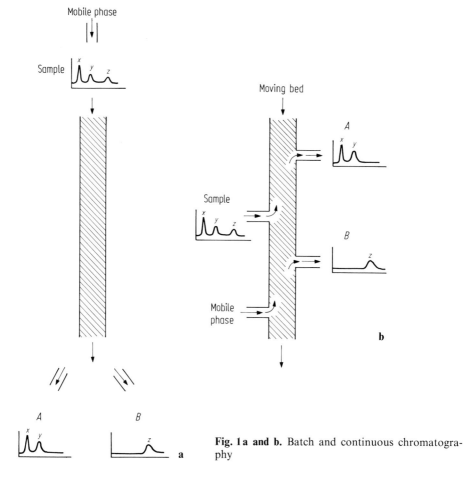

Fig. 1a and b. Batch and continuous chromatography

chromatography is the province of the chemical engineer, continuous chromatography is excluded here.

This chapter deals with batch chromatography, in particular with liquid chromatography, which, following the revival in 1970, led to HPLC. This period included the study of preparative separations with forced elution flow rates and brought about rapid advances in theory, optimization, apparatus, and applications.

2 Theory and Strategy

2.1 Loadability

The three main goals of any chromatographic separation are:

– Resolution;
– Separation speed;
– Capacity.

In analytical chromatography, speed and resolution are a requirement. In preparative chromatography capacity is most important.

Unfortunately, chromatographic systems have a low capacity and thus loadability. The column linear loadability is less than 1 mg/g sorbens, as demonstrated in Fig. 2 for analytical columns of 4 mm–17 mm diameter and 250 mm in length. From the figure it can be deduced that in an ideal range of sample weights/g sorbens, the column efficiency (the resolution R) changes little with sample size, and separations can be carried out with analytical separation power. Then the condition is reached, where column overload begins. At this point, with the increasing sample size, the efficiency diminishes considerably.

The overloading of the column is caused through

– excessive sample feed volume[3] or
– excessive sample mass[4].

The first type of overloading, the sample feed-overload, is usually encountered as a result of poor solubility of solutes in the mobile phase. In this case the fronts of the peaks being eluted remain at constant retention time, while retention of the peak-backs rise with increasing feed volume (comp. Fig. 3).

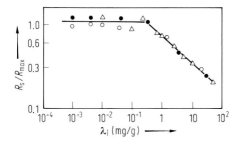

Fig. 2. Relation between relative resolution (R/R_{max}) and loadability/g sorbens (λ_i); column and sample as in Fig. 6

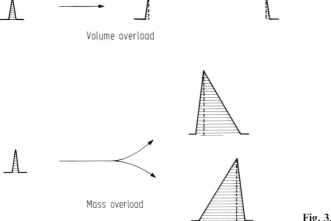

Fig. 3. Sample overload

In contrast, when mass overload occurs, the retention of both the back and the front of the peaks as well as the elution time may change (Fig. 3). These changings are caused by

– the occurrence of non-linear solute isotherm;
– deactivation of sorbent;
– change of mobile phase polarity;
– solute saturation;
– change of the nature of separation
  (In critical cases, increased feed can change the nature of the separation, e.g., from elution to displacement chromatography).

It is evident that under such circumstances any general discussion is impossible. To get practical guidelines, we therefore have to examine the case where the column is assumed to operate in the linear part of solute isotherm in which no overloading occurs. For this situation, an expression describing the dependence of feed volume and chromatographic resolution can be found. This is shown by several authors[5]. We[6] deduced the following equations:

$$V_f = \frac{V_0}{\psi} \frac{k_i'(\alpha - 1)^2}{R_s^2} \frac{(1 + k_i')^2}{N} \tag{1}$$

where

V_f = feed volume
V_0 = dead volume of the column
ψ = σ_0/V_f (a proportional factor depending on the mixing of the sample)
σ_0 = volume variance of the distribution function of the component at the beginning of the column
k_i' = the capacity factor for solute i
α = the selectivity coefficient between solute i and j
R_s = resolution between solute i and j
N = the number of theoretical plates of the column

Furthermore, when chromatographic characteristics are such that N is large, the feed volume is expressed by the equation

$$V_f = \frac{V_0}{\psi_4} \frac{k_i'(\alpha-1)}{R_s} \tag{2}$$

and the amount of sample Q_f which can be injected into the column is

$$Q_f = C_f V_f = C_f \frac{V_0}{\psi} \frac{k_i'(\alpha-1)}{R_s} \tag{3}$$

where

C_f = feed concentration

Equations (1)–(3) describe the type of separation where the resolution is so great that feed volume can be increased. Of course, a small feed volume in preparative liquid chromatography is not desirable. Consequently, to optimize a preparative separation, the product of resolution and the sample input should be a maximum. This means R_s should be minimized. Therefore, for maximal sample load:

$$Q_{f,\,max} = Q_{f,\,max} \frac{V_0}{\psi} \frac{k_i'(\alpha-1)}{R_{s,\,mini}} \tag{4}$$

and

$$Q_{f,\,max} = \lambda \cdot \frac{V_0(\alpha-1)}{R_{s,\,mini}} \tag{5}$$

where

$Q_{f,\,max}$ = maximal sample load
$C_{f,\,max}$ = maximal feed concentration
$R_{s,\,mini}$ = minimalized resolution

The expression $C_f \cdot k_i'/\psi$ has been the subject of considerable study, e.g., Cretier[7] has shown that maximum injection concentration $C_{i,\,max}$ decreases while the capacity factor k' increases and the product $C_{i,\,max} \cdot k'$ remains roughly constant. Therefore, it can be concluded that $C_i^0 k_i'/\psi$ is a system constant, and we can define this expression as the specific loadability λ.

    In consideration of these special circumstances, Eq. (5) in practical terms implies that feeding is proportional to:

– specific loadability,
– column dead volume,
– selectivity,
    and will be optimum
– at the point where the peaks begin to separate. Therefore, overloading is the rule in preparative liquid chromatography.

2.2 Throughput

The most essential criteria for a preparative separation is not the loadability Q_f, it is the amount of component i that can be separated per unit time with a chosen purity. This so-called throughput or production rate P_j is defined as

$$P_j = Q_{f,\,max}/t_{R,\,j} \tag{6}$$

where $t_{r,\,j}$ is the retention time of the last eluted compound.

The term 'throughput' was introduced by de Jong et al.[8]. This production rate and its dependence can be considered under linear and above linear isothermal (mass-overload) conditions.

2.2.1 Throughput Under Linear Isothermal Conditions (Volume-Over-Load)

In Eq. (6) it is assumed that the substance under consideration is the last eluting component and that the amount of the substance producing Q_f is injected. Thus the cycle time $t_{R,\,j}$ is given by:

$$t_{R,\,j} = \frac{L(1+k'_j)}{u_0} \tag{7}$$

and

L = column length
u_0 = linear velocity

Modifying the loadability-equation [Eq. (5)] for volume/overload conditions by means of $t_{R,\,j}$, we obtain for the corresponding throughput:

$$P_{j,\,max} = \lambda\,A\,u_0\,\frac{1}{1+k'_j}\,(\alpha-1)\,1/_{R_{S,\,mini}} \tag{8}$$

Analyzing this Equation we can state:

- The selectivity should be optimized in consideration of the dependence of throughput from the capacity factor (α_{max} at k'_{mini}).
- The sample capacity will be best on totally porous packings.
- At a certain column length, the throughput rises by increasing the cross section of the column rather than the length.
- For highest sample throughput per unit time, a velocity as high as possible should be chosen (V_{max}).
- For maximal throughput, the column must be overloaded to the point where the peaks still separate ($R_{s,\,mini}$).

These statements show that the geometrical effects (A, L), physical effects (k' and u_0), and chemical effects (α) play a main role in the throughput equation and that optimal throughput means optimizing these parameters.

2.2.2 Throughput Under Non-linear Isothermal Conditions (Mass-overload)

Non-linear isothermal conditions are caused by excessive sample mass, and as mentioned in Sect. 2.1, under such conditions no general discussion can be made, since

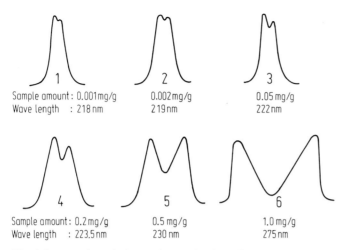

Fig. 4. Increased resolution under overload conditions

the isotherm is characteristic for a definite substance, and in a mixture the chromatographic behavior and the technique can be influenced by the sample and the load respectively. For these reasons, and because systematic studies have not been published, we have to rely on a trial-and-error method in order to decide whether overloading has a positive or negative influence.

An example which suggests overloading the column prior to repeated separations is shown in Fig. 4. The example demonstrates that this solute pair (phenylethane and ethylphenylcarbinol) separated on a reversed-phase system increased the resolution under overload conditions. The reason for this better separation may not at all be attributed to the previously-mentioned displacement effect.

Such mass-overload-facts and the recent developments in high-performance liquid chromatography with respect to column engineering and instrumentation have prompted Horwarth[9] to explore the potential of liquid chromatography in the displacement mode for preparative-scale separations with columns and precision instruments that are presently used in analytical work.

In this mode of chromatography, the column packed with a solid adsorbent is first equilibrated with a carrier solvent (mobile phase) that has a low affinity to the stationary phase. Then the feed solution containing the mixture dissolved in the carrier is introduced so that its components are adsorbed in the inlet section of the column.

Subsequently, the solution of a displacer substance having a stronger affinity to the stationary phase than any of the feed components is pumped into the column. Provided the column affinity be sufficiently long, the components of the feed arrange themselves upon the action of the displacer front moving down the column into a "displacement train" of adjoining square-wave concentration pulses of the pure substances, all moving with the same velocity. A displacement chromatogram is illustrated in Fig. 5. It is very similar to the overloaded elution chromatogram of Fig. 4 and confirms the hypothesis that in elution mode, displacement effects can be the background of peak destruction but also of peak separation. Therefore, we can state:

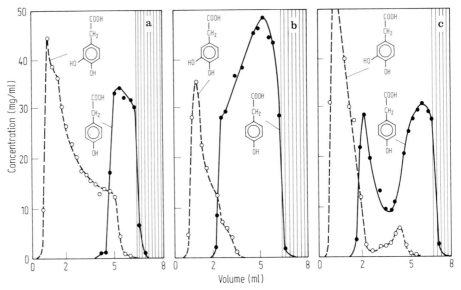

Fig. 5a–c. Effect of feed size on the separation of 3,4-dihydroxyphenylacetic acid (●) and 4-hydroxyphenylacetic acid (○) by displacement chromatography. Displacer: 0.64 M phenol in 0.10 M phosphate buffer; pH 2.1; temperature 25° C; mobile phase flow-rate 0.3 ml/min. The respective amounts of 3,4-dihydroxyphenyl acetic acid and 4-hydroxyphenylacetic acid in the feed were: **a** 114 and 57 mg; **b** 57 and 114 mg; **c** 114 and 114 mg. The column was a 250 × 4.6 mm, 10 μm Partisil ODS-2. (Reprinted from Ref.[9] with permission.)

Prior to repeated conventional separation, overload experiments should be realized because overloading respectively, displacing can increase separation and sample throughput. In other words: throughput optimization must include mass-overload experiments.

2.3 Optimization Strategy

As seen in the previous sections, high throughput is the criterion required for both routine and occasional preparative work. To reach this, several parameters must be optimized as shown in Eq. (5), Sect. 2. Such an optimization can be achieved by use of two different approaches. The most usual, but perhaps not always the most successful, is to base the choice of chromatographic conditions on knowledge and experience with similar samples. This approach is useful if all relevant factors are known and their importance in the actual case is well understood. But, if the circumstances are not fully known, e.g., if we are working in the mass-overload mode as mentioned above, or if better selectivity is required, we have to optimize and in this situation the best possible method is to seek the optimum preparative chromatographic conditions via a mathematical statistical method. To do this
1. the type of variables important in effecting throughput has to be specified and,
2. the optimization strategy that is the most appropriate to those variables has to be chosen[10].

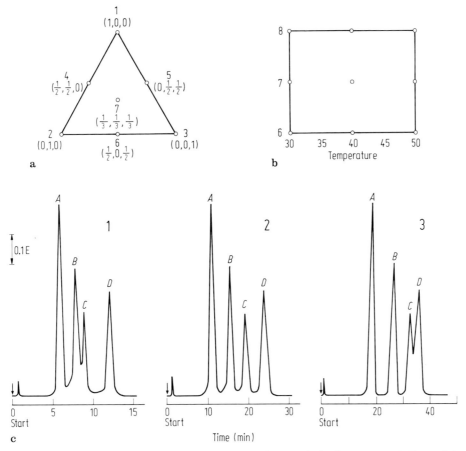

Fig. 6a. Mixture design experiments used in mobile-phase optimization strategy. **b** Fractorial design experiments used for examining temperature and pH effects. **c** Three chromatograms of a pH variation from a factorial design (1 = pH 10, 2 = pH 11, 3 = pH 12) of four ergotoxin alkaloids (A = ergocornin, B = α-ergocryptine, c = ergocristine, D = β-ergocryptine)

The more critical part of these optimizations is the selection of the variables. For this step, chromatographic experience has to be used. If too many variables are included, there will be too many experiments to be performed; on the other hand, if too few or the wrong variables are chosen, valuable information will be lost.

Within the limits of the available preparative instrumentation there are many operating parameters that can be used to change throughput. As can be deduced from Eq. (8) and as shown by Rocca[11], there are fourteen possibilities, viz., column length, column diameter, particle diameter, particle porosity, elution flowrate, mobile phase velocity and permeability, resolution, capacity factor, selectivity, reduced injection volume, injection concentration, as well as the pressure drop over the column and the run number. But within the limits of the available instrumentation only

Table 1. Results of an optimization

L	n	D	ΔP		H		R_s		R_H		$R_H/\Delta P$		R_H/D	
			a	b	a	b	a	b	a	b	a	b	a	b
cm		cm³/min	bar		μm				mg/min		mg/min·bar		mg/cm	
25	1	5	7.1	3.2	56	110	2.7	2.1	0.6	0.7	0.085	0.218	0.12	0.14
		10	14.2	6.9	62	139	2.6	1.9	1.2	1.25	0.085	0.181	0.12	0.13
		15	21.2	10.8	64	117	2.55	1.75	1.9	1.8	0.090	0.167	0.13	0.12
		20	28.3	14.9	65	197	2.5	1.64	2.5	2.3	0.080	0.154	0.13	0.12
12.5	2	5	3.8	1.9	56	110	1.95	1.57	0.6	0.6	0.157	0.315	0.12	0.12
		10	7.6	4.0	62	139	1.87	1.44	1.15	1.1	0.151	0.275	0.12	0.11
		15	11.7	6.4	64	167	1.84	1.35	1.7	1.5	0.145	0.234	0.11	0.10
		20	15.6	9.0	65	697	1.83	1.29	2.2	1.9	0.141	0.211	0.11	0.10
10	5	5	2.7		56		1.65		0.5		0.185		0.10	
		10	5.7		62		1.59		1.0		0.175		0.10	
		15	8.5		64		1.57		1.55		0.182		0.10	
		20	11.3		65		1.51		2.10		0.186		0.11	

five of these fourteen are main parameters, viz., the particle diameter, column diameter, column length, elution flow rate, and the selectivity.

These five parameters can be divided into two categories, related and discrete variables. Related variables are those which directly effect each other, and discrete variables are those which cause little direct effect on each other. The definition of these two types of variables provides the optimization strategies. Well-known are the mixture design for related variables such as mobile phase etc. and the factorial design for discrete variables.

Normally, in preparative LC, the selectivity will first be optimized with one of these designs. This means the mobile phase composition, is, e.g., varied with the mixture design; the ionic effects, the pH, and possibly the temperature are varied with the factorial design to get the maximal resolution as described in Fig. 6.

In a second design the discret variables are then chosen within the limits of the available apparatus, or are determined by the theory described. The results of such an optimization are shown in Table 1, which was published by Rocca[11].

This table shows that, as predicted in Sect. 2.2.1, the highest velocity permitting adequate resolution ($R_{s, min}$) results in the highest sample throughput. Furthermore, it demonstrates that within the pressure drop limits (ΔP_{max}) of the available instrumentation, and with a given column, better throughput will be reached by increasing the flow rate and the run number at the cost of column length or by changing the column over to a coarse-grained one.

In essence, it can be stated that:

Physical effects such as column efficiency can well be predicted by theory, whereas chemical effects or overload effects which influence the selectivity should be optimized within the limits of the available apparatus with strategies. Such processing is very powerful.

3 Apparatus for Preparative LC

As already mentioned, preparative LC is used for several purposes. We can distinguish between analytical-, laboratory-, and production-scale uses, by which we mean purification of milligram-, gram-, and kilogram quantities. This classification can be maintained when speaking about the apparatus in use.

3.1 Analytical Scale Preparative LC

In analytical preparative liquid chromatography the best possible separation power is required. In the last decade this was achieved on HPLC columns, chosen with small particle sizes (3–10 μm).

Recent studies have shown[12] that the required efficiency can also be obtained with the aid of larger diameter columns ($\varnothing$ 25 mm) and that a direct scaling-up from analytical to preparative liquid chromatography can be achieved without speed or resolution losses, as represented in Fig. 7.

Nowadays, such HPLC separations are performed with analytical instruments, permitting flow rates up to and including 50 ml/min at 300 bar. These analytical instruments allow the operation of analytical columns as well as analytical preparative columns up to diameters of 40 mm, filled with small particles up to a column length of 250–500 mm. Due to the volume gained with such scaled-up columns, and with the same capacity of solute per gram of packing material, it is possible to separate in intervals of minutes up to 200 mg of a main-solute in a not too complex mixture, or up to 2 mg of a 1% by-product in one chromatogram or one run, respectively.

In high-performance liquid chromatographs used for preparative separation the following components should be adapted to the preparative specifications of large analytical preparative columns: reservoir, pump, sample introduction, detector, and outlet devices.

Reservoir. The solvent reservoir, the tube connections, and the pump must guarantee the required flow-rate.

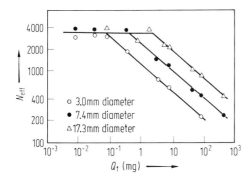

Fig. 7. Relation between relative resolution and sample amount for columns with different diameters; column: LiChrosorb RP-8, 10 μm, length 25 cm; sample: benzophenone

Table 2. Pumps that meet preparative requirements

Manufacturer	Model	Type	Max. pressure (bar)	Max. flow rate (ml/min)
Gilson Inc.	303	A	70	100
Knauer GmbH	64.00	A	100	50
Gilson Inc.	303	A	140	50
Du Pont Co.	8800	A	200	40
Bran + Lübbe	NP-31	B	285	110
Orlita KG	DMP-SK 15	B	325	60
Bran + Lübbe	NK-31	B	325	158
Kontron A.G.	414	A	400	28
Waters Ass. Inc.	590 EF	A	438	45
Haskel Co.	DSTV-122	C	600	800

A, reciprocating, piston; B, reciprocating, diaphragm; C, pneumatic amplifier

Table 3. Commercially available packed columns for analytical preparative purposes

Column length (cm)	10–15								
Column diameter (mm)	8–10			16–22			30		
Particle size (μm)	5	10	20	5	10	20	5	10	20
Packing	sbig	sbig	sbig	sbig	sbig	sbig	sbig	sbig	sbig
Manufacturer									
Bio-Rad									
Brownlee									
Chrompack									
Du Pont									
Ercatech									
Hewlett Packard									
Innovativ	+++	+++							
Knauer	+++	+++		+++			+++	+++	
Merck									
Perkin-Elmer									
Serva									
Supelco									
Varian									
Waters	++	++							
Wathmann		++							

s = silica gel; b = bonded phase; i = ion exchange; g = gel

Pump. A very important part of a preparative apparatus is the pump. The flow-rate should be variable between 1 and 100 ml/min for columns up to 3 cm in diameter; this is also necessary at inlet pressures of up to 100 bar. Pumps that meet these requirements are listed in Table 2.

Sample Introduction Device. Sample introduction can be effected with a syringe (large sample volumes can be accumulated by replicate injections made under

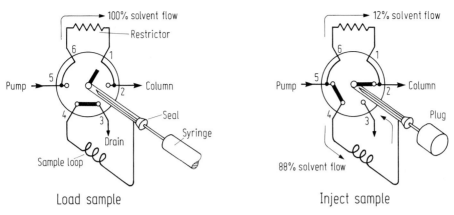

Fig. 8. Schematic diagram of the Rheodyne Model 7105 dosing valve

20–30									50								
8–10			16–22			30			8–10			16–22			30		
5 sbig	10 sbig	20 sbig	5 sbig	10 sbig	20 sbig	5 sbig	10 sbig	20 sbig	5 sbig	10 sbig	20 sbig	5 sbig	10 sbig	20 sbig	5 sbig	10 sbig	20 sbig
	++																
	++++																
	++ +		++														
++	+++		++														
++++	++++		++++						++++	++++		++++	++++				
	+ ++ +																
+++	+++		+++ +++			+++ +++											
++++	++++		+++		++	+++ +++ ++										+	
++ +	++ +			++													
+	++ +			++													
+++	+++		+		+	+++ +++			+++	+++		+++ +++			+++	+++ +++	+++ +++
++	++																
	++++									++++							
+++	+++		+++					++									
	+++			+++						+++			+++			+++	

stopped-flow conditions). A better and simpler approach is to utilize sampling with a loop valve, e.g., a Valco or a Rheodyne valve or a loop arrangement such as the Waters Assoc. inlet system.

Automatic Inlet Systems are desirable and are a necessity in the automatic batch systems as described later. Such systems are pneumatic loop-valve systems and are supplied by Valco, Rheodyne, and Hewlett-Packard. These valves can be used at pressures up to 200 bar. The loop tubes are interchangeable and the volume can be varied from microliters to milliliters. A possible arrangement is shown in Fig. 8.

Column Head. For a column with a diameter greater than 10 mm, a good distribution of the sample over the cross section is essential. Using point injection, the column will be centrally overloaded by infinite-dimater[13], and often noticeable tailing is observed. If one wishes to obtain a high loading, one has to spread the sample homogeneously over the whole cross section.

Columns. Analytical preparative columns (diameter up to 30 mm) can be filled with coarse or fine particles. Several suppliers offer columns that differ in length and diameter (Table 3).

For a particular packing, the column capacity can be improved by increasing the length or inner diameter. However, an increase in the column length results in a decrease in the column permeability and a greater operating pressure is necessary. The realizable column length, therefore, depends on the particle size.

For separation purposes, the choice of size of the packing does not play the same role in preparative chromatography as in analytical chromatography. For high and rapid throughputs, as e.g., in repetitive operating mode[14], fine-particle columns are most suitable: For many laboratory applications, one may prefer long, coarse-particle rather than short, fine-particle columns with the same separation capacity. The sorbens as well as the techniques used in the preparative mode are the same as in analytical chromatography. This means that adsorption chromatography, bonded phase chromatography, ion exchange chromatography, and size exclusion chromatography are well established; other techniques such as liquid-liquid chromatography or affinity chromatography can be used if necessary[15].

Detectors. The same principles of detection are used in both preparative and analytical work. The two detectors that have the widest range of application and are the most often used in preparative chromatography are the UV absorption detector with adjustable wavelength (suppression of too intense signals) and the refractive index (RI) detector. The detectors can be connected either directly or with by-passing; the question whether to use by-passing or not is related to resistance in the detection chamber.

The problems that can be caused by highly pressure-sensitive detector cells (e.g., RI detector cells) have been solved in different commercially available detectors by the construction of cells with larger volumes and greater outlet diameters. This is possible because the volume spreading of the peaks in the detection cell is not so critical as in analytical work.

Fraction Collection. For single-run separations, manual fraction collection or classical collection devices will suffice, but for repetitive cyclic operation, collectors such as those described in Sect. 4.2.3.1 are advantageous.

3.2 Laboratory-scale Preparative LC

Analytical preparative column liquid chromatography is the area where high-quality separations are scaled up in such a way that a quantity of substance sufficient to carry out compound identification or structure elucidation is obtained. Another area is preparative LC on the laboratory-scale, where quantities to be separated lie between several mg and a few grams.

Typical examples for laboratory-scale LC are the production of highly purified substances for chemical, physical, or biological tests.

In analytical applications, the chromatographs deliver flow velocities up to 5 cm · sec^{-1} at pressures up to 200 to 300 bars. The medium pressure used in analytical work may be 100 bars. In preparative LC, pressures of 100 bars are attained, but in most laboratories, as well as in production-scale LC, chromatography is run in a transition region between 1 and 100 bars. The apparatus in use in our and other laboratories can be divided into two groups:

– devices up to 50 bars and
– devices between 50 and 100 bars.

3.2.1 Columns and Equipment Constructed for Separations up to 50 bars

Today, many preparative liquid chromatography applications on the laboratory scale up to 50 bars are performed in glass columns of standard sizes manufactured by commercial companies. These columns withstand considerable pressure and a whole array of backing hardware has been put onto the market (injection parts,

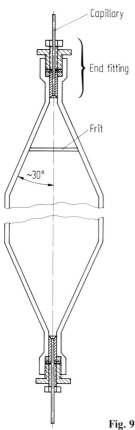

Fig. 9. Preparative LC column, glass construction

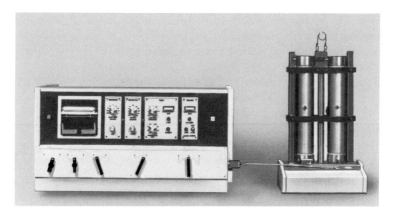

Fig. 10. Prep LC/System 500, a preparative liquid chromatograph

jackets, columns, column heads and ends, connections, etc.). These columns are best suited for filling without the suspension technique and are therefore mostly applied with stationary phases which do not give high pressure. Therefore neither high resolution nor high throughput as described in the previous chapters are used.

A glass column as used in our laboratories is described in Fig. 9[16]. The column itself is constructed of glass tubing of 2.5, 5, or 7.5 cm diameter and 60 cm in length. The maximum loading capacity of this column by LiChrosorb is 1 kg, which corresponds to a packed bed height of 500 cm.

To operate such columns, low-pressure equipment is required. Home-made systems composed of a pump, a loop injector, and a detector can be found in many laboratories. Some are combined with a low-pressure gradient device, others with a fraction collector.

Other laboratories work in this pressure range with commercial equipment. In 1975, Waters Associates introduced the Prep-LC-500[17], shown in Fig. 10. The important advantage in the design of this instrument is the use of a disposable, prepacked column with a plastic wall. This cartridge is filled in a cylindrical compression chamber and will be rapidly compressed prior to sample injection. This procedure compresses the column packing into a highly efficient chromatographic bed.

The Prep-PAK 500 silica catridges are 5.7 cm in diameter and 30 cm long. The effective column length can be increased by connecting such cartridges in series. The flow-rate is adjustable between 0.05 and 0.5 l/min to meet the selected separation conditions. The maximum operating pressure is adjustable to 30 bar. The sample is loaded directly onto the column with a syringe. The sample-introduction system permits the loading of virtually any size of sample. The detector used is a differential refractometer optimized for preparative operation.

3.2.2 Columns and Equipment Constructed for Separations Between 50 and 100 bars

When working in the region up to 50 bars, medium resolution is required in most separations. In such cases, columns filled with supports 30 μm in diameter are used.

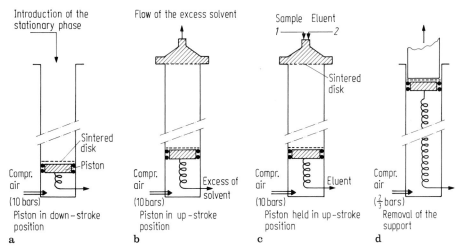

Fig. 11 a–d. Four steps to illustrate the method of packing the stationary phase in the Chromatospac-Prep 100

Such columns can be homogeneously dry-packed down to the mentioned particle size. When working with pressures higher than 50 bars, medium to high resolution is necessary and is often required in the pharmaceutical industry. This resolution can be reached by elongation of the column length or by using finer particles than 30 μm. Under the conditions mentioned, two problems have to be solved:

– all parts must withstand this pressure and
– the column has to be filled by the slurry technique.

These two problems have been solved by Jobin Yvon in the new Prepmatic Liquid Chromatograph. The key to effective performance in dealing with these two tasks lies in the design of the column (1 m × 8 cm), fitted with a piston. Upward movement of the piston compacts the column bed thoroughly and makes good column packing a relatively simple operation, as illustrated in Fig. 11. A slurry of the packing material in the mobile phase solvent is poured into the top of the open column (b) and the piston is moved upwards to compress the bed while driving the excess of solvent out through the end of the column. The final pressure on the piston is maintained throughout the chromatographic process (c).

This system has the advantage of providing homogeneous and reproducible packing to operate with elution and column extrusion[18].

3.2.3 Some Trends in Laboratory-scale LC

3.2.3.1 Repetitive Cycle Operation

In preparative chromatography the factors of sample size and retention are combined in the so-called throughput and in the production rate, respectively. The troughput is the amount of separated compounds per unit time for a given number of theoretical plates, the production rate is represented by the band broadening in a chromatogram. In a chromatographic system with a defined column geometry the qualitative rela-

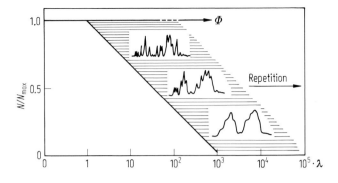

Fig. 12. Scaling up of LC

tionship between band broadening, respectively the separation efficiency R/R_{max} and the production rate xB is known and indicated in Fig. 12. The graph shown is well suited to illustrate this situation. Naturally, there is a floating border between area A (blank) and B (hatched). But each area can be extended a) by diameter elevation and b) by repeating the chromatograms, respectively the separation cycles. Both can naturally be combined and also be provided with an automatic control system. These control systems for performing programed separations can be divided into three groups:

1. time control only (system without feedback of the actual separating sequence);
2. concentration-dependent control (system in which the program is effected by the separation sequence);
3. time- and concentration-dependent control (see e.g., Gilson[19]).

The most reliable system is probably that with time- and concentration-dependent control. Figure 13 shows a block diagram for such a type of automatic batch chromatograph fitted with a control system.

   In this cycle operation chromatograph, the sample is discharged automatically into one of the injectors mentioned in the previous section. Once the sample has been inserted, an automatic program is activated in which an isocratic run or a programed separation occurs. Means for collection of the eluted compound and for re-positioning of the collection mechanisms for the next component (or to the waste position) are provided. Following the elution of the sample, the elution program is reset at the initial chromatographic conditions and the entire cycle is repeated. Such instruments operate unattended, and can be stopped after sufficient material has been collected. This technique can increase the economy of preparative liquid chromatography, insofar as it requires a minimum of human effort and interaction in order to separate adequate amounts of material in runs that may take several days or even weeks.

   As a fault in a run usually leads to contamination of the material that has already been separated and collected, all pure products obtained prior to such a fault will be lost. Therefore, the advantages of automatic over manual operation can only be realized fully when the controls are so effective that no special consideration needs to be paid to their functioning when choosing the separation conditions, and when the whole chromatography can be guaranteed to give troublefree and reliable operation.

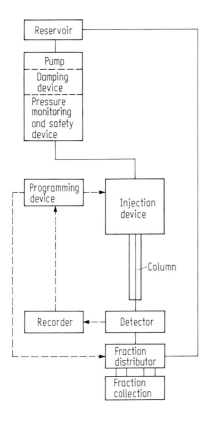

Fig. 13. Scheme of an automated batch system

3.2.3.2 Gradient Elution, Column Switching, Recycling

Techniques such as gradient elution and column switching may be interesting supplements in preparative instrumentation. In particular, gradient elution apparatus solve the general elution problem and help to recondition columns and is therefore very useful in many separations.

Recycling[20] may also help, and often overcomes difficulties in resolution. The simplest concept of such a recycling machine places the pump, the sampling system, the column, and the detector in a closed loop (Fig. 14).

The equipment for recycling operations differs from conventional liquid chromatographic equipment in that it imposes a double carrier-solvent inlet to the pump (one from the carrier reservoir and the other from the recycle), and a splitting valve at the detector exit for recycling or collecting particular fractions of a chromatogram. This additional valve is necessary because a recycling system is a closed system with a finite volume. Hence, the fast-moving material will eventually overtake the slower moving material and re-mix. This valve must be provided so that the operator can remove a portion of possibly mixed components before peak overlapping can occur. An example made by connecting the outlet of a UV detector to a Waters Assoc. pump with small-diameter tubing is shown in Fig. 15. Figure 15a shows the starting chromatogram, and the separation illustrated in Fig. 15b was obtained by using eight cycles at 9 ml/min and a 25 cm long column at 8 mm in

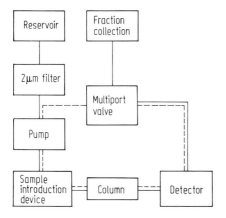

Fig. 14. Recycling arrangement

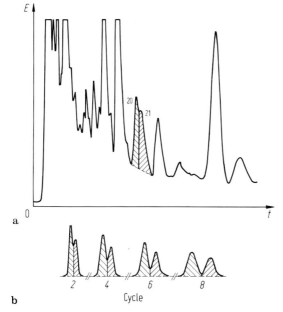

Fig. 15a and b. Separation by recycling. **a** Original chromatogram. The sample in question is an extract of metabolites in plasma. **b** Separation by recycling of substances 20 and 21

diameter. The silica used was Merckosorb Si 100, 5 μm in diameter. Because of the good separation, it was possible to collect the two compounds without any loss of material.

This last-mentioned approach cannot, however, be completely satisfactory, mainly due to peak dispersion and skewing, which occurs in the pump chamber and associated hydraulic damping circuits. Therefore, another approach to recycling a sample in gel permeation chromatography was developed by Biesenberger et al.[21]. They used an alternating pumping principle to eliminate the problems associated with the closed-loop approach. Alternating pumping has the major advantages over closed-loop pumping that the sample does not have to pass through the pump

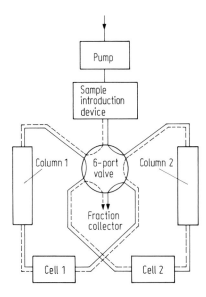

Fig. 16. Alternating pumping device as described by Biesenberger et al.

chamber and the dead volume can be kept to an absolute minimum. A diagram of the alternating pumping system is shown in Fig. 16.

3.3 Production-scale LC

In order to increase the capacity of such preparative laboratory work, cycle operation combined with progressive column diameter[22] (as shown in Fig. 12) or continuous chromatography[23] permits the scaling-up to production-scale throughputs. It is quite clear that such automatic devices, which are in use in development laboratories today, will be applied and expanded to production scale if necessity and economy are proved.

But, today, in most factories, compounds or compound groups which have to be enriched or cleaned are separated in one run. The columns are made of glass and have diameters between 2.5 cm and about 1 m. They are filled with silicagel, Al_2O_3, or cellulose or its derivatives to a height of 50 to 250 cm with particles of ~ 100 μm diameter. In most cases, the particle size is based on practical considerations, such as permeability or facilitation (e.g., dry packing) and security (hazard of inhalation) of the packing method.

Columns with this low separation efficiency (compared with those of analytical chromatography) are still in use for simple separations, as e.g., group selective separation, size separation, etc. Production-scale equipment used for this purpose is shown in Fig. 17. There are two parts shown in this block diagram, not specially mentioned in the previous sections, which play a role in preparative LC. They are:
− the solvent and column regeneration and
− the sample distribution.

Whether a solvent should be purified has to be decided by the user. Generally, solvent regeneration or purification is achieved in preparative LC in three steps:

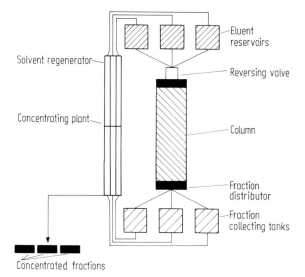

Fig. 17. Block diagram of a production-scale LC plant used for group-specific separation

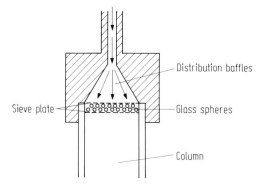

Fig. 18. Dosing distributor

1. Freeze drying,
2. Redistillation,
3. Passing through a selective column (possibly a pre-column).

Freeze drying of the recovered solvents yields, in most cases, solvents pure enough for preparative separations in laboratory or production scale. (In analytical preparative LC, further purification is very often necessary.) The necessity for cleaning and reconditioning the column packing depends on the sample introduced.

Samples that have not been prefractioned often have too large an elution volume and repetitive cyclic separation of compounds is not possible, except if the column is purified and reconditioned. The economic feasibility of purifying and reconditioning is therefore dependent upon the costs of the solvent and stationary phase. Often it is preferable to replace the column packing instead of reconditioning.

The second part mentioned, which plays a role in preparative LC, relates to the column head. If one feeds a column with a diameter larger than 20 mm, a good distribution of the sample over the cross section of the column is essential. Using point-injection, the column will be centrally overloaded by infinite-diameter and often noticeable tailing or fronting is observed. If one wishes to obtain a high loading, one has to spread the sample homogeneously over the whole cross section, as demonstrated in Fig. 18, where the distribution baffles (five metal tubes), the glass spheres, and the sieve plates allow a homogeneous distribution of the sample.

4 Applications of Preparative LC

Liquid chromatography has been widely applied for the separation of various classes of substances. It is therefore not surprising that preparative liquid chromatography has been used to isolate samples of synthetic and naturally occurring products such as plant growth regulators, biologically active molecules of natural products, or plant and animal metabolites.

The products isolated in the g to kg range were used in most cases for

– structure elucidation (analytical application);
– biological tests (laboratory-scale application);
– or as a stage in production (production-scale application).

4.1 Analytical Applications

Separation on analytical preparative high-resolution columns with subsequent purification and identification is a matter of routine in many industrial laboratories. The fractions containing the separated components are extracted or rechromatographed into chloroform or into another solvent, which then will be evaporated under nitrogen. The resulting µg–mg fractions are used for structure elucidation with NMR, IR, and mass spectrometry. The final step in the identification is a comparison of the spectral and retention data with reference data for the presumed compounds [24].

Most identification problems do not need the above-mentioned total structure elucidation equipment. Prior knowledge of the nature of the sample allows a tentative identification as well as structure confirmation, e.g., in systematic investigations of plant-, animal-, or fermentation extracts, in the control of synthetic products and edducts in metabolic studies, etc. When this can or must be done, it may suffice to collect enough material for mass spectra or NMR confirmation of the expected structure.

Because mass spectrometry needs about 1,000 times less substance for a spectrum than other spectroscopic methods, off-line LC-MS – using small bore, packed analytical columns – is applied in many cases as an analytical pilot-identification tool [25].

Figure 19 shows the chromatogram of a Claviceps purpurea fermentation extract using an analytical column (4 mm diameter) packed with LiChrosorb RP-8 and developed with gradient elution. Four major components of the sample were separated in this mode.

In order to enhance the concentration of the components, 2 µl of the fraction cut at the peak maximum are placed on a Whisker [26] with the aid of a syringe,

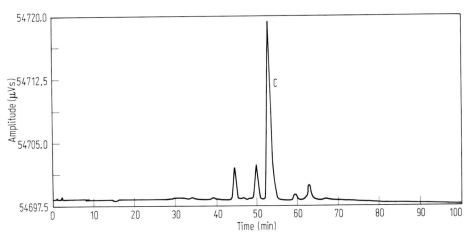

Fig. 19. Separation of the ergot main compounds from a Claviceps purpurea fermentation extract by gradient elution

Fig. 20. Placing a droplet of eluent on a whisker

as shown in Fig. 20. This droplet is then evaporated in a warm nitrogen stream to prevent oxidation prior to analysis. To get an in-beam spectrum, less than approximately 10 ng solute are necessary. This means ca. 1–10 µl of the central cut fraction condensed on the whisker are in some cases sufficient for such a MS spectrum.

It is even possible to get different MS spectra with only ng quantities of substance, e.g., EB spectra and soft ionization spectra. The EB spectra give detailed information about the structure of a pure compound, but a crude information about a mixture (Fig. 21 b). The soft ionization spectra have such high molecular ion intensities (Fig. 21 a) that mixture analysis is possible. Such spectra have, however, no fragment-ion information and are therefore not appropriate for identification purposes. This disadvantage can be mastered by so-called MS/MS analysis.

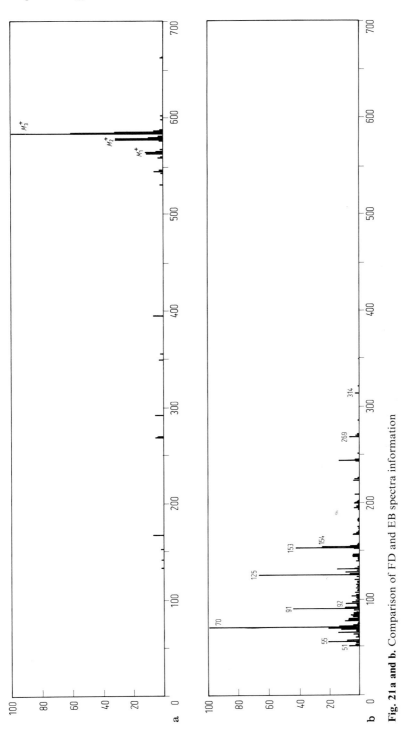

Fig. 21 a and b. Comparison of FD and EB spectra information

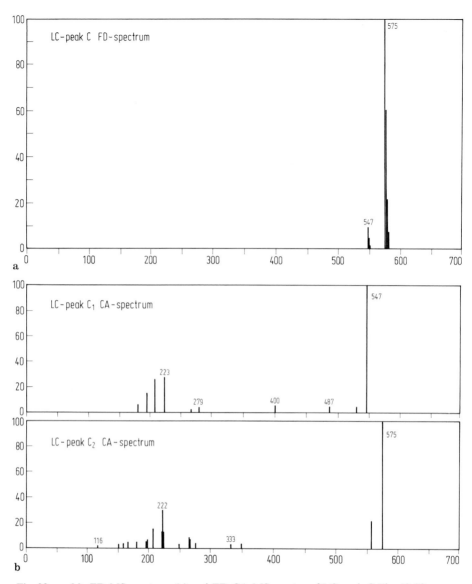

Fig. 22a and b. FD-MS spectrum (**a**) and FD-CA-MS spectra of LC peak C Fig. 19 (**b**)

In its simplest form, MS/MS analysis is a technique whereby a component-ion of a mixture is separated from the other mixture components by one stage of mass analysis, and, after reaction by collision with a target gas[25] in the mass spectrometer, identified by a second stage of mass analysis. Used in this way, MS/MS is analogous to GC/MS or LC/MS. Such MS/MS or CAMS spectra of peak C in Fig. 19 are shown in Fig. 22. The FD spectrum in Fig. 22a proves that peak C is a mixture of two compounds. Figure 22b illustrates the MS/MS spectra of the FDM$^+$ peak 1 and peak 2. As the spectra demonstrate, one gets similar fragments as in EB spectra and therefore sufficient information for a tentative identification of the various compounds. In the experiment shown, the two compounds could be identified as $C_1 =$ ergokryptin and $C_2 =$ ergosinin, two well-known ergotalkaloids. The results were veryfied by comparison with reference spectra.

This experiment demonstrates that such a MS/MS tool is of considerable importance for eluent or crude extract analysis and for mixture analysis, as the soft produced molecular ions, for example, can be selected for each compound and fragmented one after the other so that a second mixture analysis or an extrapolated higher separation power for identification is realized. It can therefore be stated that the substitution of a mass spectrometer separator giving MS/MS data provides a complementary separator-identifier system, and that such a system combined with ultra micropreparative separation is an analytical tool which will find quite a number of applications, e.g., as a pilot identification method in research and development laboratories.

4.2 Laboratory-scale Applications

Analytical preparative column liquid chromatography is employed to carry out compound identification or structure elucidation. Laboratory-scale liquid chromatography means the production of highly purified substances for chemical, physical, or biological tests in the mg to the g range.

A first example [28] illustrating such a separation is shown in Fig. 23. The example demonstrates an analytical separation (a). Chromatogram (b) shows the same system in a preparative mode using a large-diameter, low-resolution glass column operated under overloaded conditions.

These examples demonstrate that many applications of preparative lab-scale liquid chromatography allow overloading in order to reach higher capacity and greater throughput at the expenses of the resolution. Therefore, low-pressure systems with a smaller efficiency are still in demand and, as mentioned, in use in a number of laboratories.

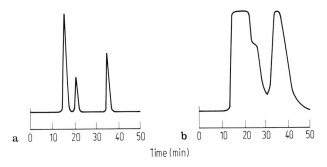

Time (min)

Fig. 23a and b. Purification of pyranosides (glucomanno- and galactopyranoside. **a** Analytical separation, re-calculated. **b** Preparative separation with overload. Column: 40×7 cm; silicagel; mobile-phase: ethylacetate/n-heptane; sample weight: 5 g. Reprinted from Ref.[16] with permission

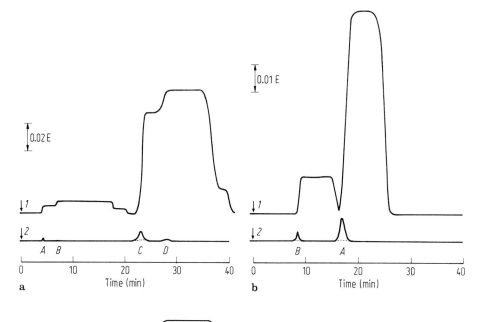

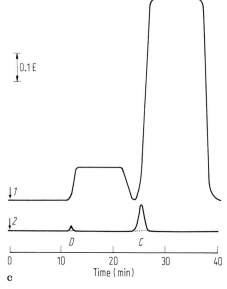

Fig. 24a–c. Enrichment of 3-nitrophenace-tin reaction products with the aid of column strategy. **a** Analytical and preparative chromatogram of nitrophenacetin mixture. Column, 250×4.6 mm and 250×16 mm; stationary phase: LiChrosorb SI-100, $d_p = 10$ µm; mobile phase: *n*-hexane-dichloromethane-acetonitrile-water ($195:780:24:0,1$); flow-rate: 2.5 and 30 ml/min. **b** and **c** Column, 250×4.6 mm and 250×16 mm; stationary phase: LiChrosorb RP-8; $d_p = 10$ m; mobile phase: water-acetonitrile-triethylamine ($870:124:6$); flow-rate: 2.5 and 30 ml/min

Another example[29] is shown in Fig. 24 and involves the preparative (1) and analytical (2) separations of four peaks. The chromatography was carried out in two steps. At first, the sample was divided into two groups of compounds as shown in the chromatograms (a) in the figure. The chromatogram (a1) demonstrates the superposition of two nearly rectangular peaks in each group on the LiChrosorb-60 system chosen. Fractions of the column effluent corresponding to the two groups of compounds were collected manually and transferred as samples to a more efficient

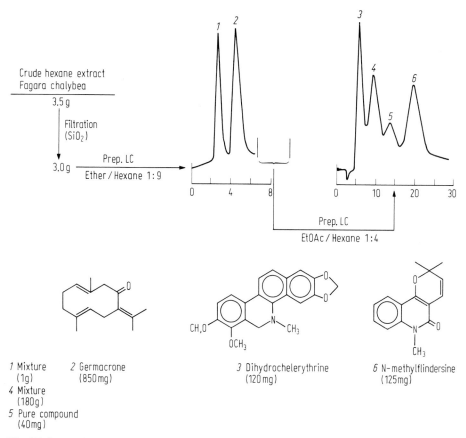

Crude hexane extract
Fagara chalybea

3.5 g
        Filtration
        (SiO₂)
3.0 g   Prep. LC
        Ether / Hexane 1:9

Prep. LC
EtOAc / Hexane 1:4

1 Mixture   *2* Germacrone   *3* Dihydrochelerythrine   *6* N-methylflindersine
   (1g)   (850 mg)   (120 mg)   (125 mg)
4 Mixture
   (180 g)
5 Pure compound
   (40 mg)

Fig. 25. Separation of a crude hexane extract of Fagara chalybea on a Waters Prep LC-500 using silicagel cartridges and forced pressure. Reprinted from Ref.[30] with permission

LiChrosorb RP-8 system as shown in chromatograms (b) and (c), so that a complete separation of all four compounds was achieved in this manner. The total enrichment and separation time for all three chromatographic separations was two hours.

Concerning such chromatographic experiments, we can state: In a chromatographic separation of a mixture, the aim is to find a column for which the smallest occurring selectivity coefficient is as large as possible. Furthermore, a good strategy has to be chosen. If this is done, as large a sample as possible should be placed on the column so as to increase the throughput per chromatogram.

The third example (Fig. 25), demonstrates that it is possible and advantageous to use forced-pressure preparative-scale liquid chromatography for the separation of pure constituents from a crude plant extract containing terpeneoids and alkaloids. This example was published by Hostettmann[30] and shows the separation of the crude hexane extract of Fagara chalybea (Rutaceae) on a Waters Prep LC-500 using silica gel cartridges (30 cm × 5 cm).

In order to avoid column contamination, the small amount of polar material was removed from the bulk of the extract by filtration through silica gel with

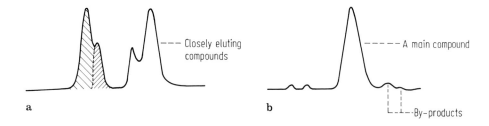

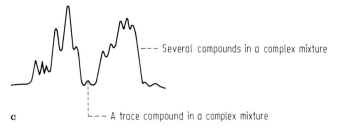

Fig. 26a–c. Chromatograms illustrating isolation problems in preparative LC

ether:hexane 1:1. As shown in the scheme, the filtrate was submitted to preparative chromatography, upon which the least polar fractions 1 (about 1 g) and 2 (850 g) were obtained in six minutes. After peak 2, the solvent system was switched to the more polar ethyl acetate/hexane 1:4 mixture, and the remainder was collected in one fraction. This fraction was concentrated, and a second stage of preparative LC with the above solvent system was carried out. This resulted in the separation of four fractions, Nos. 3 (120 mg), 4 (180 mg), 5 (40 mg), and 6 (125 mg), in 25 minutes. Considering that the whole isolation process required 2–3 hours, whereas the conventional technique needs about two weeks, we can state that such chromatography can be a great simplification of the isolation technique.

These are three examples illustrating the use of lab-scale LC in the pharmaceutical industry. In the literature, there are many other applications to be found; most of them can be classed with the examples shown in Fig. 26.

They are:

- separation of closely eluting compounds (a),
- purification of relatively well-separated main compounds or enrichment of by-products (b),
- isolation of small compounds in a complex matrix (c).

In the third case, the minor compound must first be enriched by one or more steps. This enrichment is accomplished by overloading the column and collecting fractions in the area of expected retention of the desired compound. These collected fractions can then be pooled, concentrated, and reinjected on different systems until isolation is reached.

A much simpler case is the purification of a relatively well-separated main compound or the enrichment of by-products. The best method will be to first optimize the retention in the k' range up to 10. Then the sample load has to be increased up to the point where peaks are overlapping and threaten to disturb. At this point the material will be collected.

In the last case, two or more closely eluting compounds have to be isolated. In this situation it is useful to increase the resolution (by selectivity or efficiency) or else a recycling technique can be employed.

4.3 Production-scale Application

Often larger than g quantities are required in industrial applications. Such applications can be divided into two groups:

– purification of a main compound or
– separation of relatively close-eluting compounds.

The first group is a major segment of the chemical engineering field. In this field, classical techniques such as distillation, extraction, crystallization, etc. are well established in industry. In addition to these techniques, however, there are many others. One of these, not so familiar to chemical engineers, is preparative liquid chromatography. It is well known for its large-scale extraction of ions in water[31], or for contaminants in solvents[32], and as a large-scale pre-cleaning step of drugs manufactured by fermentation processes[33].

Also larger quantities of relatively closely eluting compounds can be isolated with columns of large internal diameters (20 cm). For instance, a mixture of prostaglandins has been separated into purified components on a longitudinally compressed 100×20 cm diameter column of silicea gel (Fig. 27). Although this is not a "very pretty" chromatogram, very large amounts (150 g) of highly purified prostaglandin E were isolated in a single run[34]. Columns of this internal diameter can be used to prepare kilogram quantities of purified materials within a reasonable time.

Naturally, such isolations can be assisted by automatic preparative LC. As the last example shows, such repetitive LC can be used to isolate very large amounts of highly purified components. Figure 28 shows the repetitive separation of diastereomeric carbamates on a column of acidic alumina (about one hour required per individual run). This system separates clearly the diastereoisomers ($\alpha = 1.37$[35]).

This is an example illustrating that with repetitive separations, gram to kilogram quantities of mixtures with small selectivities can also be processed.

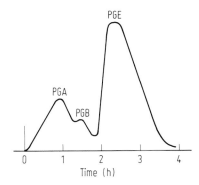

Fig. 27. Large-scale separation of prostaglandins. Column: 100×20 cm; silicagel Woelm; mobile phase: ethylacetate/n-hexane. Reprinted – Ref.[34]

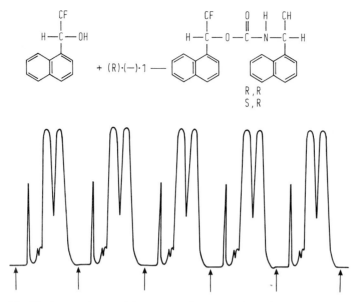

Fig. 28. Automatic repetitive preparative separation of diastereomeric carbamates. Column: 122 × 6.4 cm; acidic alumina; mobile phase: benzene; sample weight: 1 g. Reprinted from Ref.[35] with permission

5 Summary

The existing knowledge of preparative chromatography suggests that the main parameters to be optimized in this technique are: the selectivity, loadability, column diameter, the column length, and the elution flow rate. To obtain maximal throughput, these parameters have to be optimized within the limits of the available instruments according to the following rules:

- The selectivity should be optimized in consideration of the dependence of throughput from the capacity factor (α_{max} at k'_{mini}).
- The sample capacity will be best on totally porous packings.
- At a certain column length, the throughput is raised by increasing the cross section of the column rather than the length.
- For highest sample throughput per unit time, a velocity as high as possible should be chosen (V_{max}).
- For maximal throughput, the column must be overloaded to the point where the peaks still separate ($R_{s, mini}$).

In the past, only a few instrument manufacturers and accessory suppliers were active in preparative liquid chromatography. Compared with the rate of development of analytical equipment, the development of commercial preparative equipment for (HPTHC) "High-Pressure Throughput Chromatography" is less advanced (Table 4).

Table 4. State of preparative liquid chromatography

Application	Amounts required	Columns		
		$\varnothing$ [cm]	p [bar]	N
Situation				
Identification	ng–mg	0.4– 2.5	100	20,000
Tests	mg–g	2.5–10.0	50	5,000
Production	g–kg	10.0–100.0	10	1,000
Trends				
Pressure p resp. N		1 10		100 (bar)
Repetitive operation		3 30		300 (Cycles) 10 (Days)
Continuous operation		...		
Hopes Systems with specific selectivity				

Preparative applications can be classified as:

– Analytical preparative,
– Laboratory-scale, and
– Production-scale preparative.

The amounts required are nanograms to kilograms (depending on the user).

In the research and development laboratories, this technique is applied to the collection of material for structure identification or structure elucidation (requirement of substance [mg]), as e.g., when searching for new effective substances in plants or animals. For these applications, analytical instrumentation is used.

Quantities in the gram range are also separated and purified by preparative column liquid chromatography. They are collected for chemical, physical, or biological tests. To attain such amounts, apparatus, columns, and strategy have to be adjusted according to preparative requirements.

Preparative column liquid chromatography is rather seldom used in the production stage, except as a cleaning method or sometimes as a means for the separation of main components on large-diameter columns.

Considering the stormy development of instrumentation, phases, and methods, one is inclined to predict an increase in the importance of preparative liquid chromatography as shown in Table 4.

6 References

1. Barker, R.E.: In: "Developments in Chromatography", Knapman C.E.H. (ed.): Applied Science, London 1978
2. Sussman, M.V., Rathmore, R.N.S.: Chromatographia 8, 55 (1975)

3. Scott, R.P.W., Kucera, P.: J. Chromatogr. Sci. 119, 467 (1976)
4. Barford, R.A., McGraw, R., Rothbart, H.L.: J. Chromatogr. 166, 365 (1978)
5. Comp. 3 and 4
   Done, J.N.: J. Chromatogr. 125, 43 (1976)
   Spedding, F.H., Powell, J.E.: In: "Ion Exchange Technology," Nachod, F.C., Schubert, J. (eds.). Academic Press, New York 1976
6. Wehrli, A., Hermann, U., Huber, J.F.K.: J. Chromatogr. 125, 59 (1976)
7. Cretier, G., Thesis, Lyon (1979)
8. De Jong, A.W.J., Poppe, H., and Kraak, J.C.: Chromatogr. 148, 127 (1981).
   Wehrli, A.: Z. Anal. Chem. 277, 289 (1975)
9. Horwath, C., Nahum, A., Freuz, J.H.: J. Chromatogr. 218, 365 (1981)
10. Glajch, J.L., Kirkland, J.J.: Anal. Chem. 55, 319 A (1983)
11. Cretier, G., Rocca, J.L.: Chromatographia 16, 32 (1982)
12. Wehrli, A.: Z. Anal. Chem. 277, 289 (1975)
13. De Stefano, J.J.: In: "Introduction to Modern Liquid Chromatography," Snyder, L.R. and Kirkland, J.J. (eds). John Wiley, New York, 1974
14. Pirkle, W.H., Anderson, R.W.: J. Org. Chem. 39, 3901 (1974)
    Wehrli, A.: In: "Instrumentation for HPLC," Huber, J.F.K. (ed.). Elsevier Scientific Publishing, Amsterdam, pp. 93–111 (1978)
15. Turkova, J.: "Affinity Chromatography". Elsevier Scientific Publishing, Amsterdam 1978
16. Loibner, H., Seidel, G.: Chromatographic 12, 600 (1979)
17. Waters Associates, Inc. (Ed.), Milford, Massachussetts 01757, USA. Prep LC/System 500 (1978)
18. Godbille, E., Devaux, P.: J. Chromatogr. Sci. 12, 664 (1974)
19. "Modern Liquid Chromatography," Gilson Medical Electronic, Inc., Ed., West Beltine Midleton, Wisconsin 53562 U.S.A. 1983
20. Bombaugh, K.J.: In: "Modern Practice of Liquid Chromatography," J.J. Kirkland (ed.). Wiley Interscience, New York 1971, Chapt. 7
21. Biesenberger, J.A., Tan, M., Dudevani, I., Maurer, T.: Polymer Sci. B9, 353 (1971)
22. Porath, J.: Biotechnol. Bioeng. Symp. 3, 145 (1972)
23. Barker, P.E., Chuah, C.M.: Chem. Ing. Techn. 53, 987 (1981)
24. Hupe, K.P., Lauer, H.H.: Chromatographia 13, 1 (1980)
    Barker, D.R., Henry, R.A., Williams, R.C., Hudson, D.R.: J. Chromatogr. 83, 233 (1973)
25. Düblin, Th., Quiquerez, Ch., Wehrli, A.: Chromatographia 7, 414 (1982)
26. Schultern, H.R.: Intern. J. Mass Spectrom. Ion Phys. 32, 97 (1979)
27. Cocks, R.G. (ed.): "Collision Spectroscopy". Plenum Press, New York 1978
28. Little, J.N., Cotter, R.L., Prendergast, J.A., McDonald, P.D.: J. Chromatogr. 126, 439 (1976)
29. Comp. 6
30. Hostettmann, K., Pettel, M.J., Kubo, I., Nakaniski, K.: Helv. 60, 670 (1977)
31. Bartels, Ch.R., Kleinman, G., Kerzun, J.N., Irish, D.B.: Chem. Eng. Progress 54, 49 (1958)
    Becker, E.H.: Chem. Ing. Techn. 27, 579 (1955)
    Brown, P.R., Kustolovic, A.M.: In: "Separation and Purification", Perry, S.E., Weissberger, A. (eds.). John Wiley, New York 1978, p 197
32. Breck, W.D.: In: "Zeolite Molecular Sieves. John Wiley, New York 1973, Chapt. 8, p. 593
    Cassidy, H.G.: In: "Adsorption and Chromatography," Weissberger, A. (ed.). Interscience Publishing, New York 1951, p. 177
33. Schmid, G., Gölker, C.: In: „Handbuch der Biotechnologie," Präve, P. et al. (eds.). Akad. Verlagsges. Wiesbaden 1982, pp. 235–242
34. Jobin, Yvon (Ed.): Division d'Instruments S.A. 91160 Longjumeau, France (1980)
35. Pirkle, W.H., Anderson, R.W.: J. Org. Chem. 39, 3901 (1974)

Column-switching

Herwig Hulpke and Ulrich Werthmann
Bayer AG, Sparte Pflanzenschutz, Pflanzenschutzzentrum Monheim, Geb. 6100,
5090 Leverkusen-Bayerwerk/FRG

1 Introduction

For some time now applications of the technique of column switching in HPLC
have been growing steadily. The method was introduced in 1973 by Huber et al.
[1] who regarded column switching in its simplest form as a technically simple
and inexpensive alternative to gradient elution which, at that time, could only be
achieved at considerable expense. The continued development of HPLC instrumentation, however, soon enabled solvent gradients to be generated inexpensively, so
that column switching as a simplifying alternative became unnecessary.

With increasing application of HPLC in very diverse fields, especially in the
analysis of complex samples such as biological extracts, much effort is once again
being devoted to column switching – but this time from completely different aspects.

It was recognized that this technique permitted very efficient analyses to be
performed with extremely simple sample preparation. Particularly in the current
developmental phase of HPLC involving faster and faster separations, sample preparation has become a bottleneck that determines the total output and cost of analyses
in most cases.

The determination of pharmaceuticals in biological materials is but one example
of a general and more frequently occurring problem of determining traces of single
or a group of special components in very complicated matrices. Interferences pose
great difficulties in the analysis of such materials, and the usual remedy is to increase
column efficiency and/or selectivity. In most cases, however, the best solution is
the use of column switching systems and multidimensional chromatography.

Switching systems can be readily adapted to solve a variety of problems. Thus,
they function reproducibly and reliably in on-line sample preparation, sample enrichment, various cut techniques, backflushing, and multidimensional chromatography.

2 Various Switching Systems and their Applications

For a long time one of the main problems in the construction of a column switching
system was the lack of a suitable switching valve. Such a valve should be able

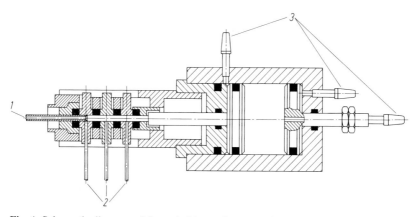

Fig. 1. Schematic diagram of the switching valve. 1 = Inlet; 2 = Outlets; 3 = Air supply

Table 1. Specifications for the high-pressure switching valve

Volume with connections	20 µL
Max. pressure	300 bar
Max. temperature	150° C
Switching frequency	10↑5
Pneumatic switching pressure	2.5–5 bar
Material	PTFE glass

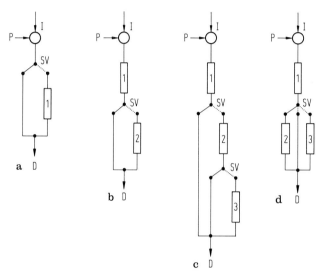

Fig. 2 a–d. Various switching arrangements of this valve. **a** Control of total elution; **b** Heart-cutting capability; **c** Heart-cutting capability with subsequent further separation on different stationary phases. P = pump; I = injector; D = detector; SV = switching valve

to operate under high pressure (up to 400 bar) and to withstand several hundred thousand switching cycles without malfunction. In addition, its total volume and that of any connections must be so small as to contribute negligibly to band broadening. It should be constructed from materials that are resistent to the eluents to be used and should exhibit no absorption properties.

In their original work, Huber et al. employed a pneumatically-driven piston valve (Fig. 1) having one inlet that could be connected to three outlets[1, 2].

The technical data for the valve are presented in Table 1. Because of the small total volume, the contribution to band broadening was negligible.

Figure 2 illustrates various switching configurations.

As mentioned in the introduction, originally an alternative to the then expensive gradient elution was sought. For this purpose configuration b of Fig. 2 could be used, for example. Such coupling of columns in series is suitable for increasing the resolution of the early-eluted peaks, while the late-eluted ones pass directly from the first column to the detector. In this way samples with a broad range

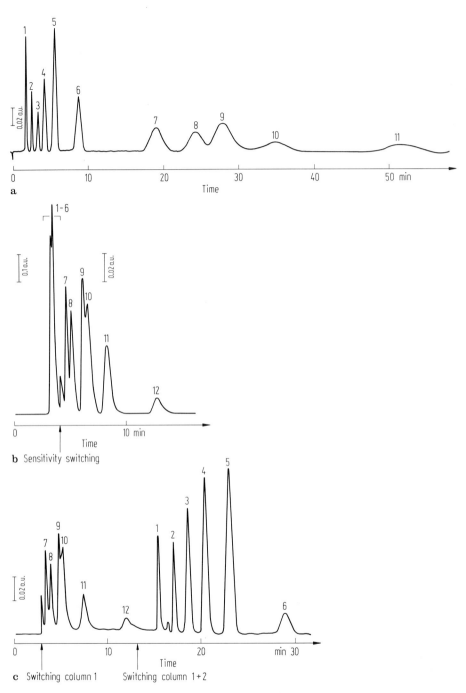

Fig. 3a–c. Analysis of sample components having a broad range of polarities. (For explanation see text)

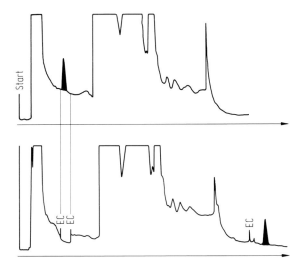

Fig. 4. An example of the removal of a component from the flank of a peak by heart-cutting. (EC = switching points)

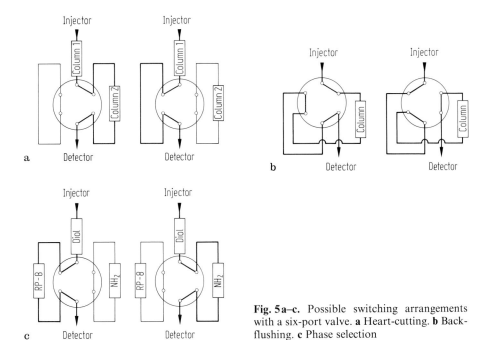

Fig. 5a–c. Possible switching arrangements with a six-port valve. **a** Heart-cutting. **b** Backflushing. **c** Phase selection

of polarities can be eluted isocratically. Thus, the components represented by the first portion of the chromatogram are held on the second column until even the late-eluted ones have emerged from first column and have been detected. Such storage of samples on a column for longer periods of time is possible because of their very small diffusion coefficients in liquids which therefore results in no observable band broadening.

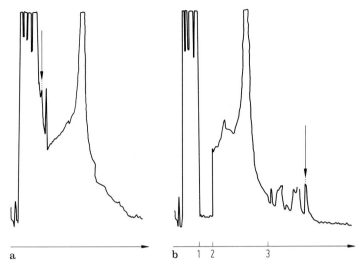

Fig. 6a and b. An example of heart-cutting. **a** Original sample without heart-cutting. **b** Use of heart-cutting. Arrow points to the component of interest. 1 = Start of the heart-cut; 2 = End of the heart-cut; 3 = Start of the elution of the cut portion

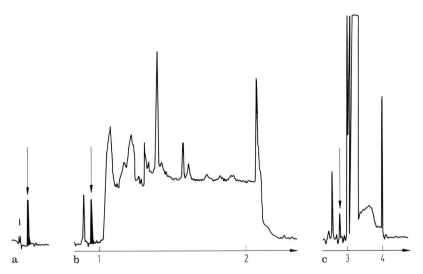

Fig. 7a–c. Illustration of the advantages of backflushing. **a** Chromatogram of a standard of the component of interest. **b** Chromatogram of the component in its matrix (here rat food). 30 min were required to flush out the nonpolar contents. **c** Use of backflushing to save time in the flushing process. In this way the flushing process required only 8 min. 1 = Start of the flushing process; 2 = End of the flushing process; 3 = Start of the backflushing; 4 = End of the backflushing

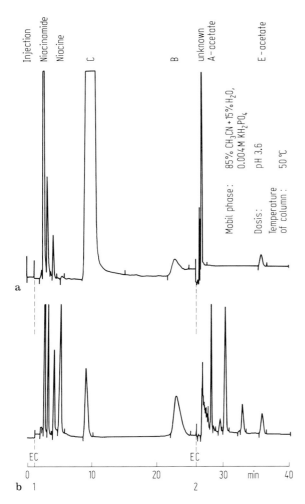

Mobil phase: 85% CH₃CN + 15% H₂O, 0.004 M KH₂PO₄

Dosis: pH 3.6

Temperature of column: 50 °C

Fig. 8a and b. Illustration of the use of phase selection with the switching system of Fig. 5c. Analysis of a vitamin tablet. Water- and fat-soluble vitamins on the same chromatogram[4]. Separation sequence: 1) Group separation of the water- and fat-soluble vitamins on a Diol phase; 2) Storage of the unretained fat-soluble vitamins on an RP-8 phase; 3) Analysis of the water-soluble vitamins on an NH₂ phase; 4) Analysis of the fat-soluble vitamins on the RP-8 phase. 1 = Separation on the NH₂ phase; 2 = Separation on the RP-8 phase

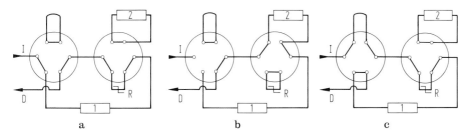

Fig. 9a–c. A switching system for heart-cutting and backflushing Valve configurations. **a** Normal position. **b** Heart-cut position. **c** Backflush position. I = from the injector; D = to the detector; R = restrictor capillary (generates the same back pressure as column 2); 1, 2 = columns

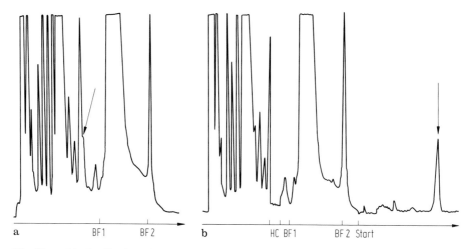

a   BF1   BF2   b   HC BF1   BF2 Start

Fig. 10a and b. Application of the switching arrangement of Fig. 9. **a** Purification by backflushing. **b** Heart-cutting of the component of interest and purification by backflushing. BF1 = Start of backflushing; BF2 = End of backflushing; HC = Region removed by heart-cutting; Start = Start of the elution of the heart-cut region

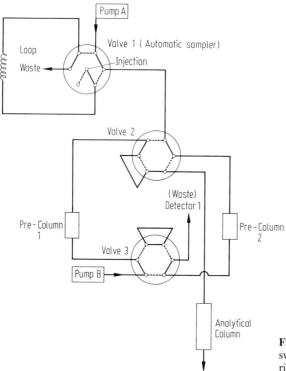

Fig. 11. Alternating pre-column switching technique for sample enrichment, exemplified with three 6-way valves[9]

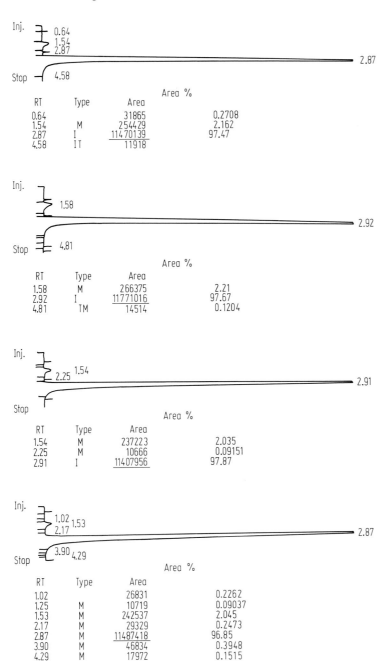

Fig. 12. Typical chromatogram of directly injected human plasma (injection volume = 150 μL), spiked with 100 ng/ml of a drug. The drug is concentrated on the pre-column and then eluted by backflushing to the analytical column. During the analysis another sample is concentrated on the second pre-column

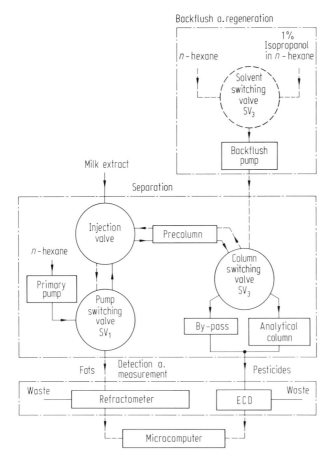

Fig. 13. Schematic diagram of the HPLC system in [11]

This system facilitated the analysis of complex samples, as the following example illustrates (Fig. 3).

Components with a high partition coefficient are separated on the column with a low phase ratio, while those with a low partition coefficient are resolved on a high phase ratio column.

In chromatogram (a) both columns were utilized; the flattening of the last peaks is clearly evident. Peak 12 was not eluted even after 70 min. Chromatogram (b) was obtained with only one column. Peaks 1–6 are very poorly resolved. Chromatogram (c) shows the results obtained with column switching. After holding the first six peaks in column 2, the valve was switched and peaks 7–12 were conveyed directly to the detector. Upon re-diverting the eluent to column 2, further resolution of the first six peaks was effected. The overall result was a greatly shortened analysis time and sensitive detection of the late-eluted peaks. Today such a separation problem would be solved by gradient elution.

This type of switching is still employed to remove interfering main components or impurities from a chromatogram, or to cut out peaks of interest or poorly resolved ones from complex chromatograms[3].

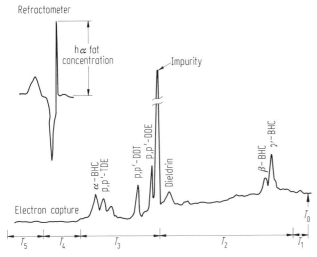

Fig. 14. Chromatogram of "spiked" milk fat using the switching system

The possibilities offered by column switching in such cases in shown in Fig. 4.

For some time now ordinary multiport valves have found widespread acceptance as switching valves for such purposes. These valves, which can also serve as an injection system, possess the necessary characteristics and are relatively inexpensive. Furthermore, they can be readily automated, which is important for routine applications.

The switching systems that can be constructed with a six-port valve are shown in Fig. 5.

Examples of practical applications of these systems are shown in Figs. 6–8.

The use of ten-port valves for such purposes has also been described [5, 6].

The various single systems shown can be coupled by combining several such switching valves to produce a very versatile and universally applicable total system. This is illustrated in Fig. 9.

In this switching arrangement heart-cutting can be combined with backflushing, which is especially effective for the analysis of very complex matrices [7, 8]. This is illustrated in the next example (Fig. 10).
A component on the flank of a peak was first removed by heart-cutting, the portion of no interest was backflushed, and then the compound of interest was eluted.

Another very significant switching arrangement that has recently gained widespread use, especially in the area of clinical analysis, is presented in Fig. 11.

Introduced by Roth et al. [9], this technique permits sample enrichment on a pre-column in alternating completely continuous operation. In most cases it reduces sample preparation to a minimum, improves the analysis, and greatly increases the total output. Figure 12 clarifies the procedure.

In this technique the sample, after enrichment on the pre-column, is backflushed onto the main column and analyzed while the next sample is being concurrently enriched on the second pre-column [10].

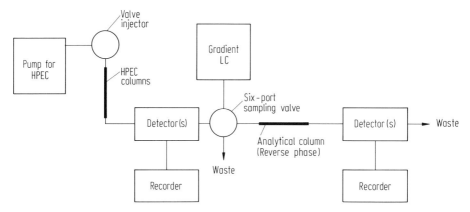

Fig. 15. Schematic diagram of a coupled column system [13]

The construction of a very complex system was described by Dolphin et al. [11] for the analysis of organochlorine pesticides in milk (Fig. 13). Sample clean-up, separation of pesticides from milk fat and quantification of both fat and pesticides are carried out on-column using column switching and backflusing techniques. The combination of liquid chromatography with electron-capture detection results in a detection limit of 0.1 ppm pesticide in milk fat. A typical chromatogram is shown in Fig. 14.

In such complex systems a microcomputer permits precise control of the column switching valves. The computer even compensates for fluctuations in the retention times which may result from column overloading.

Another possibility for multi-column operation is the coupling of GPC with other chromatographic systems, as described by Johnson et al. [13]. The scheme is presented in Fig. 15. The chromatographic determination of the pesticide malathion in a tomato extract exemplifies the advantages offered by this procedure (Fig. 16a–16d). The amount of malathion, which after a few days could no longer be quantified by GPC, could be determined unambiguously by coupling with HPLC.

A universal but too elaborate and expensive apparatus has been available for some time as a complete system. In this system up to seven valves are used. With

---

Fig. 16. a HPEC separation of tomato plant extract. Sample isolated 16 h after treatment with malathion; 20 µl injection; conditions, columns, 50 cm 2000H, 80 cm 1000H, MikroPak TSK (8 mm I.D.); flowrate, 1 ml/min; eluent, THF; UV detection an 215 nm and 0.5 a.u.f.s.; 200 µl injection. **b** RPC separation of suspected malathion fraction isolated by HPEC. Conditions, extractant, THF, 10 µl injection, column, 25 cm × 2.2 mm MikoPak MCH; flow-rate, 0.5 ml/min; gradient, acetonitrile-water (10:90, v/v) to 100% acetonitrile at 3% acetonitrile/min. UV detection at 215 nm and 0.2 a.u.f.s.; fluorescence, excitation at 254 nm, emmission at >370 nm; 5 mV full scale. **c** HPEC separation of tomato plant extract. Sample isolated 7 days after treatment with malathion. Conditions as in Fig. 16a except 200 µl injection. **d** RPC separation of suspected malathion fraction from 7-day sample. HPEC fraction concentrated 20 times before injection; 5 µl injection; other conditions as in Fig. 16b

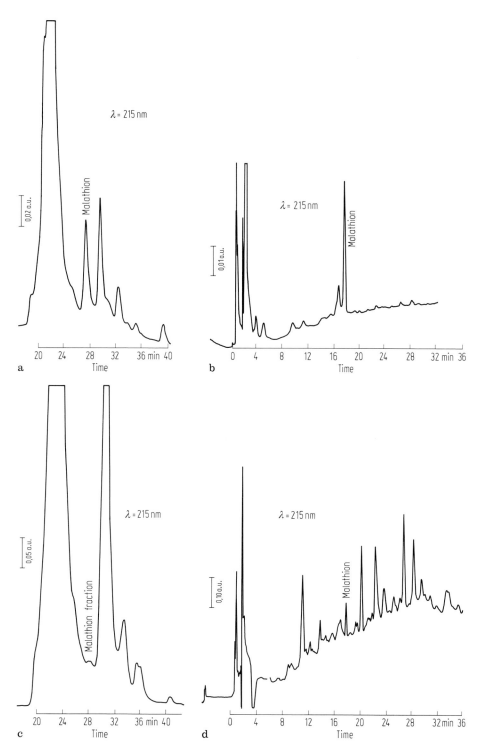

it all conceivable switching variations can be achieved without major alteration. However, since only a few of these variations are needed in the vast majority of cases, especially for routine systems, it is usually better and cheaper to assemble one's own universal system.

3 Conclusions

It is shown how extensively HPLC column switching systems can be applied in a variety of fields. Earlier, this technique had gained acceptance rather slowly, but more recently switching systems have been increasingly installed in many laboratories. This sudden breakthrough can be attributed primarily to the work of Roth et al., which is largely responsible for the increased awareness and application of this technique. The number of publications in this area has also jumped dramatically [14–32]. In many laboratories it has been recognized that there is hardly any other technique which is as useful and effective for simplifying sample preparation and improving selectivity in HPLC. Applications in routine work have demonstrated that columns, protected by pre-columns and effectively cleaned by backflushing, can be employed for hundreds of analyses, even of biological samples. In the future, combination of column switching systems with reaction detectors will afford as yet unattained selectivities [33, 34]. Coupling of HPLC with other spectroscopic or spectrometric techniques will be greatly simplified by the use of column switching.

All of this raises hopes that the column switching technique will find even more widespread application in the future.

4 References

 1. Huber, J.F.K., et al.: J. Chromatogr. 83, 267, 1973
 2. Straub, H., Ecker, E.: GIT Fachz. Lab. 19, 13, 1975
 3. Hulpke, H., Werthmann, U.: Chromatographia 12, 399, 1973
 4. Elgass, H.: Hewlett-Packard Operating Note 06, 1980
 5. Harvey, M.C., Stearns, S.D.: Am. Lab. 2, 1981
 6. Harvey, M.C., Stearns, S.D.: Am. Lab. 6, 1982
 7. Hulpke, H., Werthmann, U.: Chromatographia 13, 395, 1980
 8. Berger, G., Hulpke, H., Werthmann, U.: GIT Fachz. Lab. 25, 595, 1981
 9. Roth, W., et al.: J. Chromatogr. 222, 13, 1981
10. Riggenmann, H.J., Jürgens, U.: Labor Praxis 3, 1983
11. Dolphin, R.J.: J. Chromatogr. 122, 259, 1976
12. Willmott, F.W., Mackenzie, J., Dolphin, R.J.: J. Chromatogr. 167, 31, 1978
13. Johnson, E.L., Gloor, R., Majors, R.E.: J. Chromatogr. 149, 571, 1978
14. Fogy, I., Schmid, E.R., Huber, J.F.K.: Z. Lebensm. Unters. u. Forsch. 169, 438, 1979
15. Fogy, I., Schmid, E.R., Huber, J.F.K.: Z. Lebensm. Unters. u. Forsch. 170, 194, 1980
16. Huber, J.F.K., Fogy, I., Fioresi, C.: Chromatographia 13, 408, 1980
17. Majors, R.E.: J. Chromatogr. Sci. 18, 571, 1980
18. Lankelma, J., Poppe, H.: J. Chromatogr. 149, 587, 1978
19. Lecaillon, J.B., Souppart, C., Abadil, F.: Chromatographia 16, 158, 1982
20. Nussbaumer, K., Niederberger, W., Keller, H.P.: Journal of HRC&CC 5, 424, 1982
21. Vliet van, H.P.M., et al.: J. Chromatogr. 185, 483, 1979
22. Voelter, W., et al.: J. Chromatogr. 239, 475, 1982
23. Apffel, J.A., Alfredson, T.V., Majors, R.E.: J. Chromatogr. 206, 43, 1981

24. Alfredson, T.V.: J. Chromatogr. 218, 715, 1981
25. Benjamin, E.J., Conley, D.I.: Int. J. of Pharm. 13, 205, 1983
26. Benjamin, E.J., Conley, D.I.: J. Chromatogr. 257, 337, 1983
27. Erni, F., Frei, F.W.: J. Chromatogr. 149, 561, 1978
28. Erni, F., et al.: J. Chromatogr. 204, 65, 1981
29. Freeman, D.H.: Anal. Chem. 52, 2, 1981
30. Gfeller, J.C., Stockmeyer, M.: J. Chromatogr. 198, 162, 1980
31. Huber, R., et al.: Chromatographia 16, 233, 1982
32. Jong de, G.J., Zeeman, H.: Chromatographia 15, 453, 1982
33. Werkhoeven-Goewie, C.E., et al.: Chromatographia 16, 53, 1982
34. Werkhoeven-Goewie, C.E., et al.: J. Chromatogr. 255, 79, 1983

Sample Pretreatment and Cleanup

M. Uihlein
Hoechst Aktiengesellschaft, 6230 Frankfurt a.M. 80

Dedicated to Dr. Paul Hajdu for this 60th birthday

1 Introduction

Although one of the most versatile separation techniques, Liquid Chromatography still demands many analyte requirements, e.g.:
– clear and particle-free solutions;
– miscibility of solvent with mobile phase;
– sufficient solubility of solutes in mobile phase including any gradient program that might be necessary;
– compatibility of solvent with stationary phase and detection mode(s).

   Further restrictions arise from the fact that good isocratic separations can only be obtained within a limited k^1 range. In addition, accurate quantification of traces and main compounds in one run is limited to trace concentrations of at least 100 ppm. Real-life samples only occasionally fulfill all the requirements listed above. There is hence a need for optimum cleanup strategies in order to comply with the preconditions of Liquid Chromatography.

   The following chapters attempt to cover all aspects to be considered when developing a cleanup strategy. Nevertheless, bearing in mind the wide variety of samples and matrices, this attempt will only be a rough survey of general approaches. Although most problems may be solved by intelligent but simple combinations of the techniques presented, there will also be cases that require a more laborious and frustrating way of solution. It therefore was the author's aim not only to present helpful hints but also to increase readership interest in complex analytical strategies. At first glance, some of the topics might appear out of interest to a chromatographer but, in actuality, these are the challenges that make the scientific life of an analyst interesting.

2 Pre-laboratory Treatment of Samples

In general, analytical procedures only describe the techniques for obtaining information on the sample on the analyst's workbench. This information, however, will be completely inaccurate and misleading if the respective sample is not – or is no longer – representative of what the supplier meant to be analysed. Adequate handling of samples therefore has also to consider the history of the sample and thus may begin a long way before the analytical laboratory is reached. It includes correct sampling, use of appropriate containers and optimum sample storage and transport conditions.

2.1 Correct Sampling

It is self-evident that samples to be analysed should be representative of the lot or batch they are taken from. This prerequisite is not always easy to fulfill since only gas mixtures, true liquid mixtures and unsaturated solutions are really homogeneous. With solid samples, representative sampling requires special procedures, this being especially true in the case of biological samples which by nature are nonhomogeneous in composition. This representative sampling of solids may require taking a large number of samples from different parts of a stock, storage vessel or field, or at different times from a product being continuously produced. The samples must be chopped, ground or pulverised and thoroughly mixed. By repeated division, grinding and mixing of individual portions, followed by aliquots being taken, the sample can be reduced to a workable size (cf.[1] and [46], p. VIII-3 ff.). Respective techniques are discussed in the chapter on sample pretreatment.

2.2 Sample Storage and Transport

When samples require shipping to the analytical laboratory or cannot be analysed immediately, the possible effects of transport and storage should be evaluated.

Transportation makes the use of sample containers inevitable. Although normally not taken into account, containers may influence the analytical result, especially in trace analysis. This is self-evident with samples which are sensitive to light, where containers which protect them from light become unavoidable. In other cases, migration of plasticisers from the container into the sample may lead to interference with the determination and, finally, it has been reported that some container and seal material adsorbed the analyte[2]. In principle, it is impossible to predict whether such effects are of relevance in a given type of analysis. If there is no possibility of investigating this question, it is up to the analyst – on account of his experience – to avoid crude mistakes. In this context, the author would like to encourage his readership to communicate more about respective observations. From the above, it follows that, within a series of analyses, only standard containers of one type from a single manufacturer should be used in order to obtain at least reproducible test conditions.

Unless otherwise stated, appropriate conditions for the sample during storage or shipping are freezing or freeze-drying. In some cases, further protective measures – e.g. addition of antioxidants to the sample[3] – are necessary. It must be pointed out that, under all conditions, stability is a time-dependent property. The fact that no changes are observed after two weeks of storage is no evidence that reliable analysis can still be performed after a storage period of several months. This should particularly be borne in mind in the design of long-term studies where, quite often, a series of samples is taken and stored until the end of the entire study before being shipped to the analytical laboratory.

Under these conditions, the analyst must be prepared to make a definite statement guaranteeing his analytical results if detailed precautions with respect to shipping[4] and storage[5] are satisfied. Besides recommendations concerning containers and shipping conditions, these precautions should include information on stabilising additives requested, maximum time permissible between when the sample is taken and when it is conserved and maximum allowable storage period, i.e. the period during which, under given conditions, no alteration to the analyte has been observed.

If samples have to be shipped, the analytical laboratory should furnish the supplier with shipping instructions in order to avoid inappropriate handling of samples during shipping. It is also advisable to affix to the container an explicit comprehensive description of how it is to be handled during transport.

The sources of errors discussed so far comprise most of the so-called pre-laboratory errors. It should be stated that this type of error cannot be recognized by quality control methods.

3 Sample Pretreatment

In determinations in solids, the first step in sample processing is the destruction of the matrix structure. This has to be performed as completely as possible since the yield from the subsequent isolation procedure depends considerably on this. Besides, in special areas, further types of sample pretreatment might be necessary to enable analysis. One of the more common ones is the cleavage of drug conjugates.

3.1 Sample Disintegration

The first step in the destruction of the matrix structure is thorough grinding followed by homogenisation and blending in an appropriate solvent. These procedures may require cooling to prevent decomposition of the analyte. If possible, the solvent used should be optimum for solution of both the analyte and the matrix. Especially when analysing food, plants or microbial or animal tissues, this will not generally be possible, and special techniques may be required. Thus, when determining drug residues in animal tissues, the author has obtained excellent extraction rates where the tissue was disintegrated in methanol or acetonitrile, these being protein-precipitating agents with high solubility for the analyte.

Besides mills, blenders and mixers, ultrasonic disintegrators are often used for samples in biochemistry, enabling subcellular fractions to be obtained under very mild conditions. Nonmechanical destruction techniques are also possible here too, e.g. osmotic effects caused by the addition of three parts deionized water to one part whole blood are used for the destruction of red blood cells.

3.2 Deproteinization

For trace analysis in samples with a high protein content, precipitation of the proteins becomes necessary in order to prevent the formation of emulsions during extraction with organic solvents. Furthermore, deproteinization is required whenever there is a hint of irreversible binding of the analyte to proteins. In general, deproteinization is performed by adding deproteinizing agent such as perchloric, tungstic or trichloroacetic acid or with organic solvents such as methanol or acetonitrile.

An enormous problem connected to deproteinization is the possibilities of adsorption of the analyte on – or clustering of the analyte in – the precipitate. This effect can be minimized by not adding the deproteinizing agent to the homogenized sample, but by slowly adding the latter to a surplus of deproteinizing agent. With some compounds, a further increase in recovery was observed when the deproteinizing agent was agitated in an ultrasonic bath during addition of the sample homogenate.

The use of heat for precipitation of proteins should be avoided on account of possible deterioration of the analyte. Whenever very mild conditions are required, the use of enzymes should be considered. Subtilo-peptidase A [6] has been used for this purpose since it only affects protein-type substances [7].

Extraction by means of solid phases (cf. chapter on adsorption) does avoid formation of emulsions. It might thus become an alternative in some cases, omitting the more problematic protein-precipitation.

3.3 Cleavage of Drug Conjugates

Drug conjugates can be described as a special type of drug metabolite formed in the body. They are either esters, ethers or amides of the drug or of a primary metabolite with glucuronic acid – a sugar acid – or sulfuric acid. Direct determination of conjugates is often very difficult, since they are very polar and often poorly extractable compounds. Furthermore, the possibility of formation of several different conjugates from the mother compound alone or from one primary metabolite makes the result of analysis difficult to interpret, and synthesised standards are normally not available. Besides the possibility of direct determination by means of ion-pair chromatography (cf. [8]), it is therefore widely agreed that conjugates be determined by differential measurement, i.e. compound after cleavage of the respective conjugate(s) minus unconjugated compound. In this case, cleavage of drug conjugates is a necessary sample pretreatment. In the case of labile conjugates, it can be performed by simple pH shift in the sample prior to cleanup [9]. With stable conjugates, the use of cleaving enzymes is inevitable to release the free drug. Which type of enzyme should be used – whether the highly specific β-glucuronidase (E.C. No. 3.2.1.21 from bacteria or bovine liver) or the less specific glusulase (E.C. No. 3.1.6.1 from marine mollusc or helix pomatia) – depends on the type of conjugate and can only be established by experiment (cf. [10]). As for any enzymatic reaction optimum, pH, temperature and time are essential factors. Most cleavage procedures will require 24 h of incubation at 37° C and pH 5.0, but these conditions cannot be generalized.

4 Cleanup Procedures

Experience has taught that, besides solid material, most 'real-life' samples in solutions are also not suitable for direct submission to HPLC. The reasons might be the high complexity of the sample's composition as well as the need for trace enrichment. Thus, the sample must first undergo preseparations, so-called cleanup. Nearly all separation techniques can be considered for sample cleanup:
filtration,
sedimentation (centrifugation),
precipitation,
freeze-drying,
evaporation,
extraction of solids and liquids by liquids,
extraction of liquids by adsorption on solids.

Some of these are well-known standard laboratory procedures needing only a few comments, while others should be discussed in more detail.

Cleanup procedures, in principle, should quantitatively separate analytes from all unwanted material, a demand that normally can only be approximated by one-step procedures. Hence, in many cases, combinations of the techniques listed are used. The most simple ones are liquid-liquid extractions followed by evaporation of the extracting solvent in order to concentrate the compounds to be determined.

It pays to optimize the cleanup procedure, since it is mainly sample cleanup time which determines the workload of a laboratory (especially when the subsequent chromatographic procedure is rapid and/or completely automated).

4.1 Standard Laboratory Procedures

Precipitation, filtration, sedimentation and freeze-drying are widely used standard procedures known to most chemical laboratories. Therefore, without going into details, only those problems which might lead to a decrease in recovery of the cleanup process should be pointed out:

Whenever the analyte is a trace constituent, *precipitation* of major constituents of the sample will raise problems with regard both to adsorption on the surface of the precipitate and to occlusion by it. General methods for prevention cannot be presented, other than the possibilities that have been discussed in the chapter on deproteinization. It should be added that, in most cases, variation in recovery due to adsorption and occlusion phenomena can be compensated for by using an internal standard closely related to the analyte.

Filtration and *sedimentation* (by gravity or in a centrifuge) are assumed to be equivalent techniques for the separation of liquids and solids. In trace analysis, however, centrifugation should be preferred to filtration, since the sample does not come into contact with filter material that might turn out to be a good adsorber for the analyte and thus cause recovery and reproducibility problems.

Evaporation of solvents is widely used and considered to cause no problems. Nevertheless, prior to evaporation, it should be clarified as to whether or not the analyte is co-distilled with the solvent (e.g. whether or not it is steam-volatile). Besides, during evaporation, purging by nitrogen should be foreseen whenever oxidizable analytes are handled.

Meanwhile, a lot of devices for careful evaporation below the boiling points of the respective extraction media do exist. Some evaporate under reduced pressure and use centrifugal force in order to prevent vapor bubbles that lead to spilling of the solvent[1]. Others use nitrogen streaming through needles over the surface of the solvent (cf.[11] and Fig. 1). This not only prevents the analyte from oxidizing but also lowers the evaporation temperature.

For thermolabile compounds, *freeze-drying* should be preferred to evaporation of aqueous phases. This technique is widely used in biochemistry, especially for protein-rich liquids and homogenates. It might cause problems, however, when used for samples with high salt concentrations, e.g. concentrated urines, since its prerequisite is that the sample stays below its melting point during the whole process. A

---

1 Speed Vac Concentrator, Savant Instr. Inc., Hicksville N.Y. (USA)

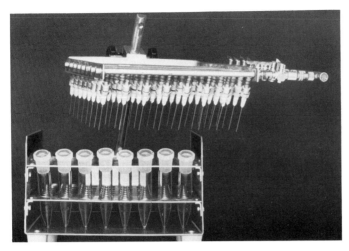

Fig. 1. Accessory to water baths for parallel concentration of a maximum of 64 extracts under nitrogen [source [11]]. Through the needles, nitrogen is blown over the surface of the extraction medium in order to prevent oxidation and to enable evaporation below atmospheric boiling point. Similar devices using alumina blocks for heating are commercially available, e.g. from Labortechnik Barkey, D-4800 Bielefeld

high salt concentration may lower the melting point so far that the sample begins to thaw immediately when put into the freeze-dryer. In these cases, it is advisable to dilute the sample with a sufficient amount of distilled water prior to freeze-drying.

4.2 Extraction

Amongst the many cleanup procedures, extraction is probably the most widely used. Despite its being usually relatively simple and rapid, extraction mostly results in high purification.

Solvents to be used should fulfill certain requirements, namely they should be stable, inert, volatile, and clear and should not be hazardous nor form emulsions. Solvents in common use are (in order of increasing polarity): hexane, petroleum ether, toluene, diethyl ether, chloroform, dichloromethane, acetone, ethyl acetate and methanol. The usual procedure is liquid-liquid extraction, normally between aqueous and nonaqueous phases. Although the solvents used should be nonmiscible with water, possible solubility of water should always be taken into account, since polar compounds may easily be transferred into the organic phase together with the dissolved water. This cannot be overcome by simple presaturation of the extraction solvent, thus the extract should be washed several times with small amounts of pure aqueous phase.

Extraction media not only separate, but can also introduce new compounds (solvent impurities) into the analytical sample. For routine analyses at the percent level, these usually have no effect, but they may severely interfere with trace analysis at the ppm and ppb level. One has to be aware that impurities at this level are normally not listed by the manufacturer. Even analytical-grade solvents may need further purification, e.g. redistillation or treatment with adsorbents. In this context, one should also remember that "impurity" is a function of the detection principle used in the final HPLC system with different purity claims for UV, fluorescence and amperomeric detection.

4.2.1 Extraction of Liquids

In *extraction of liquids*, the extracting agent is added to the liquid sample in a suitable vessel. This might be either a stoppered test tube or a separation funnel. Mixing is performed by agitating, and a number of devices – e.g. a vortex mixer or an overhead mixer – can be used for this. Intensity and duration of agitation will influence the extraction yield, thus sufficient time for extraction should be foreseen. Intense agitation is not always advisable since it can lead to formation of emulsions.

In general, the extracting solvent should have a far greater volume than the sample in order to optimize the extraction yield. This demand may easily lead to volumes that are difficult to handle, especially when batches are to be worked up. Besides, they will yield enormous amounts of solvents, thus raising waste-disposal problems and affording long evaporation times. The huge glassware needed is also difficult to rinse and, in trace analysis, problems with analyte sticking to surfaces of laboratory equipment will therefore increase. It is hence advisable to reduce the working size of a sample as far as possible. The author has found that, e.g., residue analysis based on 1 ml of milk or liver homogenate equivalent to 10 g of tissue are as accurate as – and in many cases even more accurate than – those based on $^1/_2$ l of milk or a whole liver.

The appropriate choice of the extracting agent does not only depend on its chemical and physicochemical properties. In many cases, transfer of the phase of interest is simplified if it is the upper phase in a test tube or the lower phase in a separatory funnel. In the first case, a solvent of less density than water, e.g. diethyl ether or ethyl acetate, will be more suitable and, in the latter case, a more dense solvent such as chloroform.

The *extraction yield* is directly related to the partition coefficient of the compound of interest between the two phases and thus may be optimized by use of an appropriate organic phase. Partition coefficients are in many cases pH-dependent and may even show inverse gradients due to the existence of different functional groups in the same molecule (cf.[12] and Fig. 2). It is advisable to plot log k (partition coefficient) versus pH in order to choose the most favorable pH range for extraction. Whenever this is not possible, at least the acidity of functional groups and the influence of other substituents in the molecule should be taken into account (cf. any textbook on organic chemistry). A brief indication is given by the following order of functional groups with decreasing acidity:

$$R-SO_3H > R-COOH > R-SO_2NH_2 > \text{phenolic } OH > R-N(CH_3)_2 > R-NH_2$$

Since a pH shift can be used to discriminate between closely related compounds (i.e. a drug and its metabolite[13]), it might be necessary to compromise between the optimum pH range for different compounds in order to extract them altogether (cf.[14]). In these cases, extraction might not be complete and corrections for the differing extraction rates have to be made.

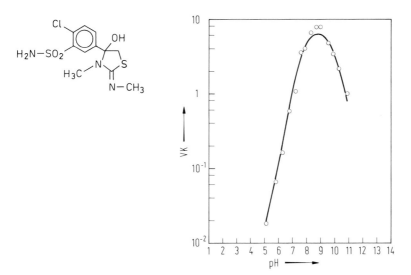

Fig. 2. pH-dependent partition of tizolemide (HOE 740) between octanol/water [source[12]]. The compound has a pK_a of 7.8 corresponding to the formation of the free base and a pK_a of 9.8 relating to deprotonation of the sulfonamide. Consequently, the coefficient of partition (VK) passes through a maximum at pH 8.7

Compounds that are *heavily extractable at any pH* do exist – mainly those which are highly polar or of amphoteric character. Here, the skill and knowledge of the analyst are challenged; success is often achieved by reducing the volume of the aqueous phase or by freeze-drying prior to extraction. Other possible solutions are salting out by the addition of NaCl or anhydrous Na_2SO_4 until the aqueous phase disappears[15]. Solvent mixtures that first lead to a homogeneous phase can also be used, the homogeneous phase then being separated by adding either n-hexane to a homogeneous water/ethanol mixture or water to a homogeneous ethanol/n-hexane mixture. The varieties of such procedures, while innumerable, must be determined empirically since they cannot be derived from theoretical knowledge.

Another very successful approach is the use of *ion-pairing agents* by means of which a neutral and well-extractable complex is formed, e.g.:

$$A^+ \text{ (analyte)} + I^- \text{ (ion-pairing agent)} = AI \text{ (neutral)}$$

This technique has, for instance, been widely used in the extraction of very polar analytes from (aqueous!) body fluids or tissue homogenates, e.g. biogenic amines[16] and tyramine[17]. The theory of ion pairing has been developed by Schill[18,19] and should be familiar to chromatographers, since ion pairing is often used in reversed-phase HPLC.

4.2.2 Extraction of Solids

Extraction of solids is usually performed by means of a Soxhlet-type extractor. Since this uses filter cartridges, the possibility of adsorption of the analyte on the filter material should again be considered. It is certainly less problematic than during normal filtration, since extraction is performed continuously over a longer period, but might become a problem when extremely low traces are to be determined.

When extracting water-containing solids, it may be useful first to homogenize them to an aqueous suspension which is then freeze-dried before extraction. This handling avoids the formation of emulsions that hinder proper separation of phases. In the case of freeze-dried material which is still difficult to extract, i.e. when grease is obtained, the extraction is ameliorated when the freeze-dried material is coated on a solid support, e.g. small glass balls. This coating can be easily obtained by adding a suitable amount of solid support prior to freeze-drying[20].

In general, it should be pointed out that extraction methods normally do not result in 100% recoveries. They will discriminate analytes with respect to extraction yield, and yields can change over the concentration range. For quantitative analysis, it is therefore necessary to construct calibration curves based on known amounts added to the sample and processed throughout the whole extraction procedure. These should cover the expected analytical range, since extrapolations from a calibration curve may be questionable. The use of internal standards does not make calibration curves obsolete, since an internal standard (even a homologue) may not necessarily show extraction characteristics identical to those of the analyte.

The extracts obtained normally far exceed the original sample volume in order to get a reasonable extraction yield. Hence, whenever the compounds of interest are minor constituents, especially in trace analysis, the extracts need to be concentrated prior to HPLC analysis. Concentration should be performed with extreme care, and excessive heat and contact with oxygen (air!) should be avoided. Apparatus for careful evaporation have already been discussed (see "Evaporation").

4.3 Automation of Extraction Procedures

Since extractions are still most widely used for sample cleanup and are also relatively simple to perform, attempts, quite naturally, were made to automatize these cleanup steps:

One of the first attempts was to use the well-established Auto-Analyzer principle, by which samples – well separated by air bubbles – are treated and extracted in a continuous flow system. The use of a moving belt further allows the extracts to be concentrated in a stream of heated nitrogen prior to HPLC separation[21].

Nevertheless, contrary to the situation in clinical chemistry, this automation principle has so far never been accepted by a relatively large number of analysts.

A totally different approach makes use of robotics, i.e. microprocessor-controlled automats that imitate the functions of human arms and hands, thus allowing the mechanization of all steps of batch-cleanup procedures[22]. Fig. 3 shows the robot arm, control station and working stations for dispensing of liquids and for extraction by microcolumns. Since this technique is relatively new, a final statement on it cannot as yet be made, but an increasing number of publications (cf.[23]) might be an indication of greater acceptance.

4.4 Adsorption

Whenever distribution problems exist, adsorption procedures are another way to get reasonably pure extracts. In principle, they seem to be of a simplicity similar to that of extraction procedures, since the analyst "only" needs an adsorbent with

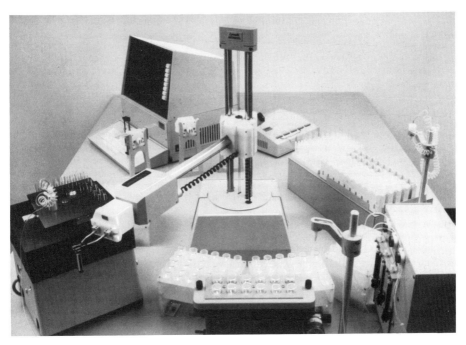

Fig. 3. Zymate® laboratory robot. Robot arm with grip hand, controller and different working stations. For details, see text. (Courtesy of Zymark GmbH, D-6240 Königstein)

suitable selectivity and a solvent by which the analyte can be quantitatively recovered. Thus, alumina had already been used in 1938 [24] for the selective extraction of catecholamines and is unexcelled for this purpose today [25]. Besides alumina, charcoal (e.g. [26]), activated charcoal (e.g. [27]) and diatomaceous earth [28] have all been used for extraction purposes.

But, experience shows that, in general, it is difficult to obtain an adsorbent with well-defined selectivity and, especially, to reproduce this from batch to batch. Furthermore, small quantities of analyte often tend to adsorb irreversibly, leading to difficulties in the quantification of trace amounts. It seems natural, therefore, that other principles and materials should also be taken into consideration. In many cases, the use of Amberlite XAD, a nonpolar resin for adsorption appears to be a reasonable solution (cf. [29]). Also, during the last few years, interest has centered on material originally designed for reversed-phase and ion-exchange HPLC and a lot of different material and disposable microcolumns have been marketed for this purpose – either "classical" silica, e.g. Extrelut® columns (cf. [30]) (E. Merck) or, as well, silica and reversed phases, e.g. Sep Pak® Cartridges (cf. [31]) (Millipore – Waters) or the whole variety of stationary phases available for HPLC, e.g. Bond Elut® columns (Analytichem International). The sample and the eluents pass through these microcolumns either due to hydrostatic pressure or centrifugal force (Extrelut, Bond Elut) or to pressure obtained by means of a syringe (Sep Pak) or are sucked through by means of a special device applying vacuum (Bond Elut, cf. Fig. 4).

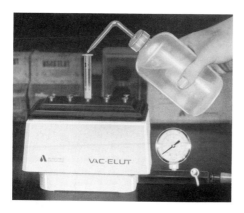

Fig. 4. Vac-Elut® station. Samples or eluents brought on top of the disposable microcolumns are sucked through by application of vacuum. Inside the apparatus, recovery tubes are placed in a rack under the columns. (Courtesy of Analytichem International)

The typical cleanup procedure is as follows (cf. [32, 33]):
1. The extracting column is washed and wetted (e.g. reversed-phase columns by means of methanol). This is of great importance, since additives and contaminants of the adsorbent – even those adsorbed from the laboratory environment during storage – can cause severe interference in the analysis[34].
2. The extracting column is equilibrated with an eluent which does not elute the analyte. With the help of this step, the sample can be adjusted to an appropriate pH.
3. The sample is applied onto the extracting column.
4. After the sample has passed through the column, the column is washed with an appropriate amount of the equilibration eluent (cf. point 2.).
5. The analyte is then eluted with a solvent of sufficient eluotropic force.

In most cases, the extraction yield of liquid-solid extraction is superior to that of liquid-liquid extraction, but the number and amount of co-extracted constituents of the analyte is also greater. Therefore, more cleanup steps than after liquid-liquid extraction might become necessary. The combination of different extraction mechanisms wherever possible is especially advisable, since it results in very clean extracts[35]. Thus, the optimization of liquid-solid extraction might be more complicated and time-consuming than that of liquid-liquid extraction. In routine analysis, however, the use of liquid-solid extraction columns may show great practical advantages such as ease of preparation, saving of time and smaller sample volumes (cf. [36]).

4.5 Automation of Adsorption Techniques

Meanwhile, a lot of concepts on how to automize liquid-solid extraction as well have been put forward. Besides the robotics already discussed, these include a specially-designed centrifuge that – while rotating clockwise – aligns the extraction column with the effluent cup. During counterclockwise rotation, the column is aligned with the recovery cup. Washing and eluting fluids are added through the rotor; the equipment allows simultaneous cleanup of 12 samples (Du Pont Prep II, cf. Fig. 5). Another concept requires all cleanup steps plus final extraction of the analyte to be performed off-line in the way described. The microcolumns which

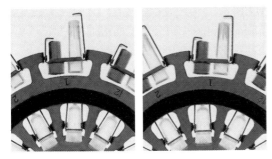

Fig. 5. Du Pont Prep® automated sample processor (detail). The figure represents a part of the rotor system with the extraction columns in the inner ring and the effluent cup and recovery cup in the outer ring. Solvents are pumped through jets in the center of the rotor; alignment of effluent cup and recovery cup is controlled by direction of rotation. (Courtesy of Du Pont Clinical & Instrument System Division, Wilmington DE 19898)

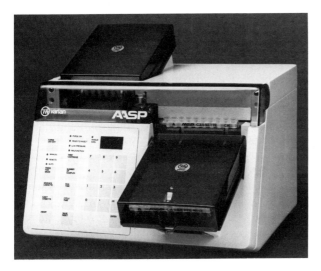

Fig. 6. Varian AASP®. The device is to be used in connection with the Vac Elut® station (cf. Fig. 4). Instead of individual microcolumns, disposable cassettes combining ten microcolumns are used. After adsorption of the analyte on the column and washing by means of a suitable solvent, the cassette is put into the AASP where the analyte is eluted by the mobile phase delivered from the chromatographic pump onto the analytical column. The microcolumn is then by-passed for HPLC chromatography. (Courtesy of Varian GmbH, D-6100 Darmstadt)

are bundled in a cartridge are then placed onto an auto-sampler that elutes directly from the microcolumn onto the analytical column of the HPLC system by means of the chromatographic pump and eluent (Varian AASP, cf. Fig. 6). At first glance, this device seems optimum compared to a chromatographer. The author's own experiments, however, have shown that, in many cases, the band-broadening of the microcolumn is very considerable. Without significant increase in chromatographic band-broadening, the AASP can therefore not be used in combination with short columns (e.g. 15 cm) and stationary phases of small particle size (e.g. 5μ) which, for sensitivity reasons, are often requested in trace analysis.

4.6 Chromatographic Techniques

The ready-to-use devices for adsorptive separations described above can be considered as an intermediate between (batch) adsorption and chromatography. Thus, it seems natural to perform parts of cleanup or even the whole analysis by combinations of chromatographic steps.

Three principles can be used:
1. Solvent programming (maintaining the column and stationary phase)
2. Column switching (maintaining the mobile phase)
3. Column and solvent switching, i.e. using one chromatographic system for a crude preseparation, and transferring the compounds of interest to another final resolving system.

Solvent programming is better known as (step) gradient programming. It allows either unwanted material first to be eluted with a solvent of low eluotropic force or the inverse, i.e. first the compounds of interest to be eluted and then flushing of all noninteresting material that has been adsorbed on the column. An interesting example was published by J.N. Little and G.J. Fallick who pumped 200 ml of river water through a reversed-phase column, enriching all the organic pollutants which were then chromatographed using a gradient [37]. In principle, innumerable variations of this technique are possible. Its biggest disadvantage, however, seems to be the fact that time-consuming re-equilibration of the system is needed after every single run, resulting in a reduction in the analytical throughput. In order to overcome this, several authors proposed not changing the mobile phase during the chromatographic run, but changing the column, thus introducing column switching. These techniques make use of the fact that different column properties, e.g. length, stationary phase, etc., lead to different elution behavior with an identical mobile phase. Mostly 'front cut' and 'heart cut' separations are described, where the compounds of interest are cut out of a complex and difficult to separate mixture for subsequent chromatographic analysis.

Using 'front cut' technique, the sample is first typically separated on a column (I) packed with low specific surface area material, resulting in low K^1 values for all compounds and thus in short elution times. All material is transferred onto a second column (II) packed with high specific area material, thus being suitable for the analysis of compounds of interest using the same mobile phase. At a set time after injection, both columns are separated by means of a switching valve. The flow and separation of the compounds of interest on column II are now stopped for a time, during which the late-eluting compounds are eluted from column I. Due to the low diffusion coefficients of solvents in solutes, the separation already obtained on column II is maintained over this time so that it can be easily continued after back-switching the valve, completing the separation of all interesting compounds. After this step, the system is ready for the next analysis (cf. [38]). The same

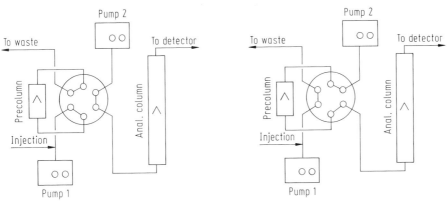

Fig. 7. Sample cleanup by column switching. Left: valve position for adsorption of analyte(s) on precolumn; right: valve position for transfer of analyte(s) to analytical column. For details, see text

procedure can be used to cut out other sections besides the front part of the elute from column I ('heart-cut technique').

A more commonly used approach is precolumn venting, using a short precolumn with packing identical to that in the main column. It makes use of the fact that, in many LC assays, the compounds of interest elute at k^1 values between 2 and 10, whereas the remainder of the sample will elute as a 'front peak' near the void volume. Particularly in trace analyses, this initial peak may tail, overlapping the peaks of interest. By means of precolumn venting, most of this disturbing material is eluted from column I into waste. When it is about to elute completely from column I, the valve is switched to allow elution onto column II. Thus, the amount of interfering material in the final separation system is reduced to the amount of compounds of interest, allowing a good separation (cf. [39]).

Being chromatographic techniques, both cleanup procedures have the advantage of easy automation, further economizing laboratory manpower. Nevertheless, both have severe restrictions:
- The eluents used must be compatible with all constituents of the analytical sample and not cause precipitation.
- Irreversible adsorption should not occur or should at least be minimal, otherwise it would lead to extensive consumption of expensive HPLC columns.

These requirements especially restrict their use for protein-containing samples of biological origin. It was people working in this area who introduced perhaps the most versatile and promising chromatographic sample-cleanup procedure to date. In 1982, Roth, Beschke et al. [40] introduced a column- and solvent-switching system for automated on-line cleanup and chromatographic separation of protein-containing samples. In principle, this combines sample cleanup by adsorption – as described above – with high-resolution HPLC. The device consists of two independently-working chromatographic systems, the first being a short, simple, often manually-packed precolumn imitating the function of a Sep-Pak or similar column (cf. Fig. 7). An HPLC pump supplies it with an eluent which, after injection of the sample, allows compounds of interest to be greatly retarded, while unwanted material such as proteins pass through independently and without precipitation. At the same time, a second pump delivers eluent – optimized for the final separation – onto the analytical column. Both systems are run in parallel and are connected by a switching valve which is switched at a suitable time, allowing eluent for final separation to flow through the precolumn and onto the analytical column. By means of this procedure, the adsorbed compounds of interest are transferred to the analytical column, where they are finally separated. Flow inversion helps minimize the band-spreading effects of the device. During the transmission period, the preseparation eluent goes either directly to waste or – by means of a more complicated arrangement using two switching valves – onto a second precolumn which is re-equilibrated for the next preseparation (cf. [40]).

Similar means of chromatographic sample cleanup have been reported by other laboratories (e.g. [41, 42]). All authors agree with our own findings that, besides their excellent practicability, column-switching cleanup procedures often result in precision and recovery data that are far better than what can be expected from conventional extraction procedures.

Nevertheless, the methods described have also problems and restrictions [43]. First of all, as already mentioned in connection with liquid-solid extractions, the precolumn will adsorb more plasma constituents than are normally extracted by liquid-liquid extraction. Thus, the following HPLC separation will become more complex, and simultaneous determinations of several analytes might even become impossible except when gradient elution is used. It is for this reason that most of the examples cited use highly specific detection, e.g. fluorescence, with which the interfering material is not seen.

Secondly, in contrast to off-line cleanup columns, the precolumn must be reused. Since the technique described is front-cutting, there is a chance that some plasma constituents are not eluted from the precolumn during the chromatographic run. In routine analysis, these are quite easily enriched by the number of samples that are chromatographed. The precolumn or the analytical column will finally become overloaded by material not eluted and will loose its separation performance, thus spoiling analysis.

Thirdly, plugging of columns by undissolved microparticles in the plasma must be taken into account. This point can be overcome by filtrating every sample through a disposable microfilter before injection.

In order to further increase sample throughput, Karger et al. [44] even divided the analysis into three consecutive chromatographic steps: By means of the first column, they eliminate plasma proteins. The remaining organic compounds are then transferred to a second chromatographic system that simply cuts out the parts of interest, transferring these only onto the final analytical column. Both preseparation steps can be performed so rapidly that the final analytical column can already be fed with the next sample before the preceding one is eluted. Due to the high degree of cleanup, no overlapping of chromatograms will occur. With a sample rate of up to 40–50 samples/hour, this so-called 'boxcar' technique can already compete with the speed of most of the automated analyzers used today in clinical pathology.

5 Analytical Strategies

So far, only different approaches to sample cleanup have been presented. They should be considered as 'building blocks', by which the analyst can obtain a suitable solution to his analytical problems. A typical example may be the sample cleanup for fenbendazole, an anthelmintic agent:

Being a drug for veterinary use, determinations have not only to be performed in plasma for the evaluation of the pharmacokinetics, but also in milk and edible tissues (meat, liver, kidney) for the evaluation of residues. An optimized HPLC system allows simultaneous determination of the mother compound and all known metabolites[45] in a single chromatographic run. Therefore, the aim was to treat all matrices in such a way that this HPLC system can be used.

The development of sample cleanup (cf. Fig. 8) started with the least complex matrix, in this case plasma. Here, a simple one-step extraction after alkalization to pH 10–11 was found to be sufficient, followed by evaporation of the diethyl ether used for extraction and take-up of the dry residue with the mobile phase.

Due to the high protein and fat content, this simple approach cannot be used for milk. Thus, in a first step, the proteins are precipitated by adding the milk to a surplus of acetonitrile. After removal of the precipitated proteins by centrifugation, the major part of the fat is eliminated by extracting the acetonitrile phase with n-hexane. The acetonitrile phase is then concentrated by evaporation. Since the residue is still too complex for chromatographic determination, it is distributed between an aqueous phase at pH 10–11 and diethyl ether. The rest of the sample cleanup is identical to the cleanup of plasma.

For meat, liver and kidney, a similar cleanup could have been used after homogenisation of the tissues in acetonitrile by means of an Ultra Turrax® mixer (method 2). This procedure, however, would be time-consuming because of the need for evaporation of great amounts of acetonitrile. Therefore, meat, liver and kidney are homogenized with 0.1N sodium hydroxide by means of an Ultra Turrax® mixer and then centrifuged. An aliquot of the supernatant equivalent to 10–20 g of tissue is then extracted by means of ethyl acetate. The extract is evaporated to dryness and the residue distributed between small amounts of acetonitrile and n-hexane. The acetonitrile phase is then evaporated to dryness and, for determinations in meat, the residue is taken up with the mobile phase for chromatography (method 1).

For liver and kidney, the result of this cleanup procedure is not yet sufficient. Thus, the residue is taken up with a mixture of chloroform and methanol and administered to a silica microcolumn (Bond Elut® Si, Analytichem International Inc.). The compounds of interest are eluted with 2 ml of the same solvent which is then evaporated to dryness. The residue is finally taken up by the mobile phase for chromatography.

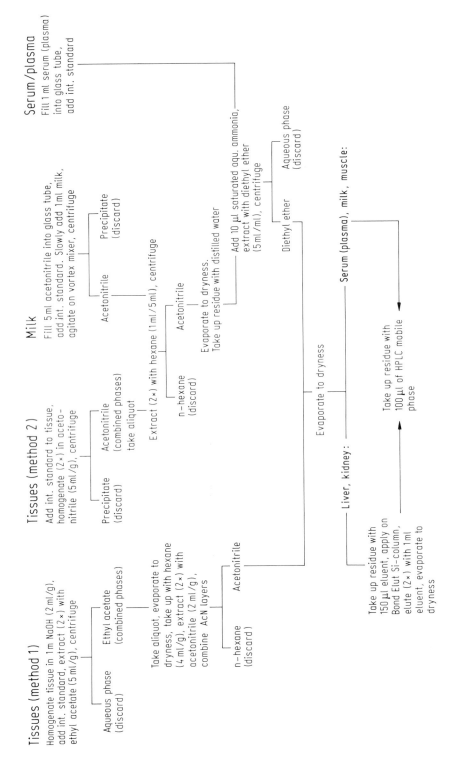

Fig. 8. Extraction scheme for fenbendazole and metabolites

All cleanup procedures described resulted in overall recoveries of less than 60%, mostly due to the impossibility of quantitative separation of phases. Therefore, in each case, an internal standard was added at the first extraction step in order to reduce imprecision due to variations in extraction yield. Besides, in order to compensate differences in extraction yield, calibration was performed by means of known amounts added to blank matrices and worked up according to the procedures described.

General solutions for optimum sample cleanup cannot be offered, bearing in mind the wide variety of problems. Most problems can however be solved in analogy to the examples described or to other examples published. In this context, compendia which describe standard procedures for special areas should also be mentioned (e.g. [46–48]).

There are still a few remarks worth mentioning: "Suitable" means "optimum" or "well-adjusted" rather then "sophisticated". Thus, it does not make sense to develop and optimize a column-switching procedure for handling only a few samples. Simple extraction would in most cases be more rapid and more appropriate, especially since analytical answers are always urgently needed. On the other hand, the continuous use of laborious daily extraction procedures should become a challenge to try some of the approaches to automate sample cleanup as described above.

Acknowledgement: The author would like to thank especially Dr. P. Hajdu for his support in preparing the material and collecting the literature, his colleagues Dr. Lehr and Dr. Schmidt for useful hints and discussions, and Miss D. Koziol and Miss G. McConaghy for editing and typing the manuscript.

6 References

1. Kienitz, H., Runge, H. in Korte, F. (ed.): Methodicum Chimicum, vol. Ia, p. 839 (1973), Thieme Verlag, Stuttgart, ISBN 3-13-480101-9
2. Steyn, J.M., Müller, F.O.: S. Afr. Med. J. 57, p. 129 (1980)
3. Verbiese-Genard, N., Hanocq, M. et al.: Analyt. Biochem. 134, p. 170 (1983)
4. Scales, B. in Reid, E. (ed.): Blood, Drugs and Other Analytical Challenges, pp. 43–54 (1978), Ellis Horwood Ltd., Chichester (Sussex), ISBN 0-85312-124-9
5. Adam, H.K., in Reid, E. (ed.): Trace-Organic Sample Handling, pp. 291–297 (1980), Ellis Horwood Ltd., Chichester (Sussex), ISBN 0-85312-187-7
6. Glaser, A.N.: J. Biol. Chem. 242, p. 433 (1967)
7. Osselton, M.D., in Reid, E. (ed.): Trace-Organic Sample Handling, pp. 101–110 (1980), Ellis Horwood Ltd., Chichester (Sussex), ISBN 0-85312-187-7
8. Fransson, B. et al.: J. Chromatogr. 125, p. 327 (1976)
9. Heptner, W., Hornke, I., Cavagna, F. et al.: Arznei.-Forsch./Drug Res. 28, p. 58 (1978)
10. Eichelbaum, M., Sonntag, B., Dengler, H.J.: Pharmacology 23, p. 192 (1981)
11. Uihlein, M.: Chromatographia 12, p. 408 (1979)
12. Sistovaris, N., Uihlein, M.: J. Chromatogr. 167, p. 109 (1978)
13. Hajdu, P., Damm, D.: Arznei.-Forsch./Drug Res. 29, p. 602 (1979)
14. Uihlein, M., Hajdu, P.: Arznei.-Forsch./Drug Res. 27, p. 98 (1977)
15. Häussler, A., Hajdu, P.: Arznei.-Forsch./Drug Res. 14, p. 710 (1964)
16. Da Prada, M., Zürcher, G.: Life Sci 19, p. 1161 (1976)
17. Causon, R.C., Brown, M.J.: J. Chromatography 310, p. 11 (1984)
18. Schill, G. in Reid, E. (ed.): Assay of Drugs and Other Trace Compounds in Biological Fluids, pp. 87–103 (1976), North Holland Publishing Company, Amsterdam, ISBN 0-7204-0584-X

19. Schill, G. et al. in Chasseaud, L.F. and Bridges, J.W. (eds.): Progress in Drug Metabolism, vol. 2, pp. 121–124 (1976), John Wiley & Sons, New York, ISBN 0-471-27702-9

20. Jones, C.R. in Reid, E. (ed.): Assay of Drugs and Other Trace Compounds in Biological Fluids, pp. 107–113 (1976), North Holland Publishing Company, Amsterdam, ISBN 0-7204-0584-X

21. Dolan, J.W., Snyder, L.R.: J. Chromatogr. 185, p. 57 (1979)

22. Little, J.N.: TrAC 2, p. 103 (1983)

23. Hawk, G. (ed.): Advances in Laboratory Automation: Robotics 1984, Zymark Corp., Hopkinton MA (USA) (1984), ISBN 0-931565-00-6

24. Shaw, F.H.: Biochem. J. 32, p. 19 (1938)

25. Mefford, I.N.: J. Neurosci. Res. 3, p. 207 (1981)

26. Adams, R.F., Vandemark, F.L.: Clin. Chem. 22, p. 25 (1976)

27. King, L.J., Parke, D.V., Williams, R.T.: Biochem. J. 98, p. 266 (1966)

28. Breiter, J.: Kontakte (Merck) 2, pp. 21–32 (1981)

29. Mule, S.J. et al.: J. Chromatogr. 63, p. 289 (1971)

30. Breiter, J., Helger, R.: Z. klin. Chem. klin. Biochem. 13, p. 254 (1975)

31. Woodbridge, A.P., McKerell, E.H. and Reid, E. (ed.): Trace-Organic Sample Handling, pp. 128–147 (1980), Ellis Horwood Ltd., Chichester (Sussex), ISBN 0-85312-187-7

32. Good, T.J., Andrews, J.S.: J. Chromatogr. Sci. 19, p. 562 (1981)

33. Sanar, W.A., Gilbert, J.A.: J. Liq. Chrom. 3, p. 1753 (1981)

34. Ende, M., Pfeifer, P., Spiteller, J.: J. Chromatogr. 183 (1980)

35. Lehr, K.H., Damm, P.: J. Chromatogr. 339, p. 451 (1985)

36. McDowall, R.D., Murkitt, G.S.: J. Chromatogr. 317, p. 475 (1984)

37. Little, J.N., Fallick, G.L.: J. Chromatogr. 112, p. 389 (1975)

38. Huber, J.F.K., Linden, R. v.d.: J. Chromatogr. 83, p. 267 (1973)

39. Wahlund, K.G., Lund, U.: J. Chromatogr. 122, p. 269 (1976)

40. Roth, W., Beschke, K. et al.: J. Chromatogr. Biomed, Appl. 222, p. 13 (1982)

41. Erni, F., Keller, H.P., et al.: J. Chromatogr. 204, p. 65 (1981)

42. Lecaillon, J.-B., Febvre, N., Souppart, C.: J. Chromatogr. 317, p. 493 (1984)

43. Uihlein, M.: Fresenius Z. analyt. Chem. 320, p. 721 (1985)

44. Karger, B.L., Giese, R.W., Snyder, L.R.: TrAC 2, p. 106 (1983)

45. Uihlein, M., Hajdu, P., Damm, D. et al.: HOECHST AG internal reports scheduled for publication.

46. Rückstandsanalytik von Pflanzenschutzmitteln, Part 7 (1984), Verlag Chemie, Weinheim

47. Sadee, W., Beelen, G.C.M. (eds.): Drug Level Monitoring, John Wiley & Sons, New York (1980), ISBN 0-471-04881-X

48. Richens, A., Marks, V. (eds.): Therapeutic Drug Monitoring, Churchill Livingstone, Edinburgh (1981), ISBN 0-443-02162-7

Liquid-liquid Chromatography

J.C. Kraak and J.P. Crombeen
Laboratory for Analytical Chemistry, University of Amsterdam,
Nieuwe Achtergracht 166, 1014 WV Amsterdam/The Netherlands

1 Introduction

In liquid-liquid chromatography the separation of the components of a mixture results from the distribution of the solutes between two immiscible liquids. One liquid is immobilized in the pores of a solid support and acts as the stationary phase. The other liquid, saturated with the stationary phase, is used as the mobile phase. Thus each phase in liquid-liquid chromatography can be considered as a bulk phase. This in contrast with bonded phase chromatography where only the mobile phase is a bulk phase.

Principally, liquid-liquid chromatography can be considered the most effective way of performing multiple extraction. This means, that the separation characteristics of extraction systems can be transferred to liquid-liquid chromatography, providing the influence of the solid support in the distribution process can be ignored. A great advantage of LLC is the almost unlimited number of phase systems, which can be composed from common liquids. This creates a vast opportunity to adjust the retention and selectivity for a specific mixture of components at hand.

Moreover, liquid-liquid systems are thermodynamically very well defined and thus exactly reproducible. This in contrast with many other forms of liquid chromatography. However, the current interest in liquid-liquid chromatograph is still small, mainly because of the precise isothermal conditions that are required to obtain stable columns. Moreover, continuous solvent gradients cannot be applied. Some of these problems have been overcome recently by applying isothermal conditions to only a limited part of the experimental set up, whereby the stationary phase is generated dynamically via the saturated mobile phase.

2 Theory of Liquid-liquid Distribution

The distribution of a solute in a liquid-liquid system is thermodynamically characterized by:

$$K_{thi} = a_i^s / a_i^m \tag{1}$$

where

K_{thi} = thermodynamic distribution coefficient of solute i.
a_i^s = activity of solute i in liquid s.
a_i^m = activity of solute i in liquid m.

When the chemical potential of the pure liquid solute $i(u_{i(T, P)})$ at temperature T and pressure P is choosen as the reference value, $K_{thi} = 1$. At infinite dilution of the solute, the distribution coefficient can be expressed in terms of mole fractions and activity coefficients of solute i in the liquid phases accordingly:

$$K_{xi} = x_i^s / x_i^m = f_i^m / f_i^s \tag{2}$$

where

K_{xi} = distribution coefficient of solute i at infinite dilution.
x_i^s and x_i^m = mole fractions of solute i in phase s and m, respectively.
f_i^s and f_i^m = activity coefficients of i in phase s and m, respectively.

In chromatography it is more convenient to use the distribution coefficient on basis of molar concentrations, which is defined as:

$$K_{pi} = c_i^s/c_i^m = K_{xi}(M_m/\rho_m)/(M_s/\rho_s) \tag{3}$$

where

K_{pi} = partition coefficient of solute i.
c_i^s and c_i^m = concentration of solute i in phase s and m, respectively.
M_s and M_m = average molecular weight of solvents s and m, respectively.
M_s/ρ_s and M_m/ρ_m = average molar Volume of the respective solvents.

The partition coefficient K_{pi} will be used in all further discussions throughout this chapter. In chromatography the capacity ratio, k_i', rather than the partition coefficient K_{pi} is preferred as retention parameter.

The relationship between k_i' and K_{pi} is given by:

$$k_i' = K_{pi} \cdot V_s/V_m \tag{4}$$

where

V_s and V_m = volume of stationary and mobile liquid, respectively.
V_s/V_m = phase ratio.

Equation 4 shows that the capacity ratio is proportional to the partition coefficient of the solute and the phase ratio. The partition coefficient is determined by the nature of the solute and the liquid phases.

The phase ratio can be varied within certain limits by the amount of the stationary phase accomodated in the pores of the support.

If retention is governed by pure liquid-liquid distribution and the phase ratio is known, then partition coefficients can be calculated from the capacity ratio. This way of estimating partition coefficients is very attractive from the point of view of speed of the determination and the less stringent purity requirements put on the samples.

According to Eq. 3 it is possible to predict partition coefficients, providing the activity coefficients in both phases and the average molar volumes of the liquids are known. There are theoretical[1–6] and emperical[7–10] expressions to calculte activity coefficients of solutes in liquids. However, up to now these expressions are not adequate for more complex solute structures for accurately predicting the retention in liquid-liquid chromatography. At present, batchwise determined partition coefficients can only be reliably transferred to column experiments.

3 Column Preparation

Several techniques have been described to prepare columns for liquid-liquid chromatography. The method of choice strongly depends on the selected phase system,
the particle size of the support and the required stationary phase loading. The
following techniques and their applicability will be considered in this paragraph.

– Solvent evaporation,
– In situ Techniques:
  – Direct coating,
  – Precipitation technique,
  – Dynamic coating technique.

3.1 Solvent Evaporation Technique

This coating technique is similar to the one used to prepare gas liquid chromatographic columns. At first the support is coated batchwise with the stationary phase
and afterwards dry-packed in a column. A measured amount of stationary phase
is dissolved in a volatile liquid and mixed gently with the dry support. Then the
solvent is gradually removed from the blend in a rotary evaporator, leaving the
stationary phase within the pores of the support. Of course this way of coating
is restricted only to non-volatile stationary phases. In order to pack the column,
small portions of the free flowing pre-coated support are poured into the column
and the column is tapped in order to provide a compact packing. This is continued
until the column is filled. The tap-filling procedure is limited to supports having
a particle size > 30 μm, since the efficiency obtained with this technique using smaller
particles has been found to be far too low. For smaller size particles, wet packing
techniques are commonly used. However, this has not been found successful, since
an effective suspension-liquid will usually dissolve the stationary phase, whereas
less solvating suspension-liquids do not produce a stable, dispersed system. To prepare liquid-liquid columns with small particles, so called in situ techniques have
been developed[11–14], whereby the stationary phase is applied in a pre-packed column.

3.2 Direct Coating

A straight-forward way to load a pre-packed column can be accomplished by pumping the stationary phase through the column. After the column has been completely
filled with stationary phase, it is eluted by saturated mobile phase to remove the
excess of stationary phase between the particles. This technique is obviously limited
to stationary phases with low viscosities (< 1.5 cP). The removal of excess stationary
phase from the interstitial space requires a large volume of mobile phase in order
to obtain a stable chromatographic system and is therefore rather time consuming.
Moreover, this coating technique provides a maximum attainable loading of the
support and thus precludes the use of the phase ratio as a parameter for adjustment
of the capacity ratio. However, this disadvantage can be overcome by applying
the so-called precipitation technique[14].

3.3 Precipitation Technique

A solution of the stationary phase in an organic solvent is pumped through a pre-packed column, which has been pre-equilibrated with the same organic solvent. When the column is filled with a solution of the stationary phase, another solvent (completely miscible with the diluent but not with the stationary phase) is pumped through the column. In this way the stationary phase "precipitates" and fills the pores of the support. Next, the column is equilibrated with the saturated mobile phase. The amount of stationary phase, and thus the phase ratio, can be adjusted by choosing the proper concentration of the stationary phase in the diluent.

For liquid-liquid systems in which the stationary phase shows a significantly larger affinity for the support and dissolves to some extent in the mobile phase (about 0.5 mol.%), the dynamic coating technique can be successfully applied[15-24].

3.4 Dynamic Coating Technique

The mobile phase, saturated with the stationary phase, is continuously pumped through a pre-packed column. Because of its large affinity for the support, the stationary phase is withdrawn from the mobile phase and is preferentially adsorbed on the support. When feeding the column with fresh saturated mobile phase, multi-layers of stationary phase are gradually formed on the surface of the support. Capillary forces may fill the entire internal volume of the support, depending on pore diameter.

The generation of the stationary phase on the support as a function of the number of column volumes of saturated mobile phase pumped through, is shown in Fig. 1 for the LL system dimethylsulfoxide/cyclohexane.

With this technique the stationary phase loading, and thus the phase ratio cannot be varied, since a maximum attainable loading is obtained at equilibrium. However, it has been shown that the phase ratio can be adjusted within certain limits by applying a small temperature difference between the stock reservoir and the column[25]. A great advantage of dynamically generated liquid-liquid systems is their intrinsic stability. Any change in the stationary phase loading is quickly restored by the saturated mobile phase. In principle, the dynamic coating technique can

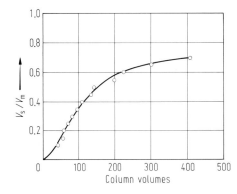

Fig. 1. Phase ratio versus (V_s/V_m) volume of mobile phase (column volumes) pumped through the column. Support: Zorbax BP-Sil. Phase System: dimethylsulfoxide/cyclohexane. Column and reservoir temperature 22.0 and 20.0° C respectively

also be used to generate stationary phases which do not dissolve easily in the mobile phase. However in that case, very large volumes of saturated mobile phase ($>1,000$ column volumes) have to be pumped through the column before a steady state is reached. It can be noted, that with some phase systems[12-13, 26] the speed of loading can be increased considerably by injecting small portions of the stationary phase into the column.

4 Selection of Solid Support

In order to create an efficient LLC system, the stationary phase has to be immobilized homogenously on the solid support with a large interfacial area. A large exchange surface is necessary to achieve a fast mass transfer between the two bulk phases. This means, that the solid support must have significant adsorptive properties for the stationary phase in order to hold it fixed under conditions of flow. In other words: the solid support must be wetted better by the stationary than by the mobile phase. On the other hand the interaction of solutes with the solid support must be avoided to provide pure liquid-liquid distribution. This demand however, is sometimes in conflict with the conditions to immobilize the stationary phase on the support.

4.1 Nature of the Surface of the Solid Support

LLC can be performed either in the normal- or reversed-phase mode, depending on wether a polar or non-polar stationary phase is used. It will be obvious, that for both LLC modes different solids have to be applied.

If a polar stationary phase (normal phase LLC) is used, the surface of the solid support must be polar and for a non-polar stationary phase (reversed phase LLC) the surface must be non-polar. In all cases the wettability of the surface for the stationary phase is larger than for the mobile phase. Currently, porous silica and alkylmodified silica are the supports of choice for immobilizing the stationary normal and reversed phases, respectively.

Apart from the requirements put on the nature of the surface of the solid support, other properties such as surface area, pore diameter and pore volume are important parameters to be considered in LLC.

4.2 Pore Diameter and Pore Volume of the Support

As mentioned before, the stationary phase must be homogeneously spread out over the support surface. The amount of stationary phase, which can be accommodated in the porous support is limited by the pore volume, yet it also depends on the pore diameter and surface area and on the nature of the stationary phase. The pore diameter is related to the specific surface and decreases with increasing surface area[27]. The pore volume does not show such a strong relationship with the surface area of the support[27]. Loading of the support is maximal if the pores are completely

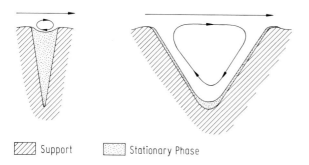

Support   Stationary Phase

Fig. 2. Schematical representation of the coating of low
viscous stationary liquids on
small and wide pore support

filled. However, the actual pore filling to maximum capacity differs with the nature
of the stationary phase (e.g. viscosity) and the pore diameter.

This is illustrated in Fig. 2, showing a schematic representation of a particle
with small pores (thus large surface area) and one with large pores (thus small
surface area).

Let us now consider a high-viscous ($> = 10$ cP) and a low-viscous ($< = 2$ cP)
stationary phase. The low-viscous stationary phases can be loaded on a pre-packed
column by either solvent precipitation or by the dynamic coating technique. When
using the dynamic coating technique the low-viscous stationary phases adsorb as
several liquid layers on the support and ultimately might fill small pores due to
capillary forces. In large pores these forces are weak or absent and the surface
in the pores is covered only with a limited number of liquid layers. In principle,
the wide pores can be filled to a larger extent by using the solvent precipitation
technique. However, low-viscosity stationary phases will be stripped out of the pores
by the flowing mobile phase.

A different situation occurs when a highly viscous stationary phase is applied.
Coating can be accomplished by solvent evaporation (and dry-packing) or solvent
precipitation after wet-packing. In contrast to low-viscosity phases, very stable systems are obtained for the wide pore support as well[28-29]. This is caused by the
high surface tension of the stationary phase, which can withstand the shear forces
of the mobile phase flow in the wide pores.

4.3 Specific Surface Area of the Support

To perform pure liquid-liquid chromatography, the solid support must be inert
toward the solutes with concomitant adsorptive properties for the stationary phase.
Since these requirements are rather contradictory, one has to be aware that in practice competitive adsorption on the support between the solute and stationary phase
might occur. Adsorption on the other hand can have either a positive or negative
effect on the selectivity, yet it diminishes the advantages of thermodynamic reproducibility.

The extent of solute adsorption on the solid depends on the nature and surface
area of the support and on the amount of stationary phase loading. In principle
the capacity ratio of a solute is a mixed adsorption and absorption process, that

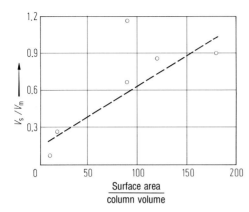

Fig. 3. Dependance of the phase ratio (V_s/V_m) on the surface area of the support per column volume. Phase system: 0.05M formic acid and 0.05M ammonium formate in water/methanol/chloroform = 40:250:710(v/v/v). Supports: Hypersil, LiChrosorb Si 100, Zorbax BP-Sil and LiChrospher Si 100, 500 and 1000. Temperature of column and support 22.0 and 20.0° C respectively

can be approximately described as:

$$k'_i = K_{adi} \cdot A_s/V_m + K_{pi} \cdot V_s/V_m \qquad (5)$$

Substitution of $V_m = V_0 - V_s$ in Eq. 5 gives:

$$k'_i = K_{adi} \cdot A_s/(V_0 - V_s) + K_{pi} \cdot V_s/V_m \qquad (5a)$$

where

k'_i = overall capacity ratio.
K_{adi} = adsorption coefficient.
K_{pi} = liquid-liquid distribution coefficient.
A_s = accessible surface area of support.
V_s = volume of the stationary phase.
V_0 = sum of interstitial and pore volume.

The first term on the right of Eq. 5 and e.g. 5a stands for the residual adsorption, which is proportional to the surface area of the support and is inversely related to the volume of mobile phase. From these considerations one would choose supports with a low surface area to prevent residual adsorption. The denominator, $(V_0 - V_s)$, ranges in practice from V_0 (at zero stationary phase loading) to $0.5\ V_0$ (maximum stationary phase loading). If the accessible surface area for the solutes is independent of the amount of stationary phase loading, then residual adsorption will be doubled at maximum loading because of the decrease of V_m. However, in practice the product $K_{adi} \cdot A_s$ approaches zero as soon as bulk solvent (stationary phase) is adsorbed into the pores with a concomitant rise of the other term, which stands for liquid-liquid distribution. The residual adsorption on high surface area silica has been found smaller than 5% when used with highly polar stationary phases[30].

Therefore, one can expect for low-viscosity stationary phases that the amount of stationary phase loading and thus the phase ratio is approximately proportional to the surface area as can be seen from Fig. 3.

Figure 4 demonstrates the change in retention occurring when going from a pure adsorption system to a solvent generated LL system on the same column[24].

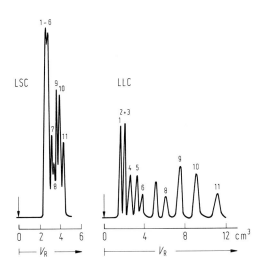

Fig. 4. Change of the retention characteristics at the transition from LSC to LLC by change of the mobile phase composition. Support: Si 100 C-8. Temperature 25.0° C (ref. [24])

	ACN	EtOH	2,2,4-TMP
LSC	90.0	3.0	7.0
LLC	87.90	2.35	9.75
			(more polar liquid phase of system V)

Test compounds in order of elution: 1 Testosterone; 2 Chrysene; 3 Fluorene; 4 Toluene; 5 Propyl Benzene; 6 Butyl Benzene; 7 Hexyl Benzene; 8 Heptyl Benzene; 9 Octyl Benzene; 10 Nonyl Benzene; 11 Decyl Benzene

Residual adsorption can be estimated from:

1. correlation of k_i' with the phase ratio[25, 31, 32],
2. correlation of k_i' with liquid-liquid partition coefficients obtained batchwise[13, 30, 33].

For columns coated by solvent evaporation and dry-packed, the first method can be applied, while the second method is suited for dynamically generated systems. Another indication of residual adsorption can be obtained by comparing the experimentally determined void volume with the sum of the stationary and mobile phase volume[30] calculated according to Method 1 or 2.

5 Selection of Liquid-liquid Systems

Theoretically, all liquid combinations, which separate into two phases can be applied as phase system in LLC. However, in practice the number of liquids is smaller because of practical reasons, e.g. an unacceptably high viscosity, detection response or chemical reactivity. In practice, a difference in specific density of 0.4% is already sufficient to obtain a settled two-phase system in the stock reservoir. Besides these obvious reasons, the polarity difference between the two liquid phases is of crucial importance in order to obtain selective and stable LLC columns. For instance, if the polarity difference is very large, then solutes will be either unretained or too strongly retained. The selectivity on the other hand will generally be high. A polarity difference which is too low favors the solubility of the solutes, yet gives rise to unstable chromatographic systems. Moreover, in this case selectivity will be rather low. Therefore, in practice a determined polarity difference must exist between the two phases in order to obtain moderate retention and stable LLC columns. The polarity of a liquid can be expressed[9] as its so called overall solubility parameter δ.

Table 1. Effect of mutual solubility on polarity difference and liquid-liquid partition coefficient (K_{pi}). $T = 22.0 \pm 0.1°$ C

Phase system (B/A)	Dimethylsulfoxide/ Cyclohexane		Acetonitrile/ Cyclohexane		2-Methoxyethanol/ Cyclohexane	
Solubility (vol.%)	A in B 5.40	B in A 0.39	A in B 13.6	B in A 2.15	A in B 29.4	B in A 0.99
Difference in polarity index P'						
Assuming pure components		7.00		5.60		5.25
Accounting for mutual miscibility		6.59		4.72		3.69
Solute				K_{pi}		
Butylbenzene		0.134		0.338		0.472
Anisole		1.06		0.672		1.42
m-Chloro-nitrobenzene		7.8		4.8		3.12
Acetophenone		8.0		6.6		3.24
Chrysene		9.8		1.71		2.56
p-Chloro-nitrobenzene		10.0		5.4		3.54
Nitrobenzene		11.8		7.7		3.87
p-Nitro-biphenyl		15.4		5.8		3.84
Benzonitrile		15.8		10.1		4.71
o-Chloro-nitrobenzene		19.9		8.7		4.84

When using alkanes as the non-polar liquid $(7 < \delta < 8)$, a stable binary LL-system can be obtained with a more polar liquid with a value greater than 12. For combinations of more polar liquids, e.g. chlorinated hydrocarbons $(9.3 < \delta < 10.2)$, liquids with an overall solubility parameter of at least 17 is required in order to meet the demands of small mutual solubility and a stable LLC system. In short, this points to liquid-liquid combinations with a 50% and 80% difference in the solubility parameter for alkanes and chlorinated alkanes, respectively. Although the polarity difference between such liquid phases is of the same order, the polarity-span of various binary LL systems, might be quite different. The result is that only a limited range of solute polarities can be handled with a particular binary LL system. Fortunately, a large number of binary LL systems, widely differing in polarity range, can be composed so that the LL system can be adapted easily to the solute mixture at hand. The magnitude of the distribution coefficient of a solute, and thus k'_i, is primarily determined by the polarity of the solute and by the polarities of the two phases. The distribution coefficient of polar solutes, for instance, increases with increasing polarity between the two phases. After blending the components of a LL system, the resulting polarity difference is smaller than might be expected from the original polarity difference of the pure components because of their mutual solubilities.

This is shown in Table 1, where Snyder's polarity index[8,9] was used for binary mixtures to calculate the actual polarity difference from their mutual solubilities.

It can be seen, that the partition coefficients of these test solutes increase with increasing polarity difference. For some solutes even a linear relationship is observed

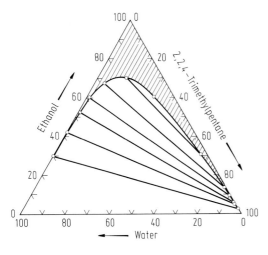

Fig. 5. Ternary phase diagram of water/ethanol/2,2,4-trimethyl-pentane at 25° C

Table 2. Binary liquid-liquid systems and applied separations

Stationary phase	Mobile phase
Normal-phase	

Di-2-cyanoethylether (ODPN) — *2,2,4-Trimethyl-pentane, di-n-butylether, n-heptane or n-hexane*

– hydroxylated aromatics[11], polychlorinated dimethoxy-benzenes and chlorinated aromatic amines[14], alkylated phenol derivatives[29], chlorinated pesticides[31, 51, 52], urea herbicides[46], 2,4-dinitro-phenyl hydrazone derivatives of 17-ketosteroids[49], phthalate esters[53] and polyaromatic hydrocarbons[52]

1,2,3-tris(2)-(cyanoethoxy)- propane (Fraktonitrile III) — *n-Hexane*

– polyaromatic hydrocarbons and m-oligo-phenylenes[47] and 2,4,-dinitro-phenyl hydrazone derivatives of carbonyl compounds[48]

Formamide, water (+buffer) — *n-Hexane, dichloromethane or ethyl acetate*

– free steroids[15, 25, 44], methylated phenols[15, 44] and urea herbicides[18, 54]

Dimethylsulfoxide acetonitrile or γ-butyrolacton — *Cyclohexane*

– polyaromatic hydrocarbons[21, 30, 43], benzoate esters[30] and mono- and polychlorinated nitrobenzenes[30]

Reversed-phase	

Tri-n-butylphosphate, tri-n-octylphos- phine or tri-n-octylamine — *Water + mineral acids or Pi-buffer*

– derivatives of phenyl carboxylic acids[23, 50, 57], phenol, benzene- and naphthalene sulfonic acids[50] and urinary free porfyrines[58–60]

Di-(2-ethylhexyl)-phosphoric acid — *Water + mineral acids*
– radionuclides[37]

n-Octanol or oleyl alcohol — Water + buffers
– 1,4-benzodiazepines and other N-containing compounds[26, 32]

Table 3. Multicomponent liquid-liquid systems and applied separations

Stationary phase	Mobile phase

Normal-phase

Dimethylsulfoxide or N,N-dimethylform-   *i-Octane or tetra (or dichloromethane) +*
amide + Carbowax or Carbowax 200–400   *i-octane*

– aromatic alcohols, free steroids and benzodiazepines[31], polycyclic aromatic hydrocarbons[61,62], polymers of alkylphenyl glycol[63]

Polar phase of   *Apolar phase of*
water (plus/or acetonitrile)/ethanol or n-pentanol/i-octane

– metal chelates[12], free steroids[13,25,55], glycosides[19], polycyclic aromatic hydrocarbons[24], thioridazines and metabolites[56]

Polar phase of   *Apolar phase of*
water/acetic acid(+aceton)/di-i-propylether or n-heptane

– dansyl amino acids[15] and urinary free porfyrines[22]

Polar phase of   *Apolar phase of*
water(+buffer)/methanol or 2-chloro-ethanol or propanol-2/dichloromethane or chloroform

– dansyl amino acids[15], free steroids[17], urea herbicides[18], carboxylic- and sulfonic acids, nucleobases and nucleosides[25], barbiturates and benzodiazepines[20,54]

Polar phase of   *Apolar phase of*
formamide/chloroform/n-hexane   *formamide/chloroform/n-hexane*

– barbiturates[25,43] and chlorinated phenols[43]

0.1 or 0.2N Sulfuric acid   *n-Bentanol or t-pentanol/chloroform/*
                                      cyclohexane

– (hydroxy)-phenyl carboxylic acids, alifatic mono- and di-carboxylic acids[65,66])

Reversed-phase

Apolar phase of   *Polar phase of*
ethanol/acetonitrile/i-octane   *ethanol/acetonitrile/i-octane*

– alkylbenzenes and polycyclic aromatic hydrocarbons[24]

---

between K_{pi} and the difference in polarity index between the two phases. However, for other solutes such a strong correlation is not found, which indicates that selectivity changes occur due to enhanced specific interactions when changing the nature of one of the phases. The strategy of keeping the polarity difference constant, yet changing the nature of one of the phases has also been found valuable in order to affect the selectivity in other forms of chromatography[9].

A very elegant way to adjust the polarity difference between two phases and influencing the selectivity, can be accomplished by applying a ternary system[13,33,34]. The three component liquids are: a non-polar, a polar and an intermediately polar one, which is completely miscible with the first two. Depending on the amount of the third component, the mutual solubility of the non-polar and polar liquid is either increased or decreased with a concomitant decrease or increase of polarity difference between the resulting phases.

Figure 5 shows a ternary phase diagram consisting of water, ethanol and 2,2,4-tri-methyl-pentane(i-octane). Blending the liquids in the ratio given by the overall com-position on one of the tie lines will cause the system to separate into two phases, the compositions of which are given by the end-points of the tie line. At point P, also called "critical mixing point", only one phase exists. All overall compositions in the shaded part above the two-phase area, represent one phase. From Fig. 5 it can be seen, that a great number of LL systems can be prepared, covering a broad scale of polarity differences.

Apart from the selection of liquids to compose LL systems use can be made of chemical equillibria in the distribution process; these include pH, ion-pairing, complexation, etc. for adjusting the retention and selectivity (see pp. 201–214).

Tables 2 and 3 describe several liquid-liquid systems which have been applied to date in normal- and reversed-phase modes.

6 Performance of LLC Systems

The performance of LLC systems compared to commonly used adsorption systems has been investigated by several authors[16, 17, 35–37]. The theoretical plate height in LLC is well described by a four-term polynomial as derived by Huber[38]. Depend-ing on the nature of the stationary phase (viscosity in particular), the diffusion coefficient in the pores of the particles in LLC might differ markedly with that in the mobile phase. This in contrast to LSC, where the intra-particle volume contains the mobile phase as well. Therefore, the decrease in diffusion coefficient will only result from the obstructive factor characteristic for the support in question.

Since diffusion is the driving force in the speed of mass transfer, differences in efficiency can be expected between LSC and LLC systems, yet between LLC systems as well. Moreover, the performance of LLC systems is also affected by the uniformity of the stationary phase layer. As mentioned before, the stationary phase must be spread out over the surface of the solid support as a thin layer, preferentially of equal thickness. This enhances a rapid mass transfer between the two liquid phases. The presence of thick layers or droplets of stationary phase in the column detoriates the column efficiency. This might happen if the support is coated by the evaporation or precipitation technique with a highly viscous station-ary phase. The risk of a non-homogeneous film does not occur with the solvent generated coating technique. Karger et al.[29] investigated the influence of stationary phase viscosity on the theoretical plate height. They found, that the theoretical plate height doubles roughly with a ten-fold increase of stationary phase viscosity. Crombeen et al.[30] measured the efficiency of adsorption and low-viscous solvent generated LLC systems on the same column.

It must be noted that the in situ coating technique allows the unambigeous comparison between LSC and LLC since the geometry of the column remains identi-cal.

Figure 6 shows the $H/<v>$ curves for a pure LSC and a LLC system at two different phase ratios. At lower velocities the adsorption system is somewhat better whereas the reverse holds for higher velocities.

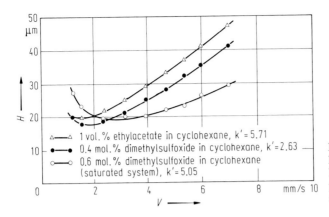

Fig. 6. Theoretical plate height (H) versus linear velocity (v). Solute: Acetophenone. Conditions and support: see Fig. 1 (ref. [30])

Others[23] have also shown that with low-viscous LLC systems (up to 4 cP for the stationary phase) similar efficiencies can be obtained. Some authors state, that the linear dynamic range of the distribution isotherm of LLC systems is superior to that in LSC systems. A large dynamic range is favorable for preparative separations. However, until now this assumption has not been verified by proper experimental support.

In a previous section it was mentioned, that the stock reservoir and the column must be well thermostated to prevent spurious mutual mixing of the phases. Unfortunately an axial and radial temperature gradient is generated in the column by frictional heat from the mobile phase flow[39].

This phenomenon acts upon the performance and retention of LSC[40-42] in general and LLC in particular, since the mutual miscibility of the phases in the latter case is increased by raising the temperature. Under these conditions the phase ratio in axial and radial direction of the column might change, which harms the column efficiency. However, in practice these effects were found to be of minor importance[43] with the commonly used 5 μm particles. However, these phenomena will become more prominent when using progressively smaller particle diameters.

7 Determination of Liquid-liquid Partition Coefficients from Retention Data

As mentioned earlier the partition coefficients of solutes in LL systems can be determined from retention data providing the phase ratio is known and residual adsorption is absent. Unfortunately, the phase ratio in the column is usually unknown unless solvent evaporation has been used to coat the support[36,44]. However, the retention of solutes with known partition coefficients allows the estimation of the phase ratio.

The retention volume of a solute i in a pure LL system is given by:

$$V_{Ri} = V_m + K_{pi} \cdot V_s \qquad (6)$$

where: V_{Ri} = retention volume of solute i.

$\quad\quad V_m$ = volume of mobile phase.

$\quad\quad K_{pi}$ = liquid-liquid partition coefficient of solute i.

$\quad\quad V_s$ = volume of stationary phase.

By plotting V_{Ri} versus the known partition coefficients of some test solutes V_s can be calculated from the slope of this correlation and V_m from the intercept. The validity of these values can be verified [24, 30] by comparing the sum of V_s and V_m with the experimentally determined void volume (V_0). If residual adsorption occurs, then $V_s + V_m$ will be greater than V_0. The equivalent of Eq. 6 expressed in retention time (t_{Ri}) and phase ratio (q) is obtained by dividing Eq. 6 by the volumetric velocity (V_v) and rewriting V_v in the second term on the right:

$$t_{Ri} = t_{Ro} + t_{Ro} \cdot K_{pi} \cdot q \tag{7}$$

where: t_{Ro} $\quad\quad$ = retention time of unretained.

$\quad\quad q = V_s/V_m$ = phase ratio.

It can be noted, that using Eq. (6) or (7) has the advantage that an absolutely unretained solute is not required. Such a component is not always readily found in liquid-liquid systems.

The standard error of estimate for data fitted according to Eq. (6) (or (7)) is strongly influenced by the residual adsorption of just one solute. In practice [30] this means, that a correlation coefficient of at least 0.99 must be found for five data points to exclude considerable residual adsorption (5% of highest capacity ratio). According to the equations mentioned above so-called dynamic partition coefficients can be calculated by interpolation between proper standards. As noted

Table 4. Accuracy of determining liquid-liquid partition coefficients (K_{pc}) from retention data using polycyclic aromatic hydrocarbons as reference compounds. ΔK_i = deviation(%) from static liquid-liquid partition coefficients

Compound	Phase system								
	1			2			3		
	K_{pi}	K_{pci}	ΔK_i	K_{pi}	K_{pci}	ΔK_i	K_{pi}	K_{pci}	ΔK_i
Anisole	2.19	2.16	−1.4	2.29	2.23	−2.6	2.90	2.80	−3.4
m-Chloro-nitrobenzene	7.8	7.9	+1.3	4.8	4.7	−2.1	8.4	8.1	−3.6
Acetophenone	8.0	8.4	+5.0	6.6	7.5	+13.6	9.7	10.0	+3.1
p-Chloro-nitrobenzene	10.0	10.1	+1.0	5.4	5.3	−1.9	10.2	9.7	−4.9
Nitrobenzene	11.8	12.1	+2.5	7.7	7.9	+2.6	12.6	12.5	−0.8
p-Nitro-biphenyl	15.4	15.6	+1.3	5.8	5.8	0	18.6	17.6	−5.4
Benzonitrile	15.8	16.6	+5.1	10.1	10.4	+3.0	16.4	16.0	−2.4
o-Chloro-nitrobenzene	19.9	20.6	+3.5	8.7	8.8	+1.1	17.4	18.3	+5.2

1: Dimethylsulfoxide/Cyclohexane, 2: Acetonitrile/Cyclohexane, 3: γ-Butyrolactone/Cyclohexane

before this way of estimating partition coefficients does not put such stringent requirements on the purity of solutes in contrast to the batch extraction method and its variants. For instance, free steroids[33] and chlorinated pesticides[45] thus have been determined in a ternary system with great precision (<1% deviation). The accuracy of dynamically determined partition coefficients of some solutes in three different binary systems[30] has been compiled in Table 4. The deviation is not greater than 5% except for acetophenone, which shows a deviate behavior in the system acetonitrile/cyclohexane.

Another feature, which is particularly interesting for workers in the field of pharmacology, is the use of n-octanol as stationary phase coated on a reversed phase support. With these columns coated dynamically in situ log P values can be reliably estimated either by the use of oleyl alcohol[32] or n-octanol[26].

8 Applications of LLC

At the beginning of the development of HPLC, normal phase LLC was the most popular technique. In that time liquids commonly used as stationary phase in gas chromatography were applied as stationary phases in LLC. Most of these systems used di-2-cyanoethyl ether (ODPN), 1,2,3-tris(2'-cyanoethoxy)propane (Fraktonitrile III) or polyethylene glycol (Carbowax 200–400) as stationary phase in combination with some alkane or di-n-butylether as the mobile phase.

The above mentioned stationary phases are rather viscous at ambient temperature (10 cP or more) and were usually coated on the solid support by the solvent evaporation technique and then dry-packed in the column[28, 29, 35, 36, 46–50]. These systems

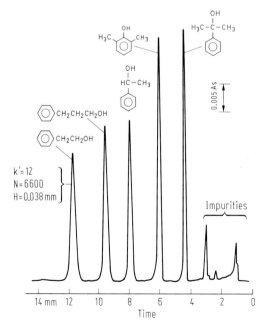

Fig. 7. Liquid-liquid chromatography with porous microspheres. 30% by weight of stationary phase (di-2-cyanoethylether) on support (silicagel). Temperature 27° C (ref. [11])

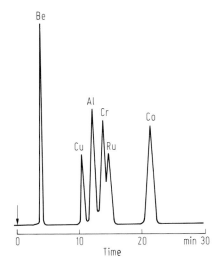

Fig. 8. Separation of six metal acetylacetonates. Support: diatomaceous earth (ref. [12])

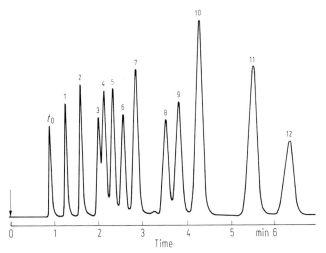

Fig. 9. Rapid separation of twelve polycyclic aromatic hydrocarbons. Phase system: dimethylsulfoxide/cyclohexane. Conditions and support: see Fig. 1. Solutes: 1 Acenaphthene; 2 Fluorene; 4 Anthracene; 5 Phenanthrene; 6 Pyrene; 7 Fluoranthene; 8 Benz[a]anthracene; 9 Chrysene; 10 Benzo[a]Pyrene; 11 Benzo[g,h,i]perylene; 12 Dibenz[a,h]anthracene (ref. [43])

have been successfully applied for the separation of various types of solutes like chlorinated pesticides [31, 51, 52], phthalate esters [53], phenols [29], polycyclic and hydroxylated aromatics [11, 47, 52], urea herbicides [46] and 2,4-dinitro-phenyl-hydrazone derivatives of ketones and aldehydes [51, 52].

The performance of the ODPN/n-hexane system for the separation of aromatic alcohols is shown in Fig. 7. In later years less viscous stationary phases such as water and formamide combined with an alkane or dichloromethane were found

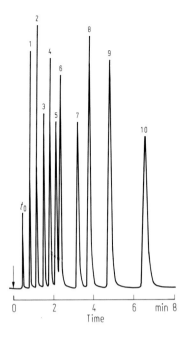

Fig. 10. Separation of some mono- and polychlorinated nitrobenzenes. Phase system: γ-butyrolacton/cyclohexane. Conditions and support: see Fig. 1. Solutes: 1 Chlorobenzene; 2 Biphenyl; 3 3,4,5-trichloro-Nitrobenzene; 4 2,3,4,5-tetrachloro-Nitrobenzene; 5 2,3,5-trichloro-Nitrobenzene; 6 2,4,5-trichloro-Nitrobenzene; 7 m-chloro-Nitrobenzene; 8 p-chloro-Nitrobenzene; 9 Nitrobenzene; 10 o-chloro-Nitrobenzene (ref. [30])

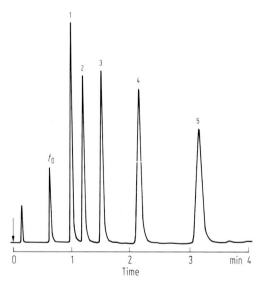

Fig. 11. Rapid separation of some esters. Phase system: acetonitrile/cyclohexane. Conditions and support: see Fig. 1. Solutes: 1 Butylbenzoate; 2 Propylbenzoate; 3 Ethylbenzoate; 4 Methylbenzoate; 5 Benzylacetate (ref. [30])

useful for the separation of phenols[15, 44], urea herbicides[18, 54] and free steroids[15, 44]. The LLC columns were usually prepared by an in situ coating technique. The same procedure or dry-packing after pre-coating was used by Huber to make LLC columns with a ternary phase, composed of water, ethanol and i-octane for

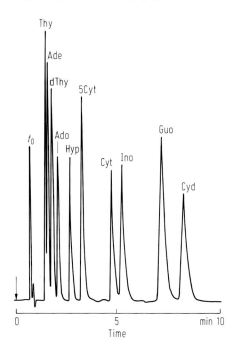

Fig. 12. Separation of nucleobases and nucleosides. Support: Hypersil. Phase system: 0.05M formic acid and 0.05M ammonium formate in water/methanol/chloroform = 40:250:710(v/v/v). Column and reservoir temperature 35 and 22° C respectively. Solutes: Thymine (Thy); Adenine (Ade); Thymidine (dThy); Adenosine (Ado); Hypoxanthine (Hyp); 5-methyl-Cytosine (5Cyt); Cytosine (Cyt); Inosine (Ino); Guanosine (Guo); Cytidine (Cyd) (ref. [25])

the separation of free steroids[55], thioridazines and metabolites[56] and metal chelates[12].

Figure 8 shows an example of the separation of metal chelates on this system. With the system water, n-pentanol, acetonitrile and i-octane, efficient separations for a number of digitalis glycosides have been obtained[19].

Recently it has been demonstrated, that very efficient binary and multicomponent LL systems can be solvent-generated on prepacked columns[25, 30, 44]. These systems, consisting of dimethylsulfoxide, γ-butyrolactone or acetonitrile combined with cyclohexane were found very useful for the separation of polycyclic aromatic hydrocarbons[30, 43], chlorinated nitrobenzenes and benzoate esters[30] as can be seen from Figs. 9–11.

The solvent generated LL system composed of aqueous formate buffer, methanol and chloroform is very well suited for a fast and selective separation of nucleobases and nucleosides (Fig. 12).

Until now, reversed-phase LL has been used little compared to normal phase LLC; however, there are some interesting applications. Thus tri-n-octylamine and tri-n-octylphosphine oxide have been applied as stationary phase combined with aqueous buffers for the separation of phenols[50], aromatic carboxylic acids[57] and aromatic sulfonic acids[50]. Tri-n-butyl phosphate/aqeous buffer has been applied for the separation of uroporfyrine isomers[58–60]. n-Octanol (or oleyl alcohol)/water(+buffer) has been tested for the estimation of log P values of a great variety of compounds[26, 32].

9 References

1. Locke, D.C., Martire, D.E.: Analyt. Chem. 39, 921–925 (1967)
2. Barton, A.F.M.: Chem. Rev. 75, 731–753 (1975)
3. Pierotti, R.A.: Chem. Rev. 76, 718–726 (1976)
4. Tijssen, R., Billiet, H.A. and Schoenmakers, P.J.: J. Chrom. 122, 185–203 (1976)
5. Karger, B.L., Snyder, L.R.: J. Chrom. 125, 71–88 (1976)
6. Tewari, Y.B., Martire, D.E., Wasik, S.P. and Miller, M.: J. Sol. Chem. 11, 435–445 (1982)
7. Rohrschneider, L.: Analyt. Chem. 45, 1241–1247 (1973)
8. Snyder, L.R.: J. Chrom. 92, 223–230 (1974)
9. Snyder, L.R.: J. Chrom. Sci. 16, 223–234 (1978)
10. Rekker, R.F.: J. Chrom. 300, 109–125 (1984)
11. Kirkland, J.J.: J. Chrom. Sci. 10, 593–599 (1972)
12. Huber, J.F.K. and Kraak, J.C., Veening, H.: Analyt. Chem. 44, 1554–1559 (1972)
13. Huber, J.F.K. and van der Linden, R. and Ecker, E. and Oreans, M.: J. Chrom. 83, 267–277 (1973)
14. Kirkland, J.J. and Dilks, C.H. Jr.: Analyt. Chem. 45, 1778–1781 (1973)
15. Engelhardt, H., Asshauer, J., Neue, U. and Weigand, N.: Analyt. Chem. 46, 336–340 (1974)
16. Berry, L.V. and Engelhardt, H.: J. Chrom. 95, 27–38 (1974)
17. Parris, N.A.: J. Chrom. Sci. 12, 753–757 (1974)
18. Gonnet, C. and Rocca, J.L.: J. Chrom. 109, 297–303 (1975)
19. Lindner, W. and Frei, R.W.: J. Chrom. 117, 81–86 (1976)
20. Gonnet, C., Rocca, J.L.: J. Chrom. 120, 419–433 (1976)
21. Durand, J.P. and Petroff, N.: J. Chrom. 190, 85–95 (1980)
22. Nordlov, H., Jordan, P.M., Burton, G. and Scott, A.I.: J. Chrom. 190, 221–225 (1980)
23. Wahlund, K.-G. and Edlen, B.: J. Liq. Chrom. 4, 309–323 (1981)
24. Huber, J.F.K., Pawlovska, M., Markl, P.: Chromatographia 17, 653–663 (1983)
25. Crombeen, J.P., Heemstra, S. and Kraak, J.C.: J. Chrom 282, 95–106 (1983)
26. Mirrlees, M.S., Moulton, S.J., Murphey, C.T. and Taylor, P.J.: J. Med. Chem. 19, 615–619 (1976)
27. Unger, K.K. (ed.): Porous Silica, Journal of Chromatography Library, vol. 16, Elsevier Pub. Comp., Amsterdam, Oxford, New York 1979, pp. 37, 48, 240
28. Halasz, I., Engelhardt, H., Asshauer, J. and Karger, B.L.: Analyt. Chem. 42, 1460–1461 (1970)
29. Karger, B.L., Conroe, K. and Engelhardt, H.: J. Chrom. Sci. 242–250 (1970)
30. Crombeen, J.P., Heemstra, S., Kraak, J.C.: Chromatograpia 19, 219–224 (1984)
31. Viricel, M. and Lemar, M.: J. Chrom. 116, 343–352 (1976)
32. Hulshoff, A. and Perrin, J.H.: J. Chrom. 129, 263–276 (1976)
33. Huber, J.F.K., Meijers, C.A.M. and Hulsman, J.A.R.J.: Analyt. Chem. 44, 111–116 (1972)
34. Huber, J.F.K.: J. Chrom. Sci. 9, 72–76 (1971)
35. De Stefano, J.J. and Beachell, H.C.: J. Chrom. Sci. 8, 434–438 (1970)
36. Rossler, G. and Halasz, I.: J. Chrom. 92, 33–46 (1974)
37. Horwitz, E.P., Bloomquist, C.A.A. and Delphin, W.H.: J. Chrom. Sci. 15, 41–46 (1977)
38. Huber, J.F.K.: Ber. Bunsenges. Phys. Chemie 77, 179–184 (1977)
39. Poppe, H., Kraak, J.C., Huber, J.F.K. and van den Berg, J.H.M.: Chromatographia 14, 515–523 (1981)
40. Snyder, L.R.: Principles of Adsorption Chromatography, Marcel Dekker Inc., New York 1968, pp. 335–343
41. Kowalczyk, J.S. and Herbut, G.: J. Chrom. 196, 11–20 (1980)
42. Sisco, W.R., Gilpin, R.K.: J. Chrom. Sci. 18, 41–45 (1980)
43. Crombeen, J.P., Heemstra, S. and Kraak, J.C.: J. Chrom. 286, 119–129 (1984)
44. Engelhardt, H. and Weigand, N.: Analyt. Chem. 45, 1149–1154 (1973)
45. Huber, J.F.K., Alderlieste, E.T., Harren, H. and Poppe, H.: Analyt. Chem. 45, 1337–1343 (1973)
46. Kirkland, J.J.: J. Chrom. Sci. 7, 7–12 (1969)

47. Randau, D. and Schnell, W.: J. Chrom. 57, 373–381 (1971)
48. Papa, L.J. and Turner, L.P.: J. Chrom. Sci. 10, 747–750 (1972)
49. Fitzpatrick, F.A. and Siggia, S., Dingman, J., Sr.: Analyt. Chem. 44, 2211–2216 (1972)
50. Kraak, J.C.: Thesis, Amsterdam 1974
51. Waters, J.L., Little, J.N. and Horgan, D.F.: J. Chrom. Sci. 7, 292–296 (1969)
52. Vermont, J., Deleuil, M., De Vries, A.J. and Guillemin, C.L.: Analyt. Chem. 46, 1329–1337 (1975)
53. Fishbein, L. and Albro, P.W.: J. Chrom. 70, 365–412 (1972)
54. Rassi, Z.El, Gonnet, C. and Rocca, J.L.: J. Chrom. 125, 179–201 (1976)
55. Huber, J.F.K., Hulsman, J.A.R.J. and Meijers, C.A.M.: J. Chrom. 62, 79–91 (1971)
56. Muusze, R.G. and Huber, J.F.K.: J. Chrom. 83, 405–420 (1973)
57. Stuurman, H.W. and Wahlund, K.-G.: J. Chrom. 218, 455–463 (1981)
58. Mundschenk, H.: J. Chrom. 37, 431–452 (1968)
59. Mundschenk, H.: J. Chrom. 38, 106–119
60. Mundschenk, H.: J. Chrom. 40, 393–409 (1969)
61. Jentoft, R.E. and Gouw, T.H.: Analyt. Chem. 40, 923–927 (1968)
62. Jentoft, R.E. and Gouw, T.H.: Analyt. Chem. 40, 1787–1790 (1968)
63. Huber, J.F.K., Kolder, F.F.M. and Miller, J.M.: Analyt. Chem. 44, 105–110 (1972)
64. Rickett, F.E.: J. Chrom. 60, 356–360 (1972)
65. Morot-Gaudry, J.F., Nicol, M.Z. et Jolivet, E.: J. Chrom. 100, 206–210 (1974)
66. Morot-Gaudry, J.F., Fiala, V., Huet, J.C. and Jolivet, E.: J. Chrom. 117, 279–284 (1974)

Ion Pair Liquid Chromatography

B.A. Persson
Department of Bioanalytical Chemistry, AB Hässle, S-431 83 Mölndal/Sweden

B.L. Karger
Barnett Institute of Chemical Analysis and Department of Chemistry,
Northeastern University, Boston, MA 02115/U.S.A.

1 General Introduction

Modern liquid chromatography has been developing very fast during the last ten years by significant progresses in both column technology and instrumentation. However, the increased interest and efforts in chemical manipulation of the phase systems have also to a great extent contributed to this tremendous development.

Bonded phase liquid chromatography is today a dominant technique not least in the rapidly growing field of separation of bioactive molecules. These are to their nature most often ionic or ionizable and require careful choice of phase system. Separation of bioactive substances has become one of the important fields of application for ion pair liquid chromatography.

2 Principles for Ion Pair Extraction

Ionic compounds can be extracted from an aqueous solution into an organic phase as ion pairs

$$A^+_{aq} + B^-_{aq} = AB_{org} \tag{1}$$

A and B may be aprotic charged ions or ionized protolytes. The extent of extraction of AB is dependent on the nature of the organic phase and on the properties and concentrations of A and B (Fig. 1). Equation (1) represents the simple model for ion pair extraction. Side reactions for A, B and AB in the aqueous and organic phases will influence the overall distribution of the compounds [1]. Such side reactions include protolysis of A and B, association and ion-pair formation in the aqueous phase. In the organic phase dissociation and association of the ion pairs may occur and the distribution may also be influenced by adduct formation. The equilibrium given above (1) may be expressed by an extraction constant

$$K_{ex} = (AB)_{org}/((A^+)_{aq}(B^-)_{aq}) \tag{2}$$

which is a formal representation of the extraction process. The distribution ratio for A in such a simple process is given by

$$D_A = (AB)_{org}/(A^+)_{aq} = K_{ex}(B^-)_{aq} \tag{3}$$

More generally the distribution ratio is given in terms comprising total concentra-

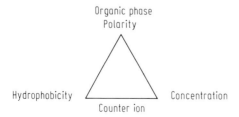

Additional: hydrogen bonding between ions and solvent
        ionization degree in aqueous phase

Fig. 1. Factors influencing ion pair extraction

tions in the two phases

$$D_A = C_{Aorg}/C_{Aaq} = K_{ex}^x \cdot C_{Baq} \tag{4}$$

where K_{ex}^x is a conditional extraction constant including all side reactions.

The hydrophobicity of the components forming the ion pair determines to a great extent the distribution ratio, however the nature of the two species may also influence distribution, e.g. if hydrogen bonding or steric effects are superimposed onto the electrostatic interaction.

The properties of the organic phase is, of course, significant in the control of the extraction of ion pairs. Alkanes are inert and lack extraction capacity for polar ion pairs, they can only act as a diluent for solvating organic agents. The polarity or dielectric constant of the solvents often controls the ion pair extraction ability but hydrogen accepting or donating properties will in many instances also be of importance.

3 Modes of Ion Pair Chromatography

In liquid chromatography the ion pair distribution technique can be used in both the liquid-liquid and liquid-solid mode. The ion pairs are either transported along the column in a liquid mobile organic phase or as in reversed-phase systems retained by distribution to a liquid stationary organic phase or by adsorption onto a solid phase [2, 3].

3.1 Normal Phase

3.1.1 Liquid-Liquid Chromatography

Ion pair liquid-liquid or partition chromatography is not much used today, being largely replaced by bonded phases because of the latter's ease of manipulation. The counter ion in the aqueous phase is coated onto a solid phase (support) in the normal-phase mode or is present in the mobile aqueous phase in reversed-phase systems. In both cases the counter ion forms ion pairs in the organic phase with ions of opposite charge imposing elution and retention, respectively [4]. Close temperature control is normally required for stability and prior to use the mobile phase must be equilibrated with the stationary liquid phase. The major advantages offered by liquid-liquid systems are the reproducibility of phase systems and the possibility to predict retention from batch experiment or physical-chemical data, this being to some extent impaired by the lack of inert supports for the stationary phase. The disadvantage is the difficulty to conduct gradient elution, particularly the slower equilibration times.

Normal-phase partition chromatographic retention of a cation A^+ is described by

$$k' = V_s/(K_{ex} \cdot (B^-)_{aq} \cdot V_m) \tag{5}$$

provided that it is a pure liquid-liquid distribution process and side reactions can be disregarded.

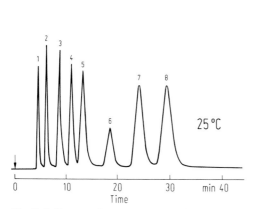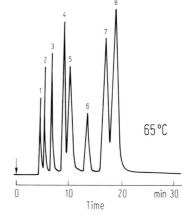

Fig. 2. Influence of temperature on the separation time and resolution of aromatic carboxylic acids[6]. Sample: (1) 4-aminobenzoic acid; (2) 4-hydroxybenzoic acid; (3) 2,4-dihydroxybenzoic acid; (4) benzoic acid; (5) 4-nitrobenzoic acid; (6) 2-hydroxybenzoic acid; (7) 4-methylbenzoic acid; (8) cinnamic acid. Column: trioctylamine, 0.04 g per gram of solid support; 0.05 M $HClO_4$, pH 1.5

In 1965 kiselguhr was used as support by Schill and co-workers[5] who found good correspondence with bulk equilibria. With the introduction of modern liquid chromatography Kraak and Huber[6] showed efficient separations with microparticles of kiselguhr as a support with low surface area and adsorption activity (Fig. 2). Later microparticles of silica with different surface area were introduced as solid phase[7-12] and also diol modified silica[13].

The stationary aqueous phase can be mixed with the support prior to packing of the column[6,14] or, more conveniently injected in portions or pumped through the packed column pure[9,11,12] or diluted with organic solvent[7,8]. The degree of coating is dependent on the surface area of the solid phase. A high degree of coating is desirable in order to promote a liquid-liquid distribution and to minimize the influence of the support. The aqueous phase is often in the range of 20–100% of the solid phase, it is buffered to a suitable pH, where the compounds are in ionic form, and contains the counter ion in a concentration of 0.01–1 M to ensure constant distribution conditions. Perchlorate[7-11] and sulfonates[12] have been employed as counter ions for the separation of ammonium ions (protonated amines) and aliphatic quaternary ammonium ions as counter ions for the separation of carboxylates[15] and anions of other organic acids[4]. The use of strong acid salts as counter ions gives a free choice of pH which then can be adapted to the protolytic properties of solutes.

An option with normal-phase ion pair partition chromatography is to use a counter ion with high detector response in the stationary phase. The ion pairs in the mobile phase passing through the detector will then give a response independent of the solute itself. This has been shown with napthalene sulfonate[12,16] and picrate[14] as counter ions for ammonium ions and with N,N-dimethylprotriptyline as counter ion for different anionic compounds[13] (Fig. 3).

Mobile phases are often halogenated hydrocarbons such as dichloromethane and chloroform in mixtures with alcohols such as butanol and pentanol, sometimes

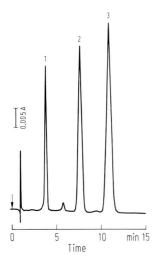

Fig. 3. Separation of aliphatic carboxylic acids[13]. Solid phase: LiChrosorb DIOL, 10 μm. Liq stationary phase: N,N-dimethylprotriptyline in phosphate buffer pH 6.2. Mobile phase: Chloroform + 1-propanol (9 + 1 by volume). Wavelength: 254 nm. Sample: (1) propionate, (2) acetate, (3) formiate

with alkanes as nonsolvating diluents. The combination of mobile organic phase and counter ion in the stationary phase will determine the retention for a specific solute. A more lipophilic counter ion with a higher extraction constant will require a mobile phase with less ion pair solvating ability. The mobile phase is equilibrated with the aqueous phase and should contain the counter ion in low or very low concentration. This latter point is only critical using detector sensitive counter ions when too high a background is not desirable. A more careful choice of solvent components in the mobile phase is then required [16].

The sample components such as amines and carboxylic acids are most often injected in uncharged form and not necessarily as ion pairs with the counter ion in the system. The long-term stability is often very good provided that the system is thermostatted and the liquid phases are properly equilibrated. If the phases are not equilibrated, stripping off of stationary phase by dissolution in the mobile phase or elution of droplets due to overloading may occur.

3.1.2 Liquid-Solid Chromatography

Normal-phase ion pair chromatography in the liquid-solid mode is a very useful technique although not widely employed. The border between this kind of chromatography and liquid-liquid or partition chromatography is however not very distinct. As mentioned above there is often no pure liquid-liquid distribution process but adsorption effects from the support are also involved. By increasing the content of aqueous solution in the mobile organic phase the extent of dynamic coating of the solid phase will increase. However, this dynamic coating will not make a defined bulk stationary phase of the same amount as one separately applied. In studies on ion pair adsorption chromatography on silica typical mobile phases have consisted of low amounts of aqueous perchloric acid in methanol added to dichloromethane [17]. Other acids can also be used, other alcohols, acetonitrile or similar solvents and, as the bulk phase, other organic solvents such as ethers [18, 19]. These ion pair systems are durable and have been used for the separation of amines in biological samples as an alternative to systems containing organic bases in the mobile

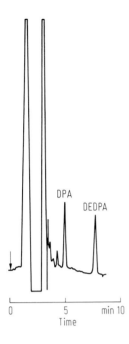

DPA

DEDPA

0   5   min 10
Time

Fig. 4. Disopyramide (DPA) and its N-deisopropyl metabolite (DEDPA) from a plasma sample[17]. Solid phase: LiChrosorb SI 60, 7 μm. Mobile phase: 1 M (aq.) $HClO_4$ + methanol + dichloroethane (1 + 9 + 90, v/v). Wavelength: 265 nm. Sensitivity: 0.01 a.u.f.s. Sample: 250 μl of an extract from 1 ml of plasma spiked with 250 pmol/ml (90 ng/ml) of each amine

organic phase. It has been possible to inject directly an aliquot of the organic phase from an extracted plasma or urine sample onto the column, thus simplifying the assay (Fig. 4).

Separation systems based on ion pair adsorption on silica have also been utilized for the resolution of enantiomers. Pettersson and Schill used (+)-10-camphorsulfonic acid as ion pairing agent for the chiral separation of amino alcohols[20]. For the chromatographic separation of enantiomers of acids quinine and quinidine were used as chiral counter ions[21, 22]. Mobile phases have included dichloromethane with addition of pentanol or similar polar solvent with control of the water content. Separation factors up to 1.6 have been achieved with these systems.

3.2 Reversed-Phase

3.2.1 Liquid-Liquid Chromatography

Reversed-phase ion pair chromatography in the liquid-liquid mode has not been used very much. The advantages have not been obvious and the requirements on temperature control and equilibrated mobile phases have restricted its use. Wahlund and Beijersten used dynamic coating of butyronitrile and pentanol onto bonded phases by passing saturated aqueous mobile phases through the columns[23, 24]. Measuring the adsorption isotherm showed that when approaching complete saturation of pentanol in the mobile phase a strong increase in the amount of adsorbed alcohol was obtained[25]. The retention mechanism was treated as a combination of adsorption to the solid phase and liquid-liquid partition to the bulk phase of adsorbed pentanol.

The systems with completely saturated mobile phases seem to offer limited stability and this is even more true if the support is loaded with pentanol by the injection technique since droplets of organic phase occasionally will occur in the eluting mobile phase. However, tributylphosphate can be applied as a stationary phase by this technique and such systems have been used for separation of biogenic amines and their metabolites [26, 27]. Tributylphosphate is a strong hydrogen bond acceptor with low solubility in the aqueous phase compared to pentanol. The systems with tributylphosphate are well suited for separation of extracts from biological samples and show good stability without close temperature control.

The retention properties of these systems can be regulated by the pH and counter ion concentration of the aqueous mobile phase and also by addition of an organic solvent such as methanol as shown by Tjaden and co-workers [26, 27]. The authors indicate that supersaturation of the mobile phase may be responsible for stability problems encountered in liquid-liquid chromatography.

3.2.2 Liquid-Solid Chromatography

Reversed-phase chromatography on bonded phase silica packings is today the dominant separation technique. In the pH region used, 2–8, organic bases are mainly present in cationic form and retained by ion pair adsorption. Acids are either uncharged or present in anionic form and ion pair separation systems for these often use quaternary ammonium compounds as counter ions allowing a flexible pH selection. Tetrabutylammonium (TBA^+) and other symmetrical quaternaries are available [28, 29] but also other with a long alkyl chain e.g. cetrimide [30, 31]. Ionized amines will act as counter ions for anionic compounds but the useful pH range is then more limited. The mobile aqueous phase, buffered to a sufficient pH stability, contains a counter ion e.g. TBA^+, in a concentration range of 0.01–0.05 M. The pH is chosen so that the solutes are present mainly in ionic form but for selectivity reasons other pH's may be considered. Acetonitrile, methanol or another organic solvent is used to regulate the retention which of course is also influenced by the type and concentration of the counter ion.

Separation of cationic compounds, often containing amino groups, has attracted most interest. Counter ions, within a wide range of hydrophobicity, are used depending on the properties of the sample components. For hydrophobic cationic solutes hydrophilic counter ions from pH buffering substances or from strong acids, e.g. dihydrogen phosphate, trifluoroacetate, perchlorate and bromide may be used to provide suitable retention [32–34].

Organic sulfonates and sulfates are frequently used as counter ions. Hydrophobic anions such as dodecyl-, decyl- and octylsulfate are often needed for cationic hydrophilic solutes to provide a sufficient retention [35–37] but these species have also found more general use. The counter ion concentration is usually in the range of 0.001–0.05 M. For a specific solute appropriate retention in many cases can be achieved by combining either a liphophilic counter ion and high content of organic modifier or a more hydrophilic counter ion and a lower content of the modifier. The latter system is preferable but in terms of selectivity, the impurity pattern in the sample may require other considerations to be made.

Ion pair systems for the reversed-phase separation of ammonium compounds seem rather straight forward. An organic or inorganic counter ion 0.01 M in buffer pH 2–3 and methanol or acetonitrile as organic modifiers is used. However, with respect to peak shape the chromatographic performance is in many instances not as good as anticipated[38–41]. The problem was referred to free non-reacted silanol groups and "end-capping" was one solution. Through the years a lot of efforts have been made by manufacturers to produce bonded phase material with minimal free silanols. In recent times special columns for separation of amines have been introduced but even though progress has been made precautions are still required. The chromatographic behaviour can be dependent on minor variations in the properties of the packing material. It is advisable to use a composition of the mobile aqueous phase not more complex than necessary but which determines in which form the solutes are retained and enables separations to be reproducible from one batch of support to another.

Efficient separations of ammonium compounds on bonded-phases by ion pair chromatography most often require the presence of a modifier in the aqueous mobile phase[38–41]. Neutral modifiers such as acetonitrile, butanol or other organic solvents are tools to regulate retention but are in general not able to improve peak shape. Ionic ammonium modifiers are then preferred and di- and trimethylsubstituted amines and ammonium ions such as N,N-dimethyloctylamine and N,N,N-trimethyloctylammonium have been found most effective. The importance of the structure of the amine modifier is to some extent dependent on the properties of the solutes and the bonded-phase packing used.

Non-reacted silanol groups may contribute to the presence of adsorption sites of different activity on the surface of the packing material. Some of the sites show high activity and are neutralized by the amine modifier having high affinity for those sites in particular[42–44]. Besides the influence on chromatographic performance, increased concentration of amine modifier will decrease retention of the amine solutes by competition for binding sites on the solid phase. The adsorption of amine modifier onto the bonded phase will increase with increased hydrophobicity and concentration of the counter ion. The isotherm will approach a maximum level, where the retention of the solutes will correlate to an ion-exchange mechanism[45].

Bonded-phase systems with mobile phases containing both ion pairing agents, neutral and ionic modifiers will offer a number of possibilities to regulate the selectivity by changing proportions and properties of the mobile phase components. In this connection it may be appropriate to note that the composition of the sample injected may influence the chromatographic performance. This is not specific for ion pair chromatography but distortion of peaks or ghost peaks may appear if the sample solution is not compatible with the mobile phase.

The use of regular unmodified silica as solid phase with aqueous mobile phases in ion pair chromatography is today an established technique[46–48]. The variation in properties of bonded-phase material made this kind of system interesting. The mobile phases are usually free from acetonitrile, methanol or other neutral organic modifiers in order to give sufficient retention. As expected the range of capacity factors is much smaller than for bonded-phases and the retention seems to be as much correlated to hydrogen bonding as lipophilic properties (Fig. 5). In the absence of modifiers strongly retained ammonium compounds are often not that well chromatographed showing tailing.

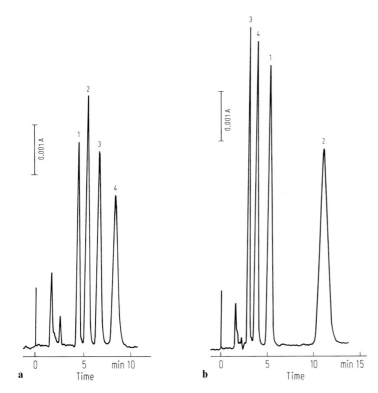

Fig. 5a and b. Separation of (1) alprenolol (2) propranolol (3) metoprolol (4) acebutolol on silica[48]. Solid phase: LiChrosorb SI 60, 7 μm. Mobile phase: Aqueous phosphate buffer pH 2.2 with **a** $1.5 \cdot 10^{-3}$ M N,N,N-trimethyloctylammonium bromide. **b** $4 \cdot 10^{-2}$ M N,N,N-trimethyloctylammonium bromide and $1 \cdot 10^{-2}$ M 3,5-dimethylcyclohexylsulfate

The extent of retention and the separation pattern can be affected by the intro-duction of ionic modifiers in the mobile phase. Cetrimide in an aqueous mobile phase at pH 6.5 adsorbed by ion-exchange onto the silica forms a stationary phase available for hydrophobic interaction similar to a bonded phase[47]. The presence of both a hydrophobic counter ion, dimethylcyclohexylsulfate, and an ammonium modifier, trimethyloctylammonium, gives at high concentrations rise to a stationary ion pair phase with hydrophobic properties and a change in selectivity[48].

4 Considerations for Using Ion Pair Liquid Chromatography

4.1 Retention

In ion pair chromatography the counter ion provides a means to regulate solute retention, increased hydrophobicity of the counter ion will increase retention in reversed phase chromatography and decrease retention in the normal phase mode.

For hydrophilic solutes ion pair formation is utilized to induce a sufficient degree of retention, for hydrophobic compounds the improvement in chromatographic performance (e.g. peak width) may be the major reason for adopting this approach.

Regulation of retention by control of the pH of the aqueous phase is in many instances an alternative approach to the ion pair technique but this approach requires that the pH buffer capacity of the system be high enough to create a linear partition isotherm. It is often easier to add a counter ion in a concentration sufficient for stable conditions at a pH of 2–3 which often is suitable for protonated amines. This approach will give stable conditions and be favourable for the chromatographic performance (i.e. sharp peaks).

4.2 Selectivity

The ion pair separation concept provides an alternative to distribution in uncharged form for compounds that can be present in anionic or cationic form. It can also influence selectivity and, as noted above, is well suited for separation of amino compounds on bonded silica in view of the available pH region.

The options to use liquid-liquid versus liquid-solid chromatography in the normal or reversed phase mode are possible in ion pair chromatography as in other forms of liquid chromatography. The choice of stationary phases and mobile phase components give numerous possibilities to change the selectivity pattern. In ion pair chromatography the type and concentration of the counter ion and the nature of the organic solvent components are powerful tools to affect selectivity. The use of ionic and neutral modifiers and metal chelating additives is another way to influence the separation. Separation of enantiomers is an example of ion pair chromatography illustrating the capability of this technique[20]. Electrostatic bidentate attachment to metal chelates is suggested as the retention mechanism when amino acids as dansyl derivatives are resolved into their optical isomers with a high degree of selectivity[49].

4.3 Detectability

The first applications of ion pair extraction were directed towards the possibility to use dye molecules as counter ions for photometric measurement of ionizable compounds with low inherent absorptivity[1].

The use of detector sensitive counter ions as a means to increase the response of sample components without sufficient detectability was applied in ion pair chromatography. Picrate ions of high absorptivity were used as counter ions in the stationary phase in a normal phase low efficiency liquid-liquid system for the analysis of choline and other quaternary ammonium ions[14]. With microparticles of silica as the support naphtalene sulfonate for cationic compounds[50] and N,N-dimethyl-protriptyline for anionic solutes[13] have been employed as counter ions. The systems have to be carefully thermostatted and a reasonably low background absorbance is favourable, which may influence the choice of mobile phase composition.

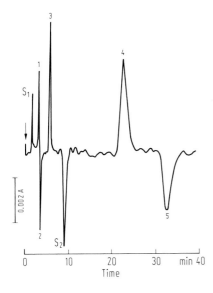

Fig. 6. Detection of non-UV-absorbing ions[52]. Mobile phase: naphthalene-2-sulfonate $4 \cdot 10^{-4}$ M in phosphoric acid 0.05 M. Solid phase: μ-Bondapak Phenyl. Wavelength: 254 nm. Sample: (1) pentanesulfonate, (2) di-isopropylamine, (3) hexanesulfonate, (4) heptylamine, (5) octanesulfonate. S_1 and S_2 are system peaks

The increased use of reversed phase chromatography with hydrocarbonaceous supports has focussed interest on the detection of non-responding ionic compounds. Counter ions are introduced into mobile aqueous phases as probes to provide detector response to such sample components[51−54]. Aromatic carboxylates and quaternary ammonium ions are examples of the probes used. Positive or negative peaks are obtained in the chromatogram depending on the charge of the sample ion relative to the probe and the retention of the solutes relative to the system peaks i.e. the peaks with constant retention and characteristic for the system[52]. The eluted peaks are a consequence of effects in the injected zone which are transferred along the column cf[55, 56] and the response is dependent on the capacity factor, the highest value of the peak area being obtained for compounds eluting close to the system peak. Both photometric and fluorometric detection mode have been employed with this probe technique (Fig. 6). An alternative and quite useful technique of general application to ionic solutes with too low detector response was developed by Brinkman, Frei and coworkers. They use counter ions with high absorptivity or fluorescence in an on-line post-column ion pair extraction system[57, 58]. The counter ion is added to the aqueous mobile phase eluting from the column and the ion pairs are extracted with chloroform for detection.

Karger, Vouros and Kirby[59, 60] used a similar approach and introduced on-line ion pair extraction as an interface between liquid chromatography and mass spectrometry. In RPLC, they found it advantageous to transfer ionic compounds to an organic phase which is compatible with a moving-belt transport system. Useful mass spectra of the ion pairs were obtained in both the EI and CI mode with less background from the solvent. Interestingly, the use of aromatic counter ions minimized the fragmentation pattern of the species for ease of structural and quantitative analysis of the solute ions. Figure 7 shows the effect of aromatic counter ions in the LC/MS of α-methylparnate.

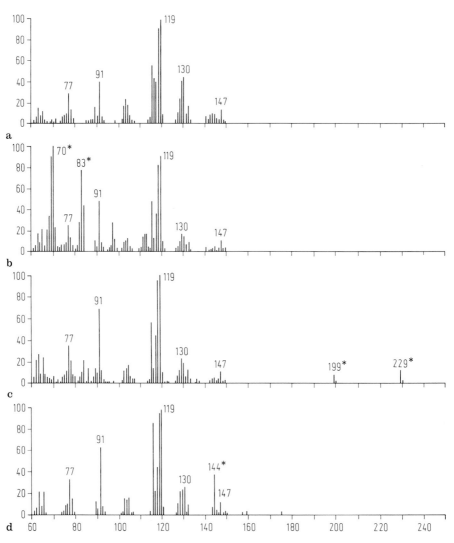

Fig. 7. a EI mass spectrum of α-methylparnate. **b** EI mass spectrum of α-methylparnate/C alkylsulfate ion pair. **c** EI mass spectrum of α-methylparnate/picrate ion pair. **d** EI mass spectrum of methylparnate/naphthalenesulfonate ion pair. All spectra recorded off-line with Finnigan 4,000 mass spectrometer [60]

Impurities in detector sensitive counter ions may disturb the ion pair extraction process by increasing the background and giving rise to extra peaks in the chromatogram. Impurities in variable concentrations with much higher extraction capability than the counter ion may also affect the linearity of a calibration plot. Purification processes include recrystallization but also a systematic extraction procedure as devised by Schill and coworkers [61] may be used.

5 Conclusion

The ion pair concept can be applied on the different modes of liquid chromatography. For ionic or ionizable compounds such as most bioactive compounds, ion pair liquid chromatography is today the major separation technique due to the dominant position of bonded phase chromatography. Chemical manipulation of the phase systems allows many ways to control retention selectivity and is a means to take advantage of the high separation efficiency of modern liquid chromatography.

6 References

1. Schill, G., in: Marinsky, J.A. and Marcus, Y. (Eds.): Ion Exchange and Solvent Extraction, Vol 6, 1. New York, Marcel Dekker 1974
2. Hearn, M.T.W., in: Giddings, J.C., Grushka, E., Cazes, J. and Brown, P.R. (Eds.): Advances in Chromatography, Vol. 18, Ch. 2., 59. New York, Marcel Dekker 1980
3. Tomlinson, E., Jefferies, T.M. and Riley, C.M.: J. Chromatogr. 159, 315 (1978)
4. Fransson, B., Wahlund, K.-G., Johansson, I.M., Schill, G.: J. Chromatogr. 125, 327 (1976)
5. Schill, G., Modin, R., Persson, B.-A.: Acta Pharm. Suecica 2, 119 (1965)
6. Kraak, J.C., Huber, J.F.K., J. Chromatogr. 102, 333 (1974)
7. Persson, B.-A., Karger, B.L.: J. Chromatogr. Sci. 12, 521 (1974)
8. Karger, B.L., Su, S.C., Marchese, S., Persson, B.-A.: J. Chromatogr. Sci. 12, 678 (1974)
9. Knox, J.H., Jurand, J.: J. Chromatogr. 103, 311 (1975)
10. Persson, B.-A., Lagerström, P.-O.: J. Chromatogr. 122, 305 (1976)
11. Westerlund, D., Nilsson, L.B., Jaksch, Y.: J. Liq. Chromatogr. 2, 373 (1979)
12. Crommen, J.: J. Chromatogr. 193, 225 (1980)
13. Hackzell, L., Denkert, M., Schill, G.: Acta Pharm. Suecica 18, 271 (1981)
14. Eksborg, S. and Schill, G.: Anal. Chem. 45, 2092 (1973)
15. Lagerström, P.-O.: Acta Pharm. Suecica 13, 213 (1976)
16. Crommen, J.: Acta Pharm. Suecica 16, 111 (1979)
17. Lagerström, P.-O., Persson, B.-A.: J. Chromatogr. 149, 331 (1978)
18. Flanagan, R.J., Storey, G.C.A., Holt, D.W.: J. Chromatogr. 187, 391 (1980)
19. Eriksson, B.-M., Persson, B.-A., Lindberg, M.: J. Chromatogr. 185, 575 (1979)
20. Pettersson, C., Schill, G. Chromatographia 16, 192 (1982)
21. Pettersson, C., No, K.: J. Chromatogr. 282, 671 (1983)
22. Pettersson, C.: J. Chromatogr. 316, 553 (1984)
23. Wahlund, K.-G.: J. Chromatogr. 115, 411 (1975)
24. Wahlund, K.-G., Beijersten, I.: J. Chromatogr. 149, 313 (1978)
25. Wahlund, K.-G., Beijersten, I.: Anal. Chem. 54, 128 (1982)
26. de Jong, J., Schouten, J.P., Muusze, R.G., Tjaden, U.R.: J. Chromatogr. 319, 23 (1985)
27. de Jong, J., van Valkenburg, C.F.M., Tjaden, U.R.: J. Chromatogr. 322, 43 (1985)
28. Tilly-Melin, A., Askemark, Y., Wahlund, K.-G., Schill, G.: Anal. Chem. 51, 976 (1979)
29. Eksborg, S., Ekqvist, B.: J. Chromatogr. 209, 161 (1981)
30. Knox, J.H., Laird, G.R.: J. Chromatogr. 122, 17 (1976)
31. Terweij-Groen, C.P., Heemstra, S., Kraak, J.C.: J. Chromatogr. 161, 69 (1978)
32. Knox, J., Jurand, J.: J. Chromatogr. 125, 89 (1976)
33. Johansson, I.M., Wahlund, K.G., Schill, G.: J. Chromatogr. 149, 281 (1978)
34. Bishop, C.A., Harding, D.R.K., Meyer, L.J., Hancock, W.S., Hearn, M.T.W.: J. Chromatogr. 192, 222 (1980)
35. Horvath, Cs., Melander, W., Molnar, I., Molnar, P.: Anal. Chem. 49, 2295 (1977)
36. Knox, J.H., Hartwick, R.A.: J. Chromatogr. 204, 3 (1981)
37. Bartha, A., Vigh, Gy., Billiet, H.A.H., de Galan, L.: J. Chromatogr. 303, 29 (1984)
38. Wahlund, K.-G., Sokolowski, A.: J. Chromatogr. 151, 299 (1978)

39. Sokolowski, A., Wahlund, K.-G.: J. Chromatogr. 189, 299 (1980)
40. Melander, W., Stoveken, J., Horváth, Cs.: J. Chromatogr. 199, 35 (1980)
41. Bidlingmeyer, B.A.: J. Chromatogr. Sci. 18, 525 (1980)
42. Jansson, S.-O., Andersson, I., Persson, B.-A.: J. Chromatogr. 203, 93 (1981)
43. Goldberg, A.P., Nowakowska, E., Antle, P.E., Snyder, L.R., J. Chromatogr. 316, 241 (1984)
44. Persson, B.-A., Jansson, S.-O., Johansson, M.-L., Lagerström, P.-O., J. Chromatogr. 316, 291 (1984)
45. Jansson, S.-O.: J. Liq. Chromatogr. 5, 677 (1982)
46. Crommen, J.: J. Chromatogr. 186, 705 (1979)
47. Hansen, S.H.: J. Chromatogr. 209, 203 (1981)
48. Jansson, S.-O., Andersson, I., Johansson, M.-L.: J. Chromatogr. 245, 45 (1982)
49. Lindner, W., LePage, J.N., Davies, G., Seitz, D.E. and Karger, B.L.: J. Chromatogr. 185, 323 (1979)
50. Crommen, J., Fransson, B., Schill, G.: J. Chromatogr. 142, 283 (1977)
51. Denckert, M., Hackzell, L., Schill, G., Sjögren, E.: J. Chromatogr. 218, 31 (1981)
52. Hackzell, L., Schill, G.: Chromatographia 15, 437 (1982)
53. Bidlingmeyer, B.A., Warren jr., F.V.: Anal. Chem. 54, 2351 (1982)
54. Barber, W.E., Carr, P.W.: J. Chromatogr. 316, 211 (1984)
55. McCormick, R.M. and Karger, B.L.: J. Chromatogr. 199, 259 (1980)
56. Melander, W.R., Erard, J.F. and Horvath, Cs.: J. Chromatogr. 282, 229 (1983)
57. Lawrence, J.F., Brinkman, U.A.Th., Frei, R.W.: J. Chromatogr. 171, 73 (1979)
58. Smedes, F., Kraak, J.C., Werkhoven-Goewie, C.F., Brinkman, U.A.Th., Frei, R.W.: J. Chromatogr. 247, 123 (1982)
59. Karger, B.L., Kirby, D.P., Vuoros, P., Foltz, R.L., Hidy, B.: Anal. Chem. 51, 2324 (1979)
60. Vuoros, P., Lankmayr, E.P., Hayes, M.J., Karger, B.L., McGuire, J.M.: J. Chromatogr. 251, 175 (1982)
61. Borg, K.O., Modin, R., Schill, G.: Acta Pharm. Suecica 5, 299 (1968)

Application of HPLC in Inorganic Chemistry

Hermann J. Möckel
Hahn-Meitner-Institut für Kernforschung Berlin, Bereich Strahlenchemie,
1000 Berlin 39, FRG

1 Introduction

HPLC and its various techniques has been the domain of organic chemistry for a long time. Only in the last few years has it been recognized that HPLC can equally well be utilized to separate inorganic compounds. It has even turned out that separations, which were definitely unfeasible before, can easily be effected now using HPLC techniques. The natural limitations of HPLC applicability will be valid also in the field of inorganic chemistry of course. That means that only those compounds are amenable to HPLC investigations which are soluble in a solvent being compatible with HPLC columns.

There are two distinctly different classes of soluble compounds in inorganic chemistry. First, there are ions – anions as well as cations – which in general are easily dissolved in aqueous solutions. Second, there are molecular compounds which more readily are soluble in organic solvents. Accordingly, two different branches of HPLC have come into use in inorganic chemistry. For the separation of ions in aqueous solution the technique of ion chromatography (IC) has been developed. The non-dissociated molecular compounds can be separated in reversed-phase systems using the well-known bonded alkyl silica phases and mostly organic eluents like methanol or acetonitrile (RPLC). In a few cases also normal-phase LC on silica or alumina has been used.

2 Ion Chromatography

2.1 The Chromatographic Separation of Anions

Before the advent of modern ion chromatography, the separation and determination of anions had been laborious and frequently very difficult. Every student who has had to face the problem of analyzing for Cl^-, J^-, CN^-, and SCN^- in awkward mixtures, qualitatively or even quantitatively, knows about the uncertainties involved in this work. Now, with the aid of ion chromatography, the answer to the "what" and "how much" is given within a few minutes.

The separation of anions can be effected according to one or more of the following three principles: first, ion exchange chromatography (IXC), second, ion pair chromatography (IPC), third, ion exclusion chromatography (ICE).

2.1.1 Ion Exchange Chromatography

The most common type of ion chromatography (IXC) makes use of an ion exchange mechanism[1] for the separation of ions. The stationary phase consists of a low capacity, highly efficient ion exchange resin of the $-NR_3^+$ type. The elution process is carried out isocratically. The retention of the ions to be separated is determined by their respective affinity to the ion exchange active sites on the stationary phase, and the strength of the eluent. The higher the affinity to the exchange sites, the

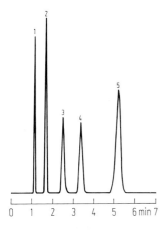

Fig. 1. Chromatogram of 1) F^-, 2) Cl^-, 3) PO_4^{3-}, 4) Br^-, 5) SO_4^{2-}, 5×10^{-5} M each. Sample 0.1 ml. Column anion exchanger HPIC AS4. Eluent 2.8 mM HCO_3^- and 2.24 mM CO_3^{2-} in H_2O. Flow 2 ml/min. Temperature 20° C. Conductivity detection with suppressor, range 10 μS

higher is the retention of the respective ionic solute. In order to make the mass transfer fast enough to avoid excessive peak-broadening, several custom-tailored exchange resins – mostly of the pellicular type – have been developed and are commercially available.

Using one or another anion exchanger and an aqueous eluent containing some millimoles of sodium carbonate and sodium bicarbonate, a great variety of inorganic anions can be separated. The absolute and relative concentrations of Na_2CO_3 and $NaHCO_3$ determine the strength of the eluent. On a typical IXC commercial column (e.g., HPIC-AS4 from Dionex) the following anions can be separated using an eluent containing 2.8 mM HCO_3^- plus 2.24 mM CO_3^{2-} (see Ref. [2]):

SiF_6^{2-}, F^-, Cl^-, Br^-, HPO_4^{2-}, HPO_3^{2-}, HPO_2^{2-}, NO_2^-, NO_3^-, S^{2-}, SO_3^{2-}, SO_4^{2-}, ClO^-, ClO_3^-, BrO_3^-, $HAsO_4^{2-}$, N_3^-, CN^-, SeO_3^{2-}, SeO_4^{2-}.

Figure 1 shows the separation of five anions, each having a concentration of 5×10^{-5} M.

On a slightly different column (HPIC-AG4 from Dionex) and using a somewhat stronger eluent (6–8 mM CO_3^{2-}), some more anions can be analyzed [2]:

I^-, IO_3^-, SCN^-, ClO_4^-, $S_2O_3^{2-}$, CrO_4^{2-}, WO_4^{2-}, MoO_4^{2-}, ReO_4^{2-}, TaO_4^{2-}, VO_4^{2-}.

The pyrophosphate $P_2O_7^{4-}$ and the triphosphate $P_3O_{10}^{5-}$ anions need a 0.05 mM HNO_3 eluent, while boric acid H_3BO_3 is eluted with NaOH, Na_2CO_3, and an excess of mannitol.

It has to be pointed out that each of the above-mentioned groups of anions needs a special stationary phase if optimal resolution is to be achieved. The main differences between those exchange resins are the degree of polymerization, which controls pore size and permeability, and the chemical character of the exchange sites. Details on IXC columns and their usefulness for particular separations should be obtained from the manufacturers.

2.1.2 Ion Pair Chromatography

An alternative separation mechanism for the anions listed above is given by ion pair chromatography (IPC). This techniques also allows to analyze:

$S_2O_4^{2-}$, $S_2O_8^{2-}$, $S_2O_6^{2-}$, $S_3O_6^{2-}$, $S_4O_6^{2-}$, $S_5O_6^{2-}$, $S_6O_6^{2-}$, BF_4^-, $Au(CN)_2^-$, $Au(CN)_4^-$; $Co(CN)_6^{3-}$, $Fe(CN)_6^{3-}$, and $Fe(CN)_6^{4-}$ [2,3].

In IPC the anion is coupled to a quaternary ammonium base like tetrabutylammonium hydroxide in an equeous sodium carbonate solution, which contains fairly large amounts of a typical reversed-phase eluent modifier like acetonitrile or methanol (about 10–50% v/v). The separation is carried out on a nonpolar, uncharged stationary phase like a RP C18 or, preferably because of the wider pH range, on a nonpolar porous polymer. The actual mechanism of the separation is not yet known with all certainty. The simplest way to think of it is as the formation of an ion pair, the dissociation of which is almost completely suppressed by the presence of the modifier. The ion pairs then are thought to behave like common solutes in RPLC, being subject to the separation mechanisms active in bonded reversed-phase systems. This model has been promoted by Horvath et al.[4,5]. Hoffmann and Liao[6] as well as Kissinger[7] prefer the idea that the ion-pairing reagent (tetraalkylammonium base) is adsorbed at the stationary phase, acting like an ion exchanger after being sorbed. From that point of view, postulating the formation of an ion exchanger, the IPC technique is not much different from IXC. As there is quite a variety of ion-pairing bases, an IPC system can easily be adjusted to the specific nature of the inorganic ion to be analyzed.

It must be added that other authors[8,9] have found that neither ion pair formation nor in situ formation of an ion exchanger can account for IPC retention completely.

2.1.3 Ion Exclusion Chromatography

The third of the above-mentioned techniques is ion exclusion chromatography (ICE). It is used to separate strong acids from weak, mainly organic acids. The stationary phase is a high capacity ion exchanger of the sulfonate type. Due to the Donnan potential (cf. Ref.[1]) strong inorganic acids are, to a large extent, excluded from the stationary phase so that they experience almost no retention. The low-strength organic acids, however, can penetrate the Donnan membrane, which in this case is represented by the boundary between the exchanger resin and the eluent. As a consequence, their concentration within the stationary phase may be even somewhat higher than in the bulk solution[10,11]. In addition to the Donnan exclusion, adsorption and steric exclusion effects are also operative in establishing the selectivity towards the various weak acids. The eluent is normally HCl in the millimolar range. One of the main applications of ICE in inorganic chemistry is the analysis of carbonate in aqueous solution. The eluent, in this case, is pure H_2O. CO_3^{2-} and HCO_3^- elute well behind all other anions.

2.1.4 Detection in Anion Chromatography

The most widely used detection system in ion exchange chromatography (IXC) is the conductometric detector[12,13]. The high background conductivity of the eluent is eliminated by the use of a "Suppressor Column" between the analytical column and the detector. This suppressor basically consists of a strongly acidic ion exchanger of the sulfonate type. It converts the carbonate to the very weak H_2CO_3 and the anions to be analyzed to the corresponding acids. The sensitivity of a conductometric detector with a suppressor is much higher than without.

Also in the ICE mode, the conductometric detector is preferred. As the ICE eluent often contains hydrochloric acid, this must be removed before the eluate enters the detector. This again is done with a suppressor which, in this case, is used in the silver-loaded form. Ag^+ eliminates Cl^-, and H^+ replaces Ag^+. Consequently, the weak acids reach the conductivity cell in an almost neutral aqueous solution. Besides the conductivity detection, several other detectors well-known from non-ion HPLC have been used more or less successfully.

2.2 The Chromatography of Cations

2.2.1 Primary Exchange Techniques

For the separation of cations, superficially sulfonated inert polymer resins are used [14-17]. The pellicular structure of the stationary phase, thin sulfonated layer, and strongly hydrophobic core, has the advantage of a fast solute transfer. With about 1 mM hydrochloric acid as eluent, the alkali metal ions $Li^+ - Cs^+$, NH_4^+, and the ions of small aliphatic amines can be eluted. As Fig. 2 shows, the analysis of the whole set of M^+ from Li^+ to Cs^+ takes less than 25 min. To make use of conductivity detection in the analysis of the univalent cations, the high background conductivity of the HCl has to be eliminated. This is done in a suppressor containing a strongly basic anion exchanger in the OH^- form, which converts the H^+ to H_2O and the ions of interest to their hydroxides.

The alkaline earth metals have a very high affinity to the cation exchanger. Thus they cannot be eluted with 1 mM hydrochlorid acid. The use of more concentrated HCl is undesirable for several reasons, the main one being the very fast

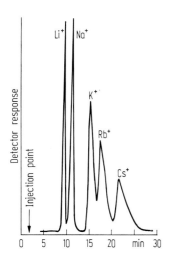

Fig. 2. Chromatogram of Li^+, Na^+, K^+, Rb^+, Cs^+, 10^{-2} M each. Sample 0.1 ml. Column 9 mm × 250 mm SSS/DVB 180–325 mesh, sp. cap. 0.016 mequiv./g. Eluent 10^{-2} M HCl. Flow 160 ml/hr. Conductivity detection with suppressor (taken from H. Hamish, T.S. Stevens, W.C. Baumann, Anal. Chem. **47**, 1801 (1975) with permission)

exhaustion of the suppressor column. A good eluent for Mg^{2+}, Ca^{2+}, Sr^{2+}, and Ba^{2+} is 2 mM m-phenylene-diamine dihydrochloride[16, 17].

2.2.2 Complexation of Cations

The direct ion exchange mechanism is not applicable to metal ions which would form insoluble hydroxides in the suppressor. Several transition metals and Pb^{2+} and Zn^{2+} belong to this group. They can be separated after complexation with oxalic and/or citric acid. For the analysis of Fe^{3+}, Cu^{2+}, Ni^{2+}, Co^{2+}, Pb^{2+}, and Fe^{2+} on a cation exchange column, the following eluent has been recommended[2]: 10 mM oxalic acid, 7.5 mM citric acid, adjustment of pH with LiOH. Lithium as the counter ion of the complexing anion is to be preferred to sodium as it has a lower affinity to the exchange resins. The pH should be between 4 and 4.6, according to the desired selectivity[18].

When anionic complexes are formed, a separation on anion exchange columns is equally well suited. In this case NaOH is sufficient for pH adjustment. The oxalate complexes of the following cations can be separated on an anion exchanger:

Pb^{2+}, Cu^{2+}, Mn^{2+}, Cd^{2+}, Co^{2+}, Zn^{2+}, and Ni^{2+}.

The separated complexes are optically detected after re-complexation in a post-column reactor with an indicator ligand. For the above-mentioned cations, 2×10^{-4} M 4-(2-Pyridylazo)-resorcinol in 3 M NH_4OH and 1 M CH_3COOH has been used[2].

Al^{3+} can be eluted from a cation exchange column with 0.01 M H_2SO_4 and 0.2 M $(NH_4)_2SO_4$. For the post-column reaction, 3×10^{-4} M Tiron (4,5-Dihydroxy-m-benzenedisulfonic acid), adjusted to pH $= 6.3$ with NaOH, has been used[2].

2.2.3 Reversed-Phase Separation of Complexes

Several complexes of cations have been separated in RP systems. Mangia et al.[19] report on the separation of dibenzo-18-crown-6 complexes of $HgCl_2$, $HgBrCl$, $HgBr_2$, and HgI_2, and HgI_2 using a Micropak CH 10 (C_{18}) column with methanol and phosphate or borate buffer as the mobile phase.

Valenty and Behnken[20] separated the bis(2,2′-bipyridyl)2,2′-bipyridyl-4,4′-dicar-boxylic acid Ruthenium II and several mono- and di-esters (anionic form of the free carboxylic groups). Depending on whether none, one, or two groups were esterified, the complexes of interest had the charge $2+$, $1+$, or 0, respectively. Separation was performed on a µ-Bondapak C_{18} column using a H_2O/THF eluent which contained acetic acid plus methanesulfonic or heptanesulfonic acid. Obviously this is an example of ion pair chromatography with a complexed cation.

Comparable is the separation of 1,10-phenanthrolin complexes of Fe^{2+}, Ni^{2+}, and Ru^{2+} reported by O'Laughlin and Hanson[21]; C_{18} and CN columns were used with an eluent of H_2O/MeOH and counterions being Me-SO_3^-H, Hept-SO_3^-, or $H_3PO_4^-$. In another investigation[22], this was extended to the separation of the tris-phenanthrolin complexes of Zn^{2+}, Cd^{2+}, Co^{2+}, and Cu^{2+}. However, in this case ClO_4^- was used as counterion and the samples were chromatographed on a

μ-Partisil-SCX cation exchanger. Thus, the separation mechanism is not clear. It is worth noting that the inverse of the net retention volume is proportional to the logarithm of the counterion concentration.

When the Fe^{2+}, Ni^{2+}, and Ru^{2+} chelates were separted on a reversed-phase column (PRP11 from Hamilton) the elution order was changed, which points to differences in the mechanisms.

The thiocyanate anion SCN^- was used in the separation of tris(2,2'-bipyridyl) Ni^{++} and Fe^{++} on a μ-Bondapak CN column[23]. The dependence on counterion concentration of the retention volume was basically the same for SCN^- and for ClO_4^- [22].

3 The Separation of Non-Ionic Inorganic Solutes at Bonded Reversed Phases

3.1 Elemental Sulfur S_6 to S_{25}

HPLC analysis methods for S_8 have been described by Cassidy[24] and Möckel[25]. Cassidy used a separation system which consisted of a porous polymer (Poragel®) with several organic eluents, mainly on the basis of THF or chloroform. With UV (254 nm) detection he reached a detection limit of less than 10 ng S_8. The selectivity towards accompanying organics and sulfur organics is reported to be unique as a consequence of "combined exclusion and adsorption effects".

Möckel separated S_8 from solutions and suspensions via extraction with CS_2 and analyzed the extract on a C_{18} column with a methanol or methanol/H_2O eluent. This method has been improved so that now by direct injection of a CS_2 extract and UV detection at 254 nm a detection limit of 1 ng has been reached. Selectivity towards other co-extracted solutes can conveniently be adjusted by a proper choice of the eluent water content. Figure 3 shows 2.4 ppm S_8 as an impurity in technical-grade methanol.

In all extraction procedures using a CS_2 solvent UV light has to be excluded, as it decomposes CS_2 causing some sulfur formation. As shown in Fig. 4, not only S_8 but also the smaller ring sizes S_6 and S_7 are produced in surprisingly high quantities.

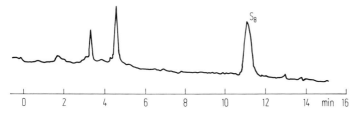

Fig. 3. S_8 as an impurity in technical methanol. 7.2×10^{-9} g S_8 in 3 μl (2.4 ppm) RCC10C18 (RCC10C18 = radially compressed column 10 cm × 8 mm i.d. with Radialpak 10 μm C18).1 ml/ min MeOH. UV 254 nm, 0.005 aufs. 25° C

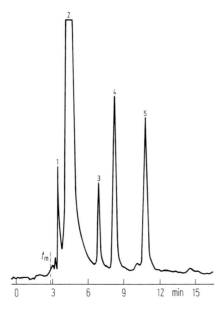

Fig. 4. Partially photolyzed CS_2 (2 h UV irradiated). Peaks 2 and 1 = solvent with unknown impurity, $3 = S_6$, $4 = S_{72}$, $5 = S_8$. Sample 10 µl undiluted. RCC5C18 (RCC5C18 = same as before, but 5 µm material). 1 ml/min MeOH. UV 254 nm, 0.01 aufs. ca. 23° C

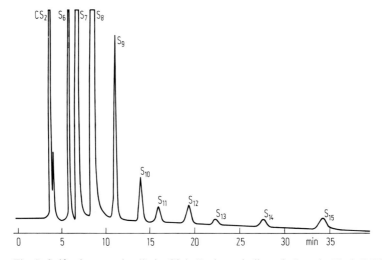

Fig. 5. Sulfur homocycles (S_x in CS_2). Peaks as indicated. Sample 20 µl. RCC5C18. 1 ml/min MeOH. UV 254 nm, 0.01 aufs. 25° C

Larger sulfur homocycles are found in sulfur melts or can be synthesized. We have separated the molecules S_6 to about S_{15} on C_{18} with a pure methanolic eluent (Fig. 5). Since for larger rings the retention becomes excessive, the polarity of the eluent has to be reduced by adding an alkane like cyclohexane or pentane. Under these conditions we have detected sulfur rings up to S_{25}, as shown in Fig. 6[26-28]. The relative concentration of the different sulfur species varies from one preparation

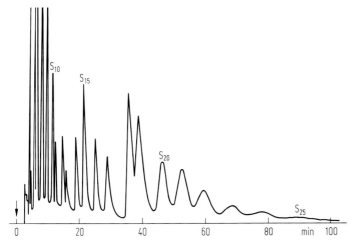

Fig. 6. Sulfur homocycles (S_x in CS_2). Peaks as indicated. Sample 20 μl. RCC5C18. 1 ml/min 75% MeOH, 25% Cyclohexane. UV 254 nm, 0.02 aufs, after S_{13} 0.005 aufs. 25° C

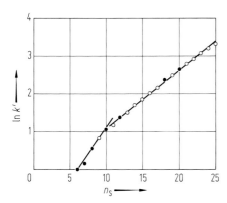

Fig. 7. Plot of logarithmic retention vs. sulfur ring size. Data taken from Fig. 6

melt to the other, but generally it appears that larger rings are less frequently found than smaller ones.

The assignment of peaks from sulfur melts to particular sulfur molecules was achieved by running S_6, S_7, S_{10}, S_{12}, S_{18}, and S_{20} under identical chromatographic conditions. The above, fairly stable homocycles had already been synthesized by Steudel et al.[28, 29].

In Fig. 7 the ln k' for the peaks in Fig. 6 are shown as a function of the ring size n_S. The triangles represent the above-mentioned calibration solutes, the dots stand for hitherto unknown ring sizes. Obviously, there are two quasi-homologous series of species, one going from S_6 to S_{10}, the other comprising the larger rings. The increase in ln k' per sulfur atom added to the molecule is larger for the smaller rings. This means that the Δln k' observed upon going from S_7 to S_8 is considerably larger than the value found between, e.g., S_{13} and S_{14}.

There are several factors influencing the retention of sulfur homocycles, the relative importance of which certainly depends on the ring size:

a) The surface area of the solute accessible to the eluent molecules. For large sulfur rings or chains this is identical with the total surface area (TSA). In smaller rings the inner part of the molecule may be inaccessible. Since for solutes, having otherwise identical properties, retention is linearly related to the accessible surface area [30-33], there must be an additional increase of ln k' until all sulfur atoms are fully exposed to the eluent. Obviously, this occurs for a ring size of the order of 10.

b) Molecular or local dipole moments, due to strongly distorted bonds and low symmetry, may cause deviations from linearity.

The sulfur homocycles exhibit some quite remarkable retention properties in RPLC. When compared to n-alkanes having about the same molecular surface area, their retention is much higher in a purely methanolic eluent. This can best be visualized by expressing retention in terms of retention indices which quote retention relative to that of n-alkanes.

In a homologous or quasi homologous series the relation between ln k' and the molecular size is linear over a fairly wide range [34, 35]. We can express this as

$$\ln k'(\text{alk}) = a(\text{alk}) + b(\text{alk}) * n(CH_2) \text{ for } n\text{-alkanes or} \tag{1}$$

$$\ln k'(S_n) = a(S_n) + b(S_n) * n(S) \text{ for sulfur rings.} \tag{2}$$

The retention index is then defined as

$$I_K(S_n) = \frac{100}{b(\text{alk})} [\ln k'(S_n) - a(\text{alk})]$$

While the retention index of n-nonane, e.g., is always 900 by definition, the I_K of S_9 is about 1480 with a pure methanol eluent. If the somewhat different molecular surface areas of n-C_9 and S_9 are taken into account and the respective ln k' are normalized to the TSA of n-C_9, the retention index of S_9 even reaches $I_K = 1610$.

If water is added to the eluent, the absolute retention values of both sulfur rings and alkanes increase, but the retention indices of the S_n approach the n-alkane values more and more, as shown in Fig. 8. If the eluent water content is high enough, the elution order is even reversed. This is demonstrated for cyclohexasulfur and n-hexane in Fig. 9. For larger sulfur rings, the eluent water content at which reversal of the retention order occurs, increases with the number of sulfur atoms in the molecule.

We have shown [33] that both enthalpy and entropy effects contribute to the high retention of S_n rings in highly methanolic eluents. Some data are given in Table 1. A more detailed investigation [36] leads to the conclusion that the sorbed state of solutes, which generally consists of C_{18} chains, solute, and eluent molecules, is more highly ordered for alkanes than for sulfur rings. Furthermore, the interaction enthalpy of alkanes with the methanol eluent is more negative than that of sulfur rings of the same size. From that point of view one has to conclude that sulfur homocycles in highly methanolic ($\sim 80-100\%$ MeOH) solvents are considerably

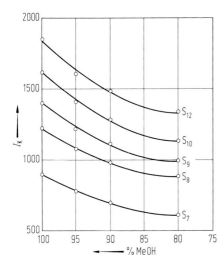

Fig. 8. Retention indices of some sulfur rings as a function of the eluent MeOH content. Column 20 cm × 4 mm i.d. SS, Nucleosil 5 μC$_{18}$. Flow 1 ml/min. 25° C

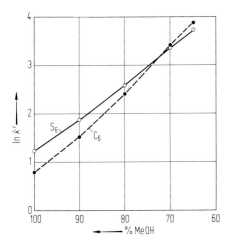

Fig. 9. Logarithmic retention of S$_6$ and n-hexane as a function of the eluent MeOH content. RCC10C18. Flow 1 ml/min. 25° C

more solvophobic than alkanes or cycloalkanes. As the extra interaction of alkane solutes with the eluent dwindles in more aqueous systems, one tends to conclude that it takes place between the methyl group of the methanol and the alkane. Thus the alkanes can no longer be regarded as the ideal solvophobic entities. This role is taken over by proton-free solutes having low or zero dipole moments, like S$_8$ and other sulfur rings, Se-rings, or P$_4$.

3.2 Polysulfide Chains

Polysulfides have the general formula R−S$_n$−R′. There are also ionic polysulfides, e.g., Na$_2$S$_n$, the separation of which via ion chromatography has not yet been completely successful[37]. Out of the non-ionic polysulfides, the dialkyl and diaryl com-

Table 1. Thermodynamic Data for the retention on a C_{18} phase of n-alkanes and sulfur homocycles

Solute	100% MeOH, 298 K, $\ln\phi = -1.531$						80% MeOH, 298 K, $\ln\phi = -3.210$				
	TSA^a	$\ln k'$	$I_K{}^b$	ΔG^c	ΔH^c	ΔS^d	$\ln k'$	$I_4{}^b$	ΔG^c	ΔH^c	ΔS^d
nC_5	138.81	−0.6127	500.1	−544	−209	1.12	0.7562	499.8	−2349	−2531	−0.61
nC_6	161.51	−0.4392	600.5	−635	−429	0.69	1.1663	600.5	−2591	−2957	−1.23
nC_9	229.62	−0.0011	899.9	−906	−1089	−0.61	2.3856	900.0	−3313	−4226	−3.06
nC_{10}	252.32	0.1524	1000.2	−977	−1309	−1.05	2.7923	999.8	−3554	−4650	−3.68
nC_{14}	343.13	0.7650	1400.6	−1360	−2190	−2.79	4.4228	1400.2	−4520	−6344	−6.12
S_6	141.69	0.0126	908.8	−914	−405	1.71	1.2809	628.7	−2659	−2686	−0.09
S_8	188.58	0.5244	1243.3	−1217	−924	0.98	2.3379	888.2	−3285	−3630	−1.16
S_9	210.89	0.8814	1476.9	−1429	−1284	0.49	2.7855	998.2	−3550	−4216	−2.23
S_{10}	235.48	1.1071	1624.2	−1562	−1548	0.05	3.3780	1143.7	−3901	−4737	−2.81
S_{12}	282.37	1.4544	1851.2	−1768	−1846	−0.26	4.1450	1332.0	−4355	−5625	−4.26

[a] TSA in $Å^2$.

[b] I_K of n-alkanes appear non-integer because they are calculated using Eq. (3).

[c] ΔG and ΔH as defined for the transition from mobile to sorbed state. Data in cal/mole.

[d] ΔS defined like ΔG and ΔH. Data in cal/K·mole.

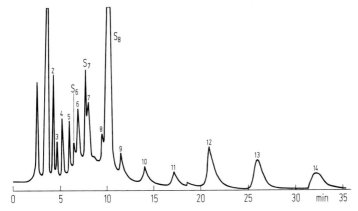

Fig. 10. Polysulfanes H_2S_n in MeOH. Numbers given with the peaks tell the respective sulfur atom number n_S. Sample 20 µl. RCC5C18. 1 ml/min MeOH. UV 254 nm, 0.2 aufs, after 18.5 min run time 0.05 aufs (base line step)

pounds have been thoroughly investigated[38]. The dihydrogen polysulfides (polysulfanes) are relatively unstable, they are very sensitive to even traces of alkali which lead to rapid decomposition into H_2S and sulfur. Only very recently could the polysulfanes be separated in a reversed-phase system[39].

3.2.1 Polysulfanes H_2S_n

In deaerated solutions of Na_2S in alcohols (methanol to pentanol), ordinary sulfur S_8 can be dissolved to give yellow solutions containing sodium polysulfide Na_2S_n. When sulfur dichloride SCl_2 is added dropwise, the color disappears and a purely white precipitate is formed. Figure 10 shows a chromatogram of a methanolic solution after the precipitate had largely settled. The column used was a radially compressed cartridge 100×8 mm filled with Radial PAK A 5 µ C_{18} (Waters), the eluent pure methanol. The peaks numbered 2 to 14 appear as a homologous series. Peaks 6, 7, and 8 are accompanied each by one peak of similar retention. These latter peaks are caused by the sulfur homocycles S_6, S_7, and S_8. A semilogarithmic plot of the peaks 2–14 is almost perfectly linear as shown in Fig. 11 on trace b. The retention values of the solutes producing peaks 2–14 do not depend on the particular kind of alcohol used as reaction solvent.

When the reaction-mixture is kept under argon for a longer time (over night), the solution yields the chromatogram in Fig. 12, which is basically identical with Fig. 5, with the one exception that an additional small peak appears before the S_6. Obviously the polysulfanes H_2S_n, to which the peaks in Fig. 10 must be attributed, have decomposed via

$$H_2S_n \rightarrow H_2S + S_{n-1} \tag{4}$$

When treated with small amounts of strong alkali (KOH), all polysulfanes in freshly prepared solutions are destroyed and only some S_6, S_7, and mostly S_8 are left.

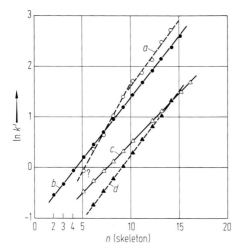

Fig. 11. Logarithmic retention vs. skeleton atom number of H_2S_n, Et_2S_n, and n-alkanes of a: S_n, b: H_2S_n, c: n-alkanes, d: Et_2S_n. Sample size 10 μl. RCC5C18. 1 ml/min MeOH, UV 254 nm, 0.2 aufs. For alkanes RI detection attenuation 8

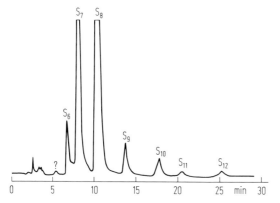

Fig. 12. Sulfur homocycles from the decay of polysulfanes in pentanol. Sample 20 μl. RCC5C18. 1 ml/min MeOH. UV 254 nm, 0.2 aufs

In addition to the H_2S_n, the S_n, n-alkane, and diethylpolysulfide retentions are also shown in Fig. 11. It is seen that the polysulfane retention is more similar to that of a sulfur homocycle having the same n_S, than to the retention of an alkane or alkylpolysulfide of approximately the same molecular size. In this connection a parallel is given to the retention behavior of n-alkanes and cyclo-alkanes whose ln k′ are very similar in methanolic eluent [26]. However, the presence of 2 SH groups seems to induce a slight increase of the energy of the solute-solvent interaction, as indicated by the somewhat greater ln k′ of the larger sulfur rings. It should be noted that the retention of a polysulfane H_2S_n is still much higher than that of an alkane C_nH_{2n+2} under identical chromatographic conditions.

From Eq. (4) the reaction $H_2S_6 \rightarrow H_2S + S_5$ may also be anticipated. We have tentatively assigned the small peak in Fig. 12, eluting before the S_6, to the cyclopentasulfur S_5.

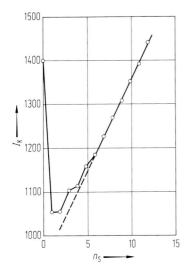

Fig. 13. Retention index versus sulfur atom number of tetradecane and all sulfides $R - S_n - R'$ with skeleton atom number 14. RCC5C18. 1 ml/min 95% MeOH, 5% H_2O. 25° C

3.2.2 Sulfur Chains with Organic Terminal Groups

Alkyl or aryl polysulfides are more stable than polysulfanes, but they are still so sensitive towards thermal or catalytic influences that GC is entirely inadequate for the analysis of this very interesting class of compounds. HPLC at reversed phases has proven to be ideally suited for the investigation of polysulfides[33, 40]. Most of these investigations have been carried out on 5 μm or 10 μm octadecylsilica columns using methanol, or methanol with up to about 30% water, as eluent. In di-*n*-propyl polysulfide, all members of the series Pr_2S_n from n = 2 to n = 21 have been identified in isocratic runs. Using a solvent gradient, polysulfides with $n_S > 30$ were found[41].

Under isocratic conditions, the polysulfides give chromatograms typical for homologous series, though there is some non-linearity in the ln k' vs. n_S functions observed with the first four or five members. These fairly small irregularities are caused by the alternating permanent dipole moment of the compounds R_2S to R_2S_5. This effect is seen in Fig. 13, which shows the retention indices of tetradecane, hexylheptylsulfide, and all polysulfides with $n_C + n_S = 14$ as a function of the number of sulfur atoms in the molecule. It is typical for the retention of aliphatic polysulfides that the introduction of one sulfur atom causes a large drop in retention when compared to the unsubstituted alkane. With increasing sulfur-atom number n_S, the I_K re-increases and eventually exceeds that of the alkane. For polysulfides with aromatic terminal groups, the initial drop in I_K is not observed.

The influence on retention of the presence of the stabilizing terminal alkyl groups dwindles gradually when the sulfur chain grows. Long-chain polysulfides show an increasing chromatographic similarity with polysulfanes or sulfur homocycles. This is demonstrated in Fig. 14. While the retention index of alkyl polysulfides with low n_S slightly increases with increasing eluent water content, the I_K of solutes with longer sulfur chains decrease. However, the effect is clearly less pronounced than in sulfur homocycles (S_8 in Fig. 14).

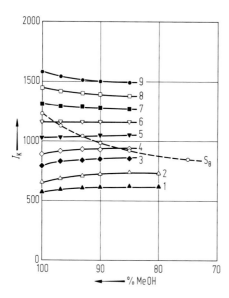

Fig. 14. Retention index versus eluent methanol content of S_8 and di-n-butylpolysulfides Bu_2S_n. Numbers at curves indicates sulfur atom number n_S. RCC5C18. 1 ml/min. UV 254 nm. 25° C

It should be pointed out that in any case the addition of terminal alkyl groups to a sulfur chain of given length lowers the retention quite remarkably, while the addition of two sulfhydryl groups incrases retention.

3.3 Selenium and Tellurium Chains

Recently, dialkyl and diaryl polyselenanes and polytelluranes have been prepared and analyzed in an ODS/MeOH system[42,43]. The retention behaviour of R_2Se_n and R_2Te_n solutes is similar to that exhibited by polysulphides, R_2S_n, as can be seen from the chromatograms of Et_2S_n, Et_2Se_n and Et_2Te_n shown in Fig. 15.

All of them are quite characteristic for homologous series. However, the retention of solutes having equal homologue numbers n_x, is distinctly higher for – Se_n – as compared to – S_n –, and again higher for – Te_n –. As can be seen from Fig. 16, also the increase of I_K per additional chalcogene atom grows in the same order.

Chalcogene chains are expected to behave chromatographically much like other proton-free, highly solvophobic entities, the retention of which in non-aqueous eluents is controlled by their respective molecular surface areas SA. From that point of view, the retention index increase per additional atom in the chain is anticipated to be proportional to the molecular surface area increase per chain atom, the proportionality factor being the same for S, Se and Te, respectively.

Consequently, Eq. (5) is expected to hold:

$$\frac{\delta I° (\text{Chalcogene 1})}{\delta I° (\text{Chalcogene 2})} = \frac{\Delta SA (\text{Chalcogene 1})}{\Delta SA (\text{Chalcogene 2})} \tag{5}$$

where $\delta I°$ is the index increment of a chalcogene atom in a hypothetical chalcogene chain carrying no terminal groups. It is found by extrapolating the retention indices

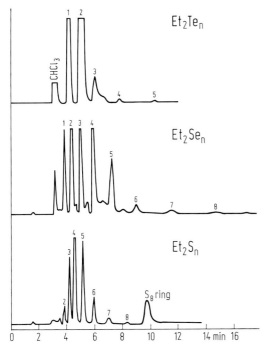

Fig. 15. Chromatograms of Et_2S_n, Et_2Se_n and Et_2Te_n. Column RCC10C18, Eluent 1 mol/min MeOH. Detection UV 254 nm for sulfides and selenides, 224 nm for tellurides

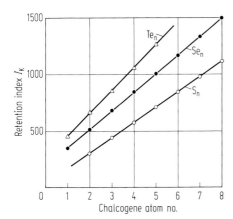

Fig. 16. Retention indices I_K versus chalcogene atom number. Data taken from Fig. 15

of dialkyl polychalcogenides to a carbon number equal zero. The following Table 2 lists the numerical values found for the ratios in Eq. (5).

The data agree well, indicating that the retention of chalcogene chains is almost completely determined by their respective surface areas. However, the actual retention index contributions in real molecules depend somewhat on the length of the terminal alkyl groups.

Table 2. Proportionality between surface area increases ΔSA and retention index increments $\delta I°$ of chalcogenes

| Chalcogene 1 | $\dfrac{\delta I° \text{ (ch.1)}}{\delta I° \text{ (ch.2)}}$ | $\dfrac{\Delta SA \text{ (ch.1)}}{\Delta SA \text{ (ch.2)}}$ |
Chalcogene 2		
Te/S	1.68	1.69
Te/Se	1.39	1.41
Se/S	1.21	1.20

Table 3. Retention index increments of n-alkanes and dialkylpolychalcogenides in an ODS/MeOH system

Solute	$\delta I°(CH_2)$	$\delta I°(CH_2 - X)$	$\delta I°(X)$	b_C''	b_S''
n-alkanes	100	0	0	0	0
R_2S_n	+91.0	−167.1	+142.2	−0.75	−0.75
R_2Se_n	+89.1	−86.5	+172.4	−1	−1
R_2Te_n	+87	−82.7	+239.0	−1	−1

Generally, the retention index I_K of dialkyl polychalcogenides R_2Xn_X with $R = H-(CH_2)_{nC}$ can be expressed in terms of index increments δI as shown in Eq. (6):

$$I_K = n_X \times [\delta I°(X) + b_X'' \times n_C] + n_C \times [\delta I°(CH_2) + b_C'' \times n_X] + 2 \times \delta I(CH_2 - X) \tag{6}$$

$\delta I°(X)$ is the above mentioned chalcogene index increment for zero carbon chain length and b_X'' is the increment decrease per carbon atom present in the terminal groups.

$\delta I°(CH_2)$ is the methylene index increment extrapolated for zero chalcogene chain length. Due to non-linearities in the retention behaviour of the first members of any homologous series, it is not identical with $\delta I(CH_2) = 100$ in n-alkanes. b_C'' is the methylene index increment decrease per atom present in the chalcogene chain.

$\delta I(CH_2 - X)$ represents the retention decrease due to the formation of local polar centers at the $CH_2 - X$ arrangements which permit a localized solvation by the surrounding MeOH eluent ("solvation patches").

The data pertinent to Eq. (6) are given in Table 3. It is seen that, e.g., a tellurium atom produces 2.4 times as much retention as a methylene group in an alkane. The "solvation patch" increment $\delta I(CH_2 - X)$ is most negative for the $CH_2 - S$ arrangement. The $CH_2 - Se$ and $CH_2 - Te$ fragments appear to be less polarised and, hence, less solvated.

Another group of solutes to be mentioned here is represented by the di-alkylthio-polyselenides $RS - Se_n - SR$ and the corresponding diaryl compounds[42]. The most remarkable observation made with this quite stable solute class is the fact that the presence of two $S - Se$ groups has only a vanishing retentional effect. There does not seem to be any appreciable solvation around these groups in chain-like solutes. The most recent development in this area appear to be the di-alkylseleno-polysulphides $RSe-S_n-SeR$ [44]. Their experimentally determined retention indices agree well with data calculated from increments deduced from the before mentioned solute classes. The di-alkylseleno-polysulphides are remarkably unstable in MeOH. Particularly, the longer chain length members decompose within a few hours after semipreparative isolation. They form shorter chain length members plus S_8. No other sulphur rings were observed in the decomposition process.

3.4 Cyclic Selenium and Selenium-Sulfur Compounds

3.4.1 Red Selenium

Red selenium is known to consist of Se_8 rings. When a sample of this selenium is dissolved in CS_2, a chromatogram like that in Fig. 17 is observed in a $C_{18}/$ methanol system. Similar results have been reported by Steudel and Strauss [45]. The indicated peaks represent Se_6, Se_7, and Se_8. No larger rings were found. After collecting the respective eluates, evaporating the solvent, and oxidizing with concentrated HNO_3, selenous acid H_2SeO_3 was obtained in each case and identified via ion chromatography.

The chromatogram also shows some S_8 present as an impurity. It is obvious that the retention of S_8 is much closer to that of Se_7 rather than Se_8. This observation is easily explained by the larger surface area of the selenium molecules, the value of Se_7 being close to that of S_8.

3.4.2 Selenium Sulfides

A melt of selenium plus sulfur, dissolved in CCl_4, yields a chromatogram as shown in Fig. 18, which exhibits a large number of partially unresolved peaks [46]. Some of these can be directly identified as Se_6 (No. 6), S_8 (No. 10), and Se_8 (No. 16). The peak sequence nos. 9, 11–15 was assumed to represent eight-membered rings

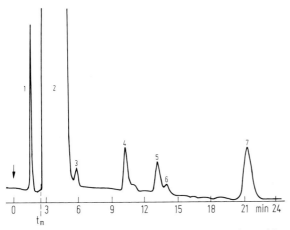

Fig. 17. Red selenium in CCl_4. Peak 1 and 3 = impurities, 2 = CCl_4, 4 = Se_6, 5 = Se_7, 6 = S_8, 7 = Se_8. Sample 5 µl. RCC10C18. 1 ml/min 95% MeOH. UV 254 nm, 0.063 aufs. 25° C

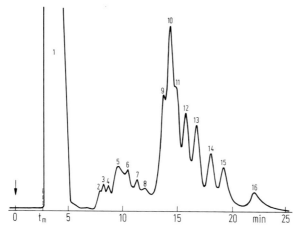

Fig. 18. Selenium-sulfur rings in CCl_4. Peak 1=solvent, 2 to 8=ring compounds with less than 8 atoms (6=Se_6), 9=Se_2S_6, 10=S_8, 11 to 15=Se_3S_5 to Se_7S, 16=Se_8. Sample 20 µl. RCC10C18. 1 ml/min 95% MeOH, 5% H_2O. UV 254 nm, 0.063 aufs. 25° C

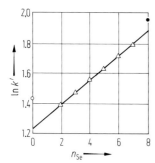

Fig. 19. Logarithmic retention versus selenium atom number in eight-membered rings. Open circle = S_8, fat dot = Se_8. Data from Fig. 16

Se_nS_{8-n} (n=1–7). Steudel et al. [7] have synthesized these cyclic selenium sulfides from SCl_2, $SeCl_2$, and KI. They found all products expected from the above formula plus Se_8. In all of these molecules, the selenium atoms are located in adjacent positions. Consequently, the selenium sulfides form a quasi-homologous series, the ln k' of which can be represented by an almost linear function of n_{Se}. An analogous linear relationship (Fig. 19) can be verified for the peaks nos. 9 and 11–15 in Fig. 18. Taking into account both the higher retention of selenium, due to a larger surface area, and the build-up of a dipole moment upon Se substitution of S atoms in S_8, one has to expect that for a low degree of selenium substitution the retention is somewhat lower than that of S_8. As observed, the Se_2S_6 elutes just before the S_8. Increasing substitution causes an almost linear increase of molecular surface area, while there are no dramatic changes of the dipole moment. The elimination of the last S, yielding Se_8, causes the dipole moment to disappear. Consequently, the ln k' of Se_8 is somewhat higher than that expected from the extrapolation of the sulfur-selenium ring values. The data show that the change in ln k' caused by the introduction or annihilation of the dipole moment is of the order of 0.1.

This is in good agreement with the ln k' changes brought about by dipole moment alterations in polysulfides[33]. The peaks at lower retention times in Fig. 18 were ascribed to smaller ring size sulfur-selenium compounds. Steudel and Laitinen[47] mention peaks observed at retention times both lower and higher than those corresponding to the eight-membered ring region. They assume that these peaks are caused by cyclic compounds of size 6, 7, and 12. Recently, Steudel and Strauss[48] have studied the retention of sulfur-selenium cyclic compounds in more detail and tried to relate them to the behavior of cyclic methylene sulfides/selenides.

3.5 Phosphorus and Phosphorus Sulfides

3.5.1 Elemental Phosphorus

White phosphorus P_4, dissolved in an organic solvent and chromatographed on a C_{18} column with 100% MeOH eluent, gives a narrow peak at a retention time lower than that of S_6 under the same conditions. A comparison of the most stable molecular forms of phosphorus, sulfur, and selenium shows that the retention of these solutes is proportional to their molecular surface area TSA:

Solute	ln k'	I_K	TSA [$Å^2$]
P_4	0.5514	989	102.96
S_8	1.0204	1277	190.06
Se_8	1.4656	1454	243.07

When a solution of white phosphorus in n-hexane is illuminated using a 125 W medium-pressure mercury lamp, a precipitate of red phosphorus is formed gradually. No oligomers are detectable via HPLC.

3.5.2 Phosphorus Sulfides

If, before illumination, some S_8 is added to the solution, several new signals appear in the chromatogram, which can be assigned to the phosphorus sulfides P_4S_{10}, P_4S_7, and P_4S_3 (Fig. 20). The chromatograms in Figs. 21 to 24 were obtained from commercial samples of P_4S_{10} and P_4S_3. Figure 21 shows P_4S_3 dissolved in CS_2. Peak 5 is the solvent CS_2, peak 7 some S_8, and peak 6 the P_4S_3. The peaks 1–3 elute within the column dead time. We assume that they represent polymeric phosphorus sulfides of unknown composition. The chromatogram in Fig. 22 was run under the same conditions as Fig. 21, but, this time, the P_4S_3 was dissolved in acetonitrile (AN). The P_4S_3 peak and the "polymer" peak are smaller now, due to the lower solubility of these compounds in the fairly polar solvent. Only 0.3 min after the dead time, a new peak is observed which coincides with the very small peak 4 in Fig. 21. We think that this is the monomer P_4S_{10} which, obviously, is much more polar than the P_4S_3, because its solubility is high an AN and the retention is quite small. This would fit with the solubility in CS_2, which is 350 times lower for P_4S_{10} as compared with P_4S_3[49]. Further evidence is gained from the chromatogram of commercial P_4S_{10} (Fig. 23) when the solvent is AN. There is a very strong

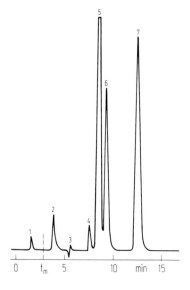

Fig. 20. Solution of P_4 and S_8 in cyclohexane after photolysis. Peak 1 = polymer phosphorus sulfides, 2 = P_4S_{10}, 3 = disturbance from solvent, 4 = P_4S_7, 5 = P_4, 6 = P_4S_3, 7 = S_8. Sample 1.5 μl RCC5C18. 1 ml/min 95% MeOH, 5% H_2O. UV 254 nm, 0.02 aufs. 25° C

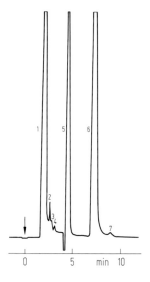

Fig. 21. Commercial P_4S_3 in CS_2. Peak 1–3 = polymer phosphorus sulfides, 4 = monomer P_4S_{10}, 5 = solvent, 6 = P_4S_3, 7 = S_8. Sample 10 μl. RCC5C18. 1 ml/min MeOH. UV 254 nm, 0.125 aufs. ca. 23° C

peak of the monomer P_4S_{10}, some "polymer", small peaks of P_4S_3 and S_8 contaminants, and another, not yet identified small impurity peak no. 3. Dissolved in CS_2, the P_4S_{10} yields chromatogram Fig. 24. As the solubility of the monomer is low in CS_2, peak no. 3 is small. Again we see the "polymer", which must be quite unpolar, represented by several peaks. Of course, it cannot be distinguished from the corresponding peaks found in P_4S_3 solutions. P_4S_3 (No. 7) and S_8 (No. 8) are present as impurities. From a comparison with Fig. 20 it can be concluded that the small peak no. 6 is caused by P_4S_7.

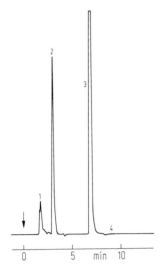

Fig. 22. Same as Fig. 19, but solvent acetonitrile. Peak 1 = polymer, 2 = monomer P_4S_{10}, 3 = P_4S_3, 4 = S_8

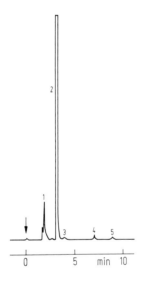

Fig. 23. Commercial P_4S_{10} in acetonitrile. Peak 1 = polymer, 2 = P_4S_{10}, 3 = unknown, 4 = P_4S_3, 5 = S_8. Conditions as in Fig. 19

3.6 Cyclic Sulfur-Nitrogen Compounds

We have used RPLC to separate tetrasulfur tetranitride S_4N_4 from reaction mixtures[6]. On a radially compressed C_{18} column with a 95% MeOH/5% H_2O eluent, S_4N_4 has a retention index of 339, as compared with $I_K(S_8) = 1124$. Apparently, the nitrogen lone electron pairs undergo a strong interaction with the eluent hydroxyl protons, which effectively shifts the free energy of sorption towards more positive values.

RPLC has also been applied to the analysis of the photolysis products of S_4N_4[51]. Depending on the nature of the solvent (alkane, alcohol, ketone) a great variety

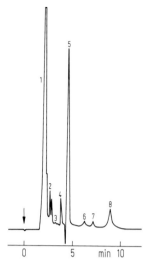

Fig. 24. Same as Fig. 21, but solvent CS_2. Peak 1–2 = polymer, 3 = P_4S_{10}, 4 = impurity?, 5 = solvent, 6 = P_4S_7, 7 = P_4S_3, 8 = S_8

of sulfur-nitrogen compounds as well as sulfur homocycles is formed. The results indicate that at the beginning of the photolysis, S_7 and S_6, possibly even S_5 are formed preferentially, while after more complete conversion of S_4N_4, S_8 is predominant.

Steudel[50] has used HPLC to separate a variety of sulfur-nitrogen cyclic compounds, mostly under normal-phase conditions using silica with 90/10 pentane/methanol or C_{18} silica with 80/20 pentane/methanol. For $S_{15}N_2$ and $S_{16}N_2$, a C_{18} phase with 30/70 pentane/methanol was used. From the experimental details given, no comments of chromatographic relevance can be made.

3.7 Phosphorus Chloronitrides

Phosphorus chloronitrides may either form cyclic compounds $(PCl_2N)_x$ of varying ring size x, chains of $R - (PCl_2N)_n - R'$, or polymers.

Figure 5 shows a chromatogram of commercial $(PCl_2N)_x$ in benzene (solvent peak no. 3). If the ln k' of peaks nos. 4–9 are plotted versus the ring size x, we find an almost linear relationship (Fig. 26). It is known[49] that the sizes x = 3 and x = 4 are most frequently found in $(PCl_2N)_x$. Consequently, the strongest peaks have been attributed to these species. Then peaks 6–9 represent the pentamer to octamer, respectively. Peak no. 1 elutes between exclusion- and dead time, and apparently is caused by polymeric $PNCl_2$. Peak no. 2 and the smaller peaks later in the chromatogram may be due to the presence of linear oligomers.

The eight-membered ring $(PCl_2N)_4$ has a retention which is considerably less than that of S_8, which indicates that it has a stronger interaction with the eluent than the sulfur ring has.

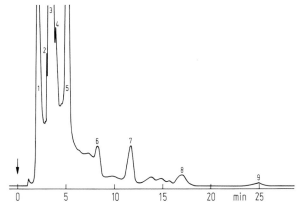

Fig. 25. Commercial phosphorus chloronitrides in benzene. Peak 1 = polymer, 2 = unknown, 3 = solvent, 4–9 cyclic $(PCl_2N)_x$ with x = 3 to 8. Small peaks in between may be linear oligomers. Sample 5 µl. RCC5C18. 1 ml/min acetonitrile. UV 254 nm, 0.063 aufs. 25° C

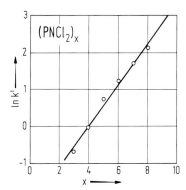

Fig. 26. Logarithmic retention of cyclic $(PCl_2N)_x$ versus ring size x. Data from Fig. 23

3.8 Cyclic Polysilanes

Brough et al.[52] have separated cyclic dimethyl polysilanes from $(Me_2Si)_5$ to $(Me_2Si)_{35}$ and describe them as an homologous series. The separation was performed on a C_{18} phase using a methanol/THF eluent. No more details are given, but from the chromatogram it appears that a multistep gradient with increasing THF concentration was applied.

4 References

1. Helfferich, F.: "Ion Exchange", McGraw Hill, New York 1962
2. Weiß, J.: Handbuch der Ionenchromatography Dionex GmbH, Weiterstadt 1985
3. Story, J.N.: J. Chromatogr. Sci. 21, 272 (1983)

 4. Horvath, C., Melander, W., Molnar, I.: Anal. Chem. 49, 142 (1977)
 5. Horvath, C., Melander, W., Molnar, I., Molnar, P.: Anal. Chem. 49, 2295 (1977)
 6. Hoffman, N.E., Liao, J.C.: Anal. Chem. 49, 2231 (1977)
 7. Kissinger, P.T.: Anal. Chem. 49, 883 (1977)
 8. Bidlingmeyer, B.A., Deming, S.N., Price, W.P., Sachok, B., Petrusch, M.: J. Chromatogr. 186, 419 (1979)
 9. Hung, C.T., Taylor, R.B.: J. Chromatogr. 202, 333 (1980)
10. Wheaton, R.M., Baumann, W.C.: Ind. Eng. Chem. 45, 228 (1953)
11. Gregor, H.P., Collins, F.C., Pope, M.J.: J. Colloid Sci. 6, 304 (1951)
12. Weiss, J.: CLB 34, 342 (1983)
13. Göbl, M.: GIT Fachz. Lab. 27, 261 (1983)
14. Fritz, J.S., Story, J.N.: J. Chromatogr. 90, 267 (1974)
15. Fritz, J.S., Story, J.N.: Anal. Chem. 46, 825 (1974)
16. Small, H., Stevens, T.S., Baumann, W.C.: Anal. Chem. 47, 1801 (1975)
17. Pohl, C.A., Johnson, E.L.: J. Chromatogr. Sci. 18, 442 (1980)
18. Weiß, J.: CLB 34, 293 (1983)
19. Mangia, A., Parolari, G., Gaetani, E., Laureri, C.F.: An. Chim. Acta 92, 111 (1977)
20. Valenti, S.J., Behnken, P.E.: Anal. Chem. 50, 834 (1978)
21. O'Laughlin, J.W., Hanson, R.S.: Anal. Chem. 52, 2263 (1980)
22. O'Laughlin, J.W.: Anal. Chem. 54, 178 (1982)
23. Mangia, A., Lugari, M.T.: J. Liq. Chromatogr. 6, 1073 (1983)
24. Cassidy, R.M.: J. Chromatogr. 117, 71 (1976)
25. Möckel, H.J., Masloch, B.: Z. Anal. Chem. 281, 379 (1976)
26. Möckel, H.J., Freyholdt, T.: In: Kaiser, U.J., Franzen, K.H. (Hrsg.); Königsteiner Chromatographietage 1980, Waters, Königstein 1980
27. Möckel, H.J., Freyholdt, T., Weiß, J., Molner, I.: In: Molner I. (ed.): "Practical aspects of HPLC". DeGruyter, Berlin 1982
28. Steudel, R., Mäusle, H.-J., Rosenbauer, D., Möckel, H.J., Freyholdt, T.: Angew. Chem. 93, 402 (1981)
29. Steudel, R., Mäusle, H.J.: Angew. Chem. 91, 165 (1979)
30. Möckel, H.J., Freyholdt, T.: Z. Anal. Chem. 368, 401 (1981)
31. Pearlman, R.S.: In: Yalkowski, S.H., Sinkula, A.A., Valvani, S.C. (eds.) "Physical Chemical Properties of Drugs". Marcel Dekker, New York 1980
32. Pearlman, R.S.: QCPE Bull. 1, 15 (1981)
33. Möckel, H.J.: J. Chromatogr. 317, 589 (1984)
34. Engelhardt, H., Ahr, C.: Chromatographia 14, 227 (1981)
35. Möckel, H.J., Freyholdt, T.: Chromatographia 17, 215 (1983)
36. Möckel, H.J.: Z. Anal. Chem. 318, 327 (1984)
37. Möckel, H.J., Weiß, J., unpublished results
38. Möckel, H.J., Molnar, I., to be published
39. Möckel, H.J.: Z. Anal. Chem. 318, 116 (1984)
40. Hiller, K.O., Masloch, B., Möckel, H.J.: Z. Anal. Chem. 280, 293 (1976)
41. Möckel, H.J., Weiß, J., unpublished results
42. Möckel, H.J., Höfler, F., Melzer, H.: Chromatographia 8, 471 (1985)
43. Möckel, H.J., Melzer, H., Höfler, F., Fojtik, A.T.: Proceedings of the 1985 Budapest Chromatography Symposium, in press
44. Schmidtke, F., Höfler, F., Asmus, K.-D., Möckel, H.J., in preparation
45. Steudel, R., Strauss, E.M.: Z. Naturforsch. 36b, 1085 (1981)
46. Möckel, H.J., Freyholdt, T., Melzer, H.: In: Franzen, K.H. (Hrsg.) „Königsteiner Chromatographietage 1982", Waters, Königstein 1982
47. Steudel, R., Laitinen, R.: Topics in Curr. Chem. 102, 177 (1982)
48. Steudel, R., Strauss, E.M.: Z. Naturforsch. 36b, 719 (1983)
49. Durrant, P.J., Durrant, B.: "Advanced Inorganic Chemistry". Longman, London 1962
50. Steudel, R., Rosenbauer, R.: J. Chromatogr. 216, 399 (1981)
51. Möckel, H.J., Melzer, H., in preparation
52. Brough, L.F., Matsumura, K., West, R.: Angew. Chem. 91, 1022 (1979)

HPLC in Forensic Chemistry

T. Daldrup, P. Michalke and S. Szathmary
Institut für Rechtsmedizin, Universität Düsseldorf,
Moorenstr. 1, 4000 Düsseldorf 1/FRG

1 Introduction

Forensic chemistry consists of chemistry in relation to the administration of justice; i.e. forensic chemistry is concerned with the processing of disputed questions of fact in criminal, civil and insurance law, which can be clarified by the application of chemical methods and knowledge. Depending on the speciality forensic chemistry is a branch of scientific criminology or of forensic toxicology. Purely analytical methods, particularly in the field of trace analysis and materials science are to the forefront in scientific criminology. Typical examples are the comparative investigations of confiscated addictive drugs to determine their origins or the investigation and classification of traces of paint after a traffic accident. In contrast it is the province of forensic toxicology, to give evidence concerning the effects of substances (poisons) on men and more rarely on animals in disputed court cases. Forensic toxicology for its part is a portion of forensic medicine alongside forensic pathology, morphology, psychiatry and serology.

The development of forensic chemistry has been closely connected with the development of analytical methods ever since its very beginning in the year 1851, when Jean Servais Stas[1] employed chemical analysis to demonstrate the presence of nicotine in the stomach of a poisoning victim and this result became the object of a legal process. Maehly and Strömberg[2] name 1950 as the year that forensic chemistry was born, since the application of new techniques such as UV and IR spectrophotometry, X-ray diffraction and paper chromatography led to an explosive development after this date. Of the chromatographic processes paper chromatography was gradually replaced by the more rapid and sensitive thin-layer chromatography, particularly after the fundamental investigations of Egon Stahl[3]. During the 1960s gas chromatography became of great importance and in the mid 1970s, particularly after the development of reversed-phase chromatography, HPLC conquered the analytical laboratories of forensic chemistry.

There is scarcely a medicament, scarcely a foreign substance that cannot play a role in a forensic problem. A forensically (toxicologically) important substance is only involved in the minority of cases. This can be emphasized by the following example:

Eleven seemingly intact tablets were discovered in the upper intestinal tract of a suspected case of homicide. It was only after a considerable expense of analytical effort, including the employment of HPLC, that is was possible to demonstrate that these tablets were harmless sugar-coated vitamin E tablets which had lost their glaze and scarcely to be implicated in the death. This example shows that the HPLC identification of vitamins can come into the province of forensic chemistry. However we wish to limit ourselves here to the areas typical of forensic chemistry and have concentrated on the analytical problems where HPLC techniques offer advantages over other chromatographic methods. These include screening investigations (general unknown) e.g. in intoxication, further investigations of dealing in addictive drugs and the consumption of addictive drugs and finally from the fields concerned with criminology, including the identification of explosives and dyestuffs. We would like at this point to refer you to the review article by Wheals and Jane[160], which covers the literature up to 1977.

2 Screening Investigations

The investigations of biological (human) materials, such as gastric contents, blood, urine, organ samples etc. as well as substances discovered at the scene of the crime, for toxicologically relevant foreign substances (poisons, medicaments which are toxic in overdose etc.) constitutes one of the most important tasks of a forensic chemist. In the majority of cases the substances to be detected are unknown at the commencement of the analysis so that broadly based screening investigations have to be carried out. Another difficulty is that the analysis must often be carried out in great haste, since life-saving measures may depend on the result of the analysis, e.g. in certain intoxications. The forensic chemist must, therefore, possess rapid analytical methods, which include a broad spectrum of substances and allow identification via data collections. Numerous proven screening systems and data collections have been developed for both gas chromatography and TLC[4−7]. HPLC is gaining more and more in importance as a third chromatographic method, because it is possible to detect numerous substances with it which are difficult or impossible to detect by other methods. Recently comprehensive tables of HPLC retention data have become available for numerous toxically relevant pharmacologicals and related compounds[6,8] (cf. Table 1).

2.1 Extraction Processes

When reversed-phase materials are employed extraction before HPLC can often be omitted. Thus deproteinated blood samples can be chromatographed directly after the addition of acetonitrile or methanol[9,10,29,66]. However, prior extraction of biological materials is not usually omitted before screening investigations, particularly if cadaveric tissue is to be investigated. It is basically possible to apply to HPLC all the extraction methods developed for employment in conjunction with thin-layer or gas chromatography. A repetition of the multiplicity of possibilities lies outside the scope of this article. It should only be mentioned here that the employment of absorption columns such as Extrelut® or Extube® has proven itself for working up investigation materials from very complex matrices including cadaveric organs[1]. The extracts exhibit a relatively low proportion of interfering substances combined with a high recovery rate for the majority of substances sought (11, own unpublished investigation). The use of hexane as an extractant has led to good results particularly with basic pharmaceuticals with low effective blood levels of a few ng/ml[12,13,22]. Extracts are obtained with good to very good recovery rates that allow the employment of the highest sensitivity of the UV detector (Fig. 1).

2.2 Data Collection and the Choice of the Separation Material

Perusal of the literature reveals that pure absorption materials (silica gel) are only rarely employed as column packings[22,23]. C_{18} reversed-phase material is primarily employed for this purpose[6,8,14−21,154]. For this reason we will limit ourselves

---

1 Extraction at pH 8 with ethyl acetate/ether 1:1

Table 1. Relative retention times of drugs and related compounds. Chromatography: isocratic conditions; eluent: 312 g acetonitrile/688 g phosphate buffer (4.8 g of 85% orthophosphoric acid and 6.66 g of KH_2PO_4 in 1 l water, pH 2.30); column: Nucleosil-C 18, 10 μm, 250 × 4 mm; temperature: 26 °C; flow: 1 ml/min; detection: UV, 220 nm; Reference compound: 5-(p-Methylphenyl)-5-phenylhydantoin (= MPPH). Sensitivity for LCII: + +: very good, +: good, 0: reasonable, S: bad. (from [8])

Compound	RRT	Sens.	Compound	RRT	Sens.
Acebutolol	0.24	++	Bevoniummetil-	0.43	++
Acecarbromal	0.27	+	sulfate		
Acefylline	-	-	Bezafibrat	1.78	++
Acetazolamide	0.25	+	Biperiden	0.72	0
Acetyldigoxine	0.40	++	Bisacodyl	0.39	++
Acetylsalicyclic			Bornaprine	1.23	0
acid	0.52	++	Brallobarbital	0.47	++
Ajmaline	0.29	+	Bromazepam	0.42	++
Alimemazine	0.84	++	Bromhexine	0.76	++
Allobarbital	0.39	+	Bromisoval	0.45	++
Allopurinol	0.19	++	Bromophosethyl	-	-
Alprenolol	0.42	++	Bromophosmethyl	-	-
Amantadine	-	-	Bromoprid	0.27	++
Ambutoniumbromide	0.31	+	Brompheniramine	0.29	++
Amiloride	0.20	++	Bumadizon	2.49 3.22	S
Aminophenazone	0.19	+	Bumetanid	2.00	++
Amiphenazole	0.23 0.52	++	Bunitrolol	0.26	+
Amitriptyline	0.90	++	Buphenine	0.39	++
Amobarbital	0.70	++	Bupivacaine	-	++
Amphetamine	0.70	S	Bupranolol	0.43	++
Amphetaminil	0.73	0	Busulfan	-	-
Antazoline	0.50	++	Butabarbital	0.47	++
Aprindine	-	-	Butalamine	0.86	++
Aprobarbital	0.41	++	Butalbital	0.54	++
Atenolol	0.18	0	Butamirate	0.61	S
Atropine	0.23	0	Butanilicaine	0.31	++
Azapetine	0.41	++	Butaperazine	0.19 0.67	++
Azapropazone	0.31	+	Butetamat	0.34	+
Azinphosethyl	6.68	+	Butizid	0.82	0
Azinphosmethyl	-	-	Cafaminol	0.23	+
Baclofen	0.23	++	Cafedrine	0.23	+
Bamethan	0.23	++	Camazepam	2.55	++
Bamifylline	0.23	+	Camylofine	0.25	0
Bamipine	0.60	++	Carbachol	-	-
Barbital	0.30	++	Carbamazepine	0.67	++
Beclamide	0.64	+	L-Carbidopa	0.18 0.29	++
Bemegride	0.35	0	Carbimazol	0.22	++
Bemetizid	1.28 1.34	++	Carbocisteine	-	-
Bencyclan	1.40	0	Carbocromen	0.23 0.32	+
Bendroflumethia-			Carbromal	0.67	+
zide	1.52	++	Carbutamid	0.50	++
Benorilate	0.24	+	Carisoprodol	-	-
Benperidol	0.41	0	Carticaine	0.25	+
Benproperine	1.34	++	CBD	-	-
Benserazide	0.17	++	CBN	-	-
Benzatropine	0.96	++	Cetobemidon	1.70	++
Benzbromaron	-	S	Chloralhydrate	-	-
Benzilic-acid-	0.41 0.50	++	Chlorambucil	Z	-
tropinester			Chlorazanile	0.39	++
Benzoctamine	0.42	+	Chlorbenzoxamine	3.26	0
Benzydamine	0.64	++	Chlordiazepoxid		
Bepheniumhydro-	0.54 1.16	++	(D)	0.30	++
xynaphthoat			Chlorfenvinfos	7.23	0
Betahistine	-	-	Chlormezanone	0.61	++

Table 1 (continued)

Compound	RRT	Sens.	Compound	RRT	Sens.
Chloroquine	0.23	+	Dilazep	1.00	++
Chlorphenamine	0.28	++	Dimetacrine	1.56	++
Chlorphenoxamine	1.09	++	Dimethoate	0.41	++
Chlorpromazine	1.14	++	Dimetinden	-	-
Chlorprothixene	1.31/1.44	+	Diphenhydramine	0.51	++
Chlorquinaldol	-	-	Diponiumbromid	-	-
Chlortalidone	0.34	++	Dipyridamol	0.39	++
8-Chlortheo-phyllin	0.26	++	Disopyramide	0.31	+
Chlorzoxazone	0.68	++	Distigmine	0.30	S
Cicloniumbromide	2.30	0	Disulfiram	-	-
Cimetidine	0.19	+	Disulfoton	-	-
Cinnarizine	3.28	S	Dixyrazine	0.51	+
Clemastine	2.04	+	Dobutamine	-	-
Clenbuterol	0.03	0	Dopamine	0.18	++
Clidinimbromide	0.39	+	Doxapram	0.34	++
Clobazam	1.36	++	Doxepine	0.56	++
Clobutinol	0.42	++	Doxylamine	0.23	+
Clofedanol	0.44	++	Drofenine	1.63	0
Clofezone	0.32	+	Droperidol	0.40	++
Clofibrat	-	-	Dropropizine	0.23	++
Clomethiazol	0.68	0	Ephedrine	0.22	++
Clomifen	-	-	Epinephrine	-	-
Clomipramine	1.31	++	Eprazinon	0.77	0
Clonazepam	0.98	++	Estramustine	-	-
Clonidine	0.23	+	Etacrynic-acid	2.44	++
Clopamid	0.39	++	Etafenon	0.85	++
Clopenthixol	0.57	++	Ethacridine	0.56	+
Clorexolon	0.82 0.96	++	Ethambutol	-	-
Cocaine	0.33	+	Ethaverine	1.27	++
Codeine	0.20	++	Ethinamate	-	-
Caffeine	0.24	++	Ethion	-	-
Crotylbarbital	0.41	++	Ethosuximide	0.33	S
Cyclobarbital	0.51	++	Ethoxzolamide	0.85	++
Cyclopenthiazide	0.28 1.47	++	Etifelmine	0.60	++
Cyclopentobarbital	0.52	++	Etilefrine	0.19	+
Cyclophosphamide	-	-	Etiroxat	1.82	++
Cyproheptadine	0.78	++	Etodroxizine	0.18	++
Cytarabine	0.17	++	Etofibrat	0.32 4.13	0
Dantrolen	0.77	0	Etomidat	0.71	0
Deanol	-	-	Etozoline	1.42	0
Demeclocycline	0.27	++	Fencamfamine	0.38	++
Desipramine	0.70	++	Fenchlorphos	0.61	S
Dextromethorphane	0.45	++	Fenetylline	0.67	++
Dextromoramide	0.87	+	Fenfluramine	0.40	0
Dextropropoxyphene	0.90	++	Fenitrothion	6.32	0
Diazepam	1.54	++	Fenoprofen	2.84	++
Diazinone	-	-	Fenoterol	0.22	++
Diazoxide	0.41	++	Fenpiverinium-bromide	0.36	+
Dibenzepine	0.35	++	Fenproporex	0.27	0
Dichlorvos	0.74	0	Fentanyl	0.52	+
Diclofenac-Na	4.33	+	Fentoniumbromide	2.15	0
Dicycloverine	-	-	Fenyramidol	0.24	+
Diethylcarbamazine	0.18	+	Floctafenine	0.30 1.17	++
Diethylpentenamid	-	-	Fluanison	0.63	++
Digitoxine	2.04	0	Flufenaminic-acid	-	
Digoxine	0.41	++	Flunarizine	4.50	0
Dihydralazine	Z	-	Flunitrazepam	1.27	++
Dihydrocodeine	0.20	++	Fluoro-Uracil		
Dihydroergotamine	0.46	++	Flupentixol	0.86	++
Dikaliumchlor-azepate	0.82	++	Fluphenazine	0.67	++
			Flurazepam	0.46	++

Table 1 (continued)

Compound	RRT	Sens.	Compound	RRT	Sens.
Flurbiprofen	3.24	0	Levorphanol	0.27	+
Fluspirilen	0.39	+	Lidocaine	0.26	+
Fominoben	0.34	++	Lidoflazine	2.6	0
Furosemid	Z		Lobeline	Z	S
Glafenine	0.24 0.53	++	Lofepramine	0.7 (Z)	++
Glibenclamide	0.99	0	Loperamide	2.1	++
Glibornuride	2.81	0	Lorazepam	0.82	++
Gliclazide	1.96	0	LSD (Lysergid)	-	
Glipizid	1.13	++	Malathion	5.18	S
Gliquidon	-		Maprotiline	0.80	++
Glisoxepid	-		Mazindol	0.33	++
Glutethimide	0.87	++	Mebendazol	0.56	+
Glycopyronium-bromide	0.46	+	Mebeverine	0.94	++
			Mebhydroline	0.59	++
Glymidine	0.53	++	Meclofenoxate	1.97	0
Guaifenesin	0.29	++	Meclozine	4.97	S
Guanethidin	-		Medazepam	0.52	++
Haloperidol	0.50 0.67	++	Mefenaminic-acid	7.41	+
Heptabarbital	0.57	++	Mefenorex	-	
Heptaminol	-		Mefrusid	1.01	++
Heroine (Diacetylmorphine)	0.34	++	Melitracen	1.23	++
Hexetidine	-		Melperon	0.40	0
Hexobarbital	0.73	++	Melphalan	-	
Hexobendine	0.29	++	Menbuton	1.28	++
Hexocyclium-methilsulfate	0.50	0	Mepivacaine	0.25	+
			Meprobamate	-	
Hexoprenaline	-		Mequitazine	2.77	S
Homofenazine	0.40	0	6-Mercaptopurine	-	
Hydrochlorothiazide	0.30	++	Mescaline	0.20 Z	++
Hydrocodon	0.22	+	Mesuximide	0.86	++
Hydromorphone	0.19	++	Metamizol	0.19	+
4-Hydroxibutyrate	-		Metasystox R	-	
Hydroxycarbamide	-		Methadone	0.98	++
Hydroxychloroquine	0.17	++	Methamphetamine	0.25	0
Hydroxyzine	0.72	++	Methanthelinium-bromide	0.73	++
Hyoscin-N-butyl-bromide	0.33	+			
			Methaqualone	1.10	++
Ibuprofen	4.17	0	Methidathion	2.62	0
Ifosfamide	-		Methocarbamol	0.31	++
Imipramine	0.78	++	Methohexital	1.57	++
Imolamine	0.29	++	Methotrexat	-	
Indapamide	0.89	++	Methyldigoxine	0.74	+
Indometacine	0.40	++	Methyldopa	-	
Ipratropiumbromide	0.26	+	Methylpentynol	-	
Isoaminil	0.41	S	Methylphenidat	0.26	++
Isoetarine	0.25	0	Methylpheno-barbital	0.70 0.83	++
Isoniazide	0.18	++			
Isoprenaline	0.24	S	Methypranol	0.27 0.39	++
Isopropamidjodide	0.19 0.44	++	Methyprylone	0.38	+
Isosorbiddinitrate	1.26	++	Methysergide	0.25 0.31	0
Isothipendyl	0.55	++	Metixen	1.06	++
Isoxsuprine	0.32	++	Metoclopramide	0.27	++
Kavaine	1.69	++	Metolazon	0.68	++
Ketamine	0.24	+	Metoprolol	0.25 ·	+
Ketoprofen	1.48	++	Metronidazol	0.24	++
Ketotifen	0.35	++	Metyrapone	0.19 0.24	++
Labetalol	0.32	++	Mevinphos	0.44	++
Lanatosid C	0.33	0	Mianserine	0.50	++
Levallorphan	0.33	+	Midodrine	-	
Levodopa	-		Mitopodozid	0.26	++
Levomepromazine	0.45 0.94	++	Mofebutazon	0.70	++
			Molsidomine	-	

Table 1 (continued)

Compound	RRT	Sens.	Compound	RRT	Sens.
Monocrotophos	-		Papaverine	0.34	++
Moperon	0.60	++	Paracetamol	0.24	++
Morazon	0.33	++	Paraoxon	1.46	++
Morphine	0.18	++	Parathioncthyl	-	
Moxaverine	0.65	++	Parathionmethyl	4.41	0
Moxisylyt	0.27	++	Pemoline	0.27	+
Nadolol	0.19	++	Penfluridol	-	
Naftidrofuryl	2.28	S	Penicillamine	-	
Naproxen	1.57	++	Pentaerythrityl-	5.8	S
Nefopam	0.36	++	tetranitrate		
Neostigmine	0.21	++	Pentazocine	0.41	++
Nicethamide	0.25	++	Pentetrazol	-	
Niclosamide	0.41	++	Pentobarbital	0.68	++
Nifedipine	1.70	++	Pentorex	0.25	0
Nifenazone	0.23	++	Pentoxifylline	0.31	++
Nifluminic-acid	3.25	0	Pentoxyverine	-	
Nifuratel	0.81	S	Perazine	0.34	++
Nicotine	-		Perhexiline	-	
Nimorazol	0.19	++	Periciazine	0.55	++
Nitrazepam	0.77	++	Perphenazine	-	
Nitroglycerine	2.18	S	Pethidine	0.31	+
p-Nitrophenol	0.60	++	Phenacetine	0.50	++
Nitroxoline	-		Phenazone	0.31	++
Nomifensine	0.30	++	Phenazopyridine	0.38	++
Noradrenaline	0.18	+	Phencyclidine	0.47	+
Norfenefrine	-		Pheniramine	0.25	+
Normethadone	0.73	++	Phenobarbital	0.46	++
Norpseudoephedrine	-		Phenolphthaleine	0.93	++
Nortriptyline	0.74	++	Phenoxybenzamine	0.78	++
Noscapine	0.34	++	Phenprobamate	0.83	+
Noxiptiline	0.66	++	Phenprocumon	4.09	+
Obidoximchloride	-		Phenylbutazone	5.59	0
Octopamine	-		Phenylephrine	0.18	++
Omethoat	-		Phenytoine	0.70	++
Opipramol	0.27	++	Pholcodine	-	
Orciprenaline	-		Pholedrine	0.20	++
Ornidazol	0.35	++	Physostigmin-	0.23 <u>0.39</u>	++
Orphenadrine	0.68	++	salicylat		
Oxazepam	0.72	++	Pindolol	0.24	++
Oxeladine	-		Pipamperon	0.19	0
Oxetoron	1.00	++	Pipazetat	0.43	++
Oxomemazine	0.34	++	Pipenzolatbromide	0.45	+
Oxophenhydroxzine	-		Piperazine	-	
Oxprenolol	0.32	+	Piperidolat	0.73	++
Oxycodone	0.22	++	Piprinhydrinate	0.64	+
Oxyfedrine	0.99	++	Piracetam	-	
Oxymetazoline	0.46	0	Pirenzepine	0.19	++
Oxypendyl	0.25	+	Piribedil	0.28	+
Oxypertine	0.53	++	Piridoxylat	-	
Oxyphenbutazone	Z		Piritramide	0.70 1.10	++
Oxyphencyclimine	0.18 0.48	0	Piroxicam	0.90	S
Oxyphenonium-	0.77	0	Primidone	0.29	++
Pitofenon	0.44	++	Probenecid	2.14	S
Pizotifen	0.78	++	Procaine	0.57	++
Polythiazide	1.48	++	Procarbazine (Z)	0.41 0.69	++
Prajmaline	0.54	+	Procyclidine	0.42 <u>0.67</u>	0
Pramiverine	0.94	0	Proglumid	-	
Prazepam	4.34	++	Prolintane	0.44	0
Prazosine	0.27	+	Promazine	0.67	++
Prenoxdiazine	0.52	++	Promethazine	0.63	++
Prenylamine	2.34	+	Propafenon	0.75	++
Pridinolbromide	0.48	+	Propallylonal	0.50	++

Table 1 (continued)

Compound	RRT	Sens.	Compound	RRT	Sens.
Propanidid	1.87	+	Tiemoniummethyl- sulfate	0.33	+
Propoxur	0.88	++	Tildine	0.34	+
Propranolol	0.39	++	Timolol	0.24	0
Propylhexedrine	-		Tinidazol	0.32	+
Propyphenazone	0.95	++	Tioguanine	-	
Proquazon	2.38	++	Tiotixen	0.41 0.51	++
Proscillaridine	0.67	0	Tolazamide	1.34	++
Prothipendyl	0.49	++	Tolbutamide	1.22	++
Protionamide	0.24	++	Toliprolol	0.30	+
Protriptyline	-		Tolmetine	1.21	++
Proxibarbal	0.25	+	Tramadol	0.28	+
Proxyphylline	0.23	++	Tramazoline	0.34	+
Pyrantel	3.17 4.73	0	Tranylcypromine	0.22	+
Pyridostigmin- bromide	0.18	+	Trazodon	0.36	++
Pyrimethamine	0.34	++	Triamiphos	0.66	+
Pyrithyldione	0.34	+	Triamteren	0.19	++
Pyritinol	0.19	+	Triazolam	1.20	++
Pyrviniumpamoat	0.24 4.61	+	Trichlormethiazide	0.58	S
Quinethazon	0.25	0	Trichlorphon	0.74	S
Quindine	0.21	++	Tridihexethyl- chloride	-	
Quinine	0.20 0.34	++	Trifluoperazine	0.78	0
Reproterol	0.18	++	Trifluperidol	0.50	++
Reserpine	1.28	0	Triflupromazine	1.64	+
Salbutamol	0.19	++	Trihexyphenidyl	0.81	0
Salicylamide	0.36	++	Trimethadione	0.39	S
Scopolamine	0.22	+	Trimipramine	-	
Scopolamin-N- methylbromide	0.23	0	Triprolidine	0.31	++
			Tritoqualine	1.63	++
Secobarbital	0.85	++	Trofosfamide	-	
Sotalol	0.20	+	Trolnitrate	-	
Sparteine	-		Trospiumchloride	0.51++	
Spironolactone	3.12	S	Triazolam	1.20	++
g-Strophanthine	2.05	S	Trichlormethiazide	0.58	S
Strychnine	0.23	+	Trichlorphon	0.74	S
Strychnine-N-Oxide	-		Tridihexethyl- chloride	-	
Sulfinpyrazone	1.69	++	Trifluoperazine	0.78	0
Sulforidazine	0.59	0	Trifluperidol	0.50	++
Sulindac	1.11	++	Triflupromazine	1.64	+
Sulpirid	0.20	++	Trihexyphenidyl	0.81	0
Sultiam	0.36	+	Trimethadione	0.39	S
Suxibuzon	3.74	S	Trimipramine	-	
Synephrine	0.18	++	Triprolidine	0.31	++
Terbutaline	0.19	+	Tritoqualine	1.63	++
Tetracaine	0.49 4.85	0	Trofosfamide	-	
Tetryzoline	-		Trolnitrate	-	
THC 1 (2)	-		Trospiumchloride	0.51	++
THC 1 (6)			Viloxazine	0.26	+
Thebacon	0.27	++	Viminol	1.51	++
Theophylline	0.21	++	Vincamine	0.35	++
Thiamazol	0.22	+	Vincristine	-	
Thietylperazine	0.79	0	Vinylbital	0.70	++
Thiobutabarbital	0.48 0.88	0	Xenytropiumbromide	1.43	0
Thiopental	1.38	0	Xipamid	1.55	S
Thioridazine	1.88	++	Xylometazoline	0.68	++
Thiotepa	0.39	0	Yohimbine	0.29	+
Thiabendazol	0.24	+			
Tiaprid	0.22	++			

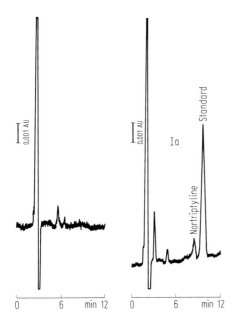

Fig. 1. Chromatograms of two sera extracted with hexane at pH 12–13. Left: normal serum; right: serum from a patient after taking nortriptyline. Standard: amitriptyline, nortriptyline concentration: 46 ng/ml. The area under the peak in the chromatogram corresponds to ca. 2 ng nortriptyline. HPLC conditions: Nucleosil C_{18} 10 μ; acetonitrile/0.09 M phosphate buffer pH 2.3, 31% w/w; 1 ml/min; 220 nm

here to a further discussion of this material. The separating column is subject to rigorous quality requirements, particularly with regard to selectivity and batch to batch reproducibility if it is to be employed for screening various compounds of very different molecular structure and polarity. The latter requirement is of particular importance for the compilation of data collections.

HPLC columns are normally tested with non-polar, neutral compounds such as aromatics, whose separation causes scarcely any problems[24]. The situation becomes more critical if the column material is tested against polar substances[24, 25]. The considerable differences between different stationary phases were revealed by our own investigations where we determined over 550 different toxicologically relevant substances on three different C_{18} phases. Many, particularly basic, pharmaceuticals could only be detected using one phase[8], while they could not be eluted or were only eluted as slow asymmetric peaks from the other phases[6].

With the aid of these copious data three substances were chosen for the testing of HPLC columns. They were 5-(p-methylphenyl)-5-phenylhydantoin (=MPPH), a relatively neutral substance that serves in the test as a standard, diazepam and diphenhydramine, a polar basic substance, which is an indicator of residual polarity particularly that caused by free silanol groups[26]. Columns sometimes exhibit considerable differences from manufacturer to manufacturer (Fig. 2) and from batch to batch so that a collection of reference retention times for the identification of foreign substances, such as the Kovats indices usual in GC, is still problematical for HPLC.

It is our experience that the test allows an adequate standardization of column materials in our own laboratory. For instance we only employ for screening investigations column material (Nucleosil C_{18}), on which diphenhydramine and diazepam

Table 2. HPLC and drug screening

Subject	Column/Detection	Mobile phase		Ref.
		Organic	Aqueous	
30 compounds; chromatographic behaviour	μ-Bondapak C_{18}/ 220 nm	MeOH: 0–80%	0.025 M phosphate buffer; pH 3–9	[17]
50 drugs of forensic interest	μ-Bondapak C_{18}/ 254 nm	MeOH: 40 0.005 M heptane sulphonate	AcOH: 1 H_2O: 59 pH 3.5	[14]
35 anticonvulsants and their metabolites	μ-Bondapak C_{18}/ 195 nm; 50° C	AcN: 21%	phosphate buffer pH 4.4	[154]
54 drugs of clinical and forensic interest	Partisil-10 ODS/ 220 nm	MeOH: 40%	phosphate buffer pH 2.3	[29]
42 commonly abused drugs	μ-Bondapak C_{18}/ 210 nm	AcN: 5–45%	phosphate buffer pH 3.5	[21]
570 compounds of clinical and forensic interest	LiChrosorb RP 18/ 220 nm	AcN: 31.2% or 31.2–90%	phosphate buffer pH 2.3 or H_2O	[6]
560 compounds of clinical and forensic interest	C-18 SIL−X−10 (Nucleosil)/220 nm	AcN: 31.2% or 31.2–90%	phosphate buffer pH 2.3 or H_2O	[8]
74 drugs of clinical and forensic interest	Mikropack MCH 10 RP 18/220 or 230 nm	AcN (azeotrope) 30–60%	0.005 M HCLO + 0.015 M $NaClO_4$	[13]

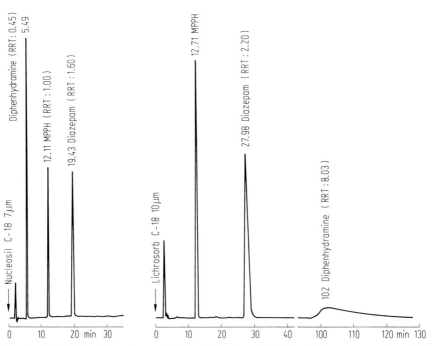

Fig. 2. DMD test: Separation of diphenhydramine, MPPH and diazepam on two different C_{18} phases. HPLC conditions acetonitrile/0.09 M phosphate buffer pH 2.3, 31.2% w/w; 1 ml/ min; 220 nm

are eluted as symmetrical peaks with RRT of 0.5 and 1.50 to 1.55 respectively in comparison with MPPH. Since this has been done there have been no significant problems concerning the reproducibility of RRT values, such as are reported in Table 1.

2.3 Detectors and Sensitivity of Detection

At the moment there does not seem to be any alternative to the UV spectrometer, usually with variable wavelength, as a detector as far as screening investigations are concerned. The coupled LC-MS that is appearing on the market more and more [30, 31, 166] has not yet found a permanent place in the forensic laboratory. It is to be expected, however, that in the future this form of detection will occupy a similarly important position to that of GC-MS in forensic analysis, particularly for the identification of the multitude of compounds that it is not possible to gas chromatograph.

The choice of wavelength usually depends upon the problem. When possible a typical maximum in the UV spectrum is chosen, e.g. 315 nm in the determination of p-nitrophenol in the serum in cases of parathion poisoning [32]. However the most commonly employed wavelengths are 195, 210, 220 or 254 nm. Some determine two wavelengths simultaneously to improve the quality of analysis [27]. Others exploit the possibility of recording the spectrum in order to identify substances [8, 29, 161]. The other types of detectors, particularly fluorescence detectors and electrochemical detectors, are mainly only employed for the solution of specific analytical problems, such as the determination of LSD in serum [33].

The sensitivity of detection of the commercial UV detectors is normally adequate for the quantitative determination of low dosage pharmaceuticals, e.g. in the few ng/ml serum range, particularly when recording in the lower UV range (195–220 nm) [12, 13]. Our own investigations have revealed that the large proportion of forensically relevant pharmaceuticals and related substances can be well or very well detected at 220 nm [8].

3 Intoxications

The most common compounds which cause death are medicaments (particularly the hypnotics) and related compounds (alkaloids, addictive drugs), plant protection agents, particularly the cholinesterase inhibitors, carbon monoxide, alcohol, heavy metals (particularly thallium, mercury and arsenic) and other inorganic agents (cyanide, fluoride) as well as various household and industrial chemicals (e.g. acids such as formic acid). The majority of these poisons can be detected by HPLC. There have been an appropriately large number of HPLC investigations concerning these topics. Thus intoxications with colchicine [34], maprotiline [38], barbiturates [39], verapamil [41], endosulfan [35], carbamates [36] and formic acid [40] have all been detected by HPLC investigations. Heavy metals can also be detected in the form of their diethyldithiocarbamate chelates by the reversed-phase technique [42]. After a fatal nickel sulphate intoxication it proved possible to detect this heavy metal in portions of the cadaver by HPLC [43].

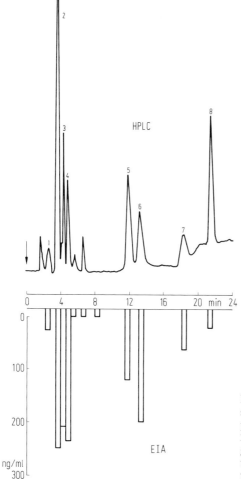

Fig. 3. Synthetic mixture of cardiac glycosides and their metabolites. Off-line coupling of HPLC with Emit. 1 digoxigenin, 2 digoxin monodigitoxose, 3 digoxin bisdigitoxose, 4 digoxin, 5 digitoxigenin, 6 digitoxin monodigitoxose, 7 digitoxin bisdigitoxose, 8 digitoxin

Another interesting possibility is the indirect coupling of HPLC with immunochemical processes, as shown for digoxin and digoxin in Figs. 2 and 3. Here it proved possible to detect the various glycoside metabolites in body fluids with high sensitivity. This method was employed to recognize signs of cardiac glycoside intoxications by post mortem organ examination[44].

4 Addictive Drugs

Investigations of addictive drugs of all types form part of the daily routine of forensic chemistry particularly since the precipitate increase in the consumption of addictive drugs in the 1960s. A distinction must be made between investigations of the drugs

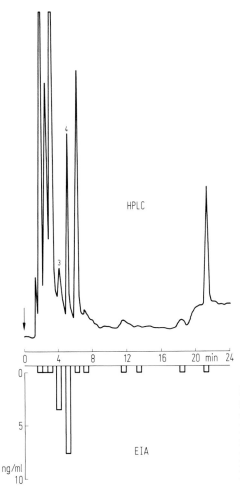

Fig. 4. Chromatogram of a cardiac blood extract from a patient treated ante-mortem with digoxin. Off-line coupling, HPLC-Emit. The high blood concentrations do not indicate an intoxication, but are caused by post-mortem diffusion of the substance from tissue into the blood. 3 digoxin bisdigitoxose, 4 digoxin

themselves, particularly their identification and the determination of their drug content, and the investigation of body fluids for evidence of drug consumption.

The majority of addictive compounds are of vegetable origin or are produced synthetically. All play a more or less important role in the illegal drug market. Only those addictive drugs most commonly employed will be dealt with here. They include cannabis products, the opiates, cocaine, LSD and others. HPLC analyses are available for all these drugs [155, 156]. The following sections are primarily devoted to problems where HPLC possesses particular advantages over other chromatographic methods.

4.1 Cannabis

The various cannabis products are amongst the least toxic addictive drugs. Severe intoxication is rarely reported. The physical addictive potential is also relatively

Table 3. HPLC of cannabinoids

Subject	Column/Detection	Mobile phase		Ref.
		Organic	Aqueous	
Cannabinoids in cannabis	Partisil 5 ODS/ 254 nm	MeOH: 4	0.02 N H_2SO_4: 1	[155]
Cannabinoids in plasma	Varian Si-10/ 273.7 nm; GC/MS	Heptane: 95–5% – DCM: 5–95% gradient		[157]
CBN in urine	Spherisorb 5 ODS resp. Partisil 10 PAC/ photochem detection	MeOH resp. isooctane-dioxane	Na_2HPO_4; 0.025 M, pH 8 resp.	[158]
THC metabolites in urine; off-line TLC-HPLC	Spherisorb silica/ 220 nm	Hexane: 97 3% MeOH in DCM: 3	–	[61]
Cannabinoids in plasma; off-line HPLC-GC/MS	Spherisorb 5 ODS/ 280 nm; GC/MS	MeOH: 50–72.5%	H_2O	[68]
Cannabinoids in urine; off-line HPLC-RIA	Hypersil 5 ODS/RIA	MeOH: 70%	Phosphate buffer pH 5.1	[69]
THC + metabolites in urine; off-line HPLC-Emit	Hypersil 5 ODS Emit	MeOH: 70%	0.025 M phospate buffer pH 5.1	[70]
Cannabinoids in cannabis resin; off-line HPLC-GC/MS	Spherisorb 5 ODS/ 220 nm; GC/MS	MeOH: 8 AcN: 9	0.02 N H_2SO_4: 7	[52]
Cannabinoids and drugs in blood; forensic cases	LiChrosorb RP 18/ 220 nm; GC/MS	AcN: 31–90% gradient	H_2O	[65]
Cannabinoids in tissue; post-column derivatization with Fast Blue Salt B	μ-Porasil or μ-Bonda-pak C_{18}/490 nm	MeOH: 85%	H_2O	[159]
THC in pharmaceutical vehicles	LiChrosorb RP 8 and RP 2/280 nm	MeOH: 78 or 73%	H_2O	[50]
THC in biological samples	μ-Bondapak-Phenyl/ electrochem. detector	AcN: 55%	0.05 M KH_2PO_4	[60]
THC and metabolites in plasma and urine; off-line HPLC-RIA-GC/MS	Spherisorb 5 ODS/ 280 nm; RIA; GC/MS	MeOH: 50–72.5%	H_2O	[59]
Cannabinoids in cannabis	LiChrosorb RP 18/ 220 nm	MeOH: 34% AcN: 37%	0.02 N H_2SO_4	[49]

small. In contrast this drug plays a not insignificant role in the road traffic situation because of its sedating and subduing effect. Traffic accidents caused by the consumption of marijuana are not infrequently observed[45,47]. The grey zone is likely to be very large, since the proof that someone was under the influence of marijuana at the time of an accident is only possible by means of very difficult analytical

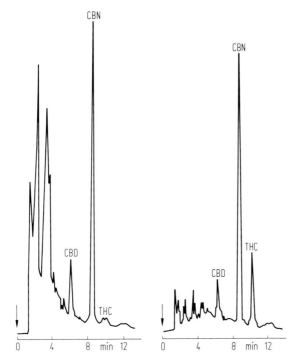

Fig. 5. Chromatogram of a marijuana extract before (right) and after (left) heating for 3 minutes to 200° C

procedures, which can only be carried out by a very few laboratories. On the other hand, it is relatively simple to demonstrate the presence of the various cannabinoids in marijuana products. In forensic chemistry these investigations are mainly undertaken with the aim of obtaining an objective measure of the amount of drug present on the basis of its tetrahydrocannabinol content and in order to come to a conclusion concerning the country where the cannabis plants were cultivated, from the quantitative distribution of the various components. The latter can be of particular importance in criminal proceedings.

More than 50 cannabinoids have so far been isolated[48]. Many of these compounds are polar and some heat labile, so that HPLC exhibits a significant advantage over GC for the determination of these substances[49, 50, 51]. Baker et al.[52] have examined a marijuana extract by HPLC before and after it has been heated, in order to demonstrate the effect on the quantitative distribution of the various substances. A particular finding was the demonstration of an appreciable increase in the THC concentration. Kanter et al.[53] recommends a three-minute heat treatment at 200° C in order to transform tetrahydrocannabinol acid to THC before the HPLC determination of the THC content (cf. Fig. 5). It is only in this way that it is possible to compare the results directly with gas chromatographic investigations.

4.1.1 Detection of Cannabinoids in Body Fluids

When cannabis products are consumed the various components are rapidly subjected to a complicated metabolization process[48, 54, 55]. The most important cannabis com-

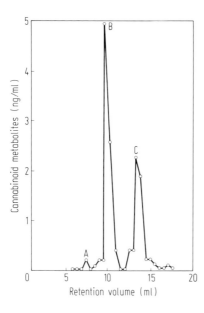

Fig. 6. Typical HPLC/RIA chromatogram obtained from a blood sample of a person who has consumed cannabis. The sample was submitted for forensic analysis: A) polar metabolites; B) Δ9-THC-11-oic acid glucuronide; and C) Δ9-THC-11-oic acid (from [162])

ponent forensically, Δ^9-THC, is hardly excreted at all unchanged in the urine and can only be detected in the blood for a very short space of time because of its extremely short residence time and low concentration and forms many metabolites of which the most important are 11-nor-Δ^9-THC-9-carboxylic acid in blood and urine and 11-hydroxy-Δ^9-THC in blood[56–58]. Pure HPLC methods for the determination of cannabinoids after passage through the body are generally beset by the lack of adequate selectivity of the detectors employed[59]. Many of the determination methods that have been described[60, 61, 62] employ conditions which seldom obtain in forensic practice. On the other hand, combination methods involving HPLC with RIA or EIA or GC-MS can be applied and will be described in further detail.

A successful method of extracting cannabinoids from blood and urine after previous hydrolysis is to repeatedly extract a 1 to 2 ml sample with diethyl ether in a silanized glass vessel after the addition of solid ammonium sulphate. Before extraction Δ^8-THC for example can be added to the sample as an internal standard[63]. A direct extraction of the blood sample with methanol is sufficient for combined HPLC-RIA or HPLC-EIA[64, 162].

A time-dependent fractionation of cannabinoids is a simple matter by HPLC. Here too surface treatment of the collection vessels is necessary to avoid bonding to glass. The eluent (methanol-water or acetonitrile-water) can be removed by freeze drying under a protective gas such as argon[59], or in a stream of nitrogen under slight heating[65], without destroying the oxidation-sensitive cannabinoids.

4.1.2 The coupling of HPLC with Immunochemical Methods

The disadvantage of all immunochemical methods lies in the fact that because of their high cross sensitivity to structurally similar compounds, single cannabinoids cannot be individually determined. One method of avoiding this problem is to perform individual immunochemical determinations on individual HPLC fractions. This indirect coupling of HPLC with RIA or EIA has been described for both blood and urine. In particular the group around Moffat has developed this method. HPLC-RIA coupling has been described for the investigation of blood samples[66 – 68], of urine samples[67, 69] and the combination of HPLC with Emit® has been described for blood and urine samples[70, 71].

4.1.3 The Coupling of HPLC with GC-MS

Here too the GC-MS detection of the cannabinoids is performed on the previously separated HPLC fractions. This method has considerable importance particularly in the forensic field, since the extracts produced from blood and urine not infrequently contain considerable amounts of accompanying substances, which interfere with the direct GC-MS determination. This micropreparative method also makes it possible to detect individual cannabinoids mass spectrometrically, when they are only present at extremely low concentrations in the body fluids. This procedure has also been amongst those methods meeting with success when it was necessary to demonstrate the presence of other addictive drugs or pharmaceuticals along with cannabis in a very few ml of blood[59, 63, 65].

4.2 Heroin, Opium Alkaloids

The largest number of investigations of HPLC methods for the detection of addictive drugs are concerned with heroin and the opium alkaloids. The aims and analytical requirements may be divided into those of investigations concerned with the examination of confiscated samples[16, 23, 72, 79, 81, 83, 84, 87, 88, 99, 103, 105, 108, 111, 112] and those of investigations which touch on this field[14, 16, 17, 20, 27, 74 – 78, 80, 85, 89, 92 – 95, 97, 101 – 104, 106, 110, 141, 148] and those which are concerned with the analysis of traces of opium alkaloids in biological material[82, 86, 90, 91, 96, 98, 107, 109].

It is only in rare cases that pure samples of heroin are submitted for analysis, more usually – as is the case with most addictive drugs – the confiscated samples (street drugs) are adulterated with more or less pharmacologically active adulterants and with impurities, which are result either of decomposition after manufacture (or of faulty synthesis) or which were present from the beginning in the form of minor alkaloids in the starting material (raw opium)[105].

HPLC possesses significant advantages over GC for the analysis without derivatization under gentle conditions (non-destructive) of both the active ingredients and the almost 50 accompanying substances that have been employed up to now[110].

In accordance with the development of HPLC stationary phases the earlier investigations employed silica or aluminium oxide columns[23, 72, 78, 80 – 84, 141]. Individual authors have also employed these principles in more recent investigations[27, 86, 89, 90, 92, 95, 100, 103, 105, 111, 112].

Considerable improvements in separation performance were obtained by the narrowing and classification of the particle size; the application spectrum of HPLC was impressively enlarged by the introduction of polar phases[73, 100, 102, 105] and above all reversed phases[14, 16, 17, 27, 83, 85, 87 – 89, 91 – 94, 96 – 99, 101, 104, 106, 107 – 110, 148]. Particular in the last mentioned group examples are to be found of the separation of a multiplicity of substances of differing polarity, whose basic and acidic character makes reversed-phase ion-pair chromatography the method of choice [14, 85, 93, 97, 110, 143].

The spectrum of HPLC methods for the analysis of opium alkaloids is completed by the ion-exchange methods, which are represented by a few examples [20, 74–77, 79].

In a very early investigation Cashman and Thornton [72] described the separation of heroin, O^6-monoacetylmorphine and morphine in synthetic mixtures on aluminium hydroxide and silica. The substances were identified by scanning with a variable wavelength detector.

Significant increases in the separation performance for the analysis of opium alkaloids and furthermore of a larger number of drugs, which are subject to abuse, were presented by Jane [80] by employing silica of uniform particle size (silica 6 μ).

While the investigations so far described did not envisage any particular sample preparation, Ziegler et al. [81] carried out a preliminary clean-up of an extract of opium on amberlite (XAD-2 column). The 6 alkaloids were then separated on a silica column with pre-column and detected UV photometrically.

Although at the time the first positive results were available concerning the separation of complex mixtures of substances on modified silica gel, Wheals [83] explicitly emphasized the advantages of employing silica (5 μ) compared with ODS, particularly for basic compounds. The separation of the following mixtures was given as an example: heroin, cocaine, methadone and morphine, together with, papaverine, narcotine, thebaine and codeine and finally the separation of the adulterants caffeine, strychnine and quinine, to be found in illegal heroin preparations. A non-adsorptive separation mechanism was postulated as the reason for the good separation performance (8,000–16,000 theoretical plates in a 25 cm column).

Huizer et al. [84] also obtained good separations of the components of illegal heroin samples such as heroin, O^6-monoacetylmorphine, codeine, acetylcodeine, caffeine and strychnine by employing a silica column with a mobile phase consisting of diethyl ether/isooctane/methanol/diethylamine. The heroin samples could be allocated to particular batches by means of quantitative analysis of the substances.

Only one investigation will be discussed here as an example of the employment of ion exchangers [20, 74–77, 79]: the analysis by Twitchett [79] of 20 authentic illegal diamorphine samples on strongly basic ion exchanger. After the separation of heroin, O^6-monoacetylmorphine, morphine, caffeine, strychnine and quinine by a pH gradient in the mobile phase the peaks were unequivocally identified in the eluates by mass spectrometry.

Trinler and Reuland [16] separated heroin, morphine, methadone and cocaine in simulated and authentic "street drugs" on RP 18 stationary phases. A graphical presentation reveals a strongly differing dependence of the RT of the materials on the composition of the mobile phase (containing acetonitrile, H_2O, ammonium carbonate).

Quantitative determinations in heroin preparations have been made by Albanbauer et al. [87]. Samples of "Hong Kong Rock" and "Grey Heroin" were separated on a RP 18 column with a weakly basic mobile phase, leading to the quantification of heroin, O^6-monoacetylmorphine, morphine and acetylcodeine. In addition narcotine, narceine, papaverine and caffeine were detected qualitatively.

After reversed-phase chromatography on ODS silica Reuland and Trinler [88] established the identity of the eluted heroin in "simulated street drugs" by IR spectroscopy of the fractionated eluates after evaporation. Morphine, monoacetyl-morphine, acetylcodeine or procaine and quinine were other components of the mixtures.

The separation of 6 opium alkaloids along with caffeine and strychnine has been described by Love and Pannell[99]. Authentic street drugs were also analysed using the same chromatographic conditions.

Baker and Gough[105] described the employment of a bound polar phase for the separation of the narcotic components of illegal heroin. They separated and quantified caffeine, heroin, acetylcodeine, O^6-monoacetylmorphine, codeine and morphine by chromatography on aminopropyl-bound SiO_2 with an acetonitrile-based mobile phase containing tetrabutylammonium phosphate (as ion-pair former). It is also possible to separate papaverine, noscapine, thebaine and strychnine which are often present as contaminants, as well as ethylmorphine and quinine. The latter is often employed as a diluent – at least in the USA.

Huizer[111] applied himself to the quantitative determination of the O^6-monoace-tylmorphine (to ca. 40%) and the smaller quantity of O^3-monoacetylmorphine (0.1 to 2%) that are found in illegal heroin samples. He attributed a certain utility for the identification of heroin samples to the latter component. Further, barbital, caffeine, acetylcodeine, codeine, strychnine and 2 unidentified substances were also separated. The employment of a silica column is striking. The relatively nonpolar mobile phase contains a small proportion of diethylamine as the basic component. Huizer[112] has also investigated 25 illegal heroin samples. The quantitative pattern of the additives noscapine, papaverine, acetylcodeine and of heroin and other sub-stances was employed to come to conclusions concerning the country of origin and, under certain circumstances, to allow the identification of the dealers. The papaverine/noscapine pattern was found to be the most informative, since the con-tent of these components varied greatly from sample to sample. The chromato-graphic conditions were identical to those employed in the previous investigation.

Bernhauer and Fuchs[108] employed two RP 18 columns of differing particle size (10 μ and 5 μ) and varying proportions of buffer in the mobile phases (aqueous acetonitrile-based) to determine the acetylthebaol, which is apparently present in all illegal heroin samples, for the purpose of sample classification. The acetylthebaol content varies from 0.003 to 0.33% and the ratio of the heroin to acetylthebaol contents varied from 200 to 2,600 in 56 street heroin samples. The additional calcula-tion of the ratio of heroin to acetylcodeine, narcotine and papaverine allows the calculation of the "exclusivity" of the sample.

A range of investigations are not directly concerned with the analysis of authentic or simulated illegal preparations, but with basic themes such as the tabulation of the analytical data of a large number of potential additives, investigations into the stability of heroin, or the separation of opium components or related substances. Reversed-phase ion-pair chromatography (PIC) has been employed in a range of such investigations[14, 85, 93, 97, 110, 143]. Separations of 21 opium alkaloids were, for example, performed by Olieman et al.[85] employing the ion-pair method (addition of n-heptane sulphonic acid to the mobile phase) with a RP 18 column.

Baker et al.[27] have reported the retention values and the extinction ratios 254/280 of 101 substances of forensic interest in three chromatographic systems, these include representatives of the morphine group (morphine, codeine, heroin), the dihydromorphine group and the morphinan group. Chromatography was performed on both a RP 18 column (aqueous methanol-based mobile phase) and a silica column (mobile phases based on methanol/ammonia and ammonia-saturated dichloromethane). According to the authors it is only possible to identify 9% of the substances on the basis of retention data alone, while inclusion of the extinction properties as well allows the identification of 95%.

Investigations have been made by Soni and Dugar[93] of the standard opiates in pharmaceutical and clandestine preparations. The mobile phase for the separation of a RP 18 column consisted of methanol together with an acidic or basic ion-pair former. Good separations could be obtained of acetylcodeine, morphine, oxymorphone, noroxymorphone, hydromorphone, nalorphine, codeine, monoacetylmorphine, oxycodone, hydrocodone, heroin and papaverine by changing the composition of the stationary phase or by the employment of a gradient. The identity of the substances could be adequately established by means of the RT value and the absorption ratios at 254 and 280 nm.

A systematic investigation of the separation performance of 22 different (laboratory-produced) separation materials, mainly with polar bound phases, has been carried out by Wheals[100] employing numerous basic substances, including 15 opium alkaloids. Two phases with an acidic surface character and unmodified SiO_2 proved the most suitable for further, more comprehensive studies. The separation mechanism was explained in terms of ion-exchange processes when employing aqueous methanolic phases (containing ammonium nitrate) at high pH.

The dependence of the retentions of heroin and 25 potential accompanying substances in clandestine preparations on the composition of the mobile phase on isocratic conditions has been investigated by Wittwer[103] employing a silica column. The mobile phases were based on cyclohexane/ammonia/methanol/chloroform (water-saturated), whereby the water content of the chloroform served to partially de-activate the polar columns. Although the ammonia component had a favourable effect on the peak shape it was, nevertheless, recommended that a volatile component be avoided, since its evaporation can cause changes in the RT value and can also be a source of erratic pump performance because of the formation of bubbles in the supply system.

Particular emphasis was placed on the possibility of separating codeine and O^6-monoacetylmorphine when the above-mentioned mobile phase contained 2% water. It was possible to separate acetylcodeine, heroin, procaine, acetylprocaine and O^6-monoacetylmorphine in ten minutes. Three authentic clandestine heroin preparations (brown, Mexican type) were quantitatively analysed by means of HPLC and GC, yielding results that were in good agreement with each other.

Lurie et al.[110] have determined the RT values of all the additives and contaminants that are likely – according to the authors – to be found in clandestine heroin samples. In order to chromatograph the 46 accompanying substances of the most varied acidity and basicity methane sulphonic acid was added to the mobile phase as an ion-pair former and two RP 18 columns from different manufacturers were employed. The appreciably different RT values obtained for some substances and the ratio of the extinction at 220 and 254 nm are said to enhance the security of identification.

Kinetic determinations of the hydrolysis of heroin to monoacetylmorphine and

Table 4. HPLC of heroin and other opium alkaloids

Subject	Column/Detection	Mobile phase		Ref.
		Organic	Aqueous	
Morphine, besides various other drugs. Evaluation of several mobile phases	Microparticulate octadecylsilane/ 220 nm	Methanol: 0–80% Changing ratios	Phosphate soln., pH 3.0–9.0	[17]
Heroin in illicit preparations, morphine, 6-mono-acetylmorphine, adulterants such as caffeine, strychnine and others	Zipax SCX (strong basic cation-exchange resin) 270 nm	Acetonitrile: 12 n-propanol: 2 pH-Gradient	Boric acid: 86	[79]
Heroin, morphine and other opium alkaloids, other substances of forensic interest	Graded silica (6 μm), (comparison with ungraded silica)/ UV-maximum of individual compound	Methanol: 27 or Methanol: 3	2 N NH_3: 2 NH_4NO_3: 1 or 0.2 N NH_4NO_3: 2	[80]
Heroin, morphine, methadone, cocaine in street drugs	Bondapak C_{18}/ Corasil/254 nm	Acetonitrile: 65 or 50	H_2O, containing $(NH_4)_2CO_3$: 35 or 50	[16]
Morphine in urine, pre-column derivatization to pseudomorphine, fluorescence detection	Partisil-7/ Fluorescence: Ex: 320 nm, Em: 436 nm	Methanol: 3	2 N ammonia: 2 1 N NH_4NO_3: 1	[82]
Heroin, street drugs, HPLC, GC, TLC	Lichrosorb Si 60, 5 μm/250 nm	Diethylether: 52.8 Isooctane: 35 Methanol: 12 Diethylamine: 0.2		[84]
Opium alkaloids and other drugs of forensic interest	Partisil 5/254 nm and 278 nm	Methanol: 27	2 N ammonium hydroxide: 2 1 N ammonium nitrate: 1	[83]
Opium alkaloids and related compounds	Bondapak C_{18}/254 nm	Methanol: 60 or 50	0.005 M n-heptane-sulfonic acid in H_2O: 40 or 50	[85]
Heroin, morphine and other opium alkaloids, besides numerous other substances of forensic interest	μ-Bondapak C_{18}/ 254 nm	Glacial acetic acid: 1, 0.005 M 1-heptane sulfonic acid in methanol: 40 (pH ca. 3.5)	Water: 59	[14]
Morphine and morphine glucuronide in blood following enzymatic hydrolysis, electrochemical detection	Silica, 7 μm/ electrochemical detector	Methanol: 9	NH_4NO_3 buffer, pH 10.2: 1	[86]
Heroin and preparations ("Hong Kong Rock", "Gray Heroin"), quantification	μ-Bondapak C_{18}/ 280 nm	Acetonitrile: 140	Water: 156 $(NH_4)_2CO_3$, 1%: 4	[87]
Heroin in simulated street drugs, off-line IR	Partisil 10 ODS/ 254 nm	Acetonitrile: 60	Water: 40	[88]

Table 4 (continued)

Subject	Column/Detection	Mobile phase		Ref.
		Organic	Aqueous	
Heroin, morphine, codeine and other opium alkaloids, besides nearly 100 other substances of forensic interest or adulterants of illicit preparations, 3 different chromatographic systems, ratio of absorbances at 254 and 280 nm as identification parameter	*System A* Bondapak C_{18} *System B* μ-Porasil *System C* μ-Porasil/ A–C: 254 nm and 280 nm	A) Methanol: 2 B) Methanol: 27 C) Dichlormethane saturated with concentr. ammonia	A) Water: 3, containing 0.025 M NaH_2PO_4, pH 7. B) 2 N ammonia: 2, 1 N NH_4NO_3: 1	27)
Opiates in pharmaceutical and illicit preparations, besides other drugs. Identification by both RT value and absorption ratio at 254 and 280 nm	μ-Bondapak C_{18}/ 254 and 280 nm	Methanol	0.01 M tetrabutyl-ammoniumphosphate, phosphoric acid, pH 7.5 or 0.01 M heptane sulfonic acid, acetic acid, pH 3 Changing ratios, gradient	93)
Heroin and hydrolysis products (kinetics of decomposition)	μ-Bondapak C_{18}/ 235 nm	Acetonitrile: 3	0.015 M KH_2PO_4, phosphoric acid, pH 3.5: 7	94)
Morphine and codeine in urine following enzymatic hydrolysis, sample preparation by use of Amberlite XAD-2	Spherisorb ODS, 5 μm/230 nm	Acetonitrile: 25	0.1 M NaH_2PO_4, pH 4.5: 75	96)
Morphine and other narcotic alkaloids, amperometric detection	μ-Bondapak C_{18}/ 254 nm and amperometric detection	Methanol: 20	50 mM tetrametyl-ammonium-hydroxide, pH 6.1: 80	97)
Illicit heroin samples, morphine, papaverine, monoacetylmorphine, codeine, acetylcodeine	μ-Bondapak C_{18}/ 280 nm	Acetonitrile: 65	0.75% ammonium-acetate: 35	99)
Heroin, morphine and numerous other basic drugs, comparison of packing materials	A) Silica B) Mercaptopropyl modified silica C) n-propyl sulfonic acid modified silica/254 nm	Methanol: 27	2 M ammonium hydroxide: 2, 1 M ammonium nitrate: 1	100)
Heroin in illicit preparations, acetylcodeine, procaine, acetylprocaine, O^6-monoacetylmorphine	μ-Porasil/254 nm	Cyclohexane: 750	Mixture: 250 ammonia: 1, methanol: 200, chloroform, water saturated: 800 3 concentrations of ammonia: 28%, 14%, 7%. 2 further mobile phases with 1% or 2% water	103)

Table 4 (continued)

Subject	Column/Detection	Mobile phase		Ref.
		Organic	Aqueous	
Narcotic components of illicit heroin: Acetylcodeine, O^6-acetylmorphine, morphine, codeine, papaverine, thebaine and others, different column packings	A) Alumina B) Octadecylsilane bonded silica C) Cyanopropyl bonded silica D) Aminopropyl bonded silica/284 nm	Acetonitrile: 85	0.005 M tetra-butylammonium-phosphate: 15	105)
Morphine in biological material, post-column derivatization to pseudomorphine, fluorescence detection	A) Partisil 10 ODS B) Zorbax ODS (8 µm) Fluorescence: Ex: 323 nm Em: 432 nm	Methanol: 12.5	0.1 M KBr: 87.5, pH 3 by addition of phosphoric acid	107)
Acetylthebaol in illicit heroin, quantification and determination of ratios, statistical evaluation of the exclusivity of street samples	A) Bondapak C_{18} B) 5 µ-Lichrosorb C_{18}/254 nm	For A: acetonitrile: 91.5, water: 8.5, 8 mg tris-(hydroxymethyl-) aminomethan/ 100 ml mobile phase. For B: acetonitrile: 53, water: 47, 23 mg ammonium carbamate/100 ml mobile phase		108)
Codeine in plasma, fluorescence detection	µ-Bondapak C_{18} ODS guard column/ Fluorescence: Ex: 213 nm Em: 320 nm	Methanol: 21 1,5 g phosphoric acid/l	Water: 79	109)
Heroin, method for quantitation in street drugs as well as for 46 adulterants commonly occurring in samples, quantitation comparison with GC	A) µ-Bondapak C_{18} B) Partisil 10 ODS-3/220 nm	Acetonitrile: 12	Water: 87, phosphoric acid: 1, 0.02 M methanesulfonic acid (adjusted to pH 2.2 with NaOH)	110)
Illicit heroin samples, comparison of samples by determination of noscapine, papaverine and acetylcodeine	Lichrosorb Si 60-7/227 nm, Fluorescence: Ex: 260 nm Em: 400 nm	Hexane: 72, dichloromethane: 20, methanol containing 0.75% V/V diethylamine: 5		112)

morphine as a function of pH and of temperature, have been carried out by Poochikian and Cradock [94], in order to determine the relevant rate constants at different temperatures. The separation in an ODS silica column was investigated with mobile phases at various pH values and the same order of elution was found in the range from pH 5 to 8.7. The separation and capacity factors were also reported. Heroin, morphine and O^6-monoacetylmorphine were separated.

Although sufficient quantities of heroin preparations or other authentic materials are normally available, so that the determination of many substances is not a matter of the limit of detection – assuming photometric activity – but rather of the separation performance, the sensitivity of detection of the same substance can be the limiting factor for its detection in biological materials. This applies, in particular, to the determination of the blood concentration which is in good correlation with the state of intoxication, but also to the substance concentrations in other media, such as liquors, organ homogenates and urine. Effective measures for increasing the sensitivity and selectivity of detection of opium alkaloids, particularly of heroin (and its metabolites) consist, alongside suitable sample preparation steps such as extraction [82, 90, 96] and conjugate cleavage [86, 107], of the employment of fluorescence detection – also in combination with derivatization [82, 106, 107, 109] – and electrochemical detection [86, 97, 98, 104].

Pre-column derivatization is employed by Jane and Taylor [82] for the analysis of morphine in urine. After extraction of the urine the extract is oxidized by treatment with potassium ferricyanide yielding fluorescent pseudomorphine and chromatographed on a silica column. Fluorescence detection allows a detection limit of 10 ng/ml urine.

Nelson et al. [107] exploited a decisive advantage of post-column derivatization, namely the possibility of simultaneous UV absorption and fluorescence detection for the detection of morphine in biological samples after similar derivatization. Normorphine, nalorphine, morphine 3-glucuronide, codeine, O^6-monoacetylmorphine, heroin, ethylmorphine, acetylcodeine, dihydromorphine and norcodeine were also detected.

The electrochemical detector is a relatively new development for increasing the sensitivity of detection in HPLC analysis. There are examples in the literature of its employment for morphine with the achievement of very low limits of detection (to 50 pg) [86, 97, 98, 104], which makes this technique seem particularly suitable for the determination of low levels in body fluids. A certain amount of selectivity results from the characteristic redox behaviour of morphine, apomorphine [104] and analogous substances with a catechol structure as compared with say codeine, which does not exhibit the redox behaviour described. Peterson et al. [97] detected morphine, oxymorphone, nalorphine, naltrexone and naloxone amperometrically.

White [86] employed an electrochemical detector for the HPLC detection of morphine and morphine 3-glucuronide (after hydrolysis) in blood and obtained a sensitivity of detection comparable with that obtained by fluorimetric detection (1 ng morphine per injection). Chromatography was performed in a silica column with a mobile phase consisting of methanol/ammoniumnitrate solution.

4.3 Cocaine

After the employment of thin-layer chromatography and gas chromatography high performance liquid chromatography has come into increasing employment since the mid-1970s for the analysis of cocaine. It has proved possible to employ it successfully both for the identification of the active agent and a range of accompanying substances – often local anaesthetics – in confiscated samples or simulated mixtures [14, 16, 17, 20, 23, 27, 73, 80, 83, 100, 110, 116, 117, 119, 120, 123, 124] as well as in the detection of cocaine and its metabolites in body fluids [115, 118, 121, 122, 125, 126].

In a relatively early investigation Trinler and Reuland [16] applied themselves to the analysis of simulated and authentic street drugs containing cocaine, heroin, morphine and methadone. Chromatography was performed on a RP 18 column with aqueous acetonitrile as the mobile phase. Trinler and Reuland[117] have also described the identification of cocaine and accompanying substances by IR spectroscopy of the evaporated HPLC eluate fractions. Alongside lactose cocaine, benzocaine, procaine, tetracaine and lidocaine were also separated employing a phenylsilica-bound phase with aqueous acetonitrile and UV photometric and refractometric detection.

Baker et al.[27] and Lurie et al.[110] have also determined the retention data of cocaine and a large number of other substances of forensic interest. The 254/280 or 220/254 extinction ratios were also employed as additional criteria of identity (see also "Heroin").

Sugden et al.[116] investigated the chromatographic behaviour of basic compounds such as cocaine, amylocaine, benzocaine and butacaine on a silica and an ODS silica column with aqueous methanol-based mobile phases in the pH range 5.5 to 8.5 at various ionic strengths.

Systematic investigations have been carried out by Wheals[100] (see "Heroin"). The separation of cocaine and three isomers has been described by Olieman et al.[119]. While RP ion-pair chromatography was employed here (ODS-silica column), Lewin et al.[120] separated cocaine, pseudococaine, allocaine and allopseudococaine on a cation exchanger.

Basic and comprehensive investigations have been made on over 300 authentic cocaine samples by Jane et al.[124]. The natural or synthetic origin of the samples were recognized by the presence or absence of cinnamoylcocaine. They determined the RT values of ca. 40 substances found as cocaine additives on a RP 2 column. The substances were identified by GC-MS.

Because cocaine is rapidly metabolized and is only present unchanged in urine to the extent of ca. 1%, the metabolites (principally benzoylecgonine) are of particular importance for the analysis of cocaine in body fluids[113,114]. While cocaine can be detected without problem by GC, this is only true of benzoylecgonine after derivatization. On the other hand both substances can advantageously be detected by HPLC without derivatization.

Graffeo et al.[115] developed a method for the determination of urinary benzoylecgonine concentrations. After direct application of 1 ml urine sample (RP 18 packing, analytical column with pre-column) the benzoylecgonine-containing HPLC fractions were collected, evaporated to dryness and trimethylsilylated. Analytical demonstration was made by GC-MS. The easy extractability of the polar benzoylecgonine makes the employment of HPLC particularly advantageous. The direct HPLC detection of cocaine, benzoylecgonine, norcocaine and benzoylnorecgonine in urine was performed by Jatlow et al.[118] using a RP 18 column. On the basis of appropriate investigations and HPLC analysis Fletcher and Hancock[123] warn of errors that can occur during the evaluation of the proportions of cocaine and benzoylecgonine after alkaline extraction by hydrolysis.

Evans and Morarity[121] investigated plasma and tissue homogenizates and detected cocaine, norcocaine and benzoylecgonine after separation on a RP 18 column. The limits of detection were under 1 µg substance/ml plasma. Garrett and Seyda[126] achieved a detection limit of 15 ng/ml plasma for the UV absorption detection of cocaine at 232 nm after chromatography through a RP 18 column during an

Table 5. HPLC of cocaine

Subject	Column/Detection	Mobile Phase		Ref.
		Organic	Aqueous	
Cocaine (and others) in street drugs (see Table 4)				[16]
Cocaine and others (see Table 4)				[17]
Cocaine and other substances of forensic interest (see Table 4)				[80]
Cocaine and other substances of forensic interest (see Table 4)				[83]
Benzoylecgonine in urine, HPLC and GC/MS	μ-Bondapak C_{18}, pre-column/254 nm	Methanol: 5–10% gradient	Water	[115]
Cocaine and numerous other substances of forensic interest (see Table 4)				[14]
Cocaine, amylocaine, butacaine benzocaine, amprolium comparison of chromatographic behaviour of silica and ODS-silica at different pH, methanol concentration and ionic strength	A) Lichrosorb Si 100 B) ODS-Lichrosorb Si 100 240 or 270 nm	Methanol: 95 or 70	Water: 5 or 30 containing ammonium formate, ammonium nitrate or sodium formate	[116]
Cocaine in simulated street drugs	Bondapak Phenyl/Porasil B/ 254 nm, IR or Refractometer	Acetonitrile: 85	Water, containing 0.1% $(NH_4)_2CO_3$: 15	[117]
Benzoylecgonine and cocaine in urine, separation of various cocaine metabolites	Partisil 10 ODS/ 220 nm or 235 nm	Acetonitrile: 17	0.25 M phosphate buffer, pH 2.7: 83	[118]
Cocaine and nearly 100 other substances of forensic interest (see Table 4)				[27]
Cocaine and other basic drugs (see Table 4)				[100]
Cocaine in human plasma (patient samples). Comparison with RT data of 22 other interfering compounds	ODS − HC SIL − X − 1/ 232 nm	Methanol: 75	0.05 M phosphate buffer, pH 6.6: 25	[122]
Cocaine and metabolites (norcocaine, benzoylecgonine) in plasma and tissue homogenates	μ-Bondapak C_{18}/ 235 nm	Acetonitrile: 1, methanol: 1 containing 1% acetic acid and 0.3 M EDTA	Water: 8	[121]

Table 5 (continued)

Subject	Column/Detection	Mobile Phase		Ref.
		Organic	Aqueous	
Cocaine, quantification and discrimination of different origins (matrices) by determination of Cinnamoyl cocaine. Further reference materials: Ecgonine, methylecgonine, cinnamoylecgonine, benzoylecgonine. HPLC, GC/MS, TLC	Lichrosorb RP 2/ 279 nm	Methanol: 2	0.1 M NH$_4$NO$_3$ pH 4.3: 3	124)
Cocaine and nearly 50 potential substances in heroin street drugs (see Table 4)				110)
Cocaine, methadone, phencyclidine and metabolites, assay of plasma levels (see Table 8)				125)
Stability evaluation of cocaine under bioassay conditions, determination of cocaine and its hydrolysis products benzoylecgonine and benzoic acid. Determination of log k profiles at 9 various temperatures	μ-Bondapak C$_{18}$/ 254 nm or 232 nm	A) Acetonitrile: 25 B) Acetonitrile: 15 C) Acetonitrile: 25	A) 0.085 M acetate buffer, pH 3.6: 75 B) 0.0002 M tetrabutylammonium hydroxide in 0.085 M acetate buffer, pH 3.63: 85 C) 0.0002 M tetrabutylammonium hydroxide in 0.085 M acetate buffer, pH 3.63: 75	126)

investigation of the stability of cocaine under the conditions of a bioassay. An even lower detection limit for cocaine (1–6 ng/ml serum) was achieved by Derendorf and Garrett [125] by fluorescence detection after ion-pair chromatography (see "Phencyclidine").

4.4 LSD

LSD is a very powerful hallucinogen and the most effective addictive drug of all in terms of the amount taken. The low effective dose of the order of ca. 1 µg/kg body weight (per os) implies particular analytical problems. Although the HPLC detection of clandestine or simulated preparations can be carried out by means of UV absorption detection[127, 129, 133, 134], a more sensitive technique e.g. fluorescence detection[33, 130] or a combination with radio immunoassay[132] is required for its detection in body fluids because of the low dosage and the rapid metabolism. Furthermore, there are some instances of the employment of fluorescence detection for LSD in preparations or in mixtures with other ergot alkaloids. The high selectivity, that is provided along with high sensitivity by fluorescence detection, is of particular value for the detection of LSD in mixtures of substances which do not belong to the ergot group of alkaloids, or in association with the 9,10-dihydrocompound (the same skeleton), since this does not fluoresce.

There are very early reports in the literature concerning the HPLC detection of LSD, which was separated from mixtures of other ergot alkaloids (some of which find medical application)[127-129]. While Wittwer and Kluckhorn[127] and Heacock et al.[128] employed silica columns Jane and Wheals[129] already employed a C_{18} column with both UV absorption and UV fluorescence detection. Again with an RP packing but in the presence of the ion-pair former 1-heptane sulphonate in the mobile phase at pH 3.5 Lurie[14] separated LSD from the other ergot alkaloids. Wheals[83], on the other hand, has described the analysis of LSD in illicit preparations under routine conditions employing a silica column (mobile phase: methanol/0.3% ammonium carbonate solution = 60:40, detection limit by fluorescence detection ca. 10 ng).

For the separation of LSD from illicit preparations Lurie and Weber[143] employed a reversed-phase ion-pair system that can also be employed for methamphetamine (cf. "Phenylalkylamines"). They identified LSD by IR and MS after isolation on a semipreparative scale.

Megges[134] only achieved a partial separation of LSD and its isomer lysergic acid methyl-propylamide (LAMPA) on a RP 18 column. Benzocaine was employed as a standard and the mobile phase consisted of acetonitrile and water containing a small proportion of 1% ammonium carbonate; detection was by UV absorption at 313 nm.

Christie et al.[130] demonstrated LSD in urine by means of fluorescence detection after extraction and chromatography on a silica column. The amounts detected in various authentic samples ranged from 0.3 to 20 ng/ml. The identity was confirmed by mass spectrometry of the fraction containing the appropriate peak after elution.

Harzer[33] achieved a detection limit of ca. 0.5 ng LSD/ml in blood, serum and urine samples employing fluorescence detection after chromatography with two different mobile phases and column switching (RP 8 and silica columns).

Twitchett et al.[132] developed a method of detecting LSD in body fluids which combines the advantages of HPLC and RIA: The samples were selected by screening with RIA and the positive samples were investigated by HPLC whereby the positive findings here were confirmed by RIA of the appropriate HPLC eluate. It is possible to detect 0.5 ng LSD/ml by HPLC with a RP 18 column and fluorescence detection (excitation wavelength 320 nm, emission wavelength 400 nm). This method formed the basis for investigations of the metabolism of LSD in animals (Sullivan et al.[150]).

Table 6. HPLC of LSD

Subject	Column/Detection	Mobile phase		Ref.
		Organic	Aqueous	
LSD and other drugs of forensic interest (see Table 4)				[83]
LSD in biological material	Partisil, 6 μm/ Fluorescence: Ex: 325 nm Em: 420 nm	Methanol: 11	0.2% NH_4NO_3: 9	[130]
LSD and numerous substances of forensic interest (see Table 4)				[14]
LSD in biological liquids, fluorescence detection, combination with RIA	A) Spherisorb-5-ODS B) Spherisorb S5W/ Fluorescence: Ex: 320 nm Em: 400 nm	A) Methanol: 65 B) Methanol: 60	A) 0.025 M Na_2HPO_4, pH 8.0: 35 B) 0.2 M NH_4NO_3: 40	[132]
LSD determination in clandestine sample; semi-preparative isolation and confirmation of identity by IR and MS. Also suitable for phenylalkylamines (Methamphetamine)	Magnum 9 Partisil-10 ODS μ-Bondapak C_{18}/254 nm	Methanol: 40, acetic acid: 1 0.005 M methanesulfonic acid, adjusted to pH 3.5	Water: 59	[143]
LSD metabolism in Rhesus monkeys after treatment with ^{14}C and ^3H labelled substance. Analysis of plasma and urine samples, fractionation by HPLC succeeded by RIA examination	Spherisorb-5-ODS/ Fluorescence: Ex: 320 nm Em: 400 nm	Methanol: 1	0.1% (w/v) $(NH_4)_2CO_3$: 1	[150]
LSD and about 20 analogues, comparison of retention data and mass spectra for 19 ergot alkaloids, TLC	A) Spherisorb S5W B) Spherisorb-5-ODS/280 nm Fluorescence: Ex: 320 nm Em: 400 nm	A) Methanol: 60 B) Methanol: 65	A) 0.2 N NH_4NO_3: 40 B) 0.025 M Na_2HPO_4, pH 8: 35	[133]
LSD in illicit preparations, determination of LAMPA isomer	μ-Bondapak C_{18}/ 313 nm	Acetonitrile: 400	Water: 572, $(NH_4)_2CO_3$, 1%: 28	[134]
LSD in blood and urine, two combined columns, fluorescence detection	A) Lichrosorb C_8 (7 μm) B) Lichrosorb Si 60 (5 μm) Fluorescence: Ex: 325 nm Em: 430 nm	A) Methanol: 50 B) Methanol: 60	A) 0.3% KH_2PO_4, H_3PO_4, pH 3: 50 B) 1% $(NH_4)_2CO_3$: 40	[33]

4.5 Psilocybin, Psilocin and Phencyclidine

While the first two hallucinogens are indole derivatives and constituents of some fungi and have to be separated chromatographically from a complex matrix of other fungus components, phencyclidine is a synthetic hallucinogen and hence confiscated samples for analysis are contaminated with impurities contingent on the synthesis.

The methods of detection of psilocybin and psilocin described in the literature are restricted to the analysis of fungal samples. White[135] separated psilocin and psilocybin from Psilocybe semilanceata on a silica column with an ammonium nitrate-containing mobile phase based on aqueous methanol at a pH of 9.7. One component that was only partially separated from psilocybin was enriched by TLC and mass spectrometrically identified as 4-phosphoryl-N-methyltryptamine (baeocystin). Homogenates of Psilocybe subaeruginosa have been chromatographed by Perkal et al.[137] at elevated column temperature on an ion exchanger with a mobile phase consisting of methanol/water (20:8) containing 0.2% ammonium phosphate and 0.1% KCl, pH 4.5; psilocybin was determined quantitatively (0.01 to 0.2%) in a range of samples by UV absorption and fluorescence detection. In addition psilocin and dimethyltryptamine were demonstrated qualitatively. Thomson[136] has reported the HPLC detection of psilocybin and psilocin in connection with cases concerned with the illegal possession of hallucinogenic fungi. Psilocin and psilocybin were separated from the other fungal components of a RP 18 column with an ion-pair former (cetyltrimethylammonium bromide) in the mobile phase.

Further methods have been developed by Beug and Bigwood[151], Christiansen et al.[152] and by Sottolano and Lurie[153]. In these methods account was taken of the ionic character of psilocybin by the inclusion of appropriate ionic or ion-pair forming additives in the mobile phase. In fact because of its ionic character psilocybin represents a good example of the advantages of HPLC over gas chromatography. The HPLC demonstration of phencyclidine together with amphetamines, ephedrine and caffeine after chromatography of simulated street drugs on silica and RP packing is described in an investigation by Trinler et al.[15]. Wittwer[127] has described a separation of phencyclidine in admixture with LSD and other ergot alkaloids with the employment of silica columns. Chan et al.[23] separated phencyclidine from other hallucinogens, THC, heroin, cocaine and other materials using a silica column and gradient elution. Lurie[14] separated phencyclidine from the related substance TCP 1-[1-(2-thienyl)cyclohexyl]-piperidine (thiophencyclidine) by employing ion-pair reversed-phase chromatography. Phencyclidine is one of a multitude of forensically interesting substances investigated on HPLC by Baker et al.[27]. The 254/280 extinction ratio and the RT obtained by two chromatographic methods were employed as additional criteria of identity.

Cook et al.[138] have studied the human bioavailability and pyrolysis behaviour of smoked (^{3}H labelled) phencyclidine by HPLC-"radiochromatography". Separations of plasma and urine samples were performed on a silica column; a plasma concentration of ca. 1 ng/ml plasma was detected 72 h after smoking a cigarette containing 100 µg PCP. The retentions of the products of a second pyrolysis investigation and an in vitro smoking trial were determined by UV absorption at 254 nm.

Table 7. HPLC of Psilocybin and Psilocin

Subject	Column/Detection	Mobile phase		Ref.
		Organic	Aqueous	
Psilocin and psilocybin from extracts of Psilocybe semilanceata mushrooms; TLC, MS	Partisil 5, 6 μm/ 254 nm	Methanol: 240	Water: 50, 1 N NH$_4$NO$_3$, pH 9.7: 10	[135]
Psilocin and psilocybin from illegal possession of hallucinogenic mushrooms	μ-Bondapak C$_{18}$/ 280 nm	Methanol: 40	0.25% Na$_2$HPO$_4$; pH 7.5: 60, cetyltrimethyl-ammoniumbromide reagent: 0.15	[136]
Psilocin and psilocybin from extracts of Psilocybe subaeruginosa; dimethyl-tryptamine	Partisil SCX-10 (ion exchange)/ Fluorescence: Ex: 267 or 260 nm Em: 335 or 312 nm	Methanol: 20	Water: 80, containing 0.2% ammoniumphos-phate and 0.1% KCl (pH 4.5)	[137]
Quantification of psilocybin and psilocin in extracts of Psilocybe baeocystis mushrooms. Additionally TLC for separation also of other indoles	μ-Bondapak C$_{18}$ (10 μm)/254 nm	Methanol: 25 or acetonitrile: 25 resp. methanol: 25	Water: 75, containing 0.05 M heptane sulfonic acid, pH 3.5 resp. water: 75, containing 0.05 M tetrabutylammo-nium phosphate, pH 7.5	[151]
Quantification of psilo-cybin in Psilocybe semilanceata	Partisil 5, 6 μm/ 254 nm, Fluorescence: Ex: 267 nm Em: 335 nm	Methanol: 220	Water: 70, 1 N NH$_4$NO$_3$, pH 9.6: 10	[152]
Quantification of psilo-cybin in dry mushroom material. Enhancement of specifity by use of ab-sorbance ratio 267: 254 nm	Partisil-10 PAC/ 267 and 254 nm	Acetonitrile: 5%	Water: 94.5%, phosphoric acid: 0.5% pH 5.5–6.0 (2 N NaOH)	[153]

A corresponding contribution employing HPLC analysis concerning the bioavailability of phencyclidine after i.v. administration has been published by Wall et al.[139]. Derendorf and Garrett[125] determined phencyclidine, methadone and cocaine in a bioassay by means of HPLC. After extraction of the serum, chromatography was performed in the presence of the fluorescent ion-pair former 9,10-dimethoxy-anthracene-2-sulphonic acid and the ion pair was detected in the eluate by UV fluorescence after on-line extraction (with chloroform). The resulting detection limit was less than 6 ng/ml plasma.

Table 8. HPLC of Phencyclidin

Subject	Column/Detection	Mobile Phase		Ref.
		Organic	Aqueous	
Phencyclidine, amphetamines, ephedrine in simulated street drugs	A) Bondapak Phenyl/ Corasil	Acetonitrile: 50	0.1% $(NH_4)_2CO_3$: 50	[15]
Phencyclidine, besides numerous other substances of forensic interest (see Table 4)				[14]
Bioavailability of phencyclidine after i.v. application of tritium labeled substance to humans, determination of phencyclidine and metabolites in plasma and urine, enzymatic conjugate cleavage	Partisil-10 PAC/ 254 nm	Ethanol: 5–35%, hexane, gradient		[139]
Bioavailability of phencyclidine by smoking tritium labeled substance, evaluation of pyrolytic behaviour	As before	As before		[138]
Phencyclidine in plasma, besides cocaine, methadone and metabolites. PIC chromatography, on-line extraction of the mobile phase with chloroform	CN-µ Bondapak, column temp. 50° C/ Fluorescence: Ex: 380 nm Em: 455 nm	Acetonitrile: 20 9,10-dimethoxyanthracene-2-sulfonic acid: 30 mg/l	0.025 M Acetate buffer. pH 3.6: 80	[125]
Fluorescence detection.				

4.6 Phenylalkylamines

Phenylethylamines and the phenylaminopropane derivatives which possess similar basic chemical structures (Table 9) are both included in this group. Pharmacologically groups of substances can be distinguished in terms of their psychostimulatory action (e.g. amphetamines), appetite-reducing action (e.g. norpseudoephedrine) and psychotomimetic action (e.g. mescaline).

There are early investigations of the separation of phenylalkylamines by adsorption or ion-exchange chromatography by, for example, Cashman et al.[140] and Chan et al.[23]. Trinler et al.[15] employed silica and reversed-phase columns for the detection of phencyclidine, ephedrine and amphetamines, that is for amphetamine itself, methamphetamine, methylenedioxy-3,4-amphetamine (MDA) and 4-methyl-2,5-dimethoxyamphetamine (DOM). Achari and Theimer[141] also employed a silica column to chromatograph amphetamines and a range of other substances. A series of stimulants of the amphetamine type have been separated by Jane[80] on silica of uniform particle size (6 µ) with ammonia/methanol containing added ammonium nitrate. Amphetamine, methamphetamine and ephedrine were amongst 30 forensically interesting substances chromatographed by Twitchett and Moffat[17] on an ODS column with aqueous methanol at differing pH values (3.0 to 9.0). However a cation exchanger proved itself more suitable for the chromatography of these basic substances (Twitchett et al.[20]). An investigation was made of the dependence of the retention time on the ionic strength, on the pH and on the methanol content and the results were presented graphically. The employment of acetonitrile instead of methanol improved the separation performed of the column.

An increase in the sensitivity of detection of amphetamine, methamphetamine and other amphetamines was achieved by Clark et al.[142] by derivatization with 4-nitrobenzoyl chloride and detection of the amides so formed by UV (254 nm). Lurie[14] chromatographed various phenylalkylamines such as phenylpropanolamine, ephedrine, amphetamine, methamphetamine and other substances along with a range of other forensically interesting substances by the ion-pair, reversed-phase method (1-heptane sulphonic acid as ion-pair former, mobile phase adjusted to pH 3.5).

Lurie and Weber[143] employed ion-pair chromatography (RP 18 column, water/methanol/acetic acid (79:20:1) as mobile phase with added pentane sulphonic acid) for the investigation of illicit preparations of methamphetamine. After the winning of phenylpropanolamine and ephedrine by a semipreparative method they separated the methamphetamine in the eluate fraction extractively from the ion-pair former with chloroform at pH 11.5 and identified by IR.

The retention behaviour of amphetamine together with a series of other substances on underivatized silica gel with various ion-pair formers in the mobile phase has been investigated by Crommen[144]. Strikingly, similar behaviour was obtained as with the chromatography on alkylated silica. Baker et al.[27] determined the RT of amphetamine, methamphetamine and mescaline on RP 18 and silica columns (cf. "Heroin" and "Phencyclidine"). The ratio of the UV absorptions of the substances at two different wavelengths was employed as an additional criterion of identity.

Wheals[100] employed three different types of column packing (silica, mercaptopropyl-modified silica and n-propylsulphonic acid-modified silica) for the determination of the chromatographic data of amphetamine, methamphetamine, ephedrine and a large number of other basic substances. The same mobile phase (methanol/2 M ammonia/1 M ammonium nitrate = 27:2:1) was employed in each case.

The retention data of a total of 8 phenylethylamine derivatives in simulated mixtures have been published by Einhellig et al.[145]. Two illicit samples containing methamphetamine as the main component were separated on a RP 18 column.

Verbiese-Genard et al.[146] demonstrated amphetamine, methamphetamine, ephedrine, fenethylline and other substances (after a rapid preliminary HPTLC screening) by HPLC with a RP 8 column (10 μ). Additional recording of the UV spectra of the eluted fractions is said to increase the certainty of assignment.

Gill et al.[147] have reported the advantages of adding amines to the mobile phase (methanol/phosphate buffer) in the reversed-phase HPLC analysis of 5 2-phenylethylamines. Alongside the improvement in the peak shape there was a dependence of the retention time on the added amine that was at times considerable.

Harbin and Lott[148] demonstrated the presence of amphetamine together with a range of other substances in genuine urine samples by HPLC. Chromatography was performed on a RP 18 column with a mobile phase composed of methanol/water

Table 9. Structure of some typical phenylethylamine derivatives according to [145]

	R_1	R_2	R_3	R_4	R_5	R_6	R_7
Phenylethylamine	H	H	H	H	H	H	H
Amphetamine	H	CH_3	H	H	H	H	H
Methamphetamine	CH_3	CH_3	H	H	H	H	H
Methylphenidate	$CH_2-CH_2-CH_2-CH_2$		$COOCH_3$	H	H	H	H
Phenmetrazine	●	CH_3	●	H	H	H	H
DOM	H	CH_3	H	OCH_3	CH_3	H	OCH_3
Mescaline	H	H	H	OCH_3	OCH_3	OCH_3	H
Ephedrine	CH_3	CH_3	OH	H	H	H	H
Norpseudoephedrine	H	CH_3	OH	H	H	H	H
	● CH_2-CH_2-O						

Table 10. HPLC of Phenylalkylamines

Subject	Column/ Detection	Mobile Phase		Ref.
		Organic	Aqueous	
Amphetamine, methamphetamine, ephedrine, besides various other drugs (see Table 4)				[17]
About 30 substances of amphetaminetype, besides numerous other drugs of forensic interest (see Table 4)				[80]
Amphetamine, methamphetamine, ephedrine, besides various other drugs (morphine, cocaine and others)	Partisil SCX (10 μm) cation exchanger/254 nm	Methanol (or acetonitrile): 0–60%.	Ammoniumdihydrogen phosphate, different pH values and ionic strengths	[20]
Amphetamine, ephedrine and phencyclidine in simulated street drugs (see Table 8)				[15]
Phenylalkylamines and other drugs of forensic interest (see Table 4)				[14]
Some phenylalkylamines and other drugs	Partisil-10 (10 μm)/254 nm	Methanol: 3, CH_2Cl_2: 1	Conc. ammonia, 1% added to the organic component of the mobile phase	[141]

Table 10 (continued)

Subject	Column/ Detection	Mobile Phase		Ref.
		Organic	Aqueous	
Diethylpropion hydrochloride in tablets	Octadecylsilane/ 255 nm	Acetonitrile: 35, methanol: 10	$(NH_4)_2CO_3$, 0.1%: 10	[149]
Phenylpropanolamine, ephedrine and methamphetamine, LSD and barbiturates (see Table 6)				[143]
Amphetamine, methamphetamine, ephedrine and various other basic drugs (see Table 4)				[100]
8 Phenylalkylamines in simulated street drugs, 2 authentic street drugs containing methamphetamine	Nucleosil 10 C_{18}/ 254 nm	Acetonitrile: 100, methanol: 100	Water: 75, $(NH_4)_2CO_3$, saturated: 3	[145]
Amphetamine in urine samples, besides other substances	μ-Bondapak C_{18}/ 254 nm	Methanol: 60	About 0.01 M KH_2PO_4 buffered to pH 6.2: 40	[148]
Ephedrine, amphetamine, methamphetamine, phentermine, amfepramone, fenethylline, UV of HPLC eluates. Screening by ion-pair, reversed-phase HPLC. Analysis of urine samples after extraction	RP 8 (10 μm)/ 215 nm			[146]
Amphetamine, dimethylamphetamine, norpseudoephedrine, phentermine, tyramine. 11 eluents containing different amines as additives were examined for influence on retention data and improvement of peak shape	ODS-Hypersil (5 μm)/250 nm	Methanol: 500 ml	Water: 4,930 ml, containing orthophosphoric acid (70 ml) and sodium hydroxide (28 g) pH 2,4 Addition of different amines resulting in a concentration of 0.05 M	[147]

(60:40) and buffered to pH 7.5. Because of the complex metabolite picture revealed by the HPLC chromatogram the authors point out the necessity for at least one other analytical parameter for the classification of genuine substances after the formation of further substances as a result of metabolization. The results were confirmed in this instance by TLC. The authors saw the employment of tables of HPLC data concerning available drugs and their metabolites as an important aid in improving the certainty of classification of a substance and its differentiation from metabolites of different origin (after combined consumption), a point of view that we, as a result of our experiences, would like to firmly endorse.

At the present moment, however, apart from a few, sometimes comprehensive data collections for pharmaceuticals, addictive drugs, etc.[6, 8, 13, 14, 17, 21, 29] appropriate data are unfortunately not available in sufficient quantity concerning metabolites.

5 Application of HPLC in Chemical Criminology

Certain differentiations can be made between forensic toxicology and chemical criminology. Although both try to supply evidence for investigation procedures, the former deals with the problems of purification and detection in biological matrices, the latter is concerned with residues and traces of the non-living milieu. It should be emphasized that this part of the chapter will not go into details, rather some interesting examples have been selected to illustrate the importance of HPLC in this field.

5.1 The Possibility of Identification by Means of Dyes

5.1.1 Dyes in the Illicit Drug Trade

The trade of illicit drugs is connected with different dye components. Presumably there is a historical background to illicit preparations containing certain colours (e.g. heroin-brown, LSD-green, amphetamine-blue or yellow) identifying them and giving them visual appeal for the drug scene. Conclusions can be drawn from the analysis of dye-components concerning location of manufacture and possibly distribution networks. A three-step process for dye analysis involving preliminary TLC, spectrophotometry and reversed-phase ion-pair HPLC has been developed by Joyce and Sanger[174]. It was possible to identify the blue dye, Green S, and the yellow dye (tartrazine) in a green LSD tablet. In a norpseudoephedrine-barbiturate preparation Patent Blue (Colour Index 42045) was responsible for the very strong blue coloration. The HPLC system employed, enabled separation of all (22 components) the food dyes currently approved in England. The water soluble azo sulfonic acid dyes with dissociable sodium ions can be detected in illicit heroin samples[173]. Fairly good separation for the seven most commonly encountered dyes could be achieved with a reversed-phased C_{18} column. The mobile phase methanol/water 58:42 contained tetrabutylammonium hydroxide to provide counter ions (Figs. 7, 8). Interferences were eliminated by employing a wool dyeing precedure as follows. Approximately 0.5 g of sample was dissolved in dilute acetic acid. A 51 mm square of wool felt was placed in the solution and allowed to stand overnight. The wool felt, with the absorbed dyes, was then removed from solution and rinsed with warm water. The wool felt square was then gently heated in dilute ammoniacal solution to dissociate the absorbed dyes from the wool[173].

5.1.2 Investigation of Documents

The investigation of documents has been an important task of forensic specialists for a long time. It is sometimes very important to know the age of a document especially if it bears no data or it is back-dated.

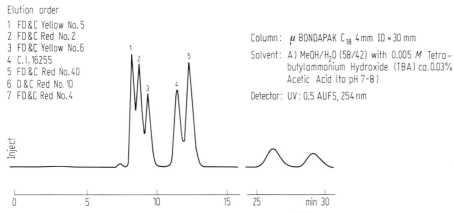

Fig. 7. Expected dye standards in illicit heroin samples (from[173])

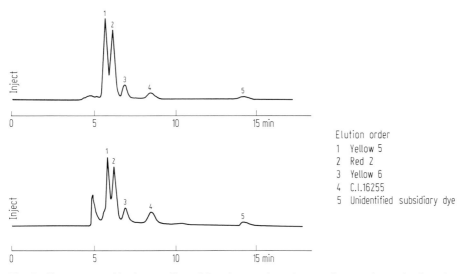

Fig. 8. Chromatographic dye profiles of heroin samples, after wool extraction, seized at the port of entry, San Ysidro, Calif. (from[173])

It was possible with HPLC to distinguish differences between the components of 25 blue ball-point pen inks. Ten plugs from written lines were punched out and extracted with 2% formamide in methanol, or with propan-2-ol/heptane. Ratios between vehicle components (resins, viscosity adjusters etc.) and dyes were calculated from the response of the spectrometer detector set at 254 or 580 nm[168]. Fingerprint chromatograms of ten ball pen-ink formulations have been reported by Lyter[169]. A reversed-phase column (μ-Bondapak C_{18}) permitted differentiation between different batches of the same formulations with a sample size as small as 0.25 μg. (this equals 10 plugs corresponding to ca. 12 mm written line).

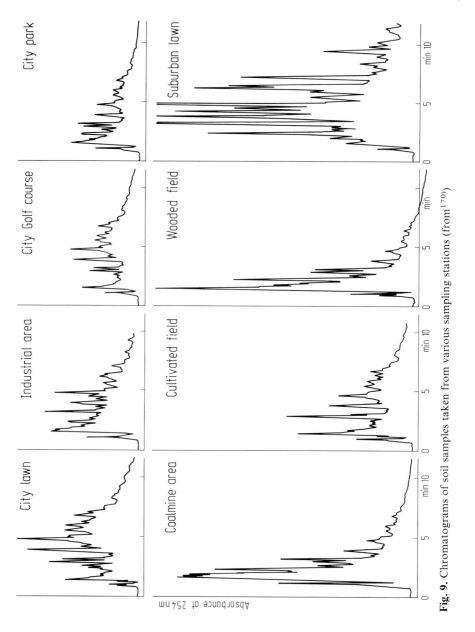

Fig. 9. Chromatograms of soil samples taken from various sampling stations (from[170])

5.1.3 Dyes in Lipstick Smears

The discriminating power of HPLC has been utilized in the identification and comparison of lipstick smears. Small quantities of lipstick on cigarette butts, glass and paper can play a role in the investigation process. Lipstick smears may be found on the clothing of a man who has assaulted a woman.

The examination procedure developed by Andrasko[171] comprises X-ray analysis for elemental composition in a scanning electron microscopy, and TLC as well as HPLC (see table 11) for the colour additives. Fats and waxes were removed by solvent extraction before the analysis. Other specialists[172] do not agree with this approach because ingredients (oils, waxes, lanolin) other than dyes can be useful in the comparision. The lipstick smears were removed on facial tissue, which was moistened with acetonitrile. The solvent program of 50% acetonitrile to 100% in water provided a good resolution. The UV detector was set at 254 nm.

5.1.4 Comparison of Soil Samples

From the forensic point of view the contribution of HPLC is useful and of interest for the comparison of different samples. The acetonitrile extracts of soils taken from different areas yielded differing but characteristic chromatograms, which remained constant over a period of several weeks and made definite identification possible[170] (Fig. 9).

5.2 Analytical Approach to Explosives by HPLC

The increasing number of labile political situations and the general increasing criminality has made the analysis of different explosives (Fig. 10) of great importance in forensic laboratories. There are two directions in which an analytical effort can be required. One is the identification of unexploded materials, which could provide information concerning their origin, the other is the collection and investigation of residues after detonation. The procedure is relatively uncomplicated in the first case because large amounts of explosives are available for exact identification. The second case is very difficult because only traces of the original material are left behind at the site of explosion. In addition the samples are always contaminated with foreign materials. These circumstances necessitate a highly specific and sensitive analytical detection.

The application of HPLC in criminal procedings has been recently reported by a Swedish group[2]. Nitroglycerine (NG), diethylene glycol dinitrate (DEGDN) and 2,4-dinitrotoluene were detected by a UV detector at 254 nm in a case concerning several burglaries by blasting. Automatic gasoline vendors were exploded by a gang of "dynamiters" in another interesting case. Among other methods the HPLC contributed to the identification of Dynamex B (a Swedish brand of dynamite) and trinitrotoluene (TNT). Krull and Champ[163] in 1980 have reviewed the HPLC application for separation of explosive nitrocompounds using various detectors. They pointed out that according to the early publication of Doali and Juhasz in 1974[164] the UV photometry is the most frequently employed technique, but that it has some drawbacks. It is difficult to employ routinely when a complex sample is composed of differently UV-absorbing components (e.g. gelatinizing, softening agents, stabilizers). Its specifity is also limited because the exact quantification demands micrograms (or hundreds of nanograms) of the components to be analysed. A new sensitive method has been presented by Lafleur and Morriseau[165] reaching down to the 10 nanogram level without interference from ancillary components such as plasticizers and stabilizers. Silica and NH_2-bound phase columns were able to separate 8 different compounds for example NG, EGDN, hexahydro-1,3,5-trinitro-s-triazine (=RDX) and octahydro-1,3,5,7-tetranitro-1,3,5,7-tetrazocine (=HMX). A

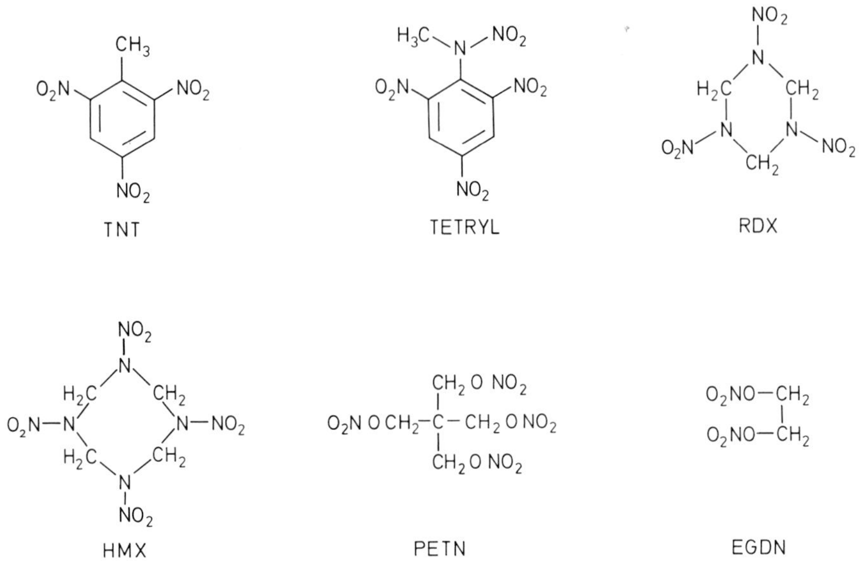

Fig. 10. Structures of some of the more commonly used explosives. For their abbreviations see table 11

Table 11. HPLC in chemical criminology

Subject	Column/Detection	Mobile Phase		Ref.
		Organic	Aqueous	
A. Identification by means of dyes				
Ball-point ink	Partisil/254 nm and 580 nm	Formamide: 2% MeOH: 98%	–	[168]
Heroin preparations	μ-Bondapak C_{18}/ 254 nm	MeOH: 58%	0.005 M tetra-butylammonium hydroxide ca. 0.03%, CH_3COOH (pH 7-8)	[173]
Identity of illicit drugs	SAS-silica/UV and visible with different wavelengths	Propanol: 28.5%	0.25% (w/v) cetrimide	[174]
Lipstick smears	μ-Bondapak C_{18}/ 254 nm	AcN: 50%–100%	H_2O	[172]
Lipstick smears	μ-Bondapak C_{18}/ 254 nm	MeOH: 67.5%	0.005 M tetra-butylammonium phosphate	[171]
Ball-point ink	μ-Bondapak C_{18}/ 546 nm	AcN: 80%	0.005 M Pic B-7 reagent	[169]

Table 11 (continued)

Subject	Column/Detection	Mobile Phase		Ref.
		Organic	Aqueous	
B. Comparison of soil samples				
Fingerprint analysis of different types of soils	Ultrasphere-ODS/ 254 nm	AcN: 90%	H_2O	[170]
C. Analytical approach to explosives				
p-NT; 2,4-DNT; TNT; RDX	Corasil II/RI and 254 nm	Hexane: 60% Methylene chloride: 40% or dioxane 10–50% in cyclohexane	–	[164]
NG; TNT; RDX; HMX (together with residue analysis after explosion)	Partisil/off-line LC/MS with CI and 254 nm	Dichloroethane or dichloroethane: 60% Heptane: 40%	–	[167]
EGDN; NG; RDX; HMX, PETN, etc.	Lichrosorb Si-60 or Lichrosorb-NH_2 TEA Analyser (nitrosylspecific detector,	EtOH: 1.5%–90% in isooctane	–	[165]
NG; EGDN; 2,4-DNT; 2,6-DNT	μ-Bondapak C_{18}/ 254 nm	MeOH: 50%	H_2O	[2]
2,4-DNT; 2-,4-,6-TNT; 2,6-DNT	μ-Porasil/254 nm	Isooctan: 70% Methylene Chloride: 30%	–	[2]
TNT; RDX; PETN; tetryl	RP-18 and RP-8/ 254 nm and on-line LC/MS with NCI (negative chemical ionization)	AcN: 50% or MeOH: 50%	H_2O	[166]

EGDN: Ethylene glycol dinitrate; HMX: Octahydro-1,3,5,7-tetranitro-s-tetrazocine;
PETN: Pentaerythritol tetranitrate; RDX: Hexahydro-1,3,5-trinitro-s-triazine (Cyclonite);
TNT: 2,4,6-trinitrotoluene; NG: nitroglycerine; p-NT: p-nitrotoluene; DNT: Dinitrotoluene

HPLC with a nitrosyl-specific detector (TEA Analyzer) was used. Combined HPLC-MS in either on-line or off-line configuration is the latest approach in the field of specific analysis. Chemical ionisation (CI) has been employed for the analysis of TNT, RDX, tetryl, and pentaerythritol tetranitrate (PETN) by Parker et al.[166].

An on-line system with a splitting ratio 1 to 100 allowed the detection of 100 ng TNT. The negative CI reagent gases were supplied by the mobile phase (acetonitrile/water 50 to 50 or methanol/water 50 to 50). The use of NH_3 as reagent gas for CI resulted in molecular peaks of high intensity in another application of the off-line configuration[167]. Efforts were made to identify some simple residues from test explosions under controlled conditions. TNT, NG, HMX and RDX residues were examined. Even though the TNT peak was surrounded by several other peaks, the component collected was pure enough to yield well defined CI spectra.

RDX was also successfully detected in exploded components, however HMX and NG could not be clearly identified.

6 References

 1. Stas, J.S.: Bull. Acad. Med. Belg. 11, 202 (1851)
 2. Maehly, A., Strömberg, L.: Chemical criminalistics. Springer, Berlin-Heidelberg-New York (1981)
 3. Stahl, E.: Dünnschicht-Chromatographie. Ein Laboratoriumshandbuch. Springer, Berlin-Göttingen-Heidelberg (1962)
 4. Clarke, E.G.C.: Isolation and identification of drugs in pharmaceuticals, body fluids and post-mortem material. The Pharmaceutical Press, London, Vol. 1 and 2 (1969/1975)
 5. Ardrey, R.E., Moffat, A.C.: J. Chromatogr. 220, 195–252 (1981)
 6. Daldrup, T., Susanto, F., Michalke, P.: Fresenius Z. Anal. Chem. 308, 413–427 (1981)
 7. Ardrey, R.E., Zeeuw, R.A. de, Finkle, B.S. et al.: Gas chromatographic retention indices of toxicologically relevant substances on SE-30 or OV-1. Chemie, Weinheim (1985)
 8. Daldrup, T., Michalke, P., Böhme, W.: Chromatogr. Newslett 10, 1–7 (1982)
 9. Rumble, R.H., Roberts, M.S., Wanwimolruk, S.: J. Chromatogr. 225, 252–260 (1981)
10. Petersdorf, S.H., Raisys, V.A., Opheim, K.E.: Clin. Chem. 25, 1300–1302 (1979)
11. Daldrup, T.: In: Symposium Psychopharmaka und Suchtstoffe. Gesellschaft Toxikologische und Forensische Chemie, Mosbach (1980)
12. Daldrup, T, Michalke, P., Schönemann, E.: In: Barz et al. (Ed.) Fortschritte der Rechtsmedizin. Springer, Berlin-Heidelberg-New York (1983)
13. Käferstein, H., Sticht, G.: Beitr. Gerichtl. Med. 41, 95–101 (1983)
14. Lurie, I.: J. Assoc. Off. Anal. Chem. 60, 1035–1040 (1977)
15. Trinler, W.A., Reuland, D.J., Hiatt, T.B.: J. Forensic Sci. Soc. 16, 133–138 (1976)
16. Trinler, W.A., Reuland, D.J.: J. Forensic Sci. Soc. 15, 153–158 (1975)
17. Twitchett, P.J., Moffat, A.C.: J. Chromatogr. 111, 149–157 (1975)
18. Draper, P., Shapcott, D., Lemieux, B.: Clin. Biochem. 12, 52–55 (1979)
19. Kabra, P.M., Koo, H.Y., Marton, L.J.: Clin. Chem. 24, 657–662 (1978)
20. Twitchett, P.J., Gorvin, A.E.P., Moffat, A.C.: J. Chromatogr. 120, 359–368 (1976)
21. Kabra, P.M., Stafford, B.E., Marton, L.J.: J. Anal. Toxicol. 5, 177–182 (1981)
22. Zeeuw, R.A. de, Westenberg, H.G.M.: J. Anal. Toxicol. 2, 229 (1978)
23. Chan, M.L., Whetsell, C., McChesney, J.D.: J. Chromatogr. Sci. 12, 512–516 (1974)
24. Müller, M., Engelhardt, H.: In: Molnar, I. (Ed.) Practical aspects of modern HPLC. De Gryter and Co, Berlin-New York (1982)
25. Engelhardt, H., Dreyer, B., Schmidt, H.: Chromatographia 16, 11–17 (1982)
26. Daldrup, T., Kardel, B.: Chromatographia 18, 81–83 (1984)
27. Baker, J.K., Skelton, R.E., Cheng-Yu, Ma.: J. Chromatogr. 168, 417–427 (1979)
28. White, P.C.: J. Chromatogr. 200, 271–276 (1980)
29. Miller, J.M., Tucker, E.: Int. Labratory may/june: 16–33 (1979)
30. Henion, J.D.: Anal. Chem. 50, 1687–1693 (1978)
31. Tsuge, S., Hirata, Y., Takeuchi, T.: Anal. Chem. 51, 166–169 (1979)
32. Michalke, P.: Z. Rechtsmed. 88, 195–202 (1982)
33. Harzer, K.: J. Chromatogr. 249, 205–208 (1982)
34. Allender, W.J.: J. Forensic Sci. 27, 944–947 (1982)
35. Demeter, J., Heyndrickx, A.: J. Anal. Toxicol. 2, 68–74 (1978)
36. Kemal, M., Imami, R., Poklis, A.: J. Forensic Sci. 27, 217–222 (1982)
37. Daldrup, T., Susanto, F.: In: Symposium Pestizide und Brände. Gesellschaft für Toxikologische und Forensische Chemie, Mosbach (1981)
38. Rejent, T.A., Doyle, R.E.: J. Anal. Toxicol. 6, 199–201 (1982)

39. Sticht, G., Käferstein, H.: Z. Rechtsmed. 85, 169–175 (1980)
40. Königshausen, T., Hort, W., Daldrup, T. et al.: Intensivmed. 18, 171–176 (1981)
41. Thomson, B.M., Pannell, L.K.: J. Anal. Toxicol. 5, 105–109 (1981)
42. Uden, P.C., Bigley, I.E.: Anal. Chim. Acta 94, 29–34 (1977)
43. Szathmary, S.: personal communication
44. Plum, J., Daldrup, T.: Z. Rechtsmed. 94, 257–272 (1985)
45. Teale, D., Marks, V.: Lancet I: 884–885 (1976)
46. Teale, J.D., Clough, J.M., King, L.J. et al.: J. Forensic Sci. Soc. 17, 177–183 (1977)
47. Owens, S.M., McBay, A.J., Cook, C.E.: J. Forensic Sci. 28, 372–379 (1983)
48. Mechoulam, R.: In: Hoffmeister, F., Stille, G. (Ed.) Handbook of experimental pharmacology Vol. 55/III. Springer, Berlin-Heidelberg-New York (1981)
49. Comparini, I.B., Centini, F.: Forensic Sci. Int. 21, 129–137 (1983)
50. Flora, K.P., Cradock, J.C., Davignon, J.F.: J. Chromatogr. 206, 117–123 (1981)
51. Smith, R.N.: J. Chromatogr. 115, 101–106 (1975)
52. Baker, P.B., Fowler, R., Bagon, K.R., Gough, T.A.: J. Anal. Toxicol. 4, 145–152 (1980)
53. Kanter, S.L., Musumeci, M.R., Hollister, L.E.: J. Chromatogr. 171, 504–508 (1979)
54. Coper, H.: In: Hoffmeister, F., Stille, G. (Ed.) Handbook of experimental pharmacology, Vol. 55/III. Springer, Berlin-Heidelberg-New York (1981)
55. Nahas, G.G.: Marihuana – Chemestry, biochemistry, and cellular effects. Springer, New York-Heidelberg-Berlin (1976)
56. Halldin, M.M., Carlsson, S., Kanter, S.L. et. al.: Drug Res. 32, 764–768 (1982)
57. Halldin, M.M., Andersson, L.K.R., Widman, M., Hollister, L.E.: Drug Res. 32, 1135–1138 (1982)
58. Hawks, R.L.: In: Hawks, R.L. (Ed.) Analysis of cannabinoids. Research monograph 42. National Institute on Drug Abuse (1982)
59. Moffat, A.C., Williams, P.L., King, L.J.: In: Hawks, R.L. (Ed.) Analysis of cannabinoids. Research monograph 42. National Institute on Drug Abuse (1982)
60. Shepard, R.M., Milne, G.M.: In: Hawks, R.L. (Ed.) Analysis of cannabinoids. Research monograph 42. National Institute on Drug Abuse (1982)
61. Kanter, S.L., Hollister, L.E., Loeffler, K.O.: J. Chromatogr. 150, 233–237 (1978)
62. Abbott, S.R., Berg, J.R., Loeffler, K.O. et al.: In: Gould, R.F. (Ed.) Cannabinoid analysis in physiological fluids. ACS symposium series. American Chemical Society (1979)
63. Daldrup, T., Matthiesen, U., Krätzschmar, T.: In: Arnold, W., Püschel, K. (Ed.) Internationales Symposium. Entwicklung und Fortschritte der Forensischen Chemie. Dr. Dieter Helm, Heppenheim (1982)
64. Peel, H.W., Perrigo, B.J.: J. Anal. Toxicol. 5, 165–167 (1981)
65. Daldrup, T.: Beitr. Gerichtl. Med. 38, 67–69 (1980)
66. Law, B., Pocock, K., Moffat, A.C.: The routine detection of cannabinoids in blood by radioimmunoassay and combined HPLC/RIA. HOCRE Report No. 388. Home Office Central Research Estabishment, Aldermaston, Reading, Berkshire (1981)
67. Law, B.: J. Forensic Sci. Soc. 21, 31–39 (1981)
68. Williams, P.L., Moffat, A.C., King, L.J.: J. Chromatogr. 155, 273–283 (1978)
69. Law, B., Moffat, A.C.: The routine determination of cannabinoids in urine by radioimmunoassay and combined HPLC/RIA. HOCRE Report No. 346. Home Office Central Research Establishment, Aldermaston, Reading, Berkshire (1980)
70. Law, B., Pocock, K., Moffat, A.C.: An evaluation of a homogeneous enzyme immunoassay (EMIT) for cannabinoid detection in biological fluids. HOCRE Report No. 365. Home Office Central Research Establishment, Aldermaston, Reading, Berkshire (1980)
71. Daldrup, T.: In: Gesellschaft für Toxikologische und Forensische Chemie (Ed.) Symposium Forensische Probleme des Drogenmißbrauchs. Dr. Dieter Helm, Heppenheim (1985)
72. Cashman, P.J., Thornton, J.I.: J. Forensic Sci. Soc. 12, 417–420 (1972)
73. Wu, C.-Y., Siggia, S., Robinson, T., Waskiewicz, R.D.: Anal. Chim. Acta 63, 393–402 (1973)
74. Hays, S.E., Grady, L.T., Kruegel, A.V.: J. Pharm. Sci. 62, 1509–1513 (1973)
75. Knox, J.H., Jurand, J.: J. Chromatogr. 87, 95–108 (1973)
76. Knox, J.H., Jurand, J.: J. Chromatogr. 82, 398–401 (1973)
77. Wittwer, J.D.: J. Forensic Sci. 18, 138–142 (1973)

 78. Verpoorte, R., Svendsen, A.B.: J. Chromatogr. 100, 227–230 (1974)
 79. Twitchett, P.J.: J. Chromatogr. 104, 205–210 (1975)
 80. Jane, J.: J. Chromatogr. 111, 227–233 (1975)
 81. Ziegler, H.W., Beasley, T.H., Smith, D.W.: J. Assoc. Off. Anal. Chem. 58/5, 888–897 (1975)
 82. Jane, J., Taylor, J.F.: J. Chromatogr. 109, 37–42 (1975)
 83. Wheals, B.B.: J. Chromatogr. 122, 85–105 (1976)
 84. Huizer, H., Logtenberg, H., Steenstra, A.J.: Bull. Narc. 29, 65–74 (1977)
 85. Olieman, C., Maat, J., Waliszewski, K., Beyerman, H.C.: J. Chromatogr. 133, 382–385 (1977)
 86. White, M.W.: J. Chromatogr. 178, 229–240 (1978)
 87. Albanbauer, J., Fehn, J., Furtner, W., Megges, G.: Arch. Kriminol. 162, 103–107 (1978)
 88. Reuland, D.J., Trinler, W.A.: Forensic Sci. 11, 195–200 (1978)
 89. Rasmussen, K.E., Tonnesen, F., Nielsen, B. et al.: Medd. Norsk. Farm. Selsk. 40, 117–125 (1978)
 90. Eriksson, B.M., Persson, B.A., Lindberg, M.: J. Chromatogr. 185, 575–581 (1979)
 91. Garrett, E.R., Gurken, T.: J. Pharm. Sci. 68, 26–32 (1979)
 92. Feher, J., Szepesy, L., Szanto, J.: Mag. Kem. Foly 85, 337–340 (1979)
 93. Soni, S.K., Dugar, S.M.: J. Forensic Sci 24, 437–447 (1979)
 94. Poochikian, G.K., Cradock, J.C.: J. Chromatogr. 171, 371–376 (1979)
 95. Vincent, P.G., Engelke, B.F.: J. Assoc. Off. Anal. Chem. 62, 310–314 (1979)
 96. Ulrich, L., Rüegsegger, P.: Arch. Toxicol. 45, 241–248 (1980)
 97. Peterson, R.G., Rumack, B.H., Sullivan, J.B., Makowski, A.: J. Chromatogr. 188, 420–425 (1980)
 98. Wallace, J.E., Harris, S.C., Peek, M.W.: Anal. Chem. 52, 1328–1330 (1980)
 99. Love, J.L., Pannell, L.K.: J. Forensic Sci. 25, 320–326 (1980)
100. Wheals, B.B.: J. Chromatogr. 187, 65–87 (1980)
101. Baker, J.K., Skelton, R.E., Riley, T.N., Bagley, J.R.: J. Chromatogr. Sci. 18, 153–158 (1980)
102. Nobuhara, Y., Hirano, S., Namba, K., Hashimoto, M.: J. Chromatogr. 190, 251–255 (1980)
103. Wittwer, J.D.: Forensic Sci. Int. 18, 215–224 (1981)
104. Smith, R.V., Humphrey, D.W.: Anal. Letters 14, 601–613 (1981)
105. Baker, P.B., Gough, T.A.: J. Chromatogr. Sci. 19, 483–489 (1981)
106. Glasell, J.A., Venn, R.F.: J. Chromatogr. 213, 337–339 (1981)
107. Nelson, P.E., Nolan, S.L., Bedford, K.R.: J. Chromatogr. 234, 407–414 (1982)
108. Bernhauer, D., Fuchs, E.F.: Arch. Kriminol. 169, 73–80 (1982)
109. Tsina, J.W., Fass, M., Debban, J.A., Matin, S.B.: Clin. Chem. 28, 1137–1139 (1982)
110. Lurie, J.S., Sottolano, S.M., Blasof, S.: J. Forensic Sci. 27, 519–526 (1982)
111. Huizer, H.: J. Forensic Sci. 28, 32–39 (1983)
112. Huizer, H.: J. Forensic Sci. 28, 40–48 (1983)
113. Fish, F., Wilson, W.D.C.: J. Pharm. Pharmacol. (Suppl.) 21, 135–138 (1969)
114. Wallace, J.E., Hamilton, H.E., Christenson, J.G. et al.: Anal. Chem. 48, 34–38 (1976)
115. Graffeo, A.P., Lin, D.C.K, Foltz, R.L.: J. Chromatogr. 126, 717–722 (1976)
116. Sugden, K., Cox, G.B., Loscombe, C.R.: J. Chromatogr. 149, 377–390 (1978)
117. Trinler, W.A., Reuland, D.J.: J. Forensic Sci. 23, 37–43 (1978)
118. Jatlow, P.J., Van Dyke, C., Barash, P., Pyck, R.: J. Chromatogr. 152, 115–121 (1978)
119. Olieman, C., Maat, L., Beyerman, H.C.: Recl. Trav. Chim. Pays-Bas 98, 501–502 (1979)
120. Lewin, A.H., Parker, S.R., Carroll, F.J.: J. Chromatogr. 193, 371–380 (1980)
121. Evans, M.A., Morarity, T.: J. Anal. Toxicol. 4, 19–22 (1980)
122. Masoud, A.N., Krupski, D.M.: J. Anal. Toxicol. 4, 305–310 (1980)
123. Fletcher, S.M., Hancock, V.S.: J. Chromatogr. 206, 193–195 (1981)
124. Jane, J., Scott, R.W., Sharpe, R.W.L., White, P.C.: J. Chromatogr. 214, 243–248 (1981)
125. Derendorf H., Garrett, E.R.: J. Pharm. Sci. 72, 630–635 (1983)
126. Garrett, E.R., Seyda, K.: J. Pharm. Sci. 72, 258–271 (1983)
127. Wittwer, J.D., Kluckhorn, J.H.: J. Chromatogr. Sci. 11, 1–6 (1973)
128. Heacock, R.A., Langille, K.R., McNeil, J.D., Frei, R.W.: J. Chromatogr. 77, 425–430 (1973)

129. Jane, J., Wheals, B.B.: J. Chromatogr. 84, 181–186 (1973)
130. Christie, J., White, M.W., Wiles, J.M.: J. Chromatogr. 120, 496–501 (1976)
131. Vandemark, F.L., Schmidt, G., Adams, R.F.: Perkin-Elmer Separation Report 123, Per-
     kin-Elmer Corp., Norwalk (1977)
132. Twitchett, P.J., Fletcher, S.M., Sullivan, A.T., Moffat, A.C.: J. Chromatogr. 150, 73–84
     (1978)
133. Ardrey, R.E., Moffat, A.C.: J. Forensic Sci. Soc. 19, 253–282 (1979)
134. Megges, G.: Arch. Kriminol. 164, 25–30 (1979)
135. White, P.D.: J. Chromatogr. 169, 453–456 (1979)
136. Thomson, B.M.: J. Forensic Sci. 25, 779–785 (1980)
137. Perkal, M., Blackman, G.L., Ottrey, A.L., Turner, L.K.: J. Chromatogr. 196, 180–184
     (1980)
138. Cook, C.E., Brine, D.R., Quin, G.D. et al.: Life Sci. 29, 1967–1972 (1981)
139. Wall, M.E., Brine, D.R., Jeffcoat, A.R. et al.: Res. Commun. Substance Abuuse 2,
     161–172 (1981)
140. Cashman, P.J., Thornton, J.I., Shelman, D.L.: J. Chromatogr. Sci. 11, 7–9 (1973)
141. Achari, R.G., Theimer, E.E.: J. Chromatogr. Sci. 15, 320–321 (1977)
142. Clark, C.R., Teague, J.D., Wells, M.M., Ellis, J.H.: Anal. Chem. 49, 912–915 (1977)
143. Lurie, I.S., Weber, J.M.: J. Liq. Chromatogr. 1, 587–606 (1978)
144. Crommen, J.: J. Chromatogr. 186, 705–724 (1979)
145. Einhellig, K., Kraatz, A., Megges, G.: Arch. Kriminol. 166, 99–106 (1980)
146. Verbiese-Genard, N., Damme, M. van, Hanocq, M., Molle, L.: Clin. Toxicol. 18, 391–400
     (1981)
147. Gill, R., Alexander, S.P., Moffat, A.C.: J. Chromatogr. 247, 39–45 (1982)
148. Harbin, D.N., Lott, P.F.: J. Liquid Chromatogr. 3, 243–256 (1980)
149. Walters, M.J., Walters, S.M.: J. Pharm. Sci. 66, 198–201 (1977)
150. Sullivan, A.T., Twitchett, P.J., Fletcher, S.M., Moffat, A.C.: J. Forensic Sci. Soc. 18,
     89–98 (1978)
151. Beug, M.W., Bigwood, J.: J. Chromatogr. 207, 379–385 (1981)
152. Christiansen, A.L., Rasmussen, K.E., Tønnesen, F.: J. Chromatogr. 210, 163–167 (1981)
153. Sottolano, S.M., Lurie, J.S.: J. Forensic Sci. 28, 929–935 (1983)
154. Kabra, P.M., McDonald, D.M., Marton, L.J.: J. Anal. Toxicol. 2, 127–133 (1978)
155. Wheals, B.B.: J. Chromatogr. 122, 85–105 (1976)
156. Gough, T.A., Baker, P.B.: J. Chromatogr. Sci. 20, 289–329 (1982)
157. Valentine, J.L., Bryant, P.J., Gutshall, P.L. et al.: J. Pharm. Sci. 66, 1263–1266 (1977)
158. Twitchett, P.J., Williams, P.L., Moffat, A.C.: J. Chromatogr. 149, 683–691 (1978)
159. Borys, H.K., Karler, R.: J. Chromatogr. 205, 303–323 (1981)
160. Wheals, B.B., Jane, I.: Analyst 102, 625–644 (1977)
161. Hodnett, C.N., Eberhardt, R.D.: J. Anal. Toxicol. 3, 187–194 (1979)
162. Law, B., Mason, P.A., Moffat, A.C., King, L.J.: J. Anal. Toxicol. 8, 19–22 (1984)
163. Krull, I.S., Champ, M.J.: Int. Labratory may/june: 15–25 (1980)
164. Doali, J.O., Juhasz, A.A.: J. Chromatogr. Sci. 12, 51–56 (1974)
165. Lafleur, A.L., Morriseau, B.D.: Anal. Chem. 52, 1313–1318 (1980)
166. Parker, C.E., Voyksner, R.D., Tondeur, Y. et al.: J. Forensic Sci. 27, 495–505 (1982)
167. Vouros, P., Petersen, B.A., Colwell, L. et al.: Anal. Chem. 49, 1039–1044 (1977)
168. Colwell, L.F., Karger, B.L.: J. Assoc. Off. Anal. Chem. 60, 613–618 (1977)
169. Lyter, A.H.: J. Forensic Sci. 27, 154–160 (1982)
170. Reuland, D.J., Trinler, W.A.: J. Forensic Sci. 18, 201–208 (1981)
171. Andrasko, J.: Forensic Sci. Int. 17, 235–251 (1981)
172. Reuland, D.J., Trinler, W.A.: J. Forensic Sci. Soc. 20, 111–120 (1980)
173. Clark, A.B., Miller, M.D.: J. Forensic Sci. 23, 21–28
174. Joyce, J.R., Sanger, D.G.: J. Forensic Sci. Soc. 19, 203–209 (1979)

Application of HPLC to the Separation of Lipids

K. Aitzetmüller[1]
1 present address: a) Feldbehnstr. 64, D-2085 Quickborn/FRG
b) Institute of Chemistry and Physics, Federal Center of Lipid Research,
Piusallee 76, D-4400 Münster/FRG

1 Introduction

It is comparatively easy to write about "separations" of lipids because, in the field of lipid HPLC, separation techniques as such have made far more progress than detector developments. Today we can separate most of the mixtures that we want to separate, but all too often we cannot detect the separated peaks, or at least we cannot quantitate them. It must be kept in mind, therefore, that in a lot of published lipid work nowadays the separation modes used were chosen primarily with an eye on the *detector* available – and not because that particular *separation* mode is the best one to separate the molecules or groups of molecules one wants to separate. Thus, often a gradient would be best but cannot be used because of an RI detector; benzene or toluene would be best for the separation on a silver nitrate column but cannot be used because one needs detection via short-wave UV, or a "group" separation and quantitation would be best but cannot be used because there is no detector with *equal* sensitivity for saturated and unsaturated molecules present together in one peak.

On the following pages the emphasis will be on separation, not detection, though the major limitations on the detector side will be pointed out where necessary. However, we still hope that someday a detector that removes the solvent prior to detection will become available. This would make most of the current limitations disappear.

2 Lipid Classes

Lipid extracts of biomedical or food samples tend to contain a large variety of substances of widely differing polarity, ranging from hydrocarbons and wax esters to highly polar, sugar- or phosphoric acid-containing glyco- and phospholipids. Their common features are usually one or more long aliphatic chains of variable chain lengths, with or without unsaturation, bound in some way to the "head group" or "rest of the molecule", e.g., via ester, ether or amide linkages.

For HPLC this means that interaction of the nonpolar chains with a nonpolar stationary phase (as in reversed-phase HPLC) or interaction of the polar "head group" with a polar stationary phase (e.g., silica) can be exploited to achieve separations. In addition, several special effects can be superimposed on the basic RP or adsorption separation. These will be discussed later.

Before the advent of HPLC, and for all preparative purposes also today, separations of total lipid extracts into "classes" on silica were very widely used. A *lipid class* then is a group of molecules that will be eluted together from a silica column, where chain-length and unsaturation differences of the nonpolar end of the molecule play little, if any, role and the polar "rest-of-the-molecule" plays the all-important role.

A few typical representatives of lipid classes are illustrated in Fig. 1.

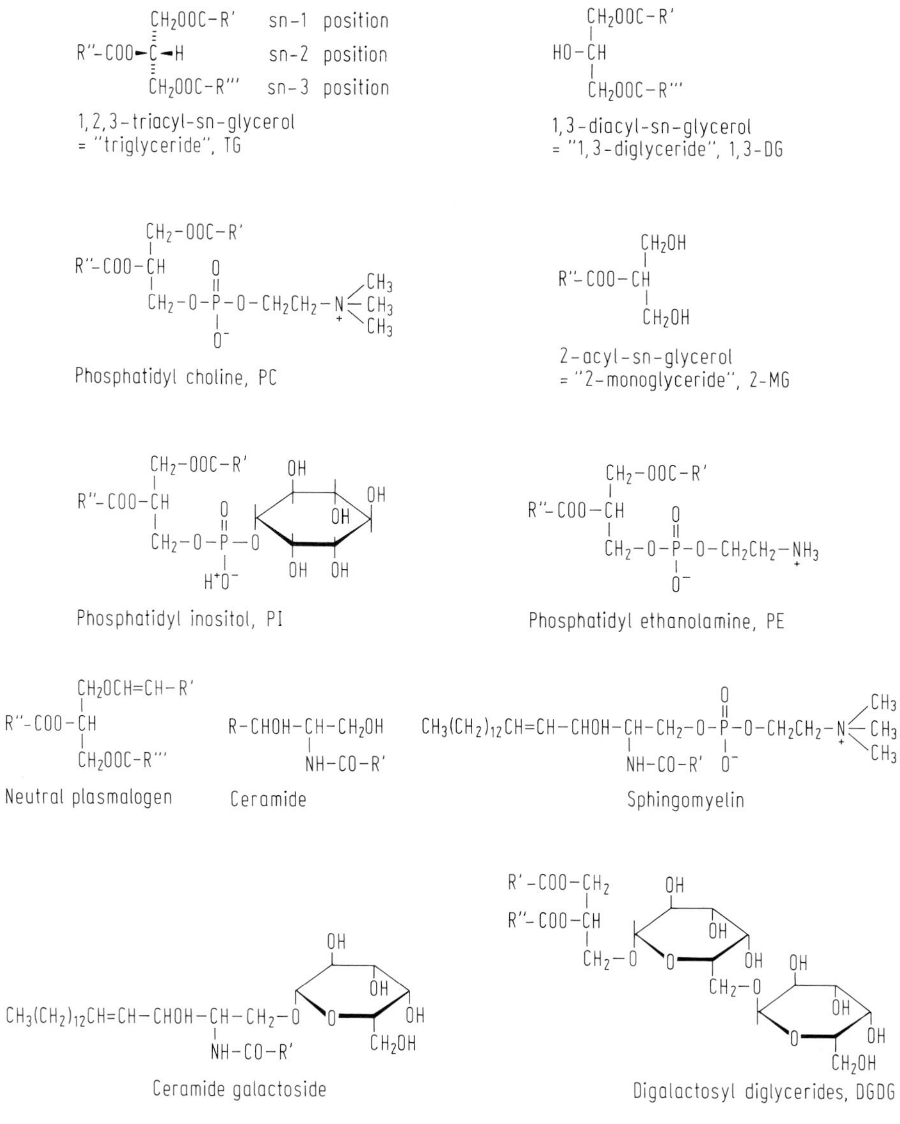

CH₂OOC–R'   sn–1 position
R''–COO–C◄H   sn–2 position
CH₂OOC–R'''   sn–3 position

1,2,3–triacyl–sn–glycerol
= "triglyceride", TG

CH₂OOC–R'
HO–CH
CH₂OOC–R'''

1,3–diacyl–sn–glycerol
= "1,3–diglyceride", 1,3–DG

CH₂–OOC–R'
R''–COO–CH   O
CH₂–O–P–O–CH₂CH₂–N⁺(CH₃)(CH₃)(CH₃)
O⁻

Phosphatidyl choline, PC

CH₂OH
R''–COO–CH
CH₂OH

2–acyl–sn–glycerol
= "2–monoglyceride", 2–MG

CH₂–OOC–R'
R''–COO–CH   O
CH₂–O–P–O   (inositol ring: OH OH OH OH OH)
H⁺O⁻

Phosphatidyl inositol, PI

CH₂–OOC–R'
R''–COO–CH   O
CH₂–O–P–O–CH₂CH₂–NH₃⁺
O⁻

Phosphatidyl ethanolamine, PE

CH₂OCH=CH–R'
R''–COO–CH
CH₂OOC–R'''

Neutral plasmalogen

R–CHOH–CH–CH₂OH
NH–CO–R'

Ceramide

CH₃(CH₂)₁₂CH=CH–CHOH–CH–CH₂–O–P(O)–O–CH₂CH₂–N⁺(CH₃)(CH₃)(CH₃)
NH–CO–R'   O⁻

Sphingomyelin

CH₃(CH₂)₁₂CH=CH–CHOH–CH–CH₂–O   (galactose ring: OH OH OH CH₂OH)
NH–CO–R'

Ceramide galactoside

R'–COO–CH₂
R''–COO–CH
CH₂–O   (sugar rings: OH OH OH OH OH CH₂–O CH₂OH)

Digalactosyl diglycerides, DGDG

Fig. 1. *A few typical representatives of "lipid classes".* Within one lipid class, individual molecules differ in the nature of the long-chain residues R, R', R'', R'''

2.1 Gradient Elution of Traditional Lipid Classes

Traditional lipid classes can be eluted as distinct *lipid class peaks* from silica columns with solvent gradients of increasing polarity. Hydrocarbons elute first, wax- and cholesterol esters, glycerol ethers, triglycerides, free sterols, diglycerides, monoglycer-

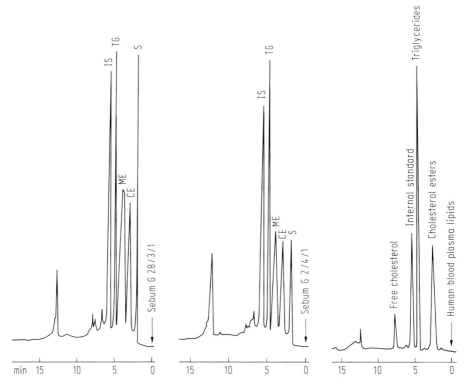

Fig. 2. *Lipid class HPLC of human hair lipids (sebum) and blood plasma lipids.* Column: LiChrosorb SI 60, 10 μ, 150 × 4.6 mm. Two concave gradients A → B → C using Ultrograd equipment (LKB) with a Milton-Roy 196–100 pump and a Pye LCM-2 detector. Solvents: A = CCl$_4$:isooctane = 34:66; B = CHCl$_3$:dioxane:hexane = 40:11:49; C = CHCl$_3$:MeOH:diisopropyl ether = 34:36:30. Flow: 1.9 ml/min. From Ref.[6] by permission of authors and journal. Peaks: S = squalene and other hydrocarbons; CE = cholesterol esters and wax esters; ME = fatty acid methyl esters (from free fatty acids originally present in sebum, after addition of diazomethane); TG = triglycerides; IS = internal standard

ides, etc. will follow. Near the end of the sequence are highly polar classes like sphingomyelins, glycolipids, and phospholipids of mainly biomedical interest.

In the early years of HPLC, intriguing lipid class separations of total lipid extracts were shown by Privett and his group[1–3], using home-made transport-flame ionization detectors for HPLC. These separations used multiple sequential solvent gradients from nonpolar solvents such as pentane up to highly polar solvents such as methanol-water-ammonia mixtures.

The lipid analyst is most often confronted with questions like "how much diglyceride is in this sample?" or "what is the PC (for phosphatidyl choline) content of this lecithin?". To the chromatographer, such questions always mean that a *lipid class* peak should be produced and quantitatively determined in a complex sample – a task that can best be carried out with a mass- or carbon-sensitive detector[4]. In particular, the detector should be *insensitive* to differences in chain unsaturation[5].

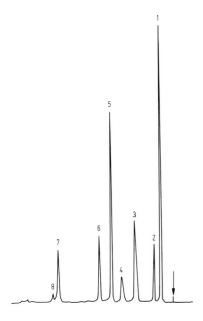

Fig. 3. *Lipid class separation of a standard mixture.* Column: 150 × 4.6 mm, LiChrosorb SI 60, 5 μ, Varian 5020 with Pye LCM-2 transport-FID. Flow: 1.5 ml/min. Gradient: A = 1,2-dichloroethane: formic acid = 997:3 (v/v); B = 1,2-dichloroethane: ethyl acetate:formic acid = 500:500:2 (v/v/v); 0–4 min: 0% B; 4–10 min: 0 → 20% B (linear); 10–12 min: 20 → 100% B (linear); 12–20 min: 100% B. Figure supplied by J.J.L. Hoogenboom[7, 8] and reproduced here by permission of Unilever Research, Vlaardingen. Peaks: 1 Tripalmitoylglycerol (Tripalmitin); 2 Palmitic acid; 3 Octadecan-1-ol; 4 Cholesterol; 5 1,3-Dipalmitoylglycerol (1,3-Dipalmitin); 6 1,2-Dipalmitoylglycerol (1,2-Dipalmitin); 7 1-Monopalmitoylglycerol (1-Monopalmitin); 8 2-Monopalmitoylglycerol (2-Monopalmitin)

Although detectors of this type are temporarily not on the market, two examples of quantitative lipid-class separations are shown here. Fig. 2 illustrates nonpolar lipid class separations of human hair lipids and human blood plasma lipids[6], and Fig. 3 shows separations of lipid classes that are of interest to the food industry[7, 8]. Class separations of phospholipids and other highly polar lipids will be discussed in Sect. 8 of this chapter.

2.2 Other Group or Class Separations

A number of other separation techniques are known where group (or class) separations other than those obtained by gradient elution from silica can be achieved. However, many of these techniques are not amenable to HPLC.

These include separations based on the acidic or basic nature of certain lipids, separations of branched vs. straight-chain lipids (e.g., on urea columns), exploitation of hydroperoxide retention on Sephadex, separation of (straight-chain) wax esters from cholesterol esters by "degree of flatness chromatography"[9] on MgO, etc. GPC separations by molecular size into groups of lipids differing by number of fatty acid residues are discussed in Sect. 3. Silver nitrate separations by double bond number are discussed in Sect. 5 and 6.1.

3 GPC Separations

Many lipid classes have a rather small "head group" and differ primarily in the number of (long-chain) fatty acids attached to this head group. If M_0 is the molecular

Table 1. Typical molecular weights of lipid classes amenable to GPC (schematic, examples only)

"Head group" (examples of)	Mol.wt. of head group M_0	Number of fatty acid residues attached to head group n	Lipid class or group name	Average mol. wt. of lipid class[a] M_n
Water	18	1	Free fatty acid	280
Glycerol	92	1	Monoglyceride	354
		2	Diglyceride	616
		3	Triglyceride	878
2 glycerols	2×92	6	Dimeric triglyceride	1756
Glycerol esterified with lactic acid	164	1 ⎫ 2 ⎬ Lactoglycerides[b] 3 ⎭		426 688 950
Propylene glycol	76	1 ⎫ Propylene glycol esters[b] 2 ⎭		338 600
Hexadecanol	242	1	Wax esters	504
Cholesterol	386	1	Cholesterol esters	648
Sphingosine	299	1 ⎫ 2 ⎬ Sphingolipids 3 ⎭		561 823 1085
Sorbitans	164	1 ⎫ 2 ⎬ Sorbitan esters[b] 3 ⎭		426 688 950
Glycerol esterified with phosphoryl-choline	257	1 2	Lyso-phosphatidylcholine[c] phosphatidylcholine[c]	519 781
Mono galactosyl-glycerol	254	1 ⎫ 2 ⎭	Monogalactosyl-mono- and -diglycerides (a simple glycolipid)	516 778

[a] Assuming an average molecular weight of 280 for the fatty acids.
[b] Typical emulsifiers.
[c] May form micelles in typical GPC solvents[10, 11]

weight of the head group, and 280 is an average molecular weight of the most widely occurring fatty acids, then we have series of molecules with molecular weights:

$$M_n = M_0 + n \cdot (\text{fatty acids}) - n \cdot (\text{water})$$
$$M_n = M_0 + 280\,n - 18\,n = M_0 + n \cdot 262$$

where n is the number of fatty acid residues attached to the head group. Table 1 gives a few examples. A series of molecules differing by multiples of 262 molecular weight units is a rather ideal case for a GPC separation. [Because n fatty acids will replace n free OH groups on the head molecule, it is also an ideal case for

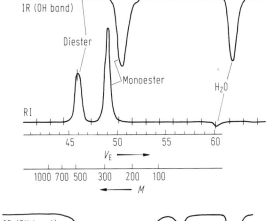

a

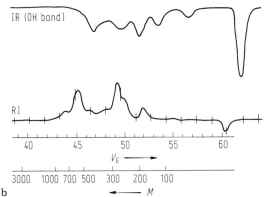

b

Fig. 4a and b. *GPC of emulsifiers with RI and IR detection.* **a** A typical propylene glycol ester. **b** A typical diacetyltartaric acid (DATA) ester. In both **a** and **b**, the upper trace is from an infrared detector set at the OH group frequency, the lower trace is from the refractive index detector. In the simple chromatogram (**a**) of the propylene glycol ester, one notices the absence of the diester peak in the IR(OH) trace, and the presence of a large water peak. In the corresponding RI trace, both ester peaks and a small negative water peak are seen. The peaks in the DATA ester chromatogram (**b**) are caused by a variety of mono- and diglycerides with or without diacetyltartaric acid residues, plus free fatty and diacetyltartaric acids. Conditions: A set of six μ-Styragel columns (300 × 7.8 mm each; 1×10^4 Å, 1×10^3 Å, 1×500 Å and 3×100 Å) in THF at 39.5° C. RI Detector: Knauer. IR Detector: Perkin Elmer 157 NaCl-IR Spectrophotometer equipped with 1 mm flow-through cells. [K. Aitzetmüller, M. Unbehend, and J. Kunze, unpublished work (1978)[14], cf. also Fig. 4 in Ref. [5]]

an adsorption separation based on the number of free OH groups (cf. Sect. 7 on partial glycerides and food emulsifiers.)]

GPC separations were used to detect dimeric triglycerides in used frying fats[12, 13] and to obtain information on lipid classes contained in food emulsifiers[5, 14]. Sometimes GPC can be used to detect micelles formed by very polar phospholipids[10, 11] in nonpolar solvents.

An example for the GPC separation of emulsifier lipids is given in Fig. 4. This also illustrates the use of an infrared detector set at the hydroxyl frequency that serves as a "hydroxyl group detector"[5, 14]. The ratio of the signals from the RI detector and from the OH-IR detector can give additional information on the separated groups of emulsifier components.

Similarly, since lipids contain a minimum of n carboxylic ester bonds, an infrared detector set at the ester carbonyl frequency[15] could give additional information on the number of ester linkages present. In a molecular size separation of lipids by GPC, the OH-IR signal will generally decrease with increasing molecular size, whereas the CO-IR signal will tend to increase with molecular size.

Experimentally, GPC is rather easy to carry out, and conventional equipment – as used for work with polymers – can be employed. There is only one solvent

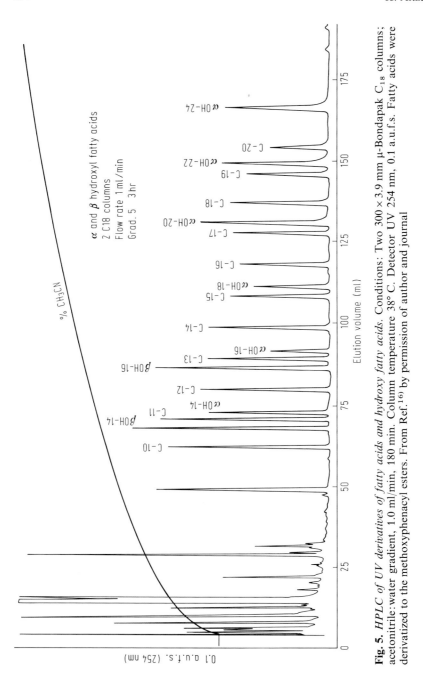

Fig. 5. *HPLC of UV derivatives of fatty acids and hydroxy fatty acids.* Conditions: Two 300×3.9 mm µ-Bondapak C_{18} columns; acetonitrile:water gradient, 1.0 ml/min, 180 min. Column temperature 38° C. Detector UV 254 nm, 0.1 a.u.f.s. Fatty acids were derivatized to the methoxyphenacyl esters. From Ref.[16] by permission of author and journal

(often tetrahydrofuran, THF) and so RI detectors can be used. Mono-, di- and triglycerides, dimeric triglycerides, and many other lipids – except the most polar ones – can be separated in THF[5,12,13]. For quantitative work, however, differences in degree of unsaturation may be a problem with RI detectors (cf. Sect. 12).

4 HPLC of Fatty Acids and Their Esters

Fatty acids are usually derivatized to UV-absorbing or fluorescent esters before HPLC to facilitate their detection. In reversed-phase HPLC, fatty acid derivatives can be separated according to both chain length or carbon number and degree of unsaturation, although with polar (e.g., hydroxy-) fatty acids[16] and prostaglandins[17] (cf. Sect. 9) the presence of polar groups will dominate the separation. An example of this is shown in Fig. 5.

Derivatization and derivatives are discussed elsewhere in this volume, but some of the most useful derivatives for lipids are listed in Table 2, which also gives some selected key references. A large number of authors have shown many interesting separations of fatty acid derivatives (cf. the references given in Table 2) by HPLC using water-acetonitrile, water-methanol, and similar gradients. The resolution of unsaturated vs. saturated fatty acids is much larger in reversed-phase HPLC than in GLC. (For solvent effects with fatty acid derivatives cf. Sect. 11 and Fig. 15).

Table 2. Reagents and derivatives often used in the HPLC of Lipids. Only a few key references are given by way of examples

Reagent or Derivative	Used for:	Ref.
p-Methoxyanilides	Fatty acids	[18]
Phenacyl esters	Fatty acids	[19–21]
p-Bromophenacyl esters	Fatty acids	[22–25]
	Prostaglandins	[17]
p-Nitrophenacyl esters	Prostaglandins	[17]
m-Methoxyphenacyl esters	Fatty acids	[16, 26–28]
9-Anthryldiazomethane	Prostaglandins	[29]
Naphthyldiazoalkanes	Fatty acids	[30]
p-Nitrobenzoates	Diglycerides	[31]
	Glycolipids	[32]
	Ceramides	[33]
Dansyl derivatives	Amino-phospholipids	[34]
Biphenylcarbonyl derivatives	Amino-phospholipids	[35, 36]
Benzoates (perbenzoyl derivatives)	Glyco- and phospholipids, ceramides, cerebrosides	[33, 36]
Iodination and bromination products	Unsaturated triglycerides	[37]
p-Nitrobenzyl esters	Fatty acids	[38]
4-Hydroxymethyl-7-acetoxy-coumarin esters	Fatty acids	[39, 40]
Phthalimidomethyl esters	Fatty acids	[41]
4-Hydroxymethyl-7-methoxy-coumarin esters	Fatty acids	[42–45]
Naphthylurethanes	Fatty alcohols, cholesterol	[46]
p-Methylthiobenzoate esters	Fatty alcohols	[47]
9-Chloromethylanthracene	Fatty acids	[48]

Free fatty acids and fatty acid *methyl esters* can also be separated according to chain length and degree of unsaturation by RP-HPLC[49, 50], but may present a considerable detection problem – at least in those cases, where only small samples are available and gradients are required.

For the analysis of fatty acids, HPLC is in direct competition with gas chromatography (GLC), and, in the author's opinion, GLC should normally be preferred[5].

However, with fluorescent derivatives HPLC detection limits are sometimes lower than those in GLC, and the analysis of very-long-chain fatty acids is no problem in HPLC – both are factors that may lead to a preference of HPLC over GLC, for example, for the analysis of long-chain fatty acids in small samples of bacterial origin[22, 23].

For compounds with very long chains, as in mycolic acids, for the highly nonpolar fatty acid cholesterol esters in blood, and in all those cases where residual triglycerides may be present in the sample to be injected, water-free mobile phases should be used (so-called "non-aqueous" reversed phase).

In the adsorption mode on silica, separation effects according to chain lengths and double bonds present are less pronounced, but hydroxy fatty acids can be separated as a group from non-hydroxy fatty acids.

Cholesterol esters in blood plasma lipids and skin surface lipids are eluted well after the triglycerides in RP systems[51, 52], but before the triglycerides on silica[6, 53]. On the other hand, fatty acid methyl esters are eluted before the corresponding triglycerides on both silica and RP columns.

5 Separations by Double Bond Number

Separations by double bond number (DB separations[1]) are best carried out by adsorption chromatography on silver nitrate-impregnated silica. This technique is most often used in TLC and is also called "argentation chromatography". In $AgNO_3$-silica adsorption TLC or HPLC, there is little or no chain-length separation (at least for the most often occurring C_{16}-C_{22} chains), so that groups differing in double bond number can be separated. It is therefore possible to separate a mixture into groups of molecules with 0, 1, 2, 3, etc. double bonds[54]. The technique is most important for triglyceride analysis (cf. Sects. 6.1 and 6.3) but works equally well for fatty acids, phospholipids, cholesterol esters, waxes, steroids, etc.

Figure 6 shows a DB separation of fatty acid methyl esters[55]. On $AgNO_3$-silica, the differences in retention between species with 1, 2, 3, 4, etc. double bonds are so large that gradient elution is usually required. *Cis* double bonds are more retained than *trans,* and on highly silver-loaded columns even separations of positional isomers of cis- and trans-C_{18}-monoenes have been reported[56].

To a lesser degree, the same effect can be achieved with Ag^+ ions in the mobile phase of RP separations[57-61]. This technique is less suitable for preparative group separations because the RP partition effects dominate. It may, however, someday become important for high resolution separations of fatty acid derivatives.

---

1 For the terminology of DB separations, CN separations, and PN separations in lipid analysis, cf. Sect. 6 and Fig. 7 where this is discussed in more detail

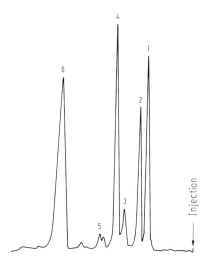

Fig. 6. *Separation of fatty acid methyl esters by double bond number.* Methyl esters of partially hydrogenated soybean oil were separated on a silica column impregnated with 10% silver nitrate. LiChrosorb SI 60, 5 μ, +10% AgNO$_3$, 150 × 4.7 mm. Solvent: toluene:petroleum ether:ethyl acetate = 60:40:1.2, 0.4 ml/min. Detection: Pye LCM-2. Supplied by M. Levacq, private communication[55]. Peaks: 1 saturated methyl esters; 2 *trans*-monoenes; 3 conjugated dienes; 4 *cis*-monoenes; 5 *cis-trans*-dienes; 6 *cis-cis*-dienes

6 Triglyceride Separations

The analysis of the triglyceride composition of an oil or fat is one of the most challenging tasks for the analytical chemist, because – counting all possible isomers – x^3 triglycerides can be formed from x fatty acids (Table 3).

Most seed fats from plants contain 5–10 different fatty acids. Animal fats and margarine compositions may be more complex (10–40 different fatty acids). At least 142 different fatty acids have been detected in butter[62]. This means that, e.g., in coconut oil with 8 fatty acids, 512 triglycerides may be expected. In butter (with 142 different fatty acids), 2863288 different triglycerides are possible[63].

For the elucidation of the triglyceride composition of an oil, there are basically three types of chromatographic separation of the intact triglycerides, as shown in a figure taken from a paper by El-Hamdy and Perkins[64] (Fig. 7). These are:

Table 3. Number of triglyceride species in a fat containing x different fatty acids

Fatty acids	Number of Triglycerides		
	All isomers	No optical isomers	No isomers
x	x^3	$\dfrac{x^3 + x^2}{2}$	$\dfrac{x^3 + 3x^2 + 2x}{6}$
4	64	40	20
8	512	288	120
10	1000	550	220
20	8000	4200	1540

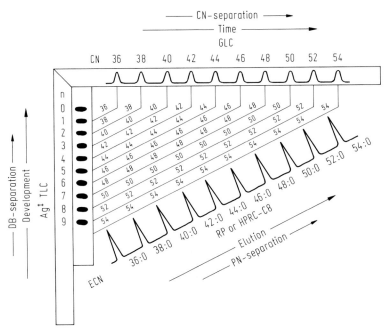

Fig. 7. *The three basic types of triglyceride separations by chromatography.* Separations by carbon number (CN) as in GLC, by double bond number (DB) as in TLC or HPLC on silver nitrate-impregnated silica, and by partition number (PN) as in reversed-phase HPLC (NARP). These are interrelated by the equation $PN = CN - 2 \cdot DB$. From El Hamdy and Perkins [64], by permission of authors and journal

Table 4. Triglycerides and their partition numbers. (Adapted from J.-P. Goiffon et al.[65])

PN	TG	CN:DB	PN	TG	CN:DB	PN	TG	CN:DB	PN	TG	CN:DB
28	CyCyLn	34:3	30	CoCyO	32:1	32	CyCyO	34:1	34	CCL	38:2
	CoCLn	3		CoCoSt	30:0		CoCO	1		CyLaL	2
	CoCyL	32:2		CoCyP	0		CoCP	32:0		CoMyL	2
	CoCoO	30:1		CoCMy	0		CoLaMy	0		CyCO	36:1
	CoCoP	28:0		CoLaLa	0		CyCyP	0		CoLaO	1
	CoCyMy	0		CyMyMy	0		CyCMy	0			
	CoCLa	0		CyCLa	0		CyLaLa	0		CLaLa	34:0
	CyCyLa	0		CCC	0		CCLa	0		CCMy	0
	CyCC	0					CoCySt	0		CoLaMy	0
			32	CyLnLn	44:6	34	CLnLn	46:6		CyCP	0
				CoLLn	42:5		CyLLn	44:5		CyCySt	0
30	CoLnLn	42:6		CCLn	38:3		CoLL	42:4		CoMyMy	0
	CyCLn	36:3		CyLaLn	3		CoOLn	4		CoLaP	0
	CoLaLn	3		CoMyLn	3		CLaLn	40:3		CoCSt	0
	CyCyL	34:2		CyCL	36:2		CyMyLn	3	36	LnLnLn	54:9
	CoCL	2		CoLaL	2		CoPLn	3		LaLnLn	48:6

Table 4 (continued)

PN	TG	CN:DB	PN	TG	CN:DB	PN	TG	CN:DB	PN	TG	CN:DB
36	CLLn	46:5	38	CLaO	40:1	42	PLLn	52:5	44	CPSt	44:0
	CyLL	44:4		CyMyO	1					CyStSt	0
	CyOLn	4		CoPO	1		MyLL	50:4			
							MyOLn	4	46	OOL	54:4
	LaLaLn	42:3		LaLaMy	38:0					StLL	4
	CMyLn	3		CMyMy	0		MyPLn	48:3		StOLn	4
	CyPLn	3		CLaP	0		LaOL	3			
	CoOL	3		CCSt	0		LaStLn	3		POL	52:3
	CoStLn	3		CyMyP	0					PStLn	3
				CyLaSt	0		MyMyL	46:2			
	CLaL	40:2		CoPP	0		LaPL	2		PPL	50:2
	CyMyL	2		CoMySt	0		COO	2		MyOO	2
	CoPL	2					CStL	2		MyStL	2
			40	LLLn	54:7						
	CCO	38:1		OLnLn	7		LaMyO	44:1		MyPO	48:1
	CyLaO	1					CPO	1		LaStO	1
	CoMyO	1		PLnLn	52:6		CyStO	1			
										MyPP	46:0
	LaLaLa	36:0		MyLLn	50:5		MyMyMy	42:0		MyMySt	0
	CLaMy	0					LaMyP	0		LaPSt	0
	CCP	0		LaLL	48:4		LaLaSt	0		CStSt	0
	CyMyMy	0		LaOLn	4		CPP	0			
	CyLaP	0					CMySt	0	48	OOO	54:3
	CyCSt	0		MyMyLn	46:3		CyPSt	0		StOL	3
	CoMyP	0		LaPLn	3		CoStSt	0		StStLn	3
	CoLSt	0		COL	3						
				CStLn	3	44	OLL	54:5		POO	52:2
							OOLn	5		PStL	2
				LaMyL	44:2		StLLn	5			
				CPL	2					PPO	50:1
38	LLnLn	54:8		CyOO	2		PLL	52:4		MyStO	1
				CyStL	2		POLn	4			
	MyLnLn	50:6								PPP	48:0
				LaLaO	42:1		PPLn	50:3		MyPSt	0
	LaLLn	48:5		CMyO	1		MyOL	3		LaStSt	0
				CyPO	1		MyStLn	3			
	CLL	46:4		CoStO	1				50	StOO	54:2
	COLn	4					MyPL	48:2		StStL	2
				CMyP	40:0		LaOO	2			
	LaMyLn	44:3		CLaSt	0		LaStL	2		PStO	52:1
	CPLn	3		CyPP	0						
	CyOL	3		CyMySt	0		MyMyO	46:1		PPSt	50:0
	CyStLn	3		CoPSt	0		LaPO	1		MyStSt	0
							CStO	1	52	StStO	54:1
	LaLaL	42:2	42	LLL	54:6						
	CMyL	2		OLLn	6		MyMyP	44:0		PStSt	52:0
	CyPL	2		StLnLn	6		LaPP	0	54	StStSt	54:0
	CoOO	2					LaMySt	0			
	CoStL	2									

PN = (Integral) Partition Number; CN = Carbon Number; DB = Number of Double Bonds; PN = CN $-$ 2 · DB.

Fatty acid abbreviations:

Co = caproic; Cy = caprylic; C = capric; La = lauric; My = myristic; P = palmitic; St = stearic; O = oleic; L = linoleic; Ln = linolenic

a) Separation by carbon number (CN), as in high-temperature gas-liquid chromatography (GLC) of triglycerides;
b) separation by double-bond number (DB) as in TLC or HPLC on silver nitrate-impregnated plates or columns, and
c) separation by partition number (PN) as in liquid-liquid partition chromatography on conventional or HPLC columns.

The partition number concept[63] defines an (integral) partition number PN as the carbon number CN minus twice the number of double bonds DB:

$$PN = CN - 2 \cdot DB$$

Table 4, modified from a paper by Goiffon et al.[65], gives the partition, carbon, and double-bond numbers for a range of common triglycerides. Since PN is the most important property of a triglyceride for HPLC, the triglycerides have been arranged in the order of increasing PN.

A discussion and a schematic description of the various elution sequences that can be obtained with a number of model triglycerides was given in an earlier publication by the author[5].

6.1 Argentation Separations

DB separations of triglycerides into classes containing 0, 1, 2, 3, ... double bonds can be carried out on $AgNO_3$-impregnated silica columns[4, 66–69]. Again, because

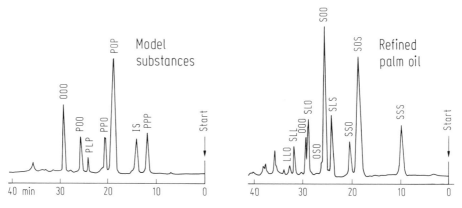

Fig. 8. *Triglyceride separations by argentation HPLC.* By gradient elution HPLC from a silica column impregnated with 11% AgNO₃ and transport-FID detection, "double bond classes" can be determined in a model mixture (left) and in a sample of refined palm oil (right). A concave gradient from hexane/toluene/ethyl acetate (72:26:2, v/v/v) to toluene/ethyl acetate (90:10, v/v) in 30 min was used. Positional isomers on the glycerol (e.g., SOS from SSO) can be separated. In the mixture of model substances (left), each peak consists of one single compound. P = palmitic, O = oleic, L = linoleic acid residues. IS = internal standard. Right hand chromatogram: Palm oil. S in this case stands for saturated fatty acids, and includes predominantly palmitic, but also stearic and myristic acid residues. O = monounsaturated (mostly oleic), L = diunsaturated (mostly linoleic) fatty acids. There is no chain length separation. The peak allocations were verified by mass spectrometry of the collected peak fractions. Figure supplied by A.C.J. Kroesen[66] and published[5] by permission of Unilever Research, Vlaardingen. Reproduced here by permission of Pergamon Press, Oxford, GB

of the large separation factors, best results are obtained by use of solvent gradients. For quantitative work, a transport-flame ionization detector is necessary[4]. Fig. 8 shows results achieved with palm oil[66] and a test mixture, including the separation of the positional isomers SOS and SSO (or POP and PPO), which is important for chocolate-type fats and cannot be achieved by reversed-phase partition HPLC.

Preparative DB separations of triglycerides are often carried out in order to obtain milligram-size fractions for further analysis by GLC, PN-HPLC, or lipase splitting.

With partially hardened fats containing *trans*-unsaturated triglycerides, DB separations on silver nitrate columns or plates give higher resolutions than reversed-phase PN separations. A zone containing triglycerides with one *trans* double bond is easily separated between the saturated and *cis*-monoene classes of triglycerides. The separation effects between molecules with *trans* or cis double bonds are much larger in argentation chromatography than in reversed-phase partition chromatography.

It is also possible to derivatize unsaturated triglycerides at the double bond, e.g., with iodine, and to separate the resulting iodized DB classes on silica[37].

6.2 Reversed-phase Separations

The technique of "non-aqueous" reversed-phase HPLC has been developed over the last few years[15, 64, 65, 70 – 90] and has been applied primarily to the separation of triglycerides and other long-chain lipids that are barely soluble in water-based mobile phases.

Reversed-phase partition is the most important HPLC technique for the analysis of triglyceride mixtures. The elution sequences are determined by the partition numbers of the various triglycerides[63] (Table 4). For PN separations – as for DB separations – ideally gradient elution and a gradient-compatible, lipid-sensitive detector *in*sensitive to unsaturation differences should be used[5, 11]. In the absence of such detectors, isocratic elution and a variety of other detectors have been used.

Again, as with DB separations, *preparative* PN separations are important. They are often used to obtain fractions (all triglycerides of identical partition number) for further analysis (cf. Sect. 6.3. below).

Certain precautions are necessary for the preparative peak-trapping of PN peaks in liquid chromatography or HPLC. These precautions have been discussed and described by Herslöf et al.[71] and particularly by Bezard and Ouedraogo[76], and are related to the fact that column type and age and mobile phase composition may cause deviations from the rule that one double bond is equal to two $-CH_2-$ units.

Hence, we can define[5, 88] an apparent partition number or apparent carbon number ACN (which then is no longer an integral number):

$$ACN = CN - n \cdot DB$$

n is then a factor for double-bond contribution.

ACN is experimentally determined by interpolation of the retention observed for a given unsaturated triglyceride between pairs of saturated triglycerides (for these, ACN = CN). ACN is equivalent to "EPN" (equivalent partition number) of Litchfield[63] and to "ECN" (equivalent carbon number) of Herslöf et al.[71].

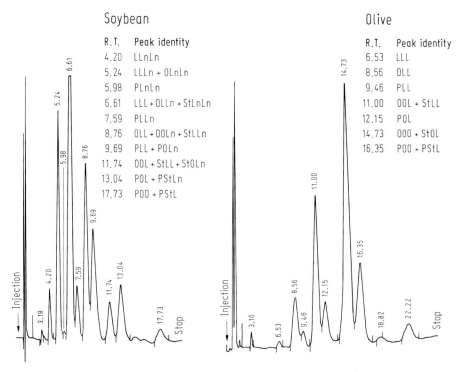

Fig. 9. *Reversed-phase HPLC separations of triglycerides in edible oils with short-wavelength UV (SWUV) detection.* Detection: short-wavelength UV at 205 nm. Separation system: aceto-nitrile:diethylether = 75:25 at 2 ml/min, LiChrosorb RP-18, 150 × 4.7 mm. (M. Levacq, private communication, April 1980[55])

We can now distinguish between two basically different types of reversed phase HPLC approaches for the analysis of triglycerides:

a) The preparative peak-trapping approach. Here one needs a column-solvent system where n is exactly equal to 2.0 to facilitate quantitative peak-trapping[76] for further analysis. When a preparative PN separation is the initial step of a triglyceride analysis strategy (cf. Sect. 6.3), a great effort must be invested to avoid a "fine structure"[76].

b) The "fingerprint" approach. A column-solvent system is needed where n ≠ 2.0 to obtain maximum resolution between saturated and unsaturated triglyceride mole-cules of the same (integral) partition number. In this way, chromatograms of the type shown in Figs. 9 and 10 are obtained, i.e., separations by PN with an "added fine structure".

Fig. 10a–d. *RP-HPLC-RI chromatograms with a flow gradient.* Four different edible oil samples were analyzed on a Spherisorb ODS-2-column (250 × 4.6 mm) in acetonitrile: acetone = 25:75 using a flow gradient from 0.6 to 1.6 ml/min in 20 min[89]. The flow gradient was produced with a Waters M660 programmer linked to a single Waters M6000 pump. Detection: Waters 401 RI at attenuation 8. **a** Coconut oil, containing short- and medium-chain triglycerides. **b** Cottonseed oil. **c** Linsed oil, containing highly unsaturated C-54 triglycerides. **d** Palm Stearine, the higher melting fraction of palm oil

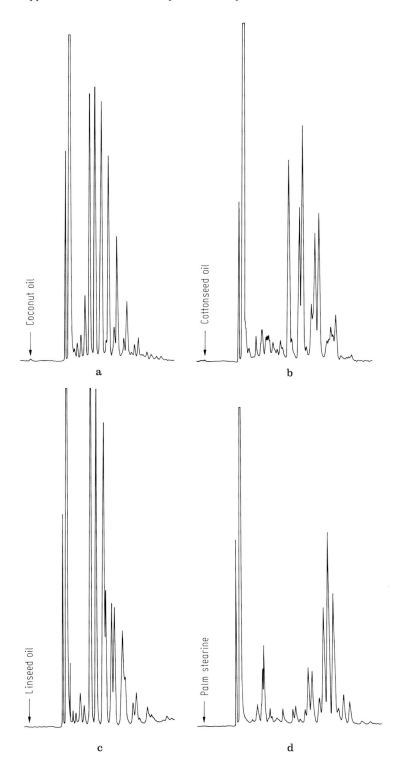

Table 5. Selected examples of conditions used for triglyceride separations by non-aqueous reversed phase HPLC

Sample type (origin of triglyceride fraction)	Class or group separation	Superimposed fine structure (n±2)	Solvent gradient Yes	Solvent gradient No	Column used	Detector used	Detector-compatible mobile phases used for NARP-HPLC	Ref.
Vegetable oils	PN	+		×	Spherisorb ODS, ODS 2	RI	Acetonitrile:acetone = 66:34, 50:50, 30:70 incl. flow gradients	[89]
Vegetable oils	PN	+		×	LiChrosorb RP-18	RI	Acetonitrile:acetone = 50:50	[65, 79]
Vegetable oils	PN	+		×	LiChrosorb RP-18	UV 205	Acetonitrile:diethyl ether = 75:25	[55]
Cod liver oil, test mixture	PN		×		Zorbax ODS	IR (CO)	Acetonitrile:THF = 60:40 → 45:55 and 100% CH_3CN → 100% of (CH_2Cl_2:THF = 47.75:52.25)	[15]
Saturated C_{12}, C_{18}, C_{36}, C_{48}	PN			×	Zorbax ODS	IR (CO)	Acetonitrile:THF = 90:10, 80:20 and 60:40	[70]
Human plasma lipoprotein TG	PN	+		×	Supelcosil-18	RI	Acetonitrile:acetone = 63.6:36.4	[77]
Olive oil, cocoa butter	PN	+		×	Spherisorb ODS-2	RI	Propionitrile	[78, 87]
Peanut and cottonseed oil	PN			×	μ Bondapak C_{18}	RI	Acetonitrile:acetone = 42:58	[76]
Coconut oil	PN			×	Partisil-ODS-2 Zorbax ODS	RI	Mixtures of 2-PrOH:acetone:MeOH:acetonitrile	[80]
Vegetable oils, test substances	PN			×	Nucleosil-5 μ C_{18}	RI	A: acetonitrile:acetone = 1:1 B: MeOH:acetone = 3:2	[71, 82]
Palm oil, soybean oil	PN	+		×	Nucleosil-5 μ C_{18}	UV 215	Mixtures of acetonitrile:2-PrOH:hexane	[81]
Model substances	PN			×	LiChrosorb RP-18	RI	Acetonitrile:THF:hexane = 224:123.2:39.6	[83]
Model substances, seed oils	PN			×	μ Bondapak C_{18}	RI	Acetonitrile:acetone = 2:1, acetonitrile	[72, 84]
Soybean oil, test substances	PN			×	7 different columns	RI, IR	Acetonitrile:acetone mixtures, Ag^+, acetonitrile:THF:CH_2Cl_2 = 60:20:20	[85]
Rapeseed and palm oil, cocoa butter	PN	+		×	Nucleosil 5 C_{18}	RI	Acetonitrile:acetone = 50:50 and 60:40	[90]
Tallow, animal fats, soybean oil	PN			×	μ Bondapak C_{18}	RI	Methanol:chloroform = 9:1	[74, 75]

With triglycerides on C-18 RP columns, type a) [n = 2.0] is sometimes more closely fulfilled in methanol-based RP mobile phases[71]. Type b) [n ≠ 2.0] is frequently achieved in nitrile-based mobile phases such as acetonitrile-acetone[65, 71, 77, 79], acetonitrile-isopropanol[51], acetonitrile-ether[55], or propionitrile[78, 87]. (The opposite has been observed with fatty acid UV derivatives[24].)

In the HPLC of triglycerides, infrared detectors based on the ester carbonyl band could become important. Parris[15, 70] has shown that they can even be used with gradient elution; however, careful control of solvent composition and purity is essential[91]. The really attractive thing with this approach is that the ester carbonyl infrared signal – like the carbon-based signal from the former Pye LCM-2 "moving-wire" detector – should be nearly independent of the number of double bonds present. Therefore, PN peaks (containing a "PN class", i.e., a mixture of triglycerides of 0, 1, 2, and more double bonds) could be quantitated with much better accuracy than is possible with short-wavelength UV (SWUV) or even RI detectors (cf. Sect. 12 for a discussion of SWUV detection).

Table 5 gives conditions for a number of reversed-phase triglyceride separations and Figs. 9 and 10 show examples of RP-HPLC short-wavelength UV and of flow-gradient RP-HPLC-RI chromatograms of a number of edible fats.

6.3 Full Triglyceride Analyses of Oils

For the full triglyceride analysis of fats and oils, chromatographic separation techniques of the *intact* triglycerides (Fig. 7), alone or in their combinations, are not sufficient. In each of the various triglyceride separation techniques, different groups of triglycerides will overlap with each other and be coeluted as one peak. To analyze for individual triglycerides, a number of additional steps must be carried out in sequence, and, therefore, *quantitative preparative* separations are necessary. Enzymatic or chemical hydrolysis or derivatization techniques must then be applied to fractions (e.g., CN groups, PN groups, or DB groups) obtained by preparative chromatography.

The products of such hydrolysis or derivatization reactions (usually diglycerides, monoglycerides, and fatty acids; cf. Sects. 4 and 7) are then further analyzed by chromatographic techniques. GLC of the fatty acid methyl esters obtained by transesterification is usually carried out in parallel, both on the original fat and all the CN-, PN-, or DB-group fractions obtained.

However, even a combination of preparative CN, PN, and DB separations of triglycerides, combined with fatty acid GLC of all the fractions obtained, is not yet sufficient for a full analysis of all the possible triglycerides, incl. isomers.

Analysis schemes, or strategies, have therefore been developed by Hammond[92], Litchfield[63], and others, to show by which combinations of techniques one can arrive at a full analysis. To illustrate this, a very simple scheme – the complete analysis of a hypothetical fat consisting of only 4 fatty acids[63] – is reproduced here in Fig. 11. Litchfield, in his book[63], calculates that for the full analysis of coconut oil, 35 preparative DB separations, 11 preparative PN separations, 49 deacylation steps, and 111 stereospecific analyses would be required.

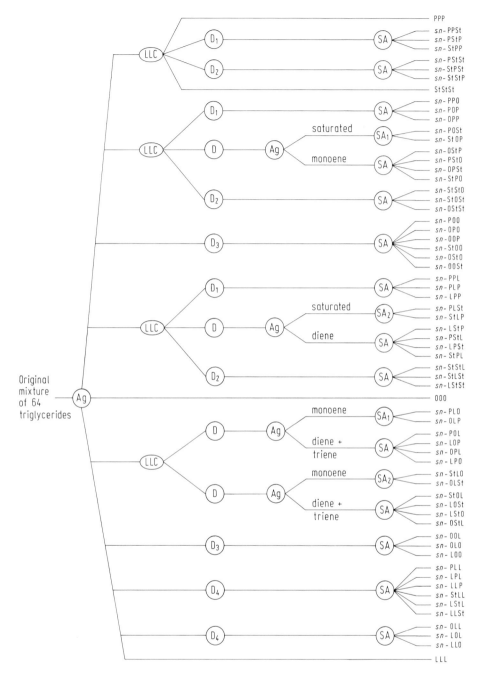

Fig. 11. *Triglyceride analysis scheme for a model triglyceride mixture formed from four fatty acids.* Theoretical scheme for the complete analysis of all triglycerides formed from palmitic (P), stearic (St), oleic (O), and linoleic (L) acids. From 4 fatty acids, 64 triglycerides (incl. stereoisomers) are formed. Ag = Separation by silver ion adsorption chromatography; LLC = Separation by liquid-liquid partition chromatography; D = Deacylation of triglycerides to representative *sn*-1,3-diglycerides; SA = Stereospecific analysis of *sn*-1,3-diglycerides. From Litchfield's book[63] by permission of author and publisher

Fortunately, in the vast majority of cases such complete analyses are not called for. With native oils (from one type of biosynthetic tissue), the fatty acid distribution theories[63] can be used to calculate triglyceride composition. The positional distribution of fatty acids over the *sn*-1, *sn*-2, and *sn*-3 positions on the glycerol is of little technical interest. In special cases, e.g., for chocolates and other fats where the melting properties in the mouth are essential, the ratio of the *sn*-2 monounsaturated vs. the *sn*-1 plus *sn*-3 monounsaturated triglycerides is important, but this can be determined either by argentation HPLC[66, 67, 68] or by lipase splitting plus monoglyceride analysis techniques.

Diglyceride and monoglyceride separations, as may become necessary during a full triglyceride analysis scheme, could also be carried out by HPLC (cf. Sect. 7). Theoretically it is even possible to devise an "all-HPLC strategy" – where all chromatographic steps (preparative and analytical; lipid class, DB and PN separations; with triglycerides, partial glycerides, and fatty acids) are carried out by HPLC.

7 Partial Glycerides and Food Emulsifiers

Mixtures of mono-, di-, and triglycerides and their derivatives with other acids (e.g., with acetic, tartaric, lactic, citric acids, etc.) are often used as food emulsifiers. They can be separated according to molecular size by GPC[5, 14] (cf. Sect. 3 and Fig. 4), according to number of polar groups by adsorption HPLC on silica[4, 6, 8, 93 – 99], and according to chain length and degree of unsaturation by RP-HPLC[49].

The group separation of partial glycerides according to number of free OH groups (cf. Figs. 2 and 3) could play a role as a micropreparative step during triglyceride analysis schemes (Sect. 6.3.). It could someday replace the usually more tedious preparative TLC step, whereas a reversed-phase partition separation of monoglycerides could perhaps replace the usual transesterification – plus GLC of fatty acid methyl esters step, for the analysis of the *sn*-2 monoglycerides formed during lipase hydrolysis.

Partial glycerides and similar emulsifiers can be derivatized at the free OH groups with a UV-absorbing reagent[31, 98], to circumvent the quantitative detection problem for lipid classes. In that case, UV detectors "seeing" only that part of the molecule orginating from the reagent can be used, but the separation as such must be changed to cope with the derivatives rather than with the original partial glycerides. In this way, unfortunately, the advantage of a clear-cut *group separation* is often lost as this is based on the predominant interaction of the original free OH groups with the silica adsorbent.

The reversed-phase separation of fatty acid esters of polyhydric alcohols was investigated by Henke and Schubert[49].

The use of IR detectors with partial glycerides and emulsifiers[5, 91] has already been discussed (cf. GPC in Sect. 3). In a few cases, when there are no double bonds present – e.g., for production control of emulsifiers made from totally hydrogenated fats – even short-wave UV detectors can be used for quantitation[98, 99].

8 Polar Lipids, incl. Glyco- and Phospholipids

As with the analysis of the mono- and diglyceride content of fats and emulsifiers, this is another area where the analytical chemist is most often confronted with "lipid class questions".

Typical examples are[1]:

a) L/S (lecithin:sphingomyelin) ratio in amniotic fluid[100];
b) PC content in high grade (pharmaceutical) "lecithin";
c) relative PE, PI, PS, PC ratios in food grade phosphatide fractions[101] or in biological tissue lipids or membranes[102–105];
d) ratio of glycosphingolipids with 1, 2, 3, and 4 hexose units in blood of patients with lipid-related diseases[36,106];
e) Content of egg yolk in food and cosmetic samples based on PC content;
f) Contents of MGDG and DGDG in wheat flour[107];
g) Lecithin content in chocolate[108], etc.

These and other questions of biochemical, medical diagnostic, food quality, and pharmaceutical interest always imply a lipid *class* separation and quantification irrespective of the fatty acid type present.

Table 6 gives experimental details for a number of HPLC systems that have been used with phospholipids. Fig. 12 shows a typical lipid class separation, whereas Fig. 13 shows an individual substance separation of a PC class sample. Recently, Phillips et al. again demonstrated the usefulness of a flame ionization detector for total lipid and phospholipid class separations[53].

With phospholipids in particular, the pH of the eluent, acids or bases impregnated on silica columns, and the presence or absence of ion pairing agents may play an important role in determining the elution sequence of lipid classes. Frequent changes of the position of PE and PC, and also of other phospholipids, have been observed[89,109].

The quantitation problem of phospholipids and lysophospholipids by "short-wavelength" UV is discussed in Sect. 12. UV and other derivatizations have frequently been used[31,34–36,124].

Glycolipids and sphingolipids are often important in the medical diagnosis area, in nervous tissue research, and in blood group active substances. Glycolipids also play a role in the baking properties of flour. A few HPLC application references for glyco- and sphingolipids, incl. ceramides, cerebrosides, gangliosides, etc. have been collected in Table 7.

Glycolipids can sometimes be eluted as a group, with changed elution sequence with regard to other lipid classes, by employing proton-donor vs. proton-acceptor solvent selectivities[131]. With glycolipids, ceramides, and other polar lipids, UV derivatization of free OH groups has also frequently been used[31,36,132].

Reviews of the HPLC separation and quantitation of phospholipids have been published[36,133].

---

1 Abbreviations used: PC = phosphatidyl choline, PA = phosphatidic acid, PI = phosphatidyl inositol, PS = phosphatidyl serine, PE = phosphatidyl ethanolamine, MGDG = monogalactosyl diglyceride, DGDG = digalactosyl diglyceride

Table 6. Selected examples of HPLC conditions used for various phospholipid applications

Sample type	Class or group separation	Individual species separation	Solvent gradient Yes	Solvent gradient No	UV derivatization	Column used	Detector used (λ)	Mobile phases used	Ref.
Immature soybeans	×		×			Silica	FID	A: pentane, B: diethyl ether, C: chloroform, D: methanolic ammonia	1)
Rat erythrocytes	×		×			Silica	FID	As above	2)
Rat kidney + liver tissue	×		×			Silica/NH₃	FID	Hydrocarbons/CH_2Cl_2/CHCl₃/methanolic ammonia	3)
Amniotic fluid	×		×			Diol	UV 203	A: MeCN, B: MeCN:H_2O=80:20	100)
Tissue lipids	×		×			Silica	UV 205	A: hexane:iPrOH = 50:50 B: hexane:iPrOH:THF:0.1 M NH₄Cl = 40:45:5:10	102)
Tissue, serum, erythrocytes	×			×		Silica	UV 203	MeCN:MeOH:H₃PO₄ = 130:5:1.5	103)
Liver cells	×		×			Silica	UV 206	Hexane:PrOH:H_2O = 6:8:0.75 → 6:8:1.5	104)
Soybean lecithin	×			×		Silica	UV 206	Hexane:iPrOH:0.2 M HOAc = 8:8:1	101)
Egg yolk	×		×			NH₂	UV 206	A: hexane:iPrOH = 5.5:8, B: hexane:iPrOH:MeOH:H_2O = 5.5:8:1:1.5	109) 109)
Chocolate, soybean lecithin	×			×		Silica	UV 210	MeCN:MeOH:H_2O = 65:21:14	108. 110)
Synthetic	×		×			NH₂	FID	A: CHCl₃, B: MeOH:H_2O = 25:4	111)
Cell membranes	×			×		SCX	UV 203	MeCN:MeOH:H_2O = 400:100:34	105)
Human erythrocyte membranes	×	×	×	×		Silica	UV 206	Hexane:iPrOH:H_2O = 6:8:1.3 (H_2O 0.75 → 1.4)	112)
Rabbit polymorphonuclear leukocytes, yeast cells	×		×			Silica	Radio-activity	CHCl₃:PrOH:HOAc:H_2O, A = 50:55:2.5:5, B = 50:55:2.5:8.75, C = 50:55:5:10	113)

Table 6 (continued)

Sample type	Class or group separation	Individual species separation	Solvent gradient Yes	Solvent gradient No	UV derivatization	Column used	Detector used (λ)	Mobile phases used	Ref.
Rat brain, liver	×			×	×	Silica	UV 280	CH_2Cl_2:MeOH:15 M NH_3 = 92:8:1 and other ratios	[35]
Egg lecithin, synthetic PC		×		×		C_{18} RP fatty acid	RI RI	MeOH:H_2O:$CHCl_3$=100:10:10 MeOH:H_2O:$CHCl_3$=70:19:10	[114]
Erythrocytes, nerve tissue	×	×	×			Silica	UV 206	Hexane:iPrOH:H_2O=6:8:0.75 → 6:1:1.4	[115]
Soybean lecithin (+ oxidation products)		×	×			C_{18} RP	UV 206 + 234	91–95% MeOH in H_2O	[116]
Soybean lecithin	×		×			Silica/NH_3	FID	$CHCl_3$ → 8% NH_4OH in MeOH	[117]
Synthetic	×			×		Silica	FID	$CHCl_3$:MeOH:NH_3 = 50:35.9:7	[118]
Rat brain	×		×		×	Silica	Fluorescence	CH_2Cl_2:MeOH:15 M NH_4OH, 91:9:1 → 70:20:5	[34]
Rat kidney, liver, serum	×		×			Silica/NH_3	FID	Hydrocarbons → MeOH/NH_3	[53]
Soy lecithins	×			×		Silica	RI	$CHCl_3$:MeOH:HOAc:H_2O=14:14:1:1	[119]
Soy lecithins	×			×		Silica	UV 206	Hexane:iPrOH:H_2O=1:4:1	[120]
Rat liver	×			×		Silica	UV 205	Hexane:iPrOH:25 mM phosphate buffer:EtOH:HOAc=367:490:62:100:0.6	[121]
Wheat roots	×		×			Silica	UV 206	Hexane:iPrOH:water = 6:8:0.5 → 6:8:1.5	[122]
Egg, brain, liver PC		×		×		C_{18} RP	UV 205	MeOH:1 mM phosphate buffer (pH 7.4)=9.5:0.5	[123]

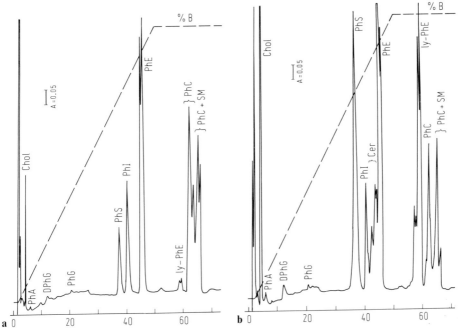

Fig. 12a and b. *HPLC of phospholipid classes from rat liver and brain.* **a** HPLC of lipid extract from rat liver. **b** HPLC of lipid extract from rat brain. Conditions: Column, Micro-Pak Si-5; solvents, hexane-isopropanol-water-sulfuric acid, A (97:3:0:0.02), B (75:24:0.9:0.1); flow-rate, 1.5 ml/min, 4% B for 2 min then increasing linearly to 100% B at 50 min. Abbreviations: Chol = Cholesterol, PhA = phosphatidic acid, DPhG = diphosphatidyl glycerol, PhG = phosphatidyl glycerol, PhS = phosphatidyl serine, PhI = phosphatidyl inositol, Cer = cerebrosides, PhE = phosphatidyl ethanolamine, LyPhE = lysophosphatidyl ethanolamine, PhC = phosphatidyl choline, SM = sphingomyelin. From Ref.[102] by permission of authors and journal

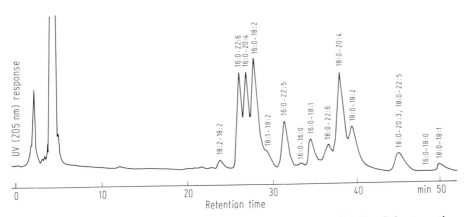

Fig. 13. *HPLC analysis of egg phosphatidyl choline.* Column: Nucleosil-5-C_{18}; Solvent: methanol − 1 mM phosphate buffer, pH 7.4, 9.5:0.5 (v/v), flow rate 1 ml/min. PC, 500 µg dissolved in dichloromethane-methanol 1:1, 10 µl, was injected. The PC species eluted were collected and analyzed. The major fatty acid composition of the PC in an individual peak is given near the peak. From Ref.[123] by permission of authors and journal

Table 7. Examples for applications of HPLC to glycolipids and sphingolipids (including ceramides, cerebrosides, sulfatides, gangliosides, and other polar lipids of the brain)

Sample type (origin of lipids)	Lipid to be analyzed	Lipid derivative separated by HPLC	Detector type	Column type	Gradient	Mobile phase	Ref.
Rat neural tissues and tumors	Neutral glycosphingolipids, ceramides	Benzoyl derivs.	UV 230	Zipax	+	Dioxane in hexane	[125]
Bovine sphingomyelins and ceramides	Sphingomyelins, ceramides	Benzoyl derivs. + native lipids	UV 230 UV 205	Silica AgNO$_3$, Nucleosil 5 C$_{18}$		MeOH:iPrOH 8:2, 9:1, 9.5:0.5, MeOH	[126]
Brain, erythrocytes	Gangliosides, neutral glycosphingolipids	Native lipids	LCM-2	Silica		CHCl$_3$:MeOH: 0.01 M HCl 60:35:5	[127]
Human plasma	Neutral glycosphingolipids	Benzoyl derivs.	UV 280	Zipax	+	2–17% AcOEt^a in hexane	[106]
Wheats, flours	Galactosyl-diglycerides (DGDG, MGDG)	Native lipids	RI	Spherisorb ODS		MeOH:H$_2$O = 60:40	[107]
Bovine + rat brain, plasma	Cerebrosides	Benzoyl derivs.	UV 280 UV 230	Zipax	+	Aqueous AcOEt in hexane, dioxane in hexane	[128]
Synthetic, egg yolk, bovine brain	Ceramides	Benzoyl and p-nitrobenzoyl derivs.	UV 240 UV 260	Lichrosorb RP-18		2–10% CHCl$_3$ in MeOH	[33]
Beef brain, human erythrocytes	Gangliosides	Native lipids	None (TLC)	Zorbax Sil	+	iPrOH:hexane: H$_2$O = 55:42:3 → 55:25:20	[129]
Blood groups active substances (from erythrocyte membranes)	Glycosphingolipids	Native lipids	None (TLC)	Iatrobeads 6RS-8010 (10 μ silica)	+	iPrOH:hexane: H$_2$O = 55:44:1 → 55:35:10	[130]

^{a} 1 part water saturated AcOEt + 5 parts dry AcOEt

9 Oxidation Products of Lipids and Prostaglandins

Lipid oxidation often leads to the formation of conjugated dienes and trienes that can be detected conveniently using UV detectors. A number of working groups[134, 135] exploited this in their work on autoxidation and lipoxigenase reactions with fatty acids. The same applies to oxidized fatty acids in the original triglyceride or phospholipid molecule[116]. The oxidized molecules contain hydroperoxy groups (or hydroxy groups derived therefrom) and show an HPLC mobility that is different from that of the parent compound. Oxidized triglycerides or phospholipids will be eluted ahead of the parent molecules in RP-HPLC, but *after* the parent molecules in adsorption HPLC.

Other oxidation products such as short-chain aldehydes and ketones can be analyzed conveniently by reversed-phase HPLC of their 2,4-dinitrophenylhydrazones.

In early work on GPC of thermally oxidized, used frying fats, an infrared OH detector was used in this laboratory in conjunction with an RI detector to obtain information about the OH-group content of di-, tri-, and polymeric triglycerides.

HPLC is frequently used for separations of prostaglandins and other arachidonic acid metabolites[17]. The subject has recently been reviewed[136]. Because prostaglandins, thromboxanes, leukotrienes, and their hydroperoxy and hydroxy intermediates are all metabolites of polyunsaturated C_{20} acids, their separation on RP columns is usually governed by the number and nature of polar functional groups present[5, 11, 17, 136]. A number of these substances contain conjugated double bonds and can be detected directly in the UV[134, 137], whereas other members of this group may have to be derivatized (Table 2).

10 Other Lipids and Minor Components of Fats and Oils

The HPLC separations of other lipids, like insect pheromones (often esters of unsaturated fatty alcohols), waxes, fatty alcohols, raw materials for detergents, sterol esters, ether lipids, etc., are basically governed by the same rules as the separations of fatty acid derivatives, triglycerides, and prostaglandins.

Fatty acid esters, like waxes and sterol esters (cf. Sect. 4), and other lipids, carotenes, chlorophylls, tocopherols and tocotrienols, and other components of the unsaponifiable are natural minor constituents of many fats. The analysis of minor components in fats by HPLC is comparatively easy when these substances can be detected by their UV absorption or fluorescence, and consequently a large number of publications exist on the HPLC of tocopherols and other fat-soluble vitamins. Most of these compounds, as well as contaminants that may include mineral oil, pesticides, and polycyclic aromatic hydrocarbons, etc., will be dealt with in more detail in other chapters of this book. Recently, HPLC was used to detect contamination of olive oil by fatty acid anilides[138 - 140].

Intentionally added minor components may include vitamins, preservatives and antioxidants, and other substances that may end up in a fat-phase extract (after having been added to a food product, for example).

One example shown here is the analysis of vitamin A esters in margarines (Fig. 14)[141].

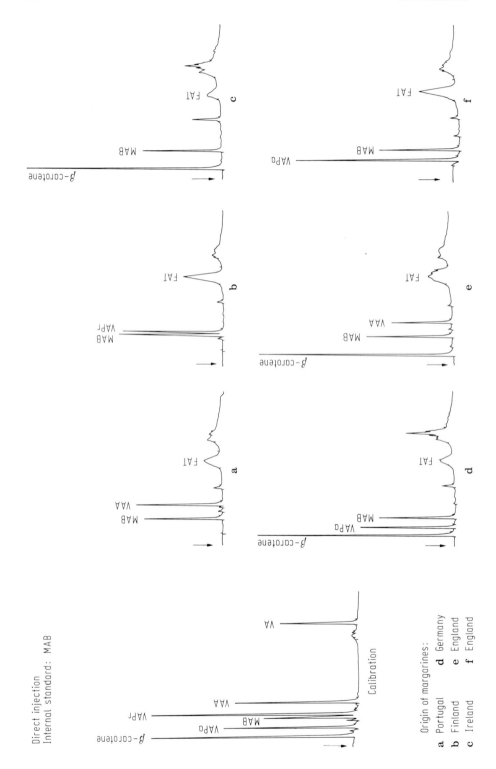

11 Special Selectivity Effects and Changes of Elution Sequence

Special selectivity effects – the results of many experimental observations that may lead to "rules of thumb" and eventually to theories of column-solvent system selectivity – can be exploited with advantage in the lipid field. A number of empirical observations were discussed by the author[141]. Changes of the elution sequence occur frequently when chemically very different molecules are involved, e.g., open-chain vs. ring compounds, etc.

By careful choice of solvent composition and column deactivation (amount of water on a silica column), it is possible to intentionally change the relative elution sequence of "triglycerides" vs. a minor or trace component.

A typical example was the shift of the "fat" peak (i.e., triglycerides) from before α-tocopherol to after γ-tocopherol in tocopherol analysis schemes[141]. Other compounds, particularly ring compounds such as phenolic antioxidants from packaging materials, etc., can be "moved" to appear before the triglyceride peak, or after the triglyceride peak. This may be important when designing an analysis scheme, because by careful choice of these "secondary solvent effects" on silica, the number of preconcentration steps required can be reduced so that savings in total analysis time can be achieved.

Of some importance is the separation of 4-demethyl-, 4-methyl-, and 4,4-dimethyl sterols vs. long-chain diglycerides[5, 8, 95].

Other changes of elution sequence have been observed with vitamin A alcohol vs. vitamin D_2, vitamin A acetate vs. α-tocopherol acetate, γ-tocopherol vs. β-tocotrienol, and with many polymer additives and antioxidants vs. long-chain triglycerides, all on silica.

In reversed-phase systems, solvent selectivity effects are important for the designing of preparative PN separations of triglycerides (cf. Sect. 6.2). Halgunset et al.[24] showed changes of elution sequence with fatty acid UV derivatives (Fig. 15), also observed by Plattner[85] for methyl esters.

Sequence changes of "minor components" vs. triglycerides can most often be achieved by changing the solvents from aliphatic to aromatic or chlorinated hydrocarbons, from proton-donor to proton-acceptor solvents, or, with silica columns, by changing from dry to water-deactivated silica columns.

The use of silver ions in stationary and mobile phases to change elution sequences of unsaturated vs. saturated compounds, and the use of acids or bases on silica to change the elution sequences of phospholipid classes have already been mentioned.

Fig. 14a–f. *Analysis of vitamin A related compounds in margarines.* Using gradient elution from a silica column and wavelength switching, β-carotene, vitamin A esters (acetate, propionate, and palmitate) and free vitamin A alcohol can be analyzed in one run. Six different 'vitamin A patterns' were found in margarines from 5 different countries (**a**: acetate only; **b**: propionate only; **c**: β-carotene only; **d**: β-carotene plus palmitate; **e**: β-carotene plus acetate; **f**: palmitate only). The margarine fat phase plus internal standard MAB was directly dissolved and injected without further sample preparation or clean-up. Conditions: Hewlett Packard 1048 B; 150×4.6 mm LiChrosorb SI 60 5 µ; 1.5 ml/min, UV at 450 and 325 nm (automatic wavelength change $450 \rightarrow 325$ nm between the β-carotene and VAPa peaks). Solvents: A = heptane, B = heptane:diisopropyl ether: $CH_2Cl_2 = 55.25:29.75:15$, all partly saturated with water. 15% B to 40% B in 8 min, then to 90% B in 4 min. Abbreviations: VA = Vitamin A-alcohol; VAPa = Vitamin A-palmitate; VAPr = Vitamin A-propionate; VAA = Vitamin A-acetate; MAB = 4-Methoxyazobenzene

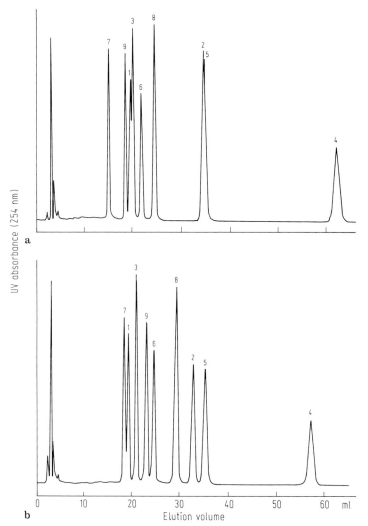

Fig. 15a and b. *Changes of elution sequence in the RP HPLC of fatty acid p-bromophenacyl esters.* In acetonitrile-water mixtures, the elution sequence is quite different from the one in methanol-water systems. Peaks: $1 = 14:0$, $2 = 16:0$, $3 = 16:1$, $4 = 18:0$, $5 = 18:1$, $6 = 18:2$, $7 = 18:3$, $8 = 20:3$, $9 = 20:4$. HPLC conditions: 250×4.6 mm Supelcosil LC-18, 5 μ; 1 ml/min, **a** acetonitrile:water $= 9:1$, **b** methanol:water $= 9:1$. Detection: UV at 254 nm. From Ref.[24] by permission of authors and journal

12 General Discussion of Detection Problems

The current detector situation for lipids is generally unsatisfactory. There is a definite need for a detector that can be used with lipids in gradient elution (e.g., partial glycerides from silica, triglycerides of a wide PN range from RP systems, blood lipid classes from silica, double bond classes from AgNO$_3$ columns, or phospholipid classes from silica) and that does not give large response differences between satu-

Table 8. Relative Short-wave UV response of saturated and unsaturated phospholipids[a]

Phosphatidyl-choline	Relative UV response (area/μmol)
di-18:0	1.00
di-18:1	21.1
di-18:2	64.0
di-18:3	91.2
di-20:4	164.8

[a] In a lipid-class separation[142] with short-wave UV detection at 203 nm, the peak areas of the phosphatidylcholine peak from a number of saturated and unsaturated phosphatidylcholines were obtained. The peak area per μmol of phospholipid is normalized to distearoylphosphatidylcholine = 1.00.
Recalculated from data obtained by F.B. Jungalwala et al. (cf. Table 1 in Ref.[142]) and reproduced by permission of Dr. Jungalwala

rated and unsaturated lipid molecules. Short-wavelength UV detection (SW-UV) cannot normally be used for quantitative work with real-life samples, except in special circumstances as, e.g., with totally hydrogenated samples[98, 99]. An interesting development in this context, however, is the recent observation of Herslöf[81] that in an RP system with acetonitrile:2-propanol:hexane mixtures the signal sizes of trilaurin and triolein are comparable in an area around 220 nm.

Quantitation by SW-UV has been most often discussed with lecithins and other phospholipids (Table 8).

Many authors[102, 105, 142] have pointed out that quantitative results cannot be obtained by integration of phospholipid class peaks obtained with short-wavelength UV detectors, because the absorbance at 200–210 nm is caused by isolated double bonds and may vary with fatty acid composition. Other authors, however, nevertheless use SW-UV peak areas to draw quantitative conclusions for particular types of samples[100, 112].

The problem is aggravated when *lyso*-phospholipids are to be determined in the same way by SW-UV, because the UV-absorbing unsaturated fatty acids occur predominantly in the *sn*-2 position of the glycerol, and may no longer be present in the corresponding much more saturated *lyso* compound[102].

The situation is similar with RI detectors, where the size of the response difference between the saturated and unsaturated lipid depends on the refractive index of the solvent used. One reason for the popularity of acetone:acetonitrile and similar mixtures in RP-HPLC of triglycerides, apart from their properties regarding triglyceride solubility, retention, and resolution, is the fact that with such mixtures the refractive index difference between solvent and solute is near a maximum.

Triglycerides composed of 2 or 3 different fatty acids are usually not available as pure reference compounds for RI or SW-UV calibration. However, because of the additivity of refractive indices from increments, it should be possible to automatically calculate – by interpolation – response factors for these "mixed triglycerides" from the response factors for the monoacid triglycerides available in a calibration solution. Most modern instruments, however, do not yet provide the necessary microprocessor or integrator software for such calibrations.

A special problem exists with fluorescence detectors (and, to a lesser extent, also with UV detectors) in the direct quantitative analysis of tocopherols and tocotrienols in vegetable oils, when only one substance – usually α-tocopherol – is available for calibration.

Firstly, there may be fluorescence quenching and enhancement effects depending not only on the type of solvents used[43], but also on whether or not invisible "triglycerides" are coeluted with the α-tocopherol peak on silica and this again will depend on mobile phase composition and column deactivation[141]. Secondly, because the excitation and emission maxima of the various tocopherols and tocotrienols do not exactly coincide, grossly erroneous quantitative results may occur unless monochromator slit widths, etc. are carefully taken into consideration. The relative response factors between α-tocopherol and other tocopherols and tocotrienols may differ by a factor of 2 or more, depending on the type of equipment used – a fact that makes it difficult to compare results between laboratories.

Fairly good quantitative results can be achieved with light-scattering detectors for lipids[143], although some doubts remain with regard to their applicability to saturated/unsaturated lipid mixtures and to solvent gradients.

Post-column reaction detection, e.g., of glycerol after enzymatic hydrolysis[144] and by a "phosphorus analyzer" for phospholipids[145, 146] may become more important in the near future. With triglyceride separations, a "glycerol detector" could directly give mole-% of triglyceride PN or DB classes, regardless of their fatty acid composition.

References

1. Privett, O.S., Dougherty, K.A., Erdahl, W.L. and Stolyhwo, A.: J. Amer. Oil Chem. Soc. 50, 516 (1973)
2. Stolyhwo, A. and Privett, O.S.: J. Chromatogr. Sci. 11, 20 (1973)
3. Phillips, F.C. and Privett, O.S.: J. Amer. Oil Chem. Soc. 58, 590 (1981)
4. Hammond, E.W.: J. Chromatogr. 203, 397 (1981)
5. Aitzetmüller, K.: Progr. Lipid Res. 21, 171 (1982)
6. Aitzetmüller, K. and Koch, J.: J. Chromatogr. 145, 195 (1978)
7. Hoogenboom, J.J.L.: Unilever Research Laboratory Vlaardingen, private communication (1979)
8. Hoogenboom, J.J.L.: Poster presentation, Symposium HPLC in Food Chemistry, Vienna, Jan. 15/16, 1980
9. Nicolaides, N.: J. Chromatogr. Sci. 8, 717 (1970)
10. Sen Gupta, A.K. and Unbehend, M.: Private communication (1975)
11. Aitzetmüller, K.: J. Chromatogr. 113, 231 (1975)
12. Aitzetmüller, K.: In: Proc. 3rd Int. Symp. Metal-Catalyzed Lipid Oxidation (IIIe S.I.O.L.), Sept. 27–30, Institut des Corps Gras, Paris, pp. 127–139 (1973)
13. Unbehend, M., Scharmann, H., Strauß, H.-J. and Billek, G.: Fette-Seifen-Anstrichmittel 75, 689 (1973)

14. Aitzetmüller, K., Unbehend, M. and Kunze, J., in preparation
15. Parris, N.A.: J. Chrom. Sci. 17, 541 (1979)
16. Bussel, N.E. and Miller, R.A.: J. Liquid Chromatogr. 2, 697 (1979)
17. Fitzpatrick, F.A.: Anal. Chem. 48, 499 (1976)
18. Hoffmann, N.E. and Liao, J.C.: Anal. Chem. 48, 1104 (1976)
19. Borch, R.F.: Anal. Chem. 47, 2437 (1975)
20. Durst, H.D., Milano, M., Kikta, E.J. Jr., Connelly, S.A. and Grushka, E.: Anal. Chem. 47, 1797 (1975)
21. Engelhardt, H. and Elgass, H.: J. Chromatogr. 158, 249 (1978)
22. Takayama, K., Qureshi, N., Jordi, H.C. and Schnoes, H.K.: J. Liquid Chromatogr. 2, 861 (1979)
23. Takayama, K., Qureshi, N., Jordi, H.C. and Schnoes, H.K.: In: Biol. Biomed. Appl. LC (G.L. Hawk, Ed.), Marcel Dekker New York, pp. 91–101 (1979)
24. Halgunset, J., Lund, E.W. and Sunde, A.: J. Chromatogr. 237, 496 (1982)
25. Jordi, H.C., Neue, U.D., Quinn, H.M. and Rausch, C.W.: In: Hawk, G.L. (Ed.) Biol./Biomed. Appl. of LC III, M. Dekker, New York, pp. 327–346 (1981)
26. Bussell, N.E., Miller, R.A., Setterstrom, J.A. and Gross, A.: In: Biol. Biomed. Appl. LC (G.L. Hawk, Ed.), Marcel Dekker, New York, pp. 57–89 (1979)
27. Miller, R.A., Bussell, N.E. and Ricketts, C.: J. Liquid Chromatogr. 1, 291 (1978)
28. Bussell, N.E., Gross, A. and Miller, R.A.: J. Liquid Chromatogr. 2, 1337 (1979)
29. Hatsumi, M., Komata, S.-I. and Hirosawa, K.: J. Chromatogr. 253, 271 (1982)
30. Matthees, D.P. and Purdy, W.C.: Anal. Chim. Acta 109, 61 (1979)
31. Batley, M., Packer, N.H. and Redmond, J.W.: J. Chromatogr. 198, 520 (1980)
32. Yamakawa, T., Handa, S., Yamazaki, T. and Suzuki, A.: 27th Int. Congr. Pure Appl. Chem. (Proc.) 1980, 351
33. Do, U.H., Pei, P.T. and Minard, R.D.: Lipids 16, 855 (1981)
34. Shi-Hua Chen, S., Kou, A.Y. and Chen, H.-H.Y.: J. Chromatogr. 208, 33 (1981)
35. Jungalwala, F.B., Turel, R.J., Evans, J.E. and McCluer, R.H.: Biochem. J. 145, 517 (1975)
36. McCluer, R.H. and Jungalwala, F.B.: In: Biol./Biomed. Appl. Liq. Chromatogr. (G.L. Hawk, Ed.), Marcel Dekker, New York, pp. 7–30 (1979)
37. Karleskind, A.: Rev. Fr. Corps Gras 24, 419 (1977)
38. Knapp, D.R. and Krueger, S.: Anal. Lett. 8, 603 (1975)
39. Tsuchiya, H., Hayashi, T., Naruse, H. and Takagi, N.: J. Chromatogr. 234, 121 (1982)
40. Tsuchiya, H., Hayashi, T., Naruse, H. and Takagi, N.: J. Chromatogr. 231, 247 (1982)
41. Lindner, W. and Santi, W.: J. Chromatogr. 176, 55 (1979)
42. Zelenski, S.G. and Huber, J.W. III: Chromatographia 11, 645 (1978)
43. Lloyd, J.B.F.: J. Chromatogr. 178, 249 (1979)
44. Lam, S. and Grushka, E.: J. Chromatogr. 158, 207 (1978)
45. Voelter, W., Huber, R. and Zech, K.: J. Chromatogr. 217, 491 (1981)
46. Wintersteiger, R., Wenninger-Weinzierl, G. and Pacha, W.: J. Chromatogr. 237, 399 (1982)
47. Willett, J.D., Brody, E.P. and Knight, M.M.: J. Amer. Oil Chem. Soc. 59, 273 (1982)
48. Korte, W.D.: J. Chromatogr. 243, 153 (1982)
49. Henke, H. and Schubert, J.: HRC + CC 3, 69 (1980)
50. Bianchini, J.-P., Ralaimanarivo, A. and Gaydou, E.M.: HRC + CC 5, 199 (1982)
51. Smith, S.L., Novotny, M., Moore, S.A. and Felten, D.L.: J. Chromatogr. 221, 19 (1980)
52. Perkins, E.G., Hendren, D.J., Bauer, J.E. and El Hamdy, A.H.: Lipids 16, 609 (1981)
53. Phillips, F.C., Erdahl, W.L. and Privett, O.S.: Lipids 17, 992 (1982)
54. Özcimder, M. and Hammers, W.E.: J. Chromatogr. 187, 307 (1980)
55. Levacq, M., Lesieur-Cotelle & Ass.: Coudekerque, France, private communication (April 1980)
56. Battaglia, R. and Fröhlich, D.: Chromatographia 13, 428 (1980)
57. Chan, H.W.-S. and Levett, G.: Chem. Ind. 578 (1978)
58. de Ruyter, M.G.M. and de Leenheer, A.P.: Anal. Chem. 51, 43 (1979)
59. Plattner, R.D.: J. Amer. Oil Chem. Soc. 58, 638 (1981)
60. Schomburg, G. and Zegarski, K.: J. Chromatogr. 114, 174 (1975)

61. Vonach, B. and Schomburg, G.: J. Chromatogr. 149, 417 (1978)
62. Jensen, R.G., Quinn, J.G., Carpenter, D.L. and Sampugna, J.: J. Dairy Sci. 50, 119 (1967)
63. Litchfield, C.: "Analysis of Triglycerides", Academic Press, New York (1972)
64. El-Hamdy, A.H. and Perkins, E.G.: J. Amer. Oil Chem. Soc. 58, 867 (1981)
65. Goiffon, J.P., Reminiac, C. and Furon, D.: Rev. Fr. Corps Gras 28, 199 (1981)
66. Kroesen, A.C.J.: Unilever Research Laboratory Vlaardingen, personal communication (cf. Fig. 8 in ref. 5)
67. Monseigny, A., Vigneron, P.-Y., Levacq, M. and Zwobada, F.: Rev. Fr. Corps Gras 26, 107 [cf. Fig. 14 on p. 117] (1979)
68. Smith, E.C., Jones, A.D. and Hammond, E.W.: J. Chromatogr. 188, 205 (1980)
69. Vigneron, P.Y., Henon, G., Monseigny, A., Levacq, M., Stoclin, B. and Delvoye, P.: Rev. franc. des Corps Gras 29, 423 (1982)
70. Parris, N.A.: J. Chromatogr. 157, 161 (1978)
71. Herslöf, B., Podlaha, O. and Töregard, B.: J. Amer. Oil Chem. Soc. 56, 864 (1979)
72. Plattner, R.D., Spencer, G.F. and Kleiman, R.: J. Amer. Oil Chem. Soc. 54, 511 (1977)
73. Plattner, R.D. and Payne-Wahl, K.: Lipids 14, 152 (1979)
74. Wada, S., Koizumi, C. and Nonaka, J.: Yukagaku 26, 95 (1977)
75. Wada, S., Koizumi, C., Takiguchi, A. and Nonaka, J.: Yukagaku 27, 579 (1978)
76. Bezard, J.A. and Ouedraogo, M.A.: J. Chromatogr. 196, 279 (1980)
77. Perkins, E.G., Hendren, D.J., Pelick, N. and Bauer, J.E.: Lipids 17, 460 (1982)
78. Schulte, E.: Lebensm. Chem. Gerichtl. Chem. 35, 99 and ibid, 36, 88 (1981/1982)
79. Goiffon, J.P., Reminiac, C. and Olle, M.: Rev. Fr. Corps Gras 28, 167 (1982)
80. El-Hamdy, A.H. and Perkins, E.G.: J. Amer. Oil Chem. Soc. 58, 49 (1981)
81. Herslöf, B.G.: HRC + CC 4, 471 (1981)
82. Herslöf, B., Herslöf, M. and Podlaha, O.: Fette-Seifen-Anstrichmittel 82, 460 (1980)
83. Jensen, G.W.: J. Chromatogr. 204, 407 (1981)
84. Lie Ken Jie, M.S.F.: J. Chromatogr. 192, 457 (1980)
85. Plattner, R.D.: J. Amer. Oil Chem. Soc. 58, 638 (1981)
86. Petersson, B., Podlaha, O. and Töregard, B.: J. Amer. Oil Chem. Soc. 58, 1005 (1981)
87. Schulte, E.: Fette-Seifen-Anstrichmittel 83, 289 (1981)
88. Aitzetmüller, K.: "Recent Progress in the HPLC of Lipids", paper given at ISF/AOCS World Congress, New York, N.Y., April 27 (1980)
89. Aitzetmüller, K. and Handt, D.: to be published
90. Petersen, B., Podlaha, O. and Töregard, B.: J. Amer. Oil Chem. Soc. 58, 1005 (1981)
91. Payne-Wahl, K., Spencer, G.F., Plattner, R.D. and Butterfield, R.O.: J. Chromatogr. 209, 61 (1981)
92. Hammond, E.G.: Lipids 4, 246 (1969)
93. Aitzetmüller, K., Böhrs, M. and Arzberger, E.: Fette-Seifen-Anstrichmittel 81, 436 (1979)
94. Aitzetmüller, K.: "Analysis of Non-ionic Food Emulsifiers by Liquid Chromatography", in: Lipids, Vol. 2 (R. Paoletti, G. Jacini, and G. Porcellati, eds.), Raven Press, New York, pp. 333–342 (1976)
95. Aitzetmüller, K.: J. Chromatogr. 139, 61 (1977)
96. Sudraud, G., Coustard, J.M., Retho, C., Caude, M., Rosset, R. Hagemann, R., Gaudin, D. and Virelizier, H.: J. Chromatogr. 204, 397 (1981)
97. Kiuchi, K., Ohta, T. and Ebine, H.: J. Chromatogr. Sci. 13, 461 (1975)
98. Brüschweiler, H.: Mitt. Geb. Lebensmittelunters. Hyg. 68, 46 (1977)
99. Riisom, T. and Hoffmeyer, L.: J. Amer. Oil Chem. Soc. 55, 649 (1978)
100. Briand, R.L., Harold, S. and Blass, K.G.: J. Chromatogr. 223, 277 (1981)
101. Nasner, A. and Kraus, Lj.: J. Chromatogr. 216, 389 (1981)
102. Yandrasitz, J.R., Berry, G. and Segal, S.: J. Chromatogr. 225, 319 (1981)
103. Shi-Hua Chen, S. and Kou, A.Y.: J. Chromatogr. 227, 25 (1982)
104. James, J.L., Clawson, G.A., Chan, C.H. and Smuckler, E.A.: Lipids 16, 541 (1981)
105. Gross, R.W. and Sobel, B.E.: J. Chromatogr. 197, 79 (1980)
106. Ullman, M.D. and McCluer, R.H.: J. Lipid Res. 18, 371 (1977)
107. Tweeten, T.N., Wetzel, D.L. and Chung, O.K.: J. Amer. Oil Chem. Soc 58, 664 (1981)
108. Hurst, W.J. and Martin, R.A.: J. Amer. Oil Chem. Soc. 57, 307 (1980)

109. Hanson, V.L., Park, J.Y., Osborn, T.W. and Kiral, R.M.: J. Chromatogr. 205, 393 (1981)
110. Press, K., Sheeley, R.M., Hurst, W.J. and Martin, R.A., Jr.: J. Agric. Food Chem. 29, 1096 (1981)
111. Kiuchi, K., Ohta, T. and Ebine, H.: J. Chromatogr. 133, 226 (1977)
112. Geurts van Kessel, W.S.M., Hax, W.M.A., Demel, R.A. and De Gier, J.: Biochim. Biophys. Acta 486, 524 (1977)
113. Blom, C.P., Deierkauf, F.A. and Riemersma, J.C.: J. Chromatogr. 171, 331 (1979)
114. Porter, N.A., Wolf, R.A. and Nixon, J.R.: Lipids 14, 20 (1979)
115. Hax, W.M.A. and Geurts van Kessel, W.S.M.: J. Chromatogr. 142, 735 (1977)
116. Crawford, C.G., Plattner, R.D., Sessa, D.J. and Rackis, J.J.: Lipids 15, 91 (1980)
117. Erdahl, W.L., Stolyhwo, A. and Privett, O.S.: J. Amer. Oil Chem. Soc. 50, 513 (1973)
118. Rainey, M.L. and Purdy, W.C.: Anal. Chim. Acta 93, 211 (1977)
119. Rhee, J.S. and Shin, M.G.: J. Amer. Oil Chem. Soc. 59, 98 (1982)
120. Nasner, A. and Kraus Lj.: Fette-Seifen-Anstrichm. 83, 70 (1981)
121. Patton, G.M., Fasulo, J.M., and Robins, S.J.: J. Lipid Res. 23, 190 (1982)
122. Ashworth, E.N., St. John, J.B., Christiansen, M.N. and Patterson, G.W.: J. Agric. Food Chem. 29, 879 (1981)
123. Smith, M. and Jungalwala, F.B.: J. Lipid Res. 22, 697 (1981)
124. Hsieh, J.Y.-K., Welch, D.K. and Turcotte, J.G.: Lipids 16, 761 (1981)
125. Chou, K.H. and Jungalwala, F.B.: J. Neurochem. 36, 394 (1981)
126. Smith, M., Monchamp, P. and Jungalwala, F.B.: J. Lipid Res. 22, 714 (1981)
127. Tjaden, U.R., Krol, J.H., van Hoeven, R.P., Oomen-Meulemans, E.P.M. and Emmelot, P.: J. Chromatogr. 136, 233 (1977)
128. Jungalwala, F.B., Hayes, L. and McCluer, R.H.: J. Lipid Res. 18, 285 (1977)
129. Kundu, S.K. and Scott, D.D.: J. Chromatogr. 232, 19 (1982)
130. Watanabe, K. and Arao, Y.: J. Lipid Res. 22, 1020 (1981)
131. Rouser, G.: J. Chromatogr. Sci. 11, 60 (1973)
132. Evans, J.E. and McCluer, R.H.: Biochim. Biophys. Acta 270, 565 (1972)
133. Porter, N.A. and Weenen, H.: Methods Enzymol. 72, 34 (1981)
134. Chan, H.W.-S. and Levett, G.: Lipids 12, 837 (1977)
135. Schieberle, P. and Grosch, W.: Z. Lebensm. Unters. Forsch. 173, 199 (1981)
136. Hamilton, J.G. and Karol, R.J.: Progr. Lipid Res. 21, 155 (1982)
137. Rabinovitch, H., Durand, J., Regaud, M., Mendy, F. and Breton, J.-C.: Lipids 16, 518 (1981)
138. Luckas, B. and Lorenzen, W.: Dt. Lebensm. Rdsch. 78, 119 (1982)
139. Wheals, B.B., Whitehouse, M.J. and Curry, C.J.: J. Chromatogr. 238, 203 (1982)
140. Beck, H., Kellert, M., Mathar, W., Tiebach, R. and Weber R.: Lebensm. Chem. Gerichtl. Chem. 36, 1 (1982)
141. Aitzetmüller, K. and Arzberger, E.: Poster presentation, V. International Symposium on Column Liquid Chromatography, Avignon, June 11–15, 1981 (to be published)
142. Jungalwala, F.B., Evans, J.E. and McCluer, R.H.: Biochem. J. 155, 55 (1976)
143. Charlesworth, J.M.: Anal. Chem. 50, 1414 (1978)
144. Hara, J., Shiraishi, K. and Okazaki, M.: J. Chromatogr. 239, 549 (1982)
145. Kaitaranta, J.K. and Bessman, S.P.: Anal. Chem. 53, 1232 (1981)
146. Kaitaranta, J.K., Geiger, P.J. and Bessman, S.P.: J. Chromatogr. 206, 327 (1981)

Manuscript received July 1983

Application of HPLC to the Separation
of Metabolites of Nucleic Acids
in Physiological Fluids

Anne P. Halfpenny and Phyllis R. Brown
Department of Chemistry,
University of Rhode Island Kingston 02881, RI/USA

1 Introduction

1.1 Background

The last decade has witnessed an explosive growth in the interest and analysis of nucleic acids and their metabolites. Purine and pyrimidine nucleotides[1], nucleosides and bases are of importance in a wide spectrum of research areas. The refinement of the instrumentation and techniques of high performance liquid chromatography (HPLC) during this decade has aided all areas of nucleic acid research[1]. As more reliable and rapid means of analyses are provided, many other research areas are benefitting, especially biomedical research[2, 3]. HPLC can be utilized in many types of research; for example, in studies of the regulatory effect of cyclic nucleotides, the determination of the composition of hydrolysates of nucleic acids, the metabolic profiling of normal and diseased subjects, and studies of disease processes. HPLC is also of great importance in the separation and purification of starting materials, intermediates, and products in the synthesis of the nucleic acids, as well as for the nucleic acids themselves. In this article we will discuss the variables involved in developing an HPLC analysis for free purine and pyrimidine compounds in biological fluids.

1.2 Structure, Chemical and Physical Properties of Nucleic Acid Metabolites

The chromatographic behavior of a compound is dictated by its structure and chemical and physical properties. The chemical behavior of nucleotides, nucleosides and bases is a function of the parent ring structure, the sugar and the degree of phosphorylation. A pyrimidine base is a heterocyclic, six-membered ring, and a purine is a pyrimidine fused with an imizadole ring (Fig. 1).

Nucleosides are composed of the parent base attached via the N_1 position of the pyrimidine or N_9 position of the purine to the C_1 position of a ribose. If the sugar is a deoxyribose, the nucleoside is referred to as a deoxyribonucleoside, and the deoxy group is usually on the 2' position of the sugar. Nucleotides are composed of nucleosides with phosphate ester groups on the 2', 3', or 5' position of the sugar (Fig. 2).

An important property of purines and pyrimidines is their ability to undergo tautomerization (Fig. 3). Tautomerization will affect the chromatographic behavior of these compounds. In addition, the acid-base character of these molecules plays an important role in their chromatographic behavior. Nucleosides are both weak

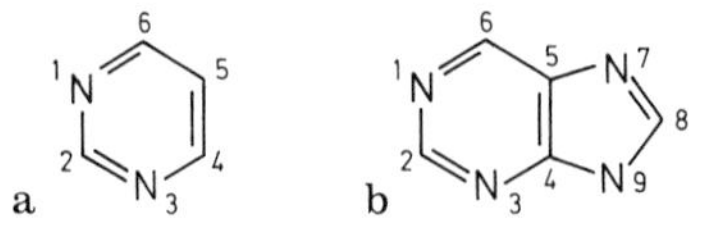

Fig. 1a and b. The structure of the purine and pyrimidine rings

1 In this chapter, the terms nucleotides and nucleosides refer to ribonucleotides and ribonucleosides

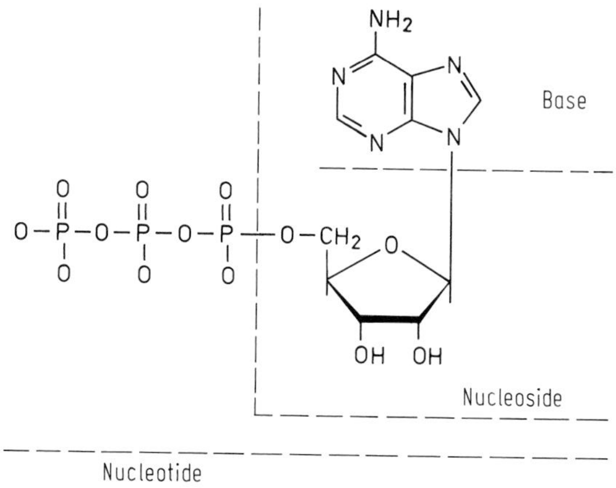

Fig. 2. The nucleotide, nucleoside and base structures of the purine adenine are shown. The nucleoside is composed of the base to which a D-ribose group is attached at the N_9 position on the imidazole ring. The mono-, di-, and triphosphate nucleotides are formed by the addition of 1 to 3 phosphate esters to 5' hydroxyl of the sugar

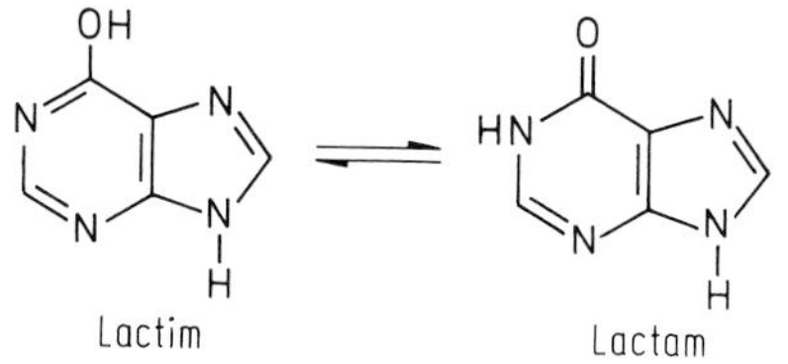

Fig. 3. The tautomeric forms of the purine base hypoxanthine

acids and bases. For example, the pK_b of adenosine is 5.7, and the pKa is 2.0. Therefore, above pH 5.7 adenosine is negatively charged and below pH 2 it is positively charged. Therefore, the retention on a reversed-phase column or on an ion-exchange column can be altered by changes in pH. Vertical base stacking behavior of the purine and pyrimidine compounds is also related to reversed-phase retention of these compounds and the degree of stacking is altered by tautomeric effects[4].

2 Sample Preparations

The primary goals of preparation techniques for physiological samples are as follows: 1) to remove protein material from the sample prior to chromatographic analysis, 2) to deactivate enzymatic activity as quickly as possible, and 3) to extract the nucleic acid constituents of interest into a suitable solvent. The procedures used for sample preparation can be the most important step in determining the sensitivity and precision of the analysis. Therefore, care must be taken not only in the execution of the sample preparation step, but also in the choice of the technique. A sample pretreatment must be chosen with the particular analysis in mind. Factors such as time, efficiency and cost must be considered. There is no single "best" method

Table 1. Deproteinization methods

Principle	Example
Dielectric change	Acetonitrile or Ethanol
Temperature change	Boiling water bath
Specific complexation or adsorption	Borate gels, pre-column concentration
pH change	TCA and PCA
Ionic strength	Ammonium Sulfate
Filtration	Ultrafiltration with membranes

for preparing a sample for chromatography; each method has its own set of advantages and disadvantages. In Table 1 are listed some of the commonly used methods of deproteinization and the principles upon which they are based.

The most common method for removal of bulk proteins and other macromolecular cells or tissue is the use of cold acid. In addition to removing the proteins, the acid will stop instantly enzymatic activity, thus preventing the enzymatic degradation of the components of interest. The most widely used acids are trichloracetic acid (TCA) and perchloric acid (PCA) [5-12]. The technique of Khym[6] involves the addition of TCA to a sample to precipitate the protein. The sample is then centrifuged and tri-octylamine in trichlorotrifluoroethane (Freon) is added. The top aqueous layer which contains the nucleic acid constituents is removed and an aliquot portion of it is chromatographed. The recovery of nucleotides from samples using this technique was studied by Chen, et al.[5], Van Haverbeke and Brown[6], and Riss[8]. The efficiency is a function of the sample matrix, the percent of TCA used and the age of the amine solution. The extraction from hepatocytes for the nucleotides range from 81.0% for GMP to 99.3% for CMP. Prior to use of this method, or any other method on a particular biological sample, the investigator is cautioned to investigate the recoveries of the analyte of interest. For example, Hartwick, et al.[11] found that the strong acid methods are unsuitable for the deproteinization of serum samples prior to analysis of nucleosides and bases. The recoveries of xanthosine, inosine and guanosine were low and varied from 54% to 67.7%.

Perchloric acid extraction followed by neutralization with KOH is another common method of protein removal, although the recoveries for nucleotides are generally lower than with the TCA methods. Other acid treatments for extraction of nucleotides from biological matrices include the use of formic and acetic acid[9]. For example, bacterial cells harvested by filtration through Millipore 0.45 micron membranes were extracted into cold 1 N acetic acid. The extracts were subjected to three freeze-thaw cycles and then lyophilized. The nucleotides were extracted with an efficiency of greater than 90% with neglible degradation. However, care must be taken in using acid for protein removal because some nucleotide derivatives are unstable in acid solution. For example, the reduced forms of NAD and NADP are subject to breakdown in acidic media[13]. Thus, if these compounds are of analytical interest, other deproteinization methods should be used.

Other commonly used techniques include precipitation by organic solvents, addition of high concentration salt solutions (salting out), ultrafiltration, and precolumn techniques. Organic solvents have been examined and found to provide poor efficiencies for nucleotides, and slightly better values for nucleosides[11, 12]. Additionally, since there is incomplete protein removal with organic solvents, column lifetimes may be shortened.

Of the non-acid methods, ultrafiltration appears to be the best choice for removal of proteins from serum[13]. This technique has a number of advantages. Because no solvent is added to the sample, impurities which could interfere with te analysis of the solutes were not introduced into the sample. In addition, the sensitivity is enhanced because there is no dilution of the sample. However, this technique is not recommended for use with cellular or tissue material, since the ultrafiltration cones will rapidly become clogged.

Pre-column techniques are a useful means of both sample deproteinization and chemical group separation. This technique with a C_{18} column was investigated for use with serum nucleosides and bases and silica for use with nucleosides[13]. These techniques are gaining popularity, although caution must be taken to avoid loss of early eluting material. The use of the reactivity between cis-diols and boronate polymers to form a pH-dependent complex is a novel and excellent pre-column technique to extract specifically ribonucleosides from biological fluids. Gerkhe's group[14] applied this technique to removal of the major and modified nucleosides from urine. A sample is passed over a column containing a borate polymer and compounds such as ribonucleosides which contain cis diols are retained on the column. The binding of the boronate to the diols can be reversed by elution with an acidic eluent, such as formic acid. Lothrop and Uziel[15] used a borate impregnated silica for both sample clean-up and group separations. To obtain the impregnated silica, a 1 ml silica column is equilibrated with 4 mM ammonium borate in 90% acetonitrile. The purine and pyrimidine bases are not retained under these conditions, while the ribonucleotides and ribonucleosides are retained. In addition, this method is used for the separation of the ribonucleosides and ribonucleotides from the corresponding deoxy analogues. Since the deoxy analogues do not certain cis diols, they are not retained by the column. After separation of the various groups, the individual members of each group can then be separated using appropriate chromatographic conditions.

Other factors may be involved in processing physiological fluids for HPLC analysis. For example, if the blood biological matrix of interest is plasma rather than serum, the effect of the anticoagulant on the separation must be assessed. Zakaria and Brown[16] investigated possible interferences caused by the anticoagulants, heparin, ethylene diamine tetraacetate (EDTA) and acid citrate dextrose (ACD) and found that EDTA and ACD contained UV absorbing impurities which interferred with the plasma nucleoside and base analysis. Thus, heparin was the anticoagulant of choice.

3 Chromatography

The HPLC separations of purine and pyrimidine compounds have been widely described in the literature[1, 2, 3]. The reversed-phase mode is generally used for the separation and analysis of nucleosides and bases, while anion-exchange is the method of choice for the ionic nucleotides. More recently, ion-pairing techniques have been developed for the reversed-phase separation of selected nucleotides. Although many of these analyses provide an excellent solution to a specific application problem, there is no universal solution for all applications. The requirements for an application must be individually determined.

3.1 Nucleotides

3.1.1 Ion Exchange

Ion exchange was the first mode of chromatography used in the separation of nucleic acid constituents. The work of Cohn at Oak Ridge[17] demonstrated the separation of nucleotides using a polymeric anion-exchanger. The development of pellicular anion exchange material by Horvath, et al.[18] brought increased speed and reproducibility to the analysis. However, it was the development of microparticulate chemically bonded silica packings that allowed the technique to grow to its present level[19]. In ion-exchange techniques the separation of the solutes is caused by selective retention of the solutes based on interactions of the negatively charged groups of the solute with the positively charged group on the stationary phase. The support materials that have been used for ion exchange include the polymeric Aminex series, the pellicular based material, and microparticulate silica. All the stationary phases used for the separation of nucleotides possess a cationic group, commonly a quaternary ammonium group.

The retention of a charged group on an anion exchange column is a function of the number of ionized groups which the solute molecule possesses. The more charged groups on a solute, the longer it is retained. Therefore, the triphosphate nucleotides are retained longer than its monophosphate analog. Separations on ion-exchange material are affected by changes in the pH and ionic strength of the mobile phase. The pH will affect the degree of ionization of the packing material, the eluent, and the solutes. With an increase in ionic strength of the eluent there is competition between the solute and the mobile phase for sites on the stationary phase. In general, high ionic strength eluents cause displacement of the solute from the stationary phase, and thus elution of that solute. In general, the higher the ionic strength of the mobile phase, the more rapid the elution of all components. Hartwick and Brown[20] were one of the first to demonstrate the use of microparticulate anion exchangers in the separation of 19 naturally occurring nucleotides. The separation of mono-, di- and triphosphate nucleotides was accomplished in under two hours on a Partisil SAX strong anion exchange column with gradient elution. These conditions were improved upon by McKeag, et al.[21]. In this latter analysis, a change in the mobile phase permitted the analysis time to be reduced to under one hour (Fig. 4).

In the past, it was difficult to obtain phosphate buffers of sufficient purity to allow the use of the high concentration of phosphate buffers required in the ion-exchange separation of nucleotides. Therefore, salts such as NaCl and KCl were added to the mobile phase to increase the ionic strength. Now HPLC grade buffers are commercially available, although there still can be UV absorbing impurities in some batches of phosphate buffers. In addition, impurities in the phosphate can form with time.

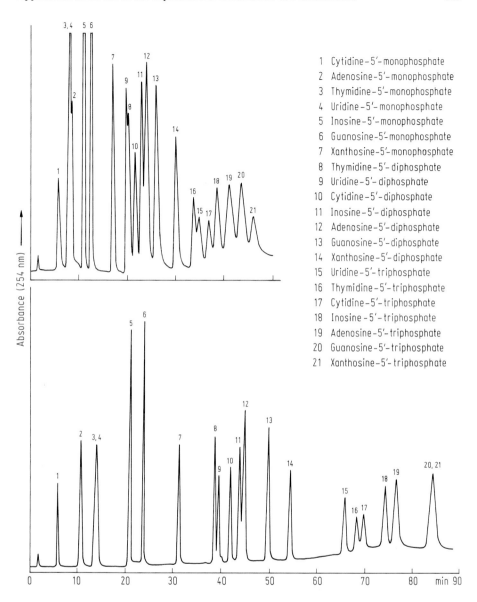

1 Cytidine-5'-monophosphate
2 Adenosine-5'-monophosphate
3 Thymidine-5'-monophosphate
4 Uridine-5'-monophosphate
5 Inosine-5'-monophosphate
6 Guanosine-5'-monophosphate
7 Xanthosine-5'-monophosphate
8 Thymidine-5'-diphosphate
9 Uridine-5'-diphosphate
10 Cytidine-5'-diphosphate
11 Inosine-5'-diphosphate
12 Adenosine-5'-diphosphate
13 Guanosine-5'-diphosphate
14 Xanthosine-5'-diphosphate
15 Uridine-5'-triphosphate
16 Thymidine-5'-triphosphate
17 Cytidine-5'-triphosphate
18 Inosine-5'-triphosphate
19 Adenosine-5'-triphosphate
20 Guanosine-5'-triphosphate
21 Xanthosine-5'-triphosphate

Fig. 4. Separation of the mono-, di-, and triphosphate ribonucleotides of cytidine, adenosine, thymidine, uridine, inosine, guanosine, and xanthosine. The nucleotides in the bottom spectra are separated on a Partisil-10-SAX column by a linear gradient using a phosphate buffer 0.007 M KH$_2$PO$_4$ to 0.25 M KH$_2$PO$_4$, 0.25 M KCl and a flow rate of 1.5 ml/min. The nucleotides in the top spectra are separated on a Partisil 10-SAX column by a linear gradient using a phosphate buffer 0.007 M KH$_2$PO$_4$, 0.007 M KCl to 0.25 M KH$_2$PO$_4$, 0.50 M KCl at a flow rate of 2.0 ml/min. Reproduced from Ref. [21]

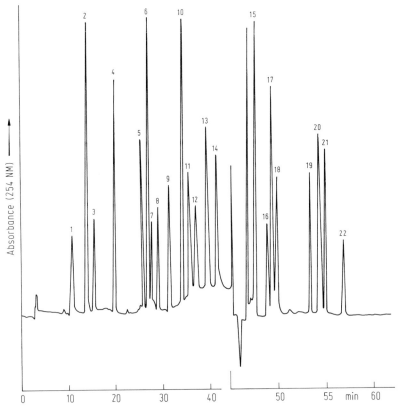

Fig. 5. Separation of a standard nucleotide, nucleoside and base mixture using column switching. The nucleotides are separated on a Partisil 10-SAX column by a non-linear gradient using a phosphate buffer from 0.005 M KH_2PO_4 to 0.5 M KH_2PO_4 at a flow rate of 2 ml/min. After elution of the XTP peak, the column is switched out of line and a radially compressed C18 column switched in line. The nucleosides are then separated by a linear gradient made up of 0.005 M KH_2PO_4 and a 30% methanol solution. Reproduced from Ref. [44]

Harmsen, et al. used Partisil-SAX to separate myocardial adenine nucleotides and creatine phosphate[22]. The mobile phase used in this analysis consisted of a phosphate buffer gradient, from 0.01 M H_3PO_4 to final conditions of 0.75 M KH_2PO_4. Ion-exchange separations have been employed in analysis of the antileukemia drug metabolite 6-thioguanosine monophosphate[23]. The separation of this compound from other constituents in erythrocytes was accomplished isocratically in six minutes. The mobile phase consisted of 0.125 M KH_2PO_4 in 0.25 M KCl.

Recently, a number of epimeric UDP-sugars were separated on a Partisil-SAX column with borate buffer as a complex-forming and eluting agent[24]. Included among the epimers separated were UDP-glucose from UDP-galactose, UDP-n-acetylglucosamine from UDP-acetylgalactosamine, and UDP-glucoronate from UDP-galacturonate. The separations were obtained isocratically at a flow rate of 2 ml/min. The mobile phases for these analyses were composed of 0.37 M boric acid, 0.15 M

Table 2. Ion-exchange mode

Compounds	Conditions	Column	Ref.

Purine ribonucleosides	A. 0.08 M $Na_2B_4O_7$
	B. 0.05 M NH_4CL, pH 9.1
	   Gradient-nonlinear
	C. 0.01 M $Na_2B_4O_7$
	   0.5 M NH_4Cl pH 9
		Aminex 29–25	[41]

Purine and pyrimidine ribonucleotides	A. 0.60 M KH_2PO_4, pH 4.2 B. 0.007 M KH_2PO_4, 2.2 M KCl, pH 4.5	Partisil SAX	[5]
Deoxyribonucleotide pools	0.4 M $(NH_4)_2HPO_4$, pH 3.25	Partisil SAX	[59]
Ribonucleotide pools	A. 0.005 M $NH_4H_2PO_4$, pH 2.8 B. 0.75 M $NH_4H_2PO_4$, pH 3.7, 0–100%, 2 ml/min	Partisil SAX	[60]
Pyrimidine monophosphate ribonucleotides	0.06 M formic acid 0.001 M Na_3PO_4, pH 4.1, 1 ml/min	Partisil SAX	[61]

Nucleobases, nucleosides, nucleotides	A. 5 mM phosphate buffer, pH 7.75
	B. 50 mM phosphate buffer pH 7.75,
	   after 3 min
	   gradient from 0%–100% B
		YEW Ax-1	[59]

Ribonucleotides	A. 0.04 M KH_2PO_4 pH 2.9
	B. 0.5 M KH_2PO_4, 0.8 M KCl, pH 2.9
	   linear gradient from A to B
	   13 min, 1 ml/min
		APS-Hypersil	[60]

disodium tetraborate, and 2.0 M glycerol, pH 6.45. Some separations by ion exchange in a number of different applications are listed in Table 2.

3.1.2 Reversed-Phase

Reversed-phase techniques are gaining use in the analysis of nucleotides, although this chromatographic mode to the nucleotides is not generally used for simultaneous analysis of the full complement of these compounds which may exist in biological fluids. In general, nucleotides are not well retained on reversed-phase systems unless methods such as ion-suppression or ion-pairing are employed.

However, Krstulović, et al.[25] showed a separation of five naturally occurring cyclic ribonucleotides using gradient elution on a C-18 μBondapak column. The system consisted of 0.02 M KH_2PO_4 and a 25-minute gradient to 25% of a 60/40 (v/v) methanolic solution of 0.02 M KH_2PO_4. More recently, Assenza, et al.[26] showed an improved procedure. In this work, an isocratic separation of cCMP, cUMP, cGMP, and cIMP was achieved in 14 minutes on a Zorbax ODS column (Fig. 6). Elution conditions consisted of 4.0 mM $(NH_4)_2PO_4$ and 4.0 mM $(NH_4)H_2PO_4$, pH 3.0. Recently, Jahngen and Rossomando[27] studied the effect of metal ions on the retention of ATP using reversed-phase. They found that separation of metal bound and free ATP could be separated isocratically using 10 mM KH_2PO_4, 4% methanol.

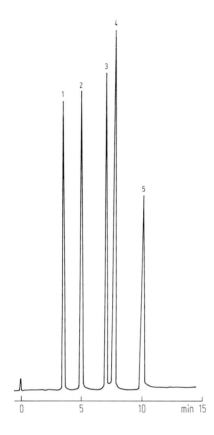

Fig. 6. Isocratic separation of the five naturally occurring 3,5 cyclic ribonucleotides. Flow rate of 2.5 ml/min with 4.0 mM $(NH_4)_2HPO_4$ and 4.0 mM $(NH_4)H_2PO_4$ pH 3.0, and a Zorbax-ODS column. Reproduced from Ref. [26]

Marinez-Valdez[28] used isocratic elution with ammonium phosphate on a C_{18} Bondapak column for the separation of adenine, ribo and deoxyribo nucleotides. At pH 5.1, the elution order is ATP, ADP, AMP, dATP, dADP, dAMP, and adenine. They applied this separation to HeLa cell extracts. Table 3 lists the compounds separated and the mobile phases used for some reversed-phase separations. Assenza, et al. also showed that deoxyribonucleosides could be separated isocratically using the reversed-phase mode[29].

3.1.3 Ion Pairing

The use of ion-pairing techniques to extend the applicability of reversed phase for nucleotide analysis was first reported by Hoffman and Liao[30]. Their separation was of 12 nucleotides and cyclic AMP, and they used tetrabutyl-ammonium (TBA) phosphate as the pairing agent. Darwish and Prichard[31] separated NAD, IDP, GDP, cAMP, cGMP and succinyl AMP by reversed-phase using a mobile phase of 65 mM KH_2PO_4, pH 3.2, to which 3.3% acetonitrile was added. They compared µBondpack C-18 with a Radialpak A column. When the radially compressed column was used, analysis time was less than 35 minutes. A radially compressed column with an ion-pairing reagent was also used by Rao, et al.[32] to study the nucleotides in platelets.

Table 3. Reverse-Phase mode

Compounds Separated	Mobile Phase	Column	Ref.
HYP, XAN, URD, KYN, PHE, INO, TRP, XAN, paraXAN	0.02 M KH$_2$PO$_4$, pH 5.7 gradient, 30 min linear 40% of 60/40 MeOH/H$_2$O	Partisil ODS-3, 10 μm	[16]
5-FU, 5-FUdR from plasma Constituents	0.01 M KH$_2$PO$_4$, pH 4.0, 2 ml/min	μBondpak C-18	[52]
UA, HYP, XAN, m-XAN	0.02 M KH$_2$PO$_4$, pH 3.65	Hypersil ODS, 3 μM	[65]
Nucleosides in urine	A. 0.01 M NH$_4$H$_2$PO$_4$ – 2.5% Methanol B. 0.01 M NH$_4$H$_2$PO$_4$ – 8% Methanol 35° C, 1 ml/min	Two 300 × 4 mm columns in series μBondpak	[53]
HYP, THY, THD, oxy-purinol allopurinol	0.05 M KH$_2$PO$_4$, 0–10% methanol pH 4.5, ambient temperature	μBondpak	[54]
5′-ADP and ATP from in-organic pyrophosphate or imidodiphosphate	A. 100 mM Triethylammonium bicarbonate, pH 6.7 (22° C) B. 100 mM triethylammonium bi-carbonate in absolute ethanol 2 ml/min 0.5% B/min gradient applied at 16 min	Spherisorb ODS (25 × 0.46 cm)	[61]
ATP, ADP, AMP, IMP, adenosine inosine, hypoxanthine, NAD, uric acid	A. 0.004 M MgSO$_4$, 0.004 M Ammediol or Tris, 0.002 M KH$_2$PO$_4$, 0.2 M NaN$_3$ pH 6.25 B. Methanol: water, 60:40 v/v 1.2 ml/min gradient 3–80% B	Zorbax ODS	[62]

3.2 Nucleosides and Bases

3.2.1 Ion Exchange

The early separations of nucleosides and bases were conducted by ion-exchange, generally cation exchange, although anion exchange was also used. However, chemically-bonded, reversed-phase packings replaced the use of ion-exchange for the separation of these compounds because of increased resolution.

3.2.2 Reversed-Phase

The reversed-phase mode is the most common method for separation of nucleosides and bases. The reversed phase mode has a number of advantages over the other modes of chromatography:

1) A variety of molecular types of molecules (ionic, nonionic and/or ionizable) can be separated. Conditions can be adjusted to permit separation of all three groups in one analysis.
2) The columns used are extremely efficient and stable. Therefore, the variation in analysis from column-to-column or day-to-day is small.

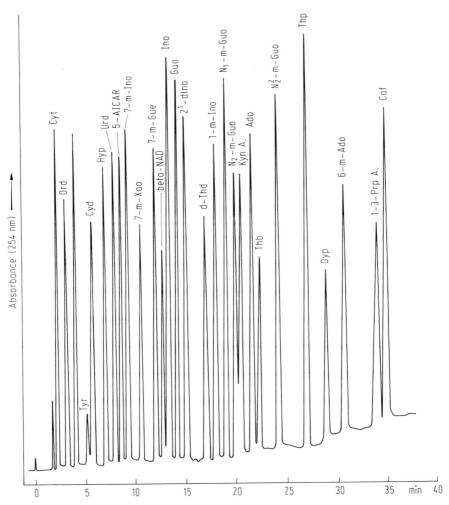

Fig. 7. Separation of 0.1 to 0.5 nmole of 28 nucleosides, bases, nucleotides, and aromatic amino acids and metabolites. Injection volume: 40 µl of a solution of 0.01 mM of each standard. Column: chemically bonded, C_{18} on 10 µm silica. Eluents: low-strength, 0.02 M KH_2PO_4: high-strength 60% methanol. Reproduced from Ref. [33]

3) Analysis time is relatively short and the column reequilibrates rapidly.

A large number of applications of reversed-phase separations to nucleoside and base separations have been reported in the literature[1]. In one of the earlier applications, Hartwick, et al.[33] optimized the conditions for separation of nucleosides, bases and other low molecular weight UV absorbing constituents in serum. Figure 8 shows the quality of separation possible for a mixture of nucleosides and bases using small particle size packings. Using high efficiency short columns (3 cm), packed with very small particle packings (3 µm), very rapid separations can be obtained (Fig. 7). Gerke and his group used HPLC as a tool for monitoring methylated

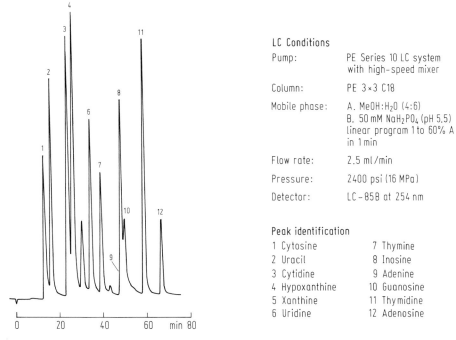

LC Conditions

Pump: PE Series 10 LC system
                 with high-speed mixer

Column: PE 3×3 C18

Mobile phase: A. MeOH:H₂0 (4:6)
                 B. 50 mM NaH₂PO₄ (pH 5.5)
                 linear program 1 to 60% A
                 in 1 min

Flow rate: 2.5 ml/min

Pressure: 2400 psi (16 MPa)

Detector: LC-85B at 254 nm

Peak identification

1 Cytosine   7 Thymine
2 Uracil   8 Inosine
3 Cytidine   9 Adenine
4 Hypoxanthine   10 Guanosine
5 Xanthine   11 Thymidine
6 Uridine   12 Adenosine

Fig. 8. The use of 3 cm columns with 3 μm particles provides good separation in shorter time. Chromatogram supplied by the Perkin Elmer Corp

nucleosides in urine[34]. Methylated nucleosides are being investigated as possible biochemical markers in cancer. More recently, this group reported on the use of RPLC as a rapid, quantitative measurement tool for nucleoside composition of tRNA's (Fig. 9)[35]. RPLC has been used to determine the base composition in hydrolysates of tRNA, DNA, and RNA by a number of groups[35–39]. Buch, et al.[38], by using a variety of gradient conditions of 0.25 M ammonium acetate (pH 6.0) and 40/60 acetonitrile/water, studied the 29 modified nucleosides of E. coli. Some examples of reversed-phase separations are given in Table 3.

3.3 Nucleotides, Nucleosides and Their Bases

3.3.1 Single Column

At times it is necessary to monitor changes in nucleotides, nucleosides and their bases in one sample. The first example of this type of analysis was shown by Floridi, et al.[40], who used an anion exchange system. Bakay, et al.[41] and Nissinen[42] also separated these compounds using the Aminex-25 series of anion exchangers and a complex gradient.

Schweinsberg and Loo[43] used reversed-phase to separate ATP, ADP, and AMP from other nucleosides, nucleotides and bases. Zakaria adapted these conditions

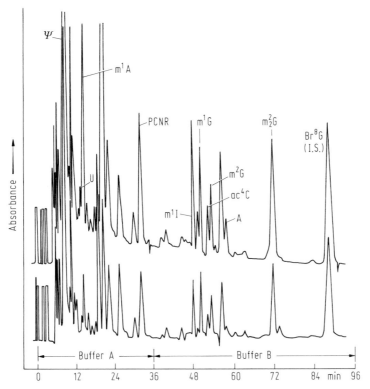

Fig. 9. RPLC step gradient separation of nucleosides in urine. Sample: Cancer Patient Urine. Column: μ Bondpak C_{18}. Conditions: Eluent A: 0.01 M $NH_4H_2PO_4$, 2.5% MeOH pH 5.3. Eluent B: 0.01 M $NH_4H_2PO_4$, 8.0% MeOH pH 5.1. Temperature: 35° C. The upper chromatogram was obtained using a wavelength of 254 nm and the lower 280 nm. Reproduced from Ref. [35]

and analyzed erythrocyte lysates for ATP, ADP, and AMP in the presence of nucleosides and bases[16].

3.3.2 Column Switching

The use of coupled columns is another way to separate groups of compounds of widely varying properties; e.g., to separate the ionic nucleotides from the non-ionic nucleosides and bases. In an analysis by Halfpenny and Brown[44], a gradient of 0.005 M KH_2PO_4 to 0.5 M KH_2PO_4 on a Partisil SAX column was used as part of a coupled column approach to separation of nucleosides, nucleotides, and bases in one analytical run. The nucleotides are introduced into the anion-exchange column where they were retained and separated, while the nucleosides and base were eluted in the void volume of the anion exchange column onto a reversed-phase column. Using automatic switching valves, the groups are first isolated. Then the nucleosides and bases separated by reversed-phase and the nucleotides are separated by anion exchange (Fig. 5).

3.3.3 Mixed Bed Columns

A rather recent approach to separate different classes of compounds is to change the nature of the stationary phase so that both reversed-phase and anion exchange properties are present in the packing. Although these packings are not presently available commercially, they can be prepared for specific applications. Using column packings that they prepared, Crowther and Hartwick[45] were able to separate both nucleotides and nucleosides in one analysis.

3.3.4 Ion Pairing

Another possible way of separating nucleotides along with their corresponding nucleosides and bases is by reversed-phase with the appropriate use of an ion-pairing reagent. The reagents and the conditions must be chosen so that the ionic reagent. The reagents and the conditions must be chosen so that the ionic nucleotides which usually elute in the void volume in a reversed-phase system are retained longer on the column without interfering with the nucleosides and bases which naturally have longer retention times. For example, several quaternary ammonium ions were evaluated for separation of the derivatives of 5-fluorouracil and the phosphorylated

Table 4. Ion pairing mode

Compounds	Conditions	Column	Ref.
5 Fluoroouracil and its deoxyribo, ribonucleotides standards and serum	0.05 M HK_2PO_4, 0.15 mM TBA, pH 5, 5% (v/v) Methanol	RCIL C-18 5 μm	46)
Ribonucleotides and cisplatine	0.05 M Phosphate buffer, pH 7.5, 4% HTAB	ODS Hypersil 5 μm	55)
Four isomers of b,g-bi-denstate Dr-ATP	20 mM MSA, pH 2.5, 1 ml/min, 23° C	Partisil-5-ODS-3 5 μm	56)
Deoxyribonucleotides from their nucleosides, ribonucleotides from ribo-nucleosides	0.05 M KH_2PO_4, pH 4.9, 7% $Mg(OH)_2$ 2.0 mM TBAP, 1 ml/min, ambient temperature	Partisil-5-C8	57)
Uracil, guanine, 5-me-thylcytosine-cytosine and thymine	5 mM HSA, 2.5 mM KH_2PO_4, pH 5.6, 2 ml/min	RP-18	58)
Adenine, Guanine	A. 30 mM TEA-Pi, pH 6.5 B. 30 mM TEA-Pi, 5 mM $MgSO$, pH 6.5 Gradient 0 to 72% B at 2%/min flow rate 1 ml/min	BioRad ODS-5S	63)
M^5d Cyd, ADE, dINO, GUO	A. 2.5% (v/v) methanol, 0.05 M KH_2PO_4, pH 4.0 B. 8.0% (v/v) methanol, 0.05 M KH_2PO_4, pH 4.0 22.5 min of A then 30 min B	Supelcosil LC-18-DB	64)

analogues of 5′-deoxy-5-fluorouridine. A separation of FU, FUR, FUdR, FUMP, 5-dFUR, FdUMP and UDPG was obtained by isocratic elution on a C-18 column [46].

3.4 Other Separations of Interest

In studies of nucleic acid metabolism, separations are needed not only for the nucleotides, nucleosides and bases, but also for the enzymes which catalyze metabolic reactions, co-enzymes which are needed for the reactions, cyclic nucleotides and the antimetabolite purine and pyrimidine reagents which inhibit the reactions. In addition, good separation techniques are needed for the nucleic acids themselves. In biotechnology HPLC has made possible separations of oligonucleotides which are the substrates, intermediates and products needed in the synthesis of DNA and RNA. However, better separations are still required, and with new developments in both instrumentation and techniques, more sensitive and rapid separations with greatly increased resolution will become possible.

The next few years will also see greatly improved separations of the nucleic acids themselves. Improvements in increased resolution and sensitivity and decreased time of analysis are needed. With more basic studies of the retention processes of oligonucleotides and nucleic acids underway, we will gain better understanding of the complex reactions causing the separations of these compounds; thus, with new and novel packings and improvements in all phases of HPLC instrumentation we should be able to optimize separations of all types of nucleic acids.

4 Detection and Identification

Presently, the most common means of detection for nucleotides, nucleosides, and bases is with a UV detector at 254 nm. All of the purine and pyrimidines possess a chromophore which absorbs at this wavelength. However, multiple techniques are required for positive identification of chromatographic peaks in physiological fluids. When the sample is of biological origin, many similar compounds may exist; thus, the use of multiple techniques is necessary for characterization and positive identification. Some of the commonly used peak identification techniques are listed in Table 5.

While retention time of a solute cannot be used as a means of positive identification, it can be used in a negative way to rule out a possible structure for the peak of interest. Therefore, retention times and co-chromatography with standards are the first steps which should be taken in the identification of a peak. While co-elution is not a conclusive test, if the substance does not co-elute, it is positive evidence that the peak is not the suspected substance. Additional information on peak identity may be obtained by chromatography on different stationary phases. The likelihood of different compounds having identical chromatographic behavior on two different stationary phases is very small; thus, either positive or negative evidence on the identity of the peak can be obtained. This approach can help to eliminate quickly certain structure possibilities.

Table 5. Commonly used methods for peak identification in HPLC separations of constituents in physiological fluids

1) Retention times

2) Co-chromatography with standards

3) With absorbance detectors:
   a) Absorbance ratios (280/254)
   b) UV-Vis scanning
   c) Chemical derivatization
   d) Enzymatic peak shifting

4) Fluorescence detectors

5) Chemical reactions specific to functional groups and structure

6) Direct interface with other techniques (MS, NMR, IR, ... etc.)

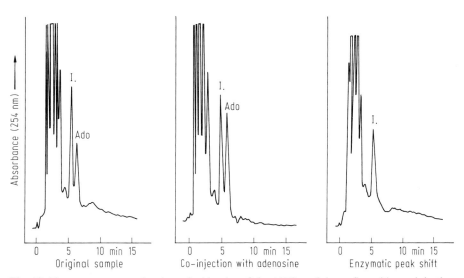

Fig. 10. Human serum sample where the identity of the ADO peak is confirmed by co-injection and the enzyme peak shift technique using adenosine deaminase which selectively converts adenosine to inosine. the nucleosides are separated on a μBondpak C_{18} column with an eluent of (10:90) anhydrous methanol: 0.007 M KH_2PO_4 at a flow rate of 2.0 ml/min. Reproduced from Ref. [47]

For the next step, on-line techniques should be used. In particular, the use of absorbance ratios at two or more wavelengths can be extremely helpful in characterizing the substance which the peak represents. In addition, a full UV spectrum of the peak can be obtained, either by the stopped-flow technique or by use of a photodiode array fast scanning UV detector. With the stopped-flow technique, the background may be stored digitally and subtracted from the background of the eluent and eluite. Once the flow is stopped, some diffusion may occur; however, this diffusion is negligible in practice. The advent of the fast scanning detectors which utilize photodiode arrays and charge coupled devices are preferable if they

Table 6. Enzymes useful for the identification of some of the low molecular weight UV-absorbing compounds found in human serum or plasma

Substrates(s)	Enzyme	Product(s)
Adenine	Adenine phosphoribosyl transferase	AMP
Adenosine	Adenosine deaminase	Inosine
	Adenosine kinase	AMP
Guanine	Guanase	Xanthine
Guanine	Phosphoribosyltransferase	GMP
Guanosine	Purine nucleoside phosphorylase	Guanine
Hypoxanthine	Xanthine oxidase	Xanthine
Xanthine		Uric Acid
Hypoxanthine, Guanine	Hypoxanthine-guanine phosphorybosyl transferase	IMP, GMP
Uric Acid	Uricase	Allantoin
Uracil	Uridine phosphorylase	Uridine
Uridine	Uridine kinase	UMP
Inosine	Purine nucleoside phosphorylase	Hypoxanthine
Tyrosine	Hydroxyphenylpyruvate oxidase	Hydroxyphenyl pyruvate
Creatinine	Creatininase	Creatine
Phenylalanine	Phenylalanine Ammonia-lyase	Trans-cinnamate
L-Tryptophan	Tryptophanase	Indole

are available and provide much information that can be used to characterize the peak.

The enzyme peak shift technique is an excellent means of peak identification for purine and pyrimidine compounds[2], since so many of these compounds are substrates for commercially available enzymes which catalyze the metabolism of a specific compound or a group of compounds. This technique is performed as follows: an aliquot of sample is chromatographed, and another aliquot of the same sample is incubated with an enzyme and its co-factors, at the optimum pH and temperature. It is important that the reaction is run under first-order conditions; that is, that an excess of the enzyme must be present. Thus, the reaction is limited by the amount of substrate present in the reaction mixture and not by the amount of enzyme present. An example of the enzyme peak shift technique is the peak shift of the nucleoside adenosine with the enzyme adenosine deaminase to form the nucleoside inosine. The reaction was performed at pH 6.88 in the presence of phosphate ions. After a sample is incubated with adenosine deaminase, the nucleoside adenosine is not present in the chromatogram, and a peak which represents the nucleoside inosine either is formed or increases in size (Fig. 10). The use of the enzyme peak shift technique is also a means of determining whether a peak is comprised of more than one component. If two compounds had co-eluted and only one of them reacts with the enzyme, the peak representing the substrate for the enzyme would decrease. We can thus say that the remaining peak is "unmasked", and its spectral properties can be obtained. Table 6 lists some of the commercially available enzymes for reactions with purines and pyrimidines.

5 Summary

HPLC is a powerful tool for use in nucleic acid research. Virtually all the metabolites of DNA and RNA can be separated and analyzed using this technique; thus, separations are possible for compounds in the nucleic acid metabolic pathway ranging from the free bases to the nucleic acids themselves. The HPLC separations are far more rapid than classical methods and provide high resolution and sensitivity. It is predicted that in the near future improvements in instrumentation, computerization, and operating conditions will improve significantly the resolution, sensitivity and time of analysis, and will make HPLC indispensible not only in studies of nucleic acid metabolism but also in biotechnology for the synthesis of genes.

References

1. "HPLC in Nucleic Acid Research" (P.R. Brown, ed.), Marcel Dekker, Inc., New York, 1984
2. "High Pressure Liquid Chromatography: Biochemical and Biomedical Applications," P.R. Brown, Academic Press, New York, 1973
3. "Reversed-Phase High Performance Liquid Chromatography: Theory, Practice, and Biomedical Applications," A.M. Krstulović and P.R. Brown, Wiley Interscience, New York, 1982
4. Brown, P.R. and Grushka, E.: Anal. Chem., 52, 1210 (1980)
5. Chen, S., Rosie, D.M. and Brown, P.R.: J. Chrom. Sci., 15, 218 (1977)
6. Van Haverbeke, D.A. and Brown, P.R.: J. of Liq. Chrom., 1, 1507 (1978)
7. Khym, J.X.: Clin. Chem., 21, 1245 (1975)
8. Riss, T.L., Zorich, N.L., Williams, M.D. and Richardson, A.: J. of Liq. Chrom., 3, 133 (1980)
9. Essigmann, J.M., Busby, W.F. and Wogan, G.N.: Anal. Biochem., 81, 384 (1977)
10. Krauss, G. and Reinboth, B.: Anal. Biochem., 78, 1 (1977)
11. Hartwick, R.A., Van Haverbeke, D., McKeag, M. and Brown, P.R.: J. of Liq. Chrom., 2, 725 (1979)
12. Hartwick, R.A., Krstulović, A.M. and Brown, P.R.: J. Chrom., 186, 659 (1979)
13. Miksic, J. and Brown, P.R.: J. Chrom. 142, 541 (1977)
14. Gehrke, C.W., Kuo, K.C. and Zumwalt, R.W.: J. of Chrom., 188, 129 (1980)
15. Lothrop, C.D. and Uziel, M.: Anal. Biochem. 109, 160 (1980)
16. Zakaria, M. and Brown, P.R.: Anal. Biochem., 120, 5 (1982)
17. Cohn, W.: Science, 109, 377 (1949)
18. Horvath, C.G., Preiss, B.A. and Lipsky, S.R.: Anal. Chem., 39, 1422 (1967)
19. Halász, I., Schmidt, H. and Vogtel, P.: J. Chrom., 126, 19 (1975)
20. Hartwick, R.A. and Brown, P.R.: J. of Chrom., 112, 651 (1975)
21. McKeag, M. and Brown, P.R.: J. of Chrom., 152, 253 (1978)
22. Harmsen, E., De Tombe, P.P. and De Jong, J.W.: J. of Chrom., 230, 131 (1982)
23. Cooley, T. and Maddocks, J.I.: J. of Chrom., 229, 121 (1982)
24. Weckbecker, G. and Keppler, D.C.R.: Anal. Biochem., 132, 405 (1983)
25. Krstulović, A.M., Hartwick, R.A. and Brown P.R.: Clin. Chem., 25, 235 (1979)

26. Assenza, S.P., Brown, P.R. and Goldberg, A.P.: J. of Chrom., 272, 373 (1983)
27. Jahngen, E. and Rossomando, E.: Anal. Biochem., 130, 406 (1983)
28. Martinez-Valdes, H., Kothari, R.M. and Taylor, M.W.: J. of Chrom., 247, 307 (1982)
29. Assenza, S.P., Goldberg, A.P. and Brown, P.R.: J. Chrom. Biomed. Applic., 277, 305 (1983)
30. Hoffman, N. and Liao, J.C.: Anal. Chem. 49, 2225 (1977)
31. Darwish, A.A. and Prichard, J.: Liq. Chrom., 4, 1511 (1981)
32. Rao, G.H.R., Peller, J.D. and White, J.D.: J. of Chrom., 226, 466 (1981)
33. Hartwick, R.A., Assenza, S.P. and Brown, P.R.: J. of Chrom., 186, 647 (1979)
34. David, J., Suits, R.D., Kuo, K.C., Gerke, C.W., Waalkes, T.P. and Borek, E.: Clin. Chem., 23, 1427 (1977)
35. Gehrke, C.W., Zumwalt, R.W., McCune, R.A. and Kuo, K.C.: Cancer Res., 84, 344 (1983)
36. Wakizaka, A., Kurosaka, K. and Okurkara, E.: J. of Chrom., 162, 319 (1979)
37. Gomes, J.D. and Chang, C.: Anal. Biochem., 129, 387 (1983)
38. Buck, M., Connick, M. and Ames, B.: Anal. Biochem., 129, 1 (1983)
39. Agris, P.F., Tompson, J.G., Gehrke, C.W., Juo, K.C. and Rice, R.H.: J. of Chrom., 194, 204 (1984)
40. Floridi, C.A., Palmerini, C.A. and Findi, C.: J. of Chrom., 138, 203 (1977)
41. Bakay, B., Nissinen, E. and Sweetman, I.: Anal. Biochem., 86, 55 (1978)
42. Nissinen, E.: Anal. Biochem., 106, 497 (1980)
43. Schweinsberg, P.D. and Loo, T.J.: J. of Chrom., 181, 103 (1980)
44. Halfpenny, A.P. and Brown, P.R.: Abstract #470, Pittsburgh Conference on Analytical Chemistry and Applied Spectroscopy, 1983
45. Crowther, J.B., Hartwick, R.A.: Chromatographia, 16, 349 (1983)
46. Gelijkens, C.F. and DeLeenHeer, A.P.: J. Chrom., 194, 305 (1980)
47. Hartwick, R.A. and Brown, P.R.: J. Chrom. Biomed. Applic., 143, 383 (1977)
48. Gelijkens, C.F. and DeLeenHeer, A.P.: J. Chrom., 194, 305 (1980)
49. Garrett, and Sasti, D.: Anal. Biochem., 99, 268 (1979)
50. Nilson, J.A., Rose, L.M. and Bennett, L.I.: Can. Res., 36, 1375 (1976)
51. Mayer, J.D. and Handschumacher, R.: Can. Res., 39, 3089 (1979)
52. Buckspitt, A.R. and Boyd, M.R.: Anal. Biochem., 106, 432 (1980)
53. Gehrke, G., Kuo, K.C. and Zumwlat, R.W.: J. of Chrom., 188, 129 (1980)
54. Taylor, G.A., Daly, P.J. and Harrap, K.R.: J. Chrom., 183, 421 (1980)
55. Riley, C.M., Sternson, L.A., Repta, A.J. and Slyter, S.: Anal. Biochem., 203 (1983)
56. Gruys, K. and Schuster, S.M.: Anal. Biochem., 125, 66 (1982)
57. Crowther, J.B., Caronia, J.P. and Hartwick, R.A.: Anal. Biochem., 124, 65 (1982)
58. Ehrlick, M. and Ehrlick, K.: J. Chrom. Sci., 17, 531 (1979)
59. Naikwadi, K.P., Rokushika, S. and Hatono, H.: J. Chrom., 280, 261 (1983)
60. Perrch, D.: Chromatographia, 16, 211 (1982)
61. Mahoney, C.W., Yount, R.G.: Anal. Biochem., 138, 246 (1984)
62. Holliss, D.G., Humphrey, S.M., Morrison, M.A. and Seelye, R.N.: Anal. Lett., 17, 2047 (1984)
63. Folley, L.S., Power, S.D. and Poyton, R.O.: J. Chrom., 281, 199 (1983)
64. Gehrke, C.W., McCune, R.A., Gamasosa, M.A., Ehrlich, M. and Kuo, K.: J. Chromatogr., 301, 199 (1984)
65. Boulieu, R., Bory, C., Baltassat, P. and Gonnet, C.: Anal. Biochem., 129, 398 (1983)

Application of HPLC to the Analysis of Natural and Synthetic Pharmaceutically Important Drugs

Gerolf Tittel
Laboratorium für Analysen und Trenntechnik Dr. Tittel GmbH
Maria-Eich-Str. 7, 8032 Gräfelfing/München/FRG

Hildebert Wagner
Institut für pharmazeutische Biologie Universität München
Karlstr. 29, 8000 München 2/FRG

1 Introduction

More than fifteen years after the introduction of HPLC into pharmaceutical analysis, a comprehensive literature survey is still lacking, and no systematic scheme of analysis has been published. Even within the scope of this handbook, only a sample of the relevant literature can be discussed, based on some especially important pharmaceuticals and classes of compounds. In view of the large number of pharmaceuticals commercially available today, this selective treatment is clearly unavoidable. Thus, the register of preparations of the German pharmaceutical industry, known as the "red list", embraces 86 areas of indication, and contains over 8900 products in more than 1100 formulations. Excluding polymeric substances, about 80% of these pharmaceuticals are suitable for separation and purification by HPLC; this provides some indication of the present and future potential of HPLC in pharmaceutical analysis.

HPLC is applied at several different levels in pharmacy and pharmacology. It is employed in production development and product control, in the isolation of pharmaceuticals, raw material control, and finally in clinical analysis, which is closely linked to the study of pharmaceuticals.

The unique value of HPLC analysis lies in the fact that a pharmaceutical must be analysed for quality (purity and content) not only before processing, but also after incorporation into various matrices (tablets, suppositories, creams, etc.), often in mixtures with other compounds.

One special analytical requirement is the identification of natural products and their determination in plant and animal extracts, i.e. in very complex mixtures. We have therefore added to the literature review a section on the possible applications and methodology of sample preparation of pharmaceuticals and plant extracts.

It has not been possible to maintain a clear distinction between synthetic pharmaceuticals and those from natural sources, because many substances originally isolated as natural products are nowadays obtained by total or partial synthesis.

The review begins with a tabulated summary of retention data of important pharmaceuticals under standardized conditions. This should serve as an aid for rapidly establishing similar separations.

Finally, a selection of literature sources is discussed, which provides information on when and how the separation conditions described in the retention tables might be improved. The discussion is limited to general improvements in apparatus and column technology, which frequently oblige the analyst to adapt and update his own chromatographic system.

2 Synthetic Pharmaceuticals

Tabulation of the retention properties of synthetic pharmaceuticals serves as an aid to rapidly evaluate mixtures of pharmaceuticals by HPLC.

2.1 Retention Data for Synthetic Pharmaceuticals

2.1.1 HPLC Separations on C18 Reversed Phase in Methanol-Water Mixtures

In 1976, Tank[1] described the retention properties of 140 synthetic pharmaceuticals on a 10 μm C18 reversed phase with various methanol-water mixtures under isocratic conditions (cf. Table 1).

Notwithstanding the recent technical developments in apparatus and column design, this summary of retention data still maintains its practical importance to rapidly determine the quality of raw materials, and as a source of parameters for simple HPLC separations.

The k′ values are listed for isocratic separations on a 25 cm, C18 column, using 20, 30, 50 and 70% methanol.

2.1.2 HPLC Separations on C18 Reversed Phase with Acetonitrile Gradients

The retention data for pharmaceuticals published in 1981 by Daltrup, Susanto and Michalke[2] represent an extension and updating of the data of Tank[1]. Retention properties were determined with a multi-step acetonitrile solvent gradient. In contrast to Table 1, the data are not recorded as absolute values, but as relative retention values (RRT*) against a standard substance. The standard was 5-(p-methylphenyl)-5-phenylhydantoin (MPH) (cf. Table 2).

Course of the gradient:

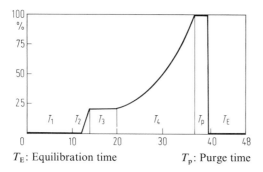

T_E: Equilibration time T_p: Purge time

2.1.3 Isocratic HPLC Separations of Synthetic Pharmaceuticals on C8 and C18 Reversed Phase

From a series of investigations on the quality control of commercially available pharmaceutical preparations, Rehm and Steinigen[3] published the list of chromatographic parameters shown in Table 3. The authors give no retention data, but they list suitable solvent systems and the appropriate wavelengths for spectrophotometric detection. The substances are not listed alphabetically, but grouped according to their pharmacological action.

The retention data in Tables 1 and 2 are a guide to the chromatographic behaviour of the substances, and are not intended to represent reproducible, absolute values. Every separation must be optimized for the individual apparatus and column.

$$* \text{ RRT} = \frac{t_{dr} \text{ Substance}}{t_{dr} \text{ Standard (MPH)}}$$

Table 1. Retention data for synthetic pharmaceuticals on a C18 reversed phase; k′-values according to Tank[1]

Substance	k′ values			
	20%	30%	50%	70%
Acetanilide	–	6.71	2.16	1.30
Acetazolamide	2.79	0.86	0.85	0.64
Acetohexamide	–	–	7.56	1.67
Acetylcarbromal	–	–	9.41	2.3
Acetylsalicylic acid	–	3.37	2.15	1.15
Acetylsalicylic acid anhydride	–	–	14.46	2.84
Aclofenac Neoston	–	–	–	1.85
Ethionamide	–	–	–	7.62
o-Ethoxybenzamide	–	15.43	2.97	2.11
Aethyl-bis-cumacetate	–	10.03	0.91	0.56
Amidopyrine	–	–	–	10.23
Anetholtrithione	–	–	–	7.87
Ascorbic acid	0.55	0.55	0.55	0.55
Barbital	–	4.72	1.82	1.09
Benzocaine	–	–	4.06	2.71
Benzoic acid	–	16.70	3.17	1.56
Benzosulfimide (= Saccharin)	0.78	0.61	0.38	0.24
Benzthiazide	–	–	4.52	1.21
Bilevon M	–	–	–	5.21
Bromvalurea	–	13.92	3.43	1.43
Butallylonal	–	–	6.67	2.05
Caffeine	–	3.71	2.1	0.95
Carbamazepine	–	–	7.16	1.84
Carbimazol	–	3.71	1.67	0.91
Carbromal	–	–	6.63	1.95
Carbutamide	–	10.80	2.20	0.89
Centalun	–	7.23	2.17	1.08
Chlordiazepoxide	–	–	–	23.54
p-Chlorphenol	–	–	6.03	2.06
Chlorthalidone	–	3.25	1.52	0.72
Chlorothiazide	2.25	1.38	0.84	0.59
Chlorpropamide	–	–	6.39	1.58
Cyclopenthiazide	–	–	6.51	1.28
Danthrone	–	–	–	6.59
Diazepam	–	–	–	2.68
Diacetylaminoazotoluene	–	–	–	3.20
Diphenan	–	–	19.23	3.26
Dipyridamol	–	–	–	13.54
Flufenaminic acid	–	–	–	11.71
Furazelidone	8.08	3.26	1.28	1.03
Furosemide	–	10.64	1.97	0.76
Glibenclamide	–	–	–	1.85
Gluthetimide	–	–	6.24	1.98
Hexobarbital	–	–	5.37	1.82
Hexyparaben	–	–	–	3.09
Hydrochlorothiazide	2.54	1.36	0.75	0.61
Hydroflumethiazide	5.37	2.21	0.92	0.79
Hydroxyvhinoline	–	–	–	11.99
Hydroxychlorphenylsulphide	–	–	–	6.93

Table 1 (continued)

Substance	k′ values			
	20%	30%	50%	70%
Isoniazid	1.85	1.18	1.01	0.82
Lactylphenetidine	–	–	3.23	1.23
Mafenaminic acid	–	–	–	12.56
Mandelic acid benzylester	–	–	16.13	2.94
Mefrusid	–	–	3.16	1.08
Metamizol	0.64	0.12	0.12	0.12
Methaqualone	–	–	11.48	2.30
Methylparaben	–	–	3.28	1.26
Methylphenobarbital	–	–	5.77	1.43
Methylthiouracil	1.87	1.22	1.02	0.89
Methyprylone	–	10.79	2.77	1.21
Metronidazole	3.51	2.07	1.29	1.02
Monophenylbutazone	–	–	8.63	1.54
Narcobarbital	–	–	8.53	1.53
Nicethamide	–	5.97	3.28	2.76
Nicotinamide	2.18	1.35	0.95	0.67
Nicotinic acid	0.95	0.90	0.85	0.69
Nitrazepam	–	–	7.68	1.55
Nitrofurantoin	6.17	2.68	1.10	0.76
Oxazepam	–	–	11.86	3.15
Oxyphenbutazone	–	–	9.35	1.87
Paracetamol	3.13	1.72	1.14v	0.74
Paraaminosalicylic acid (PAS)	3.14	2.0	1.07	0.61
Picric acid	1.01	0.69	0.46	0.25
Phenacetin	–	12.41	3.23	1.31
Phenallymal	–	–	3.86	1.21
Phenazone	–	6.69	2.06	1.06
Phenetolcarbamide	–	2.80	1.12	1.07
Phenindione (1,3)	6.18	2.91	1.40	0.51
Phenopyrazone	–	2.55	0.97	0.54
Phenprobamate	–	–	7.68	1.76
Phenylbutazone	–	–	–	4.68
Phenyl quinoline carboxylic acid	–	14.80	3.00	1.18
Phenyl salicylate	–	–	–	3.10
Phenythoin	–	–	6.01	1.44
Phtalylsulphathiazole	4.4	2.44	1.31	0.66
Physobarbital	–	12.45	3.66	1.1
Procarbazine	–	–	–	5.83
Propallylonal	–	–	3.62	1.28
Propylparaben	–	–	11.55	2.24
Propyphenazone	–	–	7.37	2.08
Protionamide	–	–	–	9.41
Pyrithylidone	–	6.22	1.99	1.18
Resorcinol	3.3	2.06	1.00	0.93
Salicylamide	–	6.2	1.83	1.09
Salicylic acid	2.69	1.82	1.36	0.92
Sorbic acid	–	9.34	3.34	1.30
Sulphacarbamide	1.47	1.02	0.75	0.67
Sulphacetamide	7.33	1.72	1.53	0.64

Table 1 (continued)

Substance	k′ values			
	20%	30%	50%	70%
Sulphaquinidine	–	3.59	1.21	0.77
Sulphadiazine	3.13	1.72	1.53	0.64
Sulphaguanidine	1.11	0.89	0.70	0.62
Sulphamerazine	6.13	2.39	1.02	0.93
Sulphafurazol	4.52	2.9	1.33	0.98
Sulphamethoxydiazine	–	3.06	1.09	0.71
Sulphamoxol	7.75	2.90	0.97	0.67
Sulphanilamide	1.30	1.01	0.70	0.61
Sulphaphenazol	–	8.95	1.69	0.92
Sulphathiazol	4.02	1.79	0.82	0.79
Sulphathiourea	1.28	0.95	0.84	0.68
Sulphinpyrazone	–	8.25	2.39	1.90
Sulphisomidine	4.28	2.13	1.29	0.79
Sulthiam	8.38	3.06	1.12	0.66
Theobromine	2.98	1.53	1.1	0.93
Theophylline	6.13	2.39	1.08	0.79
Thiaazetazone	–	5.67	1.33	0.95
Thiamazole	1.62	1.17	0.95	0.67
Tolazamide	–	–	10.61	2.21
Tolbutamide	–	–	9.15	1.32
Tribenzazoline	13.30	3.12	1.51	0.89
Tyloquinone	–	–	11.62	2.98
Vanillin	–	5.72	1.81	1.01
Xanthine	1.06	0.97	0.94	0.92

Mode of operation:
C^a: Perkin Elmer SIL-X-C18, 10 μm (250 × 4 mm ID, Stahl)
F: 1.0 ml/min
Mph: X% methanol in water
T: RT
All numbers in the table represent k′-values.
[a] Abbreviations are listed in the appendix

Table 2. Retention data for synthetic pharmaceuticals on a C18 reversed phase with solvent gradient elution [2]

Substance	RRT[a]	Substance	RRT[a]
Acecarbromal	0.63	Barbital	0.19
Acetatolamide	0.13	Bemegrid	0.26
Acetyldigoxin	0.28	Bemetizide	1.29 z
Acetylsalicylic acid	0.08	Bendroflumethiazide	0.24
Alimemazine	2.02	Benorilate	0.98 z
Allobarbital	0.28	Benproperine	0.40
Amiphenazol	0.43	Benzilic acid tropine ester	0.40
Amobarbital	0.65	Bepheniumhydroxynaphthoate	0.02
Amphetaminil	0.66	Benvoniummetil sulphate	0.01
Aprobarbital	0.31	Benzafibrate	0.18
Azapropazone	1.99	Bisacodyl	1.26
			2.48

Table 2 (continued)

Substance	RRT[a]	Substance	RRT[a]
Brallobarbital	0.36	Disulfiram (Artefact?)	2.95
Bromisoval	0.36	Disulfotone	2.96
Bumadizone	0.3	Drofenin	0.42
Bumetanide	0.06	Etacrynic acid	0.08
	0.13		0.15
Butabarbital	0.38	Ethione	3.38
Butalbital	0.45	Ethosuximide	0.21
Butaperazine	0.02	Ethoxzolamide	0.76
Butetamate	2.64	Etoroxate	0.01
	4.27	Etodroxizine	0.01
Butizide	0.16	Etofibrate	2.59
	0.73	Etozoline	1.95
Cafaminol	0.12	Fenitrothione	2.81
Caffeine	0.14	Fenoprofen	0.49
Camazepam	2.23	Floctafenin	1.76
Carbamazepine	0.63	Flufenaminic acid	0.48
L-Carbidopa	0.07	Flunitrazepam	1.22
	0.16	Fluoro-Uracil	0.07
Carbimazol	0.14	Flurazepam	1.22
Carbromal	0.65	Flurbiprofen	0.44
Carbutamide	0.34	Flurspirilen	0.30
Carticaine	0.40	Fominoben	2.71
Chlorazanil	1.16	Glibenclamide	0.79
Chlordiazepoxide (D)	4.28		0.91
Chlormezanone	0.52	Glibornuride	0.55
Chlortalidone	0.22	Gliclazide	0.58
8-Chlortheophylline	0.09	Glipizide	0.43
Chlorzoxazone	0.60	Gliquidone	2.00
Cimetidine	0.16 z	Glutethimide	0.86
Clidinium bromide	0.01	Glymidine	0.34
Clobazam	1.53	Guaifenesine	0.19
Clomethiazol	0.93	Haloperidol	0.40
Clomipramine	2.05	Heptabarbital	0.62
Clonazepam	0.92	Hexobarbital	0.67
Clopamid	0.38	Hydrochlorothiazide	0.17
Clorexolon	0.78	Ibuprofen	0.81
Cocaine	1.35	Indapamide	0.81
Crotylbarbital	0.33	Indometacin	0.33
Cyclobarbital	0.42	Isoetarin	0.15
Cyclopenthiazide	1.50	Isoniazid	0.23
Cyclopentobarbital	0.45	Isopropamide iodide	0.02
Cytarabin	0.07	Isosorbidol dinitrate	1.16
Diazepam	2.08	Kavaine	1.83
Diazinone	3.00	Ketoprofen	0.22
Diazoxide	0.31	Lanatoside C	0.21
Diclofenac-Na	0.18	Lobeline	0.73
Digitoxin	0.64	Lorazepam	0.84
	1.89	Malathione	2.72
Digoxin	0.27	Mebendazol	0.85
Dihydralazine	0.20	Meclofenoxate	1.96 z
Dilazep	0.97		
Dimethoate	0.30		
Dipotassium chloroazepate	1.42		

Table 2 (continued)

Substance	RRT[a]	Substance	RRT[a]
Mefenaminic acid	0.21	Phenyloin	0.63
Mefruside	0.98	Physostigmine salicylate	0.30
Menbutone	0.24	Pitophenone	0.40
Mesuximid	0.79		1.21
Metamizol	0.00	Polythiazide	1.38
Methaqualone	1.14	Prazepam	2.67
Methidathione	2.25	Prenoxdiazine	2.82
Methocarbamol	0.20	Primidone	0.17
Methohexital	1.67	Procaine	0.41
Methyldigoxin	0.69	Propallylonal	0.40
Methylphenobarbital	0.76	Propanidid	1.98
Methyprylone	0.28	Propoxur	0.83
Metolazone	0.23	Propyphenazone	0.94
	0.41	Proquazone	2.28
Metronidazol	0.14	Proscillaridin	0.63
Metyrapon	0.01	Proxibarbal	0.14
Mitopodozide	0.30	Proxyphylline	0.13
Morazone	0.26	Pyrantel	0.05
Naftidrofuryl	0.15	Pyrithyldione	0.23
	2.82	Pyrvinium pamoate	0.02
Naproxen	0.25	Quinethazone	0.13
Nicothamide	0.37	Salicylamide	0.26
Niclosamide	0.35	Secobarbital	0.84
Nifedipine	1.79	Spironolacetone	2.29
Nifenatone	0.21	g-Strophanthine	2.03
Nifluminic acid	0.16	Sulindac	0.14
Nimorazol	0.54	Sulphinpyrazone	0.07
Natrizepam	0.81	Sultiam	0.26
Nitroglycerin	2.03	Suxibuzone	6.75
p-Nitrophenol	0.47	Tetrahydrocannabinol (THC)	3.31
Ornidazol	0.26	Theophylline	0.13
Oxazepan	0.72	Thiamazol	0.12
Oxyfedrin	0.97	Thiobutabarbital	0.81
Paracetamol	0.13	Thiopental	1.24
Paraoxon	1.49	Timidazol	0.21
Pemolin	0.16	Tolazamide	0.50
Pentaerythrityltetranitrate	2.65	Tolbutamide	0.64
Pentobarbital	0.64	Tolmetin	0.06
Pentoxyfyllin	0.19	Triazolam	1.15
Phenacetin	0.41	Trichlormethiazide	0.42
Phenazone	0.37	Trifluperidol	0.41
Phenobarbital	0.35	Tritoqualin	0.98
Phenolphthalein	0.80	Vinylbital	0.65
Phenprobamate	0.80	Xipamide	0.31
Phenprocumone	0.27		

[a] RRT = Relative retention time against 5-(p-methylphenyl)-5-phenylhydantoin (MPH)

Mode of operation:
C: LiChrosorb RP 18, 10 μm (250 × 4 mm ID, Stahl), (Merck AG)
F: 1.0 ml/min
Mph A: acetonitrile
Mph B: 31.1% acetonitrile in water (780 g acetonitrile +1720 g water)

Table 3. HPLC conditions for the separation of various pharmaceuticals, according to Rehm and Steinigen [3]

Substance	Internal standard	Column	Acetonitrile (%)	Flow rate (ml/min)	Detection wave length (nm)
Analgetics					
Acetylsalicylic acid	Caffeine	RP18	25	2	276
(Salicylic acid)	Caffeine	RP18	25	2	234
Phenacetin	Caffeine	RP18	25	2	249
Aminophenazone	Propyphenazone	RP18	35	2	268
Propyphenazone	Aminophenazone	RP18	35	2	275
Purines					
Coffeine	Theophylline	RP18	25	2	273
Theophylline	Theobromine	RP8	12	3	270
Oxyethyltheophylline	Theobromine	RP8	12	3	270
Theobromine	Theophylline	RP8	12	3	270
Addictive drugs					
Morphine	Ethylmorphine	RP8	25	1.8	286
Codeine	Ketobemidone	RP18	25	2	285
Diacetylmorphine	Apomorphine	RP8	50	1.8	280
(Monoacetylmorphine)	Apomorphine	RP8	50	1.8	280
Apomorphine	Morphine	RP8	50	1.8	274
Ketobemidone	Codeine	RP8	30	1.8	284
Pethidine	Ketobemidone	RP8	40	5	258
Levomethadone + Fenpipramide	Normethadone	RP8	55	5	258
Normethadone	Fenpipramide	RP8	55	5	258
Digitalis glycosides					
Digitoxin	Gitoxin	RP18	50	3	220
Digitoxigenin	Gitoxin	RP18	50	3	220
Lanatoside C	–	RP18	38	3	220
Deacetyl-Lanatoside C	–	RP18	38	3	220
Digoxigenin-bis-digitoxoside	–	RP18	38	3	220
Digoxigenin-mono-digitoxoside	–	RP18	38	3	220
Digoxin	–	RP18	38/50	3	220
Digoxigenin	–	RP18	50	3	220
Corticosteroides					
Prednisolonhemisuccinate					
(Prednisolone)	–	RP8	42	2	240
Prednisolone	Dexamethasone	RP18	40	2	240
Prednisolone	Dexamethasone	RP18	40	2	240
Dexamethasone	Prednisone	RP18	40	2	240
Triamcinolone	Prednisolone	RP8	28 [a]	3	240
Antibiotics					
Tetracyclin	–	RP18	35	2	272
(Epitetracyclin)	–	RP18	35	2	258
(Epianhydrotetracyclin)	–	RP18	35	2	274
(Anhydrotetracyclin)	–	RP18	35	2	274

Table 3 (continued)

Substance	Internal standard	Col-umn	Aceto-nitrile (%)	Flow rate (ml/min)	Detec-tion wave length (nm)
Oxytetracyclin	–	RP18	35	2	270
Rolitetracyclin	–	RP18	35	2	272
Chemotherapeutic agents					
Trimethoprim	Sulphamethazine	RP18	30	2	230
Tetroxoprim	Sulphamethazine	RP18	15	3	230
Sulphamethoxazole	Sulphamethazine	RP18	30	2	270
Sulphametrole	Sulphadoxin	RP18	30	2	270
Sulphadoxin	Sulphametrol	RP18	30	2	270
Sulphamethazine	Sulphamethoxazol	RP18	30	2	270
Sulphamerazine	Sulphamethoxazol	RP18	30	2	270
Sulphanilthiocarbamide	Sulphamethoxazol	RP18	30	2	270
Sulphadiazine	Sulphamethazin	RP18	15	2	270
Sulphapyridine	Sulphamethazin	RP18	15	2	270
Sulphathiazole	Sulphamethazin	RP18	15	2	288
Sulphamoxole	Sulphadiazin	RP18	12.5	3	270
Sulphanilamide	Sulphapyridin	RP18	12.5	2	270
Sulphisomidine	Sulphapyridin	RP18	12.5	2	270
Sedatives/Hypnotics					
Barbital	Allobarbital	RP18	25	2	220
Allobarbital	Phenobarbital	RP18	25	2	220
Aprobarbital	Propallylonal	RP18	25	2	220
Phenobarbital	Phenallymal	RP18	25	2	220
Propallylonal	Pentobarbital	RP18	25	2	220
Cyclobarbital	Pentobarbital	RP18	25	2	220
Phenallymal	Hexobarbital	RP18	30	2	220
Butalbital	Hexobarbital	RP18	30	2	220
Heptabarbital	Hexobarbital	RP18	30	2	220
Butallylonal	Hexobarbital	RP18	30	2	220
Pentobarbital	Hexobarbital	RP18	30	2	220
Amobarbital	Hexobarbital	RP18	30	2	220
Hexobarbital	Allobarbital	RP18	30	2	220
Chlordiazepoxid	Medazepam	RP18	40	2	245
Medazepam	Oxazepam	RP18	40	2	231
Oxazepam	Diazepam	RP18	40	2	231
Diazepam	Oxazepam	RP18	40	2	245
Phenytoin	–	RP18	30*	4	210
Sympathomimetic agents					
Synephrine	Etilefrin	RP8	5	1.5	275
Norfenefrine	Etilefrin	RP8	5	1.5	275
Etilefrine	Synephrin	RP8	5	1.5	275
Ephedrine	Phentermin	RP18	20	1.5	258
Nor-Pseudoephedrine	Phentermin	RP18	20	1.5	258
Amphetamine	Phentermin	RP18	20	1.5	258
Methamphetamine	Ephedrin	RP18	20	1.5	258
Phentermine	Chlorphentermin	RP18	25	2	220
Chlorphentermine	Phentermin	RP18	25	2	220

Table 3 (continued)

Substance	Internal standard	Column	Acetonitrile (%)	Flow rate (ml/min)	Detection wave length (nm)
Local anaesthetics					
Procaine	Hydroxycaine	RP8	50	3	296
Hydroxycaine	Procaine	RP8	50	3	296
Benzocaine	–	RP8	60	2	292
Preservatives					
4-Hydroxybenzoic methylester	4-Hydroxybenzoic acid ethylester	RP8	45	1.8	258
4-Hydroxybenzoic propylester	4-Hydroxybenzoic acid ethylester	RP8	45	1.8	258

[a] In water instead of 0.05 M phosphoric acid; decomposition products

Mode of operation:
C: MN-Nucleosil 10-C18, or 10-C8, (250 × 4 mm ID, Stahl)
F: 1.5–5 ml/min
Mph: X% acetonitrile in 0.05 M o-phosphoric acid
T: 40° C

2.2 HPLC Separation of Selected Pharmaceuticals and Groups of Substances

In an early (1976) review, entitled "Applications of HPLC in the pharmaceutical industry", Bailey[4] treated pharmaceutical HPLC analysis under the following headings: alkaloids, antibiotics, nitrogen-containing compounds, steroids, sulphur-containing compounds, formulations and general analytical techniques. Other authors and especially commercial literature have continued to use these same subdivisions.

In view of the flood of publications in these field, we feel that a rapid search for chromatographic parameters is assisted by ordering important references alphabetically according to the pharmaceuticals they describe. However, various alternative subdivisions have been retained or introduced where similar chromatographic problems are dealt with by many analogous publications.

2.2.1 Antibiotics

There is no small group of solvent systems that is suitable for the HPLC separation of all antibiotics. Although the antibiotics are related in their pharmacological action, they represent a wide diversity of chemical structures and molecular weights.

Table 4 gives the most important references for HPLC of the chief antibiotics. The original literature must be consulted for practical details and corresponding HPLC parameters.

Table 4. HPLC separations of antibiotics

Compound(s)	Ref.[a]	Compound(s)	Ref.[a]	Compound(s)	Ref.[a]
Bacmecillinam	[5]	Doxorubicin	[49]	Penicillin	[6, 31 – 34]
Amoxycillin	[7]	Erythromycin	[18 – 21]	Polymyxin	[35]
Ampicillin	[6]	Gentamycin	[22]	Rifampicin	[36]
Anisomycin	[48]	Gramicidin	[23]	Spiramycin	[37]
Bleomycin	[8 – 10]	Kanamycin	[24]	Streptomycin	[38]
Candicidin	[11]	Levorin	[11]	Streptozocin	[39]
Cefsulphodin	[17]	Maridomycin	[25]	Tetracyclines	[40 – 46]
Cephacetrile	[50]	Mecillinam	[26]	Ticarcillin	[47]
Chephalosporines	[12, 13, 50]	Mitomycin	[27]	Trichomycin	[11]
Chloramphenicols	[14 – 16]	Nalidixic acid	[28, 29]	Vancomycin	[48]
Cycloheximide	[51]	Neomycin	[30]		

[a] See Ref. list

2.2.2 Aliphatic and Aromatic Carboxylic Acids

Polybasic and long chain carboxylic acids are frequently employed in pharmacy as additives or preservatives. As reported by Schwarzenbach[52] in 1982 in a review of the HPLC of aliphatic and aromatic mono- and polybasic carboxylic acids, this class of substances can be chromatographed both by reversed phase with addition of acids or ion pair formers to the mobile phase, and by ion exchange. For especially stereoselective separations, Schwarzenbach suggests the use of silica gel chromatography. For the chromatographic parameters, reference should be made to the original work[52], and to its 145 quoted literature sources.

2.2.3 Addictive Drugs (Opiates, Cocaine, Hallucinogens, etc.)

The narcotics and addictive drugs include the classical opiates and antidepressants, and the modern psychodelic agents.

Three fundamental studies have been published on the systematic HPLC analysis of certain addictive drugs[53 – 55]: Albanbauer et al. described the HPLC of heroin and its various preparations. As shown in Fig. 1, different preparations of heroin, or heroin from different sources can be easily differentiated by HPLC[53].

Using HPLC, Megges[54] was able to detect LSD (D-lysergic acid diethylamide) in a "trip" involving only 80–100 µg of material, and was also able to differentiate between LSD and the similarly used substitute, LAMPA (lysergic acid methylpropylamide).

Antidepressants may also be efficiently separated. Einhellig, Kraatz and Megges[55] analysed the seven most important phenylethylamine derivatives in a single separation system, which was also successfully used for the analysis of illegal amphetamine derivatives (cf. Fig. 2).

For all three series of investigations the authors employed C18 reversed phases. The mobile phases were made basic by the addition of ammonium carbonate solution. This is necessary, because all described narcotics are strongly basic compounds.

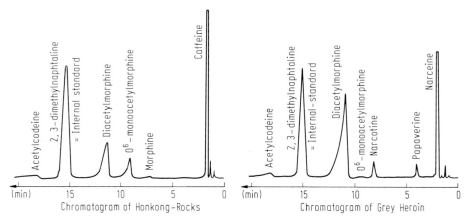

Fig. 1. HPLC separation of different types of Heroin. J. Albanbauer, J. Fehn, W. Furtner, G. Megges Arch. f. Krim. 162, 103 (1978). C: μ-Bondapak C18 (300 × 3.9 mm ID); Mph: Water-acetonitrile-ammoniumcarbonate solution (1%) = (156:140:4); F: 2.5 ml/min

Fig. 2. HPLC Analysis of Amphetamines. K. Einhellig, A. Kraatz, G. Megges, Arch. f. Krim. 166, 99 (1980). C: Nucleosil-10-C18 (300 × 4 mm ID); Mph: Acetonitrile-methanol-water-ammoniumcarbonate solution (saturated) = (100:100:75:3); F: 2.5 ml/min

This alkalinization method, which we also used earlier, has the disadvantage that the working life of the column is decreased by operation at pH 9. Alkalinization can be avoided by the addition of suitable ion pair reagents to the mobile phase.

Determination of cocaine on RP2 phase has been reported by Jane et al.[56].

2.2.4 Psychoactive Pharmaceuticals

In addition to the quality control of single psychoactive substances and the testing of pharmaceutical preparations, the chief application of HPLC in the analysis of these substances lies in the clinical field. Psychoactive pharmaceuticals are often the subject of misuse, so that their rapid and precise detection is important for the clinician as an aid to the recognition of poisoning or addiction.

The clinical application of pharmaceutical HPLC analysis is treated in another chapter. The present chapter deals only with the essential chromatographic parameters of such separations, and discusses some laboratory examples.

Psychoactive pharmaceuticals mostly belong to one of the following chemical types: barbiturates, phenothiazines, benzodiazepine derivatives and Rauwolfia alkaloids.

All the above-mentioned classes of compounds are efficiently separated on C18 reversed phases. Reports of the use of silica gel phases are found only in earlier work, e.g. the investigations of Detaevernier et al.[57] in 1976, on the separation of tricyclic antidepressants. A different system was also used by Wheals[58], who suggested a cation exchanger carrying sulphonic acid groups for the separation of phenothiazines. Wheals prepared the cation exchanger by reaction of silicic acid with 3-mercaptopropyltrimethoxysilane, followed by oxidation to the chemically bound, non-hydrolysable sulphonic acid group. For the mobile phase he used mixtures of methanol and aqueous ammonium nitrate at different ionic strengths and different pH values.

Retention data at different pH values for 31 important phenothiazines are given in Table 5.

Elution volumes in Table 5 show that the retention of all the phenothiazines is strongly pH-dependent over the range pH 4-10 and is maximal in the region pH 4-6. In addition to causing greater retention, increasing the acidity of the eluent also gives rise to minor changes in elution order.

Almost all authors use UV absorption for the detection of compounds in this group. A spectrophotometer may be used for selected wavelengths, or a fixed wavelength detector at 254 nm can be used for all compounds. An alternative method was reported by Lund, Hannisdal and Greibrook[59] who employed amperometric detection for the analysis of benzodiazepines. They established that the detection limit depended strongly on the applied reduction potential. The reported lower detection limit (3 ng for nitrazepam) is similar to that for photometric detection. Furthermore, the photometric detector is much easier to operate than the more demanding amperometric method.

Fluorimetry may also be used for detection. King[60] reported in 1981 that some psychoactive pharmaceuticals can be detected selectively by fluorimetry. The suitability of fluorimetry or spectrophotometry as detection methods, and the appropriate working wavelengths are indicated in Table 6.

In order to suppress tailing of these mostly basic compounds, basic reagents or ion pair formers are often added to the mobile phases of reversed phase separations. Thus, Wahlund and Sokolowski[61] in 1978 coated a LiChrosob RP 8 phase with 1-pentanol in situ, then chromatographed with addition of N,N,N-trimethylnonylammonium or N,N,dimethyloctylamine to the mobile phase. This technique separated antidepressive and neuroleptic amines and the related quarternary ammo-

Table 5. The influence of eluent pH on the retention of phenothiazines on an alkyl sulphonic acid strong cation exchanger [58]. Eluent: Methanol-1 M ammonium nitrate (9:1). The ammonium nitrate solution was adjusted to the pH shown before dilution with methanol. The solvent was pumped at 1 ml/min

Phenothiazines	Elution volume (ml)			
	pH 10	pH 8	pH 6	pH 4
Parent compounds				
Diethazine	–	8.7	9.8	9.5
Proquamazine	5.0	9.3	9.9	10.1
Trimeprazine	–	8.7	10.1	9.9
Pecazine	4.2	9.9	11.2	11.2
Dimethoxanate	4.8	10.8	11.7	11.9
Promazine	4.4	10.9	12.0	12.3
Methdilazine	5.2	12.0	12.6	13.2
Perazine	4.2	9.6	15.0	16.3
Derivatives of perazine				
Trifluperazine	4.0	7.3	9.2	11.9
Thiethylperazine	–	8.7	11.1	11.9
Butaperazine	–	8.0	11.1	11.9
Prochlorperazine	–	8.7	11.9	13.2
Thioperazine	3.8	9.1	12.2	13.6
Derivatives of promazine				
Triflupromazine	–	8.2	9.0	9.2
Chlorpromazine	–	9.6	10.6	11.0
Methoxypromazine	4.2	10.3	11.2	11.2
Acetylpromazine	4.4	10.9	11.8	12.0
Derivatives of phenazine				
Fluphenazine	3.8	5.7	6.7	9.0
Perphenazine	–	6.4	8.8	10.3
Carphenazine	–	6.4	8.9	9.6
Acetophenazine	–	6.7	9.6	10.9
Miscellaneous phenothiazines				
Piperacetazine	–	7.2	7.8	8.1
Pericyazine	–	7.6	9.2	9.8
Thiopropazate	–	7.5	9.6	9.6
Dimethothiazine	–	7.4	9.6	10.6
Thioridazine	4.7	9.5	9.7	10.2
Propiomazine	–	8.4	9.7	9.7
Methotrimeprazine	–	8.4	9.7	9.6
Pipamazine	–	7.9	10.4	10.9
Metopimazine	4.1	9.6	11.1	13.7
Mesoridazine	4.9	12.6	13.6	13.9

Table 6. Chromatographic and fluorescence properties of basic and neutral drugs[60]

Drugs	Fluorescence (+): yes (−): no	Excitation wavelength	Emission wavelength
Amphetamine	−		
Amtriptyline	−		
Benzocaine	+	280	397
Carbamazepine	−		
Chlordiazepoxide	+	280	weak
Chlormethiazole	−		
Chloroquine	+	280	405
Chlorpromazine	+	320	458
Cinnarizine	−		
Clomipramine	−		
Cocaine	−		
Codeine	+	280	weak
Cyclizine	−		
Desipramine	+	280	410
Desmethyldiazepam	+	280	weak
Dextromoramide	−		
Dextropropoxyphene	−		
Diamorphine	−		
Diazepam	−		
Dihydrocodeine	+	280	weak
Diphenhydramine	−		
Dipipanone	−		
Dothiepin	−		
Doxepin	−		
Fluphenazine	+	320	475
Flurazepam	−		
Haloperidol	−		
Imipramine	+	280	412
Levorphanol	−		
Lignocaine	−		
Lorazepam	+	280	weak
Maprotiline	−		
Medazepam	+	280	weak
Meprobamate	−		
Methadone	−		
Methaqualone	+	280	weak
Methotrimeprazine	+	320	456
Mianserin	+	280	412
Morphine	+	280	weak
Nitrazepam	−		
Nortripyline	−		
Orphenadrine	−		
Oxazepam	+	280	weak
Oxyprenolol	−		
Perphenazine	+	320	458
Pethidine	−		
Phenacetin	−		
Pholcodine	+	280	weak
Procaine	+	280	396

Table 6 (continued)

Drugs	Fluorescence (+): yes (−): no	Excitation wavelength	Emission wavelength
Prochlorperazine	+	320	458
Promazine	+	320	452
Promethazine	+	320	452
Propanolol	+	280	398
Protiptyline	+	280	402
Strychnine	−	−	−
Sulphathiazole	−	−	−
Temazepam	+	280	weak
Thioridazine	+	320	466
Trifluoropiperazine	+	320	476
Trimeprazine	+	320	458
Trimipramine	+	280	412

Fluorescence was measured after separation on a silicic acid column, using methanol – 2 N ammonium hydroxide – 1 N ammonium nitrat (90:7:3), or Spherisorb 5 ODS (methanol – water = 35:65) as chromatographic parameters

nium compounds. The special advantage of the method is its ability to separate cis and trans isomers (cf. Fig. 3).

The cis and trans isomers of tricyclic neuroleptic agents can, however, also be separated on silicic acid with a mixture of ethyl acetate and methanol alkalized by the addition of ammonia[62].

More recently, Sokolowski and Wahlund[63] reported the use of ion pair chromatography without the in situ coating of the stationary phase. As shown in 1977 by Harzer and Barchet[64] and later by Vilon and Vercruysse[65], the separation of benzodiazepine derivatives and their hydrolysis products (benzophenones) can be performed satisfactorily on reversed phase without additions to the mobile phase. Figure 4 and Table 7 give the separation and retention data for 11 important benzodiazepines.

Since the introduction of HPLC, various stationary phases have been tried for the chromatography of barbiturates. Reversed phase appears to be the most appropriate[66] (cf. Tables 1–3).

2.2.5 Steroids

The systematic HPLC investigation of steroids began in 1970 with a paper by Siggia and Dishman[67]. Today, all classes of steroids are analysed by HPLC and many publications deal with this subject.

In 1981, Görög[68] reviewed the HPLC analysis of steroids in pharmaceutical formulations. One year earlier Kingston[69], and two years earlier Heftman and Hunter[70] published literature surveys and general reviews on HPLC of natural products, including steroids. These publications should be consulted for literature sources.

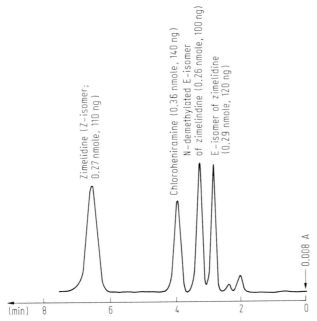

Fig. 3. Separation of cis-trans isomers of antidepressive and neuroleptic amines. K.G. Wahlund, A. Sokolowski, J. Chromatogr. 151, 299 (1978). Mph: Cyclohexylsulphamate 0.05 M, phosphate buffer (pH 2.21), H_2PO_4 0.01 M, N,N-dimethyloctylamine 0.01 M, Tetrabutyl-ammoniumhydrogen sulphate 0.03 M, 50% saturated with 1-pentanol; F: 0.78 ml/min; C: 1-pentanol, $V_s/V_m = 0.23$; support: LiChrosorb RP 8, 10 µm (150 × 3.2 mm ID)

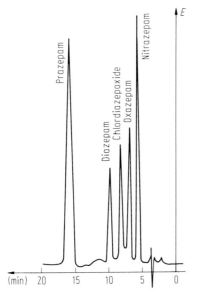

Fig. 4. Separation of Benzodiazepines. K. Harzer, R. Barchet, J. Chromatogr. 132, 83 (1977). Retention times of Benzodiazepines in different methanol-water-mixtures:

Substance	Mobile phase, Methanol-water	Retention time (min)
Chlordiazepoxide	60:40	16.8
Dipotassium chlorazepate		17.8
Nitrazepam	70:30	5.8
Lorazepam		6.8
Oxazepam		7.0
Chlordiazepoxide		8.4
Dipotassium chlorazepate		8.7
Diazepam		10.0
Prazepam	100:0	16.1
Flurazepam		9.4

C: LiChrosorb RP 18, 10 µm (250 × 4 mm ID)
F: 0.75 ml/min
D: Photometer at 254 nm wavelength

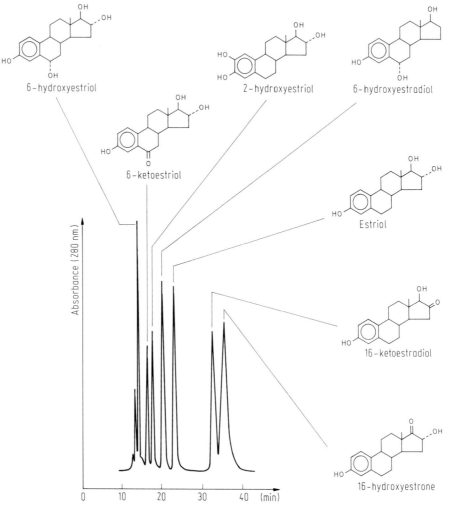

Fig. 5. Reversed phase partition chromatogram of the less polar estrogens. J.-T. Lin, E. Heftmann, J. Chromatogr. 212, 239 (1981). Mph: Acetonitrile-water = (25:75); C: Zorbax BP-ODS (250 × 4.6 mm ID)

The literature reveals that, in contrast to most other pharmaceutically important classes of compounds, the steroids are often separated on silicic acid columns. Nevertheless, reversed phase systems have become established for the routine analysis of steroids. Rehm and Steinigen[71] used reversed phase in a serial investigation of the corticoid contents of 69 creams, sprays, lotions, emulsions, foam sprays, gels and tinctures. For the UV detection of steroids a wavelength of 240 nm is normally used. The use of 280 nm and 210 mm (for estrogens) and 265 nm (for vitamin D) have also been reported. In each case a variable wavelength detector should be used, so that maximal sensitivity can be achieved by monitoring at the UV maximum.

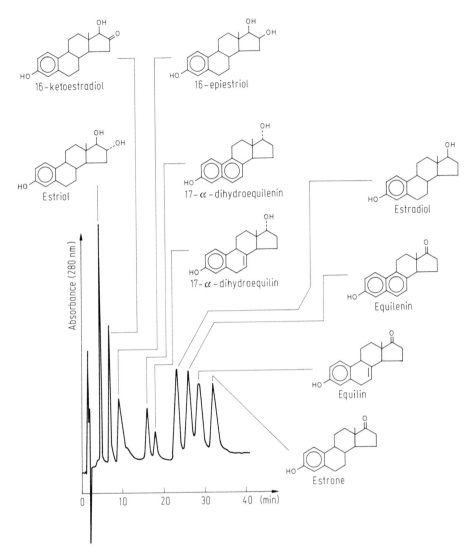

Fig. 6. Reserved phase partition chromatogram of the more polar estrogens. J.-T. Lin, E. Heftmann, J. Chromatogr. 212, 239 (1981). Mph: Acetonitrile-water = (35:65); C: Zorbax BP-ODS (250 × 4.6 mm ID)

The suitability of HPLC for the analysis of very complicated steroid mixtures is well exemplified by the investigations of Lin and Heftman[72] on the naturally occurring estrogens; separations were performed on C18 reversed phase with varying concentrations of acetonitrile in water (cf. Figs. 5 and 6).

In addition to the sex hormones, the steroids also include other pharmaceutically important groups of compounds, such as the corticoids, the various derivatives of vitamin D and the cardiac glycosides. These are later to be discussed in the chapter on natural products.

Table 7. HPLC of steroids

Compound(s)	Ref.
Corticoids	71, 73, 74)
Cardiac glycosides	see chapter on natural products
Sex hormones	72, 76, 77, 75)
Steroid glucuronides	78, 79, 253)
Vitamin D	see chapter on vitamins

Some pharmaceutically important literature sources on the HPLC separation of steroids are listed in Table 7.

Relationships between the structure and retention properties of synthetic, pharmaceutically employed steroids were investigated by Hara and Hayashi[80] on silicic acid and on chemically bound phases. Similar studies have been published by our group on naturally occurring cardiac glycosides[81].

Similar selectivity problems have been studied by Van der Wal and Huber[253] in connection with the HPLC analysis of steroid conjugates.

The work of Johnston[77] is of special pharmaceutical interest, as it describes a screening method for the detection of sex hormones contained in contraceptive preparations.

2.2.6 Sulphonamides

Today the HPLC separation of sulphonamides is performed largely by reversed phase with ion pair forming agents.

As early as 1976, Su, Hartkopf and Karger[82] studied the parameters of ion pair separation on silicic acid. They used butanol-heptane mixtures as mobile phases, and buffered the silicic acid with aqueous solutions of tetrabutylammonium sulphate. Figure 7 shows the isocratic separation of 13 important sulphonamides in this system.

At the same time, Johansson and Wahlund[83] investigated the ion pair mechanism for sulphonamides on reversed phase. Wahlund's group used a method similar to that already described[61] for the separation of antidepressants, i.e. a reversed phase coated with 1-pentanol or butyronitrile, and elution with aqueous solvents containing tetrabutyl-ammonium.

With the use of standardized ion pair forming reagents (e.g. the PIC series; Waters Assoc.) the problem free separation of sulphonamides is now possible, as evidenced by the separation shown in Fig. 8. The retention times of later eluted compounds are shortened by programming the flow rate, or by the use of solvent gradients.

Separation systems other than the ion pair technique have been suggested, but they no longer have practical importance, largely on account of their complexity or lack of resolving power. However, they do represent important stages in the development of the methodology of sulphonamide separation. Thus, Jennings and Landgraf[84] described separations on silicic acid with a five-component solvent (trimethylpentane + chloroform + propan-2-ol + acetonitrile + glacial acetic acid,

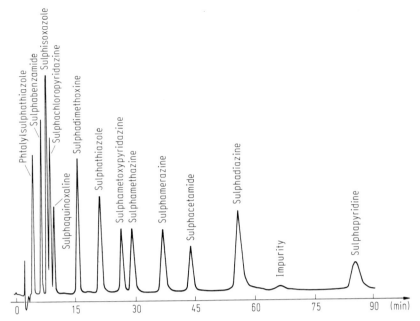

Fig. 7. Separation of sulphonamides. S. Su, A.V. Hartkopf, B.L. Karger, J. Chromatogr. 119, 523 (1976). Mph: n-butanol:n-heptane = (25:75); C: LiChrospher SI 100, 10 μm (250 × 3.2 mm ID), coated with 0.3 M Tetrabutylammoniumhydrogensulphate in 0.1 M buffer, pH 6.8; D: Photometer at 254 nm wavelength

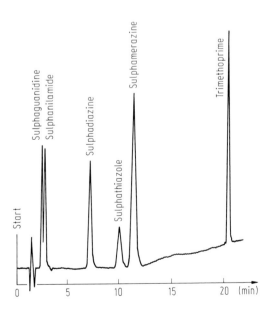

Fig. 8. Separation of some sulphonamides using a reserved phase/ion pair-system. C: MN-Nucleosil −C18-5 (125 × 4 mm ID); Mph: A: Acetonitrile-water-PIC B 6$^+$ = (500:500:10); B: Acetonitrile-water-PIC B 6$^+$ = (100:900:10); Gradient: 5 min isocratic Mph B, then in 20 min to 80% A (linear); D: Spectrophotometer at 235 nm wavelength; +: Trademark Waters Assoc. (Milford – USA)

80:9.8:4.7:4.7:0.8). Other research groups studied the suitability of different stationary phases. Thus, Umagat et al.[85] used a CN phase with a non-aqueous solvent,
while Rotsch et al.[86] investigated the behaviour of sulphonamides on porous copo-
lymers (XAD 2, XAD 7).

A reversed phase with an aqueous solvent has the advantage that methanolic
or aqueous methanolic extracts of pharmaceutical preparations can be chromatographed directly[87].

Using 10 cm columns of 1 mm internal diameter, with 5 μm – RP 18 as the
stationary phase, a good low dispersion separation of sulphonamides is possible[88].

2.2.7 Other Classes of Pharmaceutical Compounds

So far, there have been three literature reviews for the HPLC of synthetic pharmaceuticals:

Wessely and Zech[89]: 1974–1977
Gilpin[90]: 1976–1978
Gilpin, Pachala and Ranweiler[91]: 1978–1980

The reviews of Gilpin et al. are concerned generally with the analysis of pharmaceuticals, and are not restricted to HPLC.

Examples of the HPLC analysis of various pharmaceuticals are summarized
below:

Taylor and Gaudio[92] analysed *bis-substituted aminoalkyl-anthraquinones* (antitu-
mour compounds) on μ-Bondapak NH_2 and CN columns.

Das Gupta and Sachandani[93] used reversed phase for the separation of *antipyr-
ine* and *benzocaine* in ear drops.

Calcium antagonists are increasingly used besides beta blockers and nitrates in
the therapy of coronary diseases:

HPLC analysis of the six most important soluble calcium antagonists has been
reported by Eiden and Greske[94]. The chromatographic separation of the members
of this chemically heterogeneous group of compounds is especially important in
pharmaceutical formulations containing combinations of active substances. Separation is performed on an amino phase with chloroform as the mobile phase. An
example is shown in Fig. 9.

Das Gupta[96] also used reversed phase for the separation of benzoic acid and
salicylic acid in ointments. The specific H_2-receptor antagonist *cimetidine* can also
be determined in such systems[97].

Mannan et al.[98] used fluorimetric detection for the chromatographic analysis
of the antileprotic drug *dapsone* with a lower detection limit of 10 pg.

Separation of *epinephrine* in pharmaceutical preparations on a cation exchange
column by Fu and Sibley[99] represents a departure from the customary reversed
phase method.

There have been various proposals for the HPLC analysis of *isoniazid* and mixtures containing other tuberculostatic agents; both reversed phase and silicic acid
columns are used[100–102].

The HPLC analysis of *insulin, glucagon* and *somatostatin* was described by
Fischer et al. in 1978[103]. For other protein separations see the chapter on HPLC
in clinical chemistry.

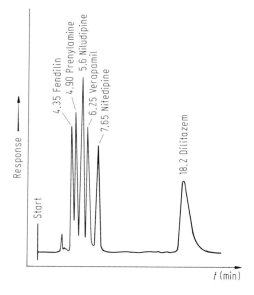

Fig. 9. Separation of the six most important, soluble calcium antagonists. K. Braatz-Greske, Dissertation Uni München (1982). C: μ-Bondapak NH$_2$ (300 × 3.9 mm ID); Mph: Chloroform; F: 1.0 ml/min; D: Photometer at 254 nm wavelength

An RP 2 phase has been proposed for the analysis of *lidocaine* in pharmaceutical preparations[104].

Propanolol and related cardioactive compounds have been separated on RP phases with the addition of ion pair forming agents to the mobile phase[105, 106].

The earliest work on the separation of *purines* (caffeine, theobromine and theophylline) used silicic acid with an ethanol-water-dichloromethane mixture as the mobile phase (Wildanger, 1975)[107]. In the meantime, this complicated separation system has been replaced by ion pair chromatography on reversed phase.

Purines like alloxanthine, hypoxanthine, allantoin and allopurinol can be separated by a methanol-water mixture on C18 solid phase[108].

The HPLC separations described so far are largely for underivatised compounds on silicic acid or reversed phases. Analytical techniques have been reported, however, which involve chemical reaction of the sample before or after separation, or modification of the chromatographic phases. Thus, the detection sensitivity may be increased by pre- or post-column derivatization[109-112], the stationary phase may be impregnated with silver nitrate[113] or silver nitrate may be added to the mobile phase[114], polyamide may be used as a stationary phase[115] and detection may be performed by the measurement of radioactivity[116]. All these techniques belong essentially to the early developmental stages of HPLC. With the exception of radioactivity detection and the post column derivatisation of amino acids, they no longer have any practical significance.

UV detection of pharmaceutically interesting compounds that lack a chromophore absorbing in the long wave UV presents a special problem in the development of routine methods of UV detection. In this connection, Van der Wal and Snyder[117] have investigated the possibility of using a detection wavelength of 185 nm. To exemplify the technique, these authors studied the chromatographic detection of polyethylene glycosides, which are used pharmaceutically as additives. These studies open up new perspectives in the HPLC analysis of additives.

3 Constituents of Plant Drugs

3.1 Introduction

HPLC detection and separation of compounds in plant extracts is incomparably more difficult than the corresponding analysis of pure synthetic pharmaceuticals in medicinal preparations.

Extracts of plant drugs are very complicated mixtures, usually containing widely varying concentrations of different classes of compounds and active principles.

The HPLC analysis of such extracts is subject to interference by the accompanying ballast substances, pigments, etc. Characterization of reference compounds (i.e. isolated or synthetic natural products) by HPLC is usually very satisfactory, whereas the separation of the same compounds in the initial crude plant extract is usually unsuccessful. In addition the establishment of the actual chromatographic parameters, the purification or enrichment of compounds in plant extracts is therefore always an important part of the analytical procedure.

3.2 Alkaloids

The basic nature of alkaloids generally necessitates the use of an alkalinized mobile phase for their HPLC separation. In earlier times, all mobile phases therefore were made alkaline with alkaline buffers, ammonium carbonate solution, or di- or triethanolamine, or by the direct addition of ammonia. In a review of the HPLC of

Table 8. HPLC separations of alkaloids

Plant genus	Constituents	Stationary phase	Ref.
Belladonna	Tropane alkaloids	Silicic acid	[124, 161]
Cactus	Phenethylamines, Tetra-hydroisoquinolines	Silicic acid	[125]
Cannabis	THC and derivatives	RP	[126 – 128]
Catharanthus	Vincristine, Vinblastine etc. (Eburane alkaloids)	RP	[129 – 134]
Chelidonium	Chelidonin etc.	RP	[268, 269]
Chondrodendron	Tubocurarine etc.	RP	[133]
Cinchona	Quinine alkaloids	Silicic acid	[254]
Colchicum	Colchicine, Colchicoside	RP	[135, 136, 223]
Maytenus	Maytansinoids	Silicic acid	[137]
Papaver	Opium alkaloids	CN/RP/NH$_2$	[138 – 141, 270 – 272]
Peganum harmala	Harman alkaloids		[142, 265]
Petasites	Pyrrolizidine alkaloids	Phenyl	[143]
Pilocarpus	Pilocarpine etc.	RP	[162 – 165]
Senecio	Pyrrolizidine alkaloids	CN/RP	[144, 145]
Symphytum	Pyrrolizidine alkaloids	NH$_2$/CN/RP	[146 – 149]
Secale	Ergot alkaloids	NH$_2$/RP	[150 – 157]
Strychnos	Strychnine etc.	Silicic acid	[158, 159]
Tussilago	Pyrrolizidine alkaloids	CN	[149]

RP = reversed phase; CN = cyano-phase; NH$_2$ = amino-phase; Phenyl = phenyl-phase

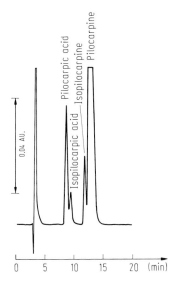

Fig. 10. Separation of pilocarpine and related compounds. J.M. Kennedy, P.E. McNamara, J. Chromatogr. 212, 331 (1981). C: μ-Bondapak Phenyl (300 × 3.9 mm ID); Mph: 5% aqueous solution of potassium dihydrogen orthophosphate adjusted to pH 2.5 with orthophosphoric acid; F: 1 ml/min; D: Spectrophotometer at 215 nm wavelength

natural products (1974–1978), published in 1979, Kingston[160] pointed out, however, that silicic acid and chemically modified phases are unstable under alkaline conditions. Thus, today, alkalinization has been replaced by the addition of ion pair forming reagents to the mobile phase. For quarternary alkaloids or nitrogen oxides it may be sufficient to use the amino phase, which possesses weak cation exchange properties. As is apparent from the review by Kingston[160], all the usual stationary phases of HPLC have been tried at one time or another for the separation of alkaloids. More recent analysis largely employ the C18 reversed phase, or the amino or cyano phase. Table 8 gives a list of published HPLC studies on the separation of the most important alkaloid groups.

The earliest fundamental investigations of the HPLC of alkaloids were reported by Verpoorte and Bearheim-Svendsen[254/255].

Pilocarpine, an alkaloid from *Pilocarpus jaborandi,* is frequently used as a pure substance in the preparation of eye drops. Earlier separations on cation exchangers[162, 163] were insufficiently sharp, and the detection by differential refractometry was lacking in sensitivity[164]. However, a practicable method now has been described by Kennedy and McNamara[165], involving a phenyl phase with spectrophotometric detection. The same method also detects the degradation products of pilocarpine, which is of great importance for pharmaceutical quality and stability control (cf. Fig. 10).

3.3 Bitter Principles

Analysis of bitter principles depended hitherto on the inaccurate and poorly reproducible organoleptic determination of indices of bitterness. HPLC permits the exact quantitative determination of individual bitter tasting compounds, thus representing a great advance in the analysis of this chemically most diverse group of pharmaceuti-

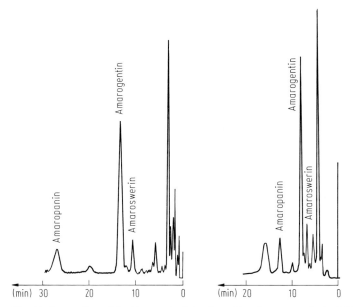

Fig. 11. Separation of the bitter principles from Gentiana purp. L. O. Sticher, B. Meier, Pharm. Act. Helv. 53, 40 (1978). Mph: Methanol – water = (40:60); Mph: Methanol – water = (50:50); F: 2.0 ml/min; F: 1.0 ml/min; D: Spectrophotometer at 233 nm wavelength; C: μ-Bondapak C18 (300 × 3.9 mm ID)

cally important substances. Investigations have been reported so far on the following four groups of plant bitter principles:

Gentiana bitter principles have been separated by Sticher and Meier[166] using a C18 reversed phase with methanolic solvent mixtures (cf. Fig. 11).

The α- and β-acids from hop resin, the so-called *hop bitter principles,* become rapidly modified by oxidation in the dried plant drug. Therefore, a rapid HPLC method for quality control of the drug is needed. For this purpose, Verzele and De Potter[167] reported an HPLC method which separates humulone and lupulone together with their early conversion products, co- and adhumulone and co- and adlupulone.

Johnson, Gloor and Majors[168] proposed coupled column chromatography (CCC) for the analysis of *limonin,* the *bitter principle* from *grapefruit* peel. They employed a first column for the enrichment of limonin by exclusion chromatography. The bitter principle was then determined by reversed phase HPLC. For the detection of limonin in grapefruit juice, Fisher[169] used a CN column and a detection wavelength of 210 nm.

In our laboratory we have investigated the determination of *quassinoids* from the bark drugs *Quassia amara* and *Picrasma excelsa*[170]. A silicic acid column with a dichloromethane-methanol mixture as the mobile phase was found to be the most suitable system for the determination of commercial quassinoid mixtures and extracts from quassia wood (cf. Fig. 12).

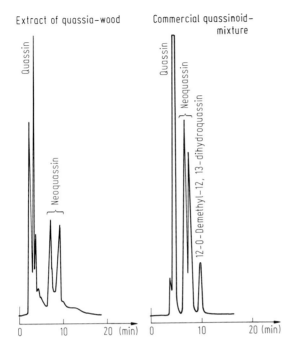

Extract of quassia-wood   Commercial quassinoid-mixture

Fig. 12. HPLC separations of quassinoids. Th. Nestler, G. Tittel, H. Wagner, Planta med. 38, 204 (1980). Mph: Dichloromethane-methanol = (98:2); C: Lichrosorb SI 100-10 μm (300 × 4 mm ID); F: 1.0 ml/min; D: Photometer at 254 nm wavelength

For the analysis of *naringin* and *naringenin,* the bitter principles from grapefruit, see Ref. 228.

3.4 Cardiac Glycosides

The earliest HPLC study of cardenolides was reported in 1974 by Evans[171]. Since then, the majority of publications on the HPLC of cardiac glycosides have been concerned with Digitalis glykosides. The first stationary phases to be used were ion exchangers. Silicic acid was introduced later. Finally, these were superseded by reversed phases[171–191]. Some authors have also reported derivatisation methods as a mean of increasing the sensitivity of detection[110].

Shimada et al.[180, 181] reported the separation of toad bufadienolides, while the separation of bufadienolides from Scilla has been described recently from our laboratory[182] and in a communication from the research group of Kubelka[183]. In addition to the HPLC of cardiac glykosides from Digitalis, corresponding studies also have been reported for Convallaria leaves[183, 190, 191], Oleander leaves[185] (cf. Fig. 13), Adonis[187] and Strophanthus seeds[187].

The relationship between the cardenolide constituents of the milkweed plant and the occurrence of cardenolides in the monarch butterfly has been studied by Enson and Seiber[193].

Present experience shows that the most suitable HPLC system for cardiac glykosides is C18 reversed phase in a solvent gradient, with UV detection. Cardenolides and bufadienolides are detected selectively at 220 nm and 300 nm, respectively.

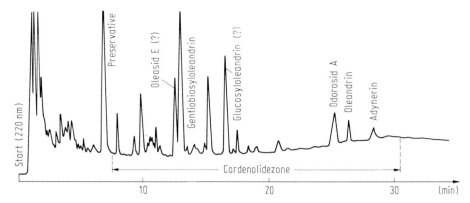

Fig. 13. HPLC-separation of cardenolides in Nerium oleander extract. G. Tittel, H. Wagner, Planta med. 43, 252 (1981). x, y, z: unidentified cardenolides; C: μ-Bondapak C18 (300 × 3.9 mm ID); Mph: Acetonitrile-water-gradient; F: 2 ml/min

Acetonitrile-water mixtures are suitable as mobile phases. The order of elution of cardenolides shows certain variations, depending on the stereochemistry of the cardiac glykoside, and on the degree of coating or residual silanol group content of the stationary phase [187].

3.5 Iridoids

Iridoid natural products occur as glycosides (iridoid glykosides, secoiridoid glyko-sides, etc.) as well as esters. Fundamental studies on the HPLC separation of these compounds were carried out by Meier and Sticher [194]. They used a C18 phase with methanolic solvent mixtures. In an extension of their work, they also reported the determination of harpagoside in *Harpagophytum procumbens* [195].

HPLC of the valepotriates (iridoid epoxide triesters from *Valeriana spp.*) was first investigated in 1978 in our laboratory [196]. There are two types of valepotriates with differing chromogenic properties, i.e. valtrate and related compounds (λ_{max} 256 nm) and didrovaltrate and related compounds (200–220 nm). In our first studies we therefore used a fixed wavelength photometer and a differential refractometer in tandem for the detection of all valepotriates in one run. The stationary phase was silicic acid [196/197] (cf. Fig. 14a).

As shown in later studies, all the valepotriates can be detected using a C18 reversed phase column with methanol-water-mixtures, and spectrophotometric detection at 220 nm wavelength.

Apart from the valepotriates, official baldrian is also characterized by the presence of valerenic acids and valerenal. Hänsel und Schulz have achieved the HPLC separation of the valerenic acids on reversed phase [266]. Freytag separated the valerenic acids, valerenal and the valepotriates on a C18 reversed phase column using gradient elution [263].

Our latest work on this subject shows, that the triwavelength detection at 220, 256 and 425 nm, using a diode array detector gives a very selective and hight sensitive

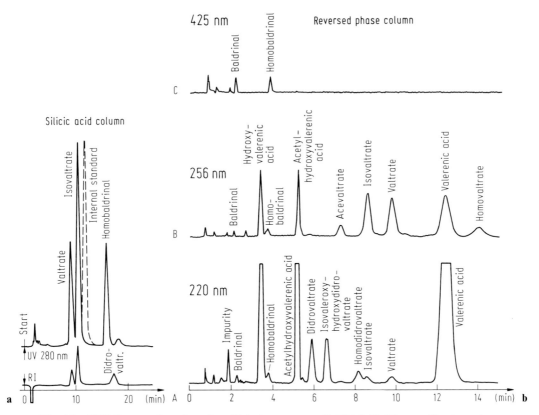

Fig. 14. a HPLC separation of mexican Valeriana extract. G. Tittel, H. Wagner, Planta med. 34, 305 (1978). **b** HPLC separation of valeriana substances (References) at different wavelengths. G. Tittel, R. Bos, to be published. C: MN-Nucleosil 5 C18 (125 × 4 mm ID); Mph: Methanol-water = 750:250; F: 1.0 ml/min

quantitative information about the composition of all important substances in valeriana products within one single separation[267]: The valerenic acids and the valepotriates of didrovaltrat-type are detected at 220 nm, the valepotriates of valtrat-type at 256 nm and the degradation products baldrinal and homobaldrinal at 425 nm (cf. Fig. 14b).

3.6 Plant Phenolics

The first important paper dealing with the separation of polyphenols by HPLC was published in 1976 by Wulf and Nagel[198]. These authors used ternary mixtures of methanol, water and acetic acid on a C18 reversed phase column to separate flavonoid aglycones and glykosides.

Below is listed one important separation method for each group of polyphenolics, selected from the great number of papers dealing with the HPLC of these compounds:

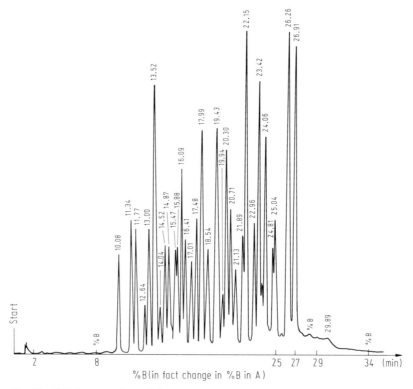

Fig. 15. HPLC separation of flavonoids. K.V. Casteele, H. Geiger, Ch.F. Van Sumere, J. Chromatogr. 240, 81 (1982)

3.6.1 Flavonoids and Related Compounds

HPLC has been used for the separation of *anthocyanidins* (Adamovics and Stermitz[199]), *xanthones* (Hostettmann et al.[200-202]), *isoflavones* (Carlson and Dolphin[203]). *procyanidins* (Lea[204/205]) and *tannins* (Okuda et al.[206]). We have published on the HPLC of *flavanolignans*[207/208] and on the separation of flavonoids in different drug preparations of Crataegus[209]. Furthermore, we have used HPLC for the standardization of flavonoids in pharmaceuticals[210].

Recently Daigle and Conkerton[211] reported retention data for 34 flavonoids in a C18 methanol-water-acetic acid HPLC system.

Casteele, Geiger and Van Sumere[212] published retention data for more than 100 flavonoids on C18 with methanol-formic acid mixtures (see Table 9) (cf. Fig. 15).

Polymethoxyflavones were separated by Rouseff and Ting[213] on a reversed phase column, while Bianchini and Gaydou[264] proposed the use of silicic acid. Own experiences show that the amino-phase is very effective as a stationary phase for these separations.

A review of these publications shows that the most widely used separation system is C18 reversed phase with acidified methanol-water or acetonitrile-water mix-

Table 9. Legend to Figure 15 in [212]

Compound	t_g (min)	Compound	t_g (min)
Kaempferol-3-sophoroside-7-glucoside	10.08	Quercetin	18.54
Quercetin-3,7-diglucoside	11.34	Genistein	19.43
Kaempferol-3-sinapoylsophoroside-7-glucoside	11.77	Kaempferol-7-rhamnoside	19.94
	12.64	Kaempferol	20.30
Kaempferol-3,7-diglucoside	13.00	Apigenin	20.71
Kaempferol-3-rutinoside-7-glucoside	13.52	3′,4′-Dimethoxy-7-hydroxy-	21.13
Quercetagitrin		flavone	
Tricetin-7-glucoside	14.04	7-Hydroxyflavone	21.89
Luteolin-3′,7-diglucoside	14.52	Pinocembrin	22.15
Myricitrin	14.87	Tricetin-3′,4′,5′-trimethyl ether	22.96
Hyperosid	15.47	Chrysin	23.42
Isoquercitrin	15.88	5-Methoxyflavone	23.73
Hesperidin	16.09	Kaempferid	24.06
Neohesperidin	16.41	2′-Methoxyflavone	24.81
Astragalin	17.01	4′-Methoxyflavone	25.04
Luteolin-5-methyl ether	17.48	5-Hydroxyflavone	26.26
Daidzein	17.99	Tectochrysin	26.91

Chromatographic parameters:

C: LiChrosorb® RP18 (250 × 4 mm ID), 10 μm
F: 2.5 ml/min
Mph A: Formic acid – water = 5:95
Mph B: Methanol

Gradient: 0– 2 min: 7% B in A (isocratic),
         2– 8 min: 7–15% B, linear
         8–25 min: 15–75% B, linear,
      25–27 min: 75–80% B, linear,
      27–29 min: 80% B isocratic.

Temperature: 35° C

tures[210]. Silicic acid has no real application in the separation of underivatized flavonoids.

The acid used for acidification of the mobile phase may be acetic, formic, phosphoric or perchloric. Fixed wavelength photometers (280 nm) or spectrophotometers are used for detection (see Table 10).

The presence of acid in the mobile phase is necessary for reversed phase HPLC as it is for TLC on silicic acid. With non-acidified solvents, the peaks become broadened and considerable tailing is observed. Increased acid concentration causes increased k′ values and hence longer retention times. The α-values therefore must be optimized by choosing a concentration of acid that decreases peak broadening without causing an unacceptable decrease in resolution (α). HPLC separations of substances with many free phenolic groups (e.g. procyanidins) require fairly high concentrations of acid in the mobile phase.

The use of perchloric acid for the HPLC of procyanidins was first described by Lea[204, 205].

Besides the acids given in Table 10 no others are normally used. The halogen acids in particular cause corrosion of the HPLC apparatus.

Table 10. HPLC systems for separation of polyphenols

Stationary phase	Mobile phase	Detection
C18 reversed phase C2 and C8 reversed phase Amino phase Cyano phase (Silicic acid)	Mostly ternary mixtures: *Component A: Methanol Acetonitrile* Tetrahydrofuran Dioxan Ethanol *Component B:* Distilled water *Component C: Acetic acid* Perchloric acid Phosphoric acid Formic acid	Photometer at fixed wavelength (280 nm) Spectrophotometer (320, 340, 366 nm, etc.)

C18 phases show varying selectivities for flavonoids, as well as for other products, depending on the supplier.

An example of this variation can be quoted from our own studies with a mixture of vitexin, vitexin-2″-rhamnoside and rutin:
All three compounds are completely separated on LiChrospher® CH 18-2 or MN-Nucleosil-C18-5. Using the same chromatographic parameters on μ-Bondapak C18, vitexin and vitexin-2″-rhamnoside migrate together in one peak. If a μ-Bondapak C18 column is conditioned with isopropanol, then all three compounds are again completely separated. Earlier production batches of μ-Bondapak C18 columns did not need conditioning with isopropanol.

A suitable choice of stationary C18 phase and the appropriate acetic acid concentration for the polarity of the mobile phase are particularly important in polyphenol analysis in order to obtain sharp peaks and the desired selectivity of separation.

One practical application of flavonoid separation by HPLC is the standardization of Camomile extract with respect to its content of apigenin and its glycosides[214].

3.6.2 Lignans

Lignans are derived biosynthetically from two phenylpropane units. Two members of this group are especially important, i.e. the podophyllotoxins which have cancerostatic properties and the eleutherosides from *Eleutherococcus senticosus*. The chief constituents of Podophyllum resin, podophyllotoxin and peltatin, as well as other lignans were determined on RP phase by Cairnes et al.[215]. These authors used acetonitrile in the mobile phase and consequently obtained only one peak for podophyllotoxin. Only recently, however, it was shown by Lim and Ayres[216] that podophyllotoxin is separated into 7 diastereoisomers by methanolic solvents. The diastereoisomers, together with other lignans, are likewise well separated by ternary solvent mixtures containing acetonitrile and methanol. A similar phenomenon was observed in our laboratory in separating flavonolignans from *Silybum marianum*, first described by us in 1977[207, 208]; resolution of the diastereoisomers of silybin and isosilybin is possible only in methanolic solvent systems, and not with acetonitrile or dioxan. These results show that the selectivity of reversed phase chromatography

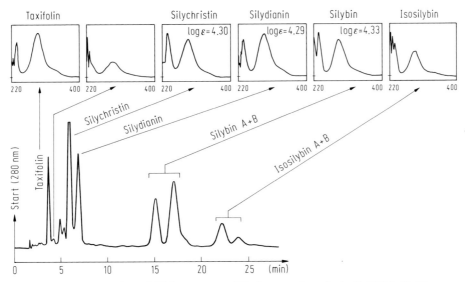

Fig. 16. HPLC separation of Silybum extract (Diode-array detection). G. Tittel, H. Wagner, J. Chromatogr. 135, 499 (1977)

depends not only on the degree of coating and the number of residual silanol groups on the stationary phase, but also on the nature of the mobile phase. Good separations of diastereoisomers therefore can be achieved with appropriate HPLC methodology.

HPLC of the liver-protective plant drug Silybum, separates the main active constituent silybin, as well as the other silymarins (silychristin, silydianin and isosilybin) and taxifolin (a flavanonol)[210] (cf. Fig. 16).

As shown by the use of a diode-array spectrophotometer, all the pharmacologically active flavanolignanes in the extract exhibit the same UV spectrum. Since the log ε values are also identical, silychristin, silydianin and isosilybin A and B can be determined quantitatively with silybin as the external standard. No further reference compounds are necessary, and the method is well suited for routine analysis[272].

Samples of *Eleutherococci radix* from different sources can be differentiated on the basis of the HPLC fingerprints of authentic *Eleutherococcus lignans* (eleutherosides)[218]. Such a facility is important for the assessment of drug quality. This reversed phase separation employs a solvent gradient.

3.6.3 Anthraquinone Derivatives

With the exception of a series of investigations on ion exchange columns[219], all published separations of anthraquinones use reversed phases. Auterhoff, Graf, et al. showed that the diastereoisomers of aloin from aloe resin can be separated by HPLC[220]. They later used this method to study the degradation of aloe extracts[221].

Frangula extracts and Frangula formulations were separated by Bonati and Forni[222] on a C8 column by gradient elution with a citrate buffer-acetonitrile mixture.

Gradient elution with the addition of buffer was also used by Hayashi et al.[224] in the analysis of Senna, whereas Komolafe[225] and Görler, Mutter and West-phal[226] proposed an ion pair separation system for the separation of anthraquinones from Senna.

Our own studies with different anthraquinone drugs and pharmaceutical preparations show that ion pair separation with PIC B-reagent[1] gives reproducible analysis[227]. All the important anthraquinones from Rheum, Aloe, Senna and Frangula can be resolved in an acetonitrile-water-PIC B gradient system; 290 nm is a suitable detection wavelength.

3.7 Saponins and Sapogenins

Saponins do not represent a chemically homogeneous group of compounds. Since most of them contain very weak chromophores, they can only be detected with very high performance spectrophotometers at wavelengths between 190 and 210 nm. Alternatively, a differential refractometer can be used for the determination of steroid sapogenins like *hecogenin, tigogenin, diosgenin, stigmasterol* and *cholesterol*. This is only possible, however, at very high saponin concentrations and in highly enriched solutions, so that the method is of no use for crude plant extracts. Higgins[229] reported the separation of similar sapogenins from Agave on RP 8 phase.

The saponins most thoroughly studied by HPLC are the active principles of ginseng, the *ginsenosides*. In 1979, Besso et al.[230] reported an HPLC separation of derivatized ginsenosides on silicic acid, whereas Sticher and Soldati[231, 232] reported the direct detection of the underivatized compounds, prepared for chromatography by extraction with butanol from methanolic extracts of the drug (cf. Fig. 17).

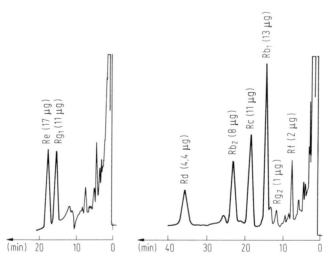

Fig. 17. HPLC separations of a standardized Ginseng extract (ginsenoside). H.R. Schulten, F. Soldati, J. Chromatogr. 212, 37 (1981). Mph: Acetonitrile-water = (18:82); Injection volume: 12 µl; F: 2.0 ml/min; Mph: Acetonitrile-water = (29:71); Injection volume: 7 µl; F: 2.0 ml/min; D: Spectrophotometer at 203 nm wavelength; C: µ-Bondapak C18 (300 × 3.9 mm ID)

---

1 Trademark Waters Assoc.

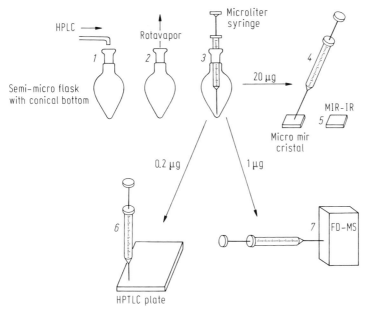

Fig. 18. Off-line methods for identifying substances separated by HPLC. H.R. Schulten, F. Soldati, J. Chromatogr. 212, 37 (1981)

Identification of the ginsenoside peaks in the HPLC separation was at first only possible from the retention times and by means of UV spectra taken in a stop-flow procedure. Later, however, Schulten and Soldati[233] employed an off-line method with micro-IR, FD-MS and HPTLC for identification purposes. In the meantime, on-line recording of UV spectra has become possible, using the photo-diode array spectrophotometer. A combination of such off-line procedures and on-line procedures enables an accurate identification of the separated peaks (cf. Fig. 18).

Kimata et al.[234] used absorption at 254 nm for determination of the triterpenoid diene-saponins from *Bupleuri radix*.

All hitherto mentioned saponin separations were optimized on RP phases. Lin, Nes and Heftmann[235], however, reported good separation of Soya sapogenins on silicic acid columns.

3.8 Further Classes of Natural Products

The possibly carcinogenic and therefore controversial *aristolochic acids* occur in several stereoisomeric forms. Good separations of these compounds could be achieved on a C18 phase in solvents containing acetic acid and methanol[236].

Using a reversed phase system, α-bisabolol can be determined in camomile flowers[237]. Thus, HPLC supplements the present GC methods for quality control of camomile by analysis of the essential oil. *Arbutin drugs* long have been used for the treatment of urinary infections. Methods for their determination (arbutin, methylarbutin, hydroquinone, etc.), which differ only in the percentage of methanol

Table 11. Ion pair HPLC of cardiotonic amines[240]. Compounds are listed in the order of their migration in TLC

	R_{t1} (min)		
1. Catecholamines			
Epinephrine	9.0		
Norepinephrine	7.0		
Dopamine	15.7		
Higenamine	31.2		

	R_{t2}	R_{t3}	R_{t4} (min), Remarks
2. Amines with a predominantly basic character			
Synephrine	5.2	–	–
Phenylephrine	5.1	–	–
Tyramine	6.4	7.3	4.2
3,4-Dimethoxy-β-phenethylamine	10.7	16.4	–
Serotonin	–	–	4.4
Norpseudoephedrine	–	19.0	5.8 Detection UV 254 nm
Ephedrine	–	20.1	6.2 Detection UV 254 nm
p-Methoxy-β-phenethylamine	18.8	23.4	6.6
Tryptamine[a]	–	–	6.5
β-Phenethylamine	–	20.6	6.6 Detection UV 240 nm
o-Methoxy-β-phenethylamine	23.8	32.2	7.7
Tetrahydroisoquinoline[a]	–	–	5.7
Carnegin	12.5	–	–

	R_{t1}	R_{t2}	R_{t3}	R_{t4}
3. Separation parameters				
Column: C18 μ-Bondapak	4×300 mm	4×300 mm	7.9×300 mm	7.9×300 mm
Solvent: $H_2O - MeOH +$	90/10	70/30	70/30	50/50
PIC B7	2 ml/min	1.8 ml/min	4 ml/min	4 ml/min

Detection: UV 280 nm, range 0.04, when no other wavelength stated.

[a] These amines have no cardiotonic activity

in the mobile phase, were developed simultaneously by the research groups of Kraus[238] and Sticher[239].

Sticher applied the method to Arctostaphylos and to species of Vaccinium, Calluna and Bergenia.

Literature on the HPLC analysis of *biogenic amines* has been reviewed by Grevel[240], together with his own method of ion pair chromatography on μ-Bondapak C18. Grevel gives the following table of retention properties for 17 cardiotonic amines (cf. Table 11):

The *diterpene esters* (phorbol esters) in pharmaceutically important members of the Euphorbiaceae are potentially carcinogenic. Detection of these esters in croton oil was reported by our research group on C18 reversed phase with an acetonitrile-water gradient (cf. Fig. 19)[244]. Some of the resulting peaks contained phorbol ester mixtures, which are not separable on C18 phases. We were able to separate such mixtures by using a mixed phase of equal parts of RP2 and RP8 (ref. Fig. 20);

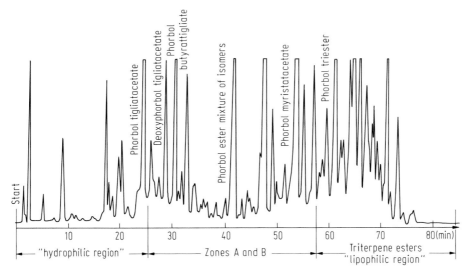

Fig. 19. HPLC. HPLC separation of Croton oil. R. Bauer, G. Tittel, H. Wagner, Planta med. 48, 10 (1983). C: μ-Bondapak C18 (300 × 3.9 mm ID); F: 2.0 ml/min; D: Spectrophotometer at 220 nm wavelength; Mph: A: Acetonitrile-water = (20:80); B: Acetonitrile; Gradient: 0–100% B in 72 min, linear

under these conditions, the shorter lipophilic chains on the stationary phase facilitate access of the ester molecules to the residual silanol groups. This procedure is important for all compounds that can be separated on silicic acid but not on a C18 phase. A mixed phase, or singly used RP2 or diol-phases have all the known advantages of chemically modified phases over silicic acid.

Curcubitacins, oxygen-rich tetracyclic triterpenes, are the active constituents of certain plant drugs which are used as laxatives and antirheumatic agents, e.g. Bryonia, Colocynthis, Luffa, Gratiola, Ecballium or Cayaponia. All curcubitacin-containing drugs can be analysed by an HPLC system established in our laboratory[241].

An HPLC method for *gossypol,* one of the spermatogenesis-inhibiting polyhydroxyphenols in Gossypium spp., was recently reported by Nomeir and Abou-Donia[242]. They used an RP phase with a methanolic solvent acidified with phosphoric acid.

According to Becker and Bingler-Timeus[243], RP8 phenyl and cyano phases are suitable for the separation of the *γ-benzopyrone derivatives khellin* and *visnagin,* and of *visnadin* from *Ammi visnaga* fruits.

HPLC analysis of the sugar substitute *stevioside* and other constituents of *Stevia rebaudiana* was investigated in detail by the research group of Dobberstein[248–250].

Separation of *terpenes* by HPLC depends largely on the presence of a suitable chromophore in the molecule. As shown by Schwanbeck and Kubeczka[251], HPLC can only serve as a supplement to GC in the analysis of terpenes, since the resolving power is too low in comparison with capillary GC.

One special advantage of HPLC, however, is its different selectivity of separation compared with GC; this is exemplified by our HPLC analysis of "spirit of Melissa"

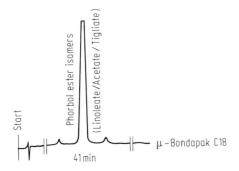

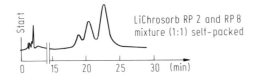

Fig. 20. HPLC separation of a mixture of phorbol esters, eluted on μ-Bondapak C18 as a single peak. R. Bauer, G. Tittel, H. Wagner, Planta med. 48, 10 (1983)

and "carmelite water"[252]. Furthermore, HPLC is a very efficient technique for the determination of phenylpropane derivatives, which accompany the terpenes in essential oils[256]. All HPLC methods reported so far for *phytohormones* rely on RP phases[245/273].

Various stationary phases (nitro, C18, silicic acid) have been described for HPLC of the *pungent principles* from pepper (piperine, etc.)[258–262].

4 Sample Preparation of Pharmaceuticals for HPLC

Pharmaceutical HPLC analysis demands more than just an optimized separation system for reference compounds! Pharmaceutical compounds must be prepared quantitatively from the matrix of the medicinal preparation, dried plant or crude plant extract, and the preparation must be free from accompanying material that may interfere with chromatography. Furthermore, sample preparation must be distinguished from sample enrichment. In the simplest case, a liquid pharmaceutical preparation with a sufficiently high concentration of active constituents can be injected directly into the liquid chromatography system, following membrane filtration. For more dilute preparations, the active constituents must be enriched by partition into an organic solvent. At this stage, the presence of sugars often leads to the production of solutions of very high viscosity. Sugars can be removed with an Extrelut® column, or liquid-liquid extraction in a separation funnel can be avoided by using the Baker-10® extraction system.

For the analysis of tablets, dragees or capsules, solvent extraction with heating is usually necessary (extraction under reflux, soxhlet extraction, etc.). The various inactive pharmaceutical additives then form an insoluble residue, which can be removed by filtration. Since the inactive additives in tablets often have strong adsorbent properties, it is absolutely necessary to monitor the course of tablet extraction.

In the case of ointments, creams, suppositories and other *fat-containing pharmaceutical preparations,* the fats must be largely removed before HPLC. The following procedures may be used:

a) The preparation is dissolved in lipophilic organic solvents, followed by partition against water or water-alcohol mixtures.

b) The preparation is suspended in water, if necessary by sonication, then loaded onto an Extrelut® column. The column can then be washed with lipophilic solvents to remove the fats, and the pharmaceutically active compounds finally eluted with suitable solvents. Similar methods of sample preparation can be performed with the Sep-Pak® or the Baker-10® extraction systems.

c) The preparation is placed on a PLC plate[1], which is developed first with ether, then with an appropriate solvent for the active constituents in question. The zones of active constituents are then eluted.

There is no universally valid procedure for the extraction of fat-containing pharmaceutical formulations. The wide variation in the type and quantity of emulsifying agents, and the type and quantity of fats means that the method of sample preparation must be devised separately for each individual case. The three methods given here serve only as guide. Fundamental studies of this problem have not been published.

Plant drug pharmaceutical preparations can be treated in the same way as all other preparations containing synthetic pharmaceuticals. In addition, a prepurification on Extrelut®, Sephadex® LH20, aluminium oxide, silicic acid or chemically modified silicic acid in small open glass columns is often necessary, because such preparations contain other plant constituents together with the usual pharmaceutical additives.

Essential oils can be separated and prepared for HPLC by steam distillation, solvent extraction with dichloromethane (or similarly polar solvents), elution from a silicic acid column with dichloromethane or elution from an Extrelut® column (for alcoholic solutions).

Alkaloids are simultaneously purified and concentrated by extraction from the matrix by acid-base partition. In many cases, however, a single methanolic extract suffices for the direct determination of alkaloids. Extraction with methanol or methanol-water should be used for *polyphenols.* The polyphenols then can usually be determined directly without further purification.

For the HPLC of *saponins,* a prior enrichment and purification by butanol extraction is absolutely necessary.

Cardiac glykosides can usually be determined directly in the alcoholic extract of the drug or pharmaceutical preparation. If, however, small quantities of cardenolides are accompanied by larger quantities of interfering material (flavonoids, tannins), then a precipitation of tannins with lead acetate is indicated. Since cardiac glykosides are also precipitated with the tannins, this method is problematic if used in quantitative analysis.

Comprehensive studies are still lacking on the potential uses, quantitative aspects and applicability of the different methods of sample preparation from pharmaceutical products for HPLC. Each pharmaceutical product therefore must be investigated

---

1 PLC: preparative layer chromatography

separately with respect to its content of active constituents and additives, in order to design an appropriate method of sample preparation. HPLC provides absolute analytical values. Thus, if the true advantage of HPLC over conventional methods of the pharmacopoeias is to be realized, sample preparation must also be quantitative. In this connection, there recently has been much discussion on the validation of this analytical method which attempts to identify the sources of error between the processes of sample preparation and chromatographic evaluation.

Additives have so far been considered only as sources of interference in the preparation of samples from pharmaceutical formulations. However, the first analyses of additives by HPLC were published recently[274].

5 Examples of the Analysis of Proprietary Pharmaceuticals

The complete HPLC analysis of proprietary pharmaceuticals is very rarely reported in the literature.

A few examples of reported analyses of preparations containing mixtures of synthetic compounds are given below:

An HPLC analysis of common analgesics was published in 1976 (Harzer and Barchet[118]. Comparison of this method with the most modern HPLC procedures, e.g. as applied in 1982 to cough mixtures (Halstead[119]), clearly reveals the improvements in resolving power that have occurred, and shows that HPLC has now acquired the status of a high performance, precise analytical method.

As early as 1977, Gupta and Ghanekar[120] and Ali[257] reported the HPLC analysis of cough mixtures.

Honigberg et al.[121] compared the HPLC of a preparation containing muscle relaxants and analgesics on silicic acid, C18 and RP phases, with and without the addition of ion pair forming reagents. Gimet and Filloux[123] reported the analysis of alkaloids in syrup-based formulations. HPLC analysis of preparations containing synthetic active constituents is nowadays more or less problem-free, whereas the analysis of preparations containing plant extracts and active plant constituents has still not been fully investigated.

From our own experience with pharmaceutical preparations containing cardenolides, bufadienolides, flavonoids or alkaloids, we can state the following:

1) HPLC is at present the most appropriate method for the analysis of complex phytopharmaceuticals, permitting the determination of individual compounds and types of compounds.

2) By monitoring the separation with the diode-array technique, in some cases different extracts can be determined simultaneously, parallel to each other in one chromatogram.

3) Fractional extraction is one of the most suitable methods for sample preparation.

4) Even for very complex preparations, standardization is possible by comparison of the chromatographic fingerprints of different charges with independently prepared laboratory Reference solutions.

6 Summary

HPLC is the most important method of pharmaceutical analysis presently available!

Routine process control, raw material analysis and stability testing are now easily and conveniently performed by this technique, which is also used very successfully in monitoring the course of synthesis of new pharmaceuticals, in the study of bioavailability, in the structural elucidation and identification of unknown natural products and in the standardization of natural products and phytopharmaceuticals.

From a review of the literature it is clear that about 80% of all separation problems can be solved by chromatography on C18 reversed phase. Appropriate mobile phases are based on either acetonitrile-water or methanol-water mixtures; depending on the compounds to be separated, these systems also contain buffer or ion pair forming agents.

The amino and cyano phases and silicic acid are now the only commonly used stationary phases. If the octadecyl phase is chosen, it should be remembered that phases from different suppliers show varying selectivities, especially in the separation of natural products. Silicic acid columns are employed in particular for the separation of these isomeric compounds that are not resolved on reversed phase, and for very lipophilic compounds.

The most appropriate detector is a spectrophotometer. Other detection techniques may be suitable for special applications, but none is universally applicable in HPLC of pharmaceutical compounds.

In the future, the method of choice will be a recording spectrophotometer with diode-array detection. During the chromatography of pharmaceuticals containing different active components and different classes of active principle, information provided by the UV spectrum largely confirms the identity of the separated peaks, and thereby satisfies the requirements of pharmaceutical analysis. For multicomponent preparations containing plant extracts, identification is inconclusive when based only on retention times, or comparison with reference compounds.

Derivatization of the sample before or after HPLC separation as a means of improving detection is not commonly practised in pharmaceutical analysis.

Isocratic elution is almost always adequate for the HPLC analysis of simple mixtures of related synthetic pharmaceutical compounds. Solvent gradients are necessary for complicated mixtures of pharmaceutical compounds of widely varying polarity. Thermostatic control of the column is not normally necessary.

Interlaboratory standardization of separation systems is urgently needed, in order to improve the future comparison of HPLC analyses of pharmaceuticals. The many stationary and mobile phases used earlier are now obsolete. We believe that practically all separations so far performed on "exotic" stationary phases, often with quarternary or even more complicated solvent mixtures, can be performed on three types of variously coated C18 phases and/or silicic acid. Amino and cyano phases are necessary only if special selectivity is demanded by the nature of the analytes.

If value is placed on the efficiency of separation rather than the speed of separation, then a 30 cm column with 10 µm particle size is still suitable for the analysis of simple pharmaceuticals.

Smaller particle sizes are essential only for the analysis of complex mixtures, or isomeric compounds, or compounds that are otherwise chemically closely related. Even for simple routine separations, the "short column" (10–15 cm) with 5 μm packing is now favoured, as it saves time and solvent; an effective argument in times of financial stringency.

Published separation methods are available for all the common pharmaceuticals in use today. In recent years, however, the technological advances in HPLC have been so rapid that, in general, a separation published four years ago as an optimized procedure now appears dated with respect to its speed and efficiency!

7 References

1. Tank, F.W.: Dissertation Uni Marburg (1976)
2. Daltrup, T., Susanto, F., Michalke, P.: Fresenius Z. Anal. Chem. 308, 413 (1981)
3. Rehm, K.D., Steinigen, M.: Pharm. Ztg. 126, 99 (1981)
4. Bailey, F.: J. Chromatogr. 122, 73 (1976)
5. Westerlund, D., Pettersson, B., Carlquist, J.: J. Pharm. Sc. 71, 1148 (1982)
6. Larsen, C., Bundgaard, H.: J. Chromatogr. 147, 143 (1978)
7. Lebelle, M.J., Wilson, W.L., Lauriault, G.: J. Chromatogr. 202, 144 (1980)
8. Sakai, T.T., Riodan, J.M.: J. Chromatogr. 178, 302 (1979)
9. Rzeszotarski, W.J., Eckelman, W.C., Reba, R.C.: J. Chromatogr. 124, 88 (1976)
10. Saki, T.T.: J. Chromatogr. 161, 389 (1978)
11. Hansen, S.H., Thomsen, M.: J. Chromatogr. 123, 205 (1976)
12. Crombez, E., Bossche, W. van den, Moerloose P. de: J. Chromatogr. 169, 343 (1979)
13. Young, M.G.: J. Chromatogr. 150, 221 (1978)
14. Ali, S.L.: J. Chromatogr. 154, 103 (1978)
15. Le Belle, M.J., Young, D.C., Graham, K.C., Wilson, W.L.: J. Chromatogr. 170, 282 (1979)
16. Vigh, Gy., Inczedy, J.: J. Chromatogr. 129, 81 (1976)
17. Granneman, G.R., Senello, L.T.: J. Pharm. Sc. 71, 1112 (1982)
18. Tsuji, K., Goetz, J.F.: J. Chromatogr. 147, 359 (1978)
19. Tsuji, K., Goetz, J.F.: J. Chromatogr. 157, 185 (1978)
20. Koch, W.L.: Analytical profiles of Drug substances, 8, 159 (1979)
21. Tsuji, K., Kane, M.O.: J. Pharm. Sc. 71, 1160 (1982)
22. Anhalt, J.P., Sancilio, F.D., Corcle, T.Mc: J. Chromatogr. 153, 489 (1978)
23. Axelsen, K.S., Vogelsang, S.H.: J. Chromatogr. 140, 174 (1977)
24. Mays, D.L., Apeldoorn, R.J. van, Lauback, R.G.: J. Chromatogr. 120, 93 (1976)
25. Kondo, K.: J. Chromatogr. 169, 329 (1979)
26. Hagel, R.B., Waysek, E.H., Hemphrey, D.: J. Chromatogr. 170, 391 (1979)
27. Srivastava, S.C., Hornemann, U.: J. Chromatogr. 161, 393 (1978)
28. Sorel, R.H., Roseboom, H.: J. Chromatogr. Biomedical appl. 162, 461 (1979)
29. Grubb, P.E.: Analytical Profiles of drug substances, 8, 371 (1979)
30. Tsuji, K., Goetz, J.F., Meter, W. Van, Gusciora, K.A.: J. Chromatogr. 175, 141 (1979)
31. Vadino, W.A., Sugita, E.T., Schnaare, R.L., Ando, M.Y., Niebegall, P.J.: J. Pharm. Sc. 68, 1316 (1979)
32. Lebelle, M., Graham, K., Wilson, W.L.: J. Pharm. Sc. 68, 555 (1979)
33. Westerlund, D., Carlquist, J., Theodorsen, A.: Act. Pharm. Suec. 16, 187 (1979)
34. Nachtmann, F.: Chromatographia, 12, 380 (1979)
35. Kalasz, H., Horvath, C.: J. Chromatogr. 215, 295 (1981)
36. Vlasakova, U., Benes, J., Zivny, K.: J. Chromatogr. 151, 199 (1978)
37. Mourot, D., Delepine, B., Boisseau, J., Gayot, G.: J. Chromatogr. 161, 386 (1978)
38. Whall, T.J.: J. Chromatogr. 219, 89 (1981)
39. Oles, P.J.: J. Pharm. Sc. 67, 1300 (1978)

40. Tsuji, K., Roberson, J.: J. Pharm. Sc. 65, 400 (1976)
41. Mack, G.D., Ashworth, R.B.: J. Chromatogr. Sc. 16, 93 (1978)
42. Know, J.H., Jurand, J.: J. Chromatogr. 186, 763 (1979)
43. Lindauer, R.F., Cohen, D.M., Munelly, K.P.: Anal. Chem. 48, 1731 (1976)
44. Mourot, D., Delepine, B., Boisseau, J., Gayot, G.: J. Chromatogr. 190, 486 (1980)
45. Eksborg, St.: J. Chromatogr. 208, 78 (1981)
46. Nelis, H.J.C.F., Leenheer, A.P. De: J. Chromatogr. 195, 35 (1980)
47. Kwan, R.H., MacLeod, S.M., Spino, M., Teare, F.W.: J. Pharm. Sc. 71, 1118 (1982)
48. Kirchmeier, R.L., Upton, R.P.: Anal. Chem. 50, 349 (1978)
49. Barth, H.G., Conner, A.Z.: J. Chromatogr. 131, 375 (1977)
50. Mangia, A., Silingardi, S., Bortesi, F., Grisanti, G., Bitetto, M. Di: J. Pharm. Sc. 68, 652 (1979)
51. Brown, L.W.: J. Pharm. Sc. 67, 669 (1978)
52. Schwarzenbach, R.: J. Chromatogr. 251, 339 (1982)
53. Albanbauer, J., Fehn, J., Furtner, W., Megges, G.: Arch. f. Krim. 162, 103 (1978)
54. Megges, G.: Arch. f. Krim. 164, 25 (1979)
55. Einhellig, K., Kraatz, A., Megges, G.: Arch. f. Krim. 166, 99 (1980)
56. Jane, I., Scott, A., Sharpe, L.W.R., White, P.C.: J. Chromatogr. 214, 243 (1981)
57. Detaevernier, M.R., Dryon, L., Massart, D.L.: J. Chromatogr. 128, 204 (1976)
58. Wheals, B.B.: J. Chromatogr. 177, 263 (1979)
59. Lund, W., Hannisdal, M., Greibrook, T.: J. Chromatogr. 173, 249 (1979)
60. King, L.A.: J. Chromatogr. 208, 113 (1981)
61. Wahlund, K.-G., Sokolowski, A.: J. Chromatogr. 151, 299 (1978)
62. Wan Po, A. Li, Irwin, W.J.: J. Pharm. Pharmakol. 31, 512 (1979)
63. Sokolowski, A., Wahlund, K.-G.: J. Chromatogr. 189, 299 (1980)
64. Harzer, K., Barchet, R.: J. Chromatogr. 132, 83 (1977)
65. Vilon, C., Vercruysse, A.: J. Chromatogr. 189, 94 (1980)
66. Riedmann, M., Hewlett-Packard: Application note AN 232-2 (1977)
67. Siggia, S., Dishman, R.A.: Anal. Chem. 42, 1233 (1970)
68. Görög, S.: Fresenius Z. Anal. Chem. 309, 97 (1981)
69. Kingston, D.G.I.: Lloydia 42, 237 (1979)
70. Heftmann, E., Hunter, J.R.: J. Chromatogr. 165, 283 (1979)
71. Rehm, K.D., Steinigen, M.: Pharm. Ztg. 127, 888 (1982)
72. Lin, J.-T., Heftmann, E.: J. Chromatogr. 212, 239 (1981)
73. Walters, M.J., Dunbar, W.E.: J. Pharm. Sc. 71, 446 (1982)
74. Ballerini, R., Chinol, M., Ghelardoni, M.: J. Chromatogr. 193, 413 (1980)
75. Hunter, I.R., Walden, M.K., Heftmann, E.: J. Chromatogr. 176, 485 (1979)
76. Capitano, G., Tscherne, R.: J. Pharm. Sc. 68, 311 (1979)
77. Johnston, M.A.: J. Chromatogr. 216, 269 (1981)
78. Hermansson, J.: J. Chromatogr. 194, 80 (1980)
79. Hermansson, J.: J. Chromatogr. 152, 437 (1978)
80. Hara, S., Hayashi, S.: J. Chromatogr. 142, 689 (1977)
81. Tittel, G., Habermeier, H., Wagner, H.: Planta med. 45, 207 (1982)
82. Su, S.C., Hartkopf, A.V., Karger, B.L.: J. Chromatogr. 119, 523 (1976)
83. Johansson, I.M., Wahlund, K.-G.: Acta Pharm. Suec. 14, 459 (1977)
84. Jennings, E.C., Landgraf, W.C.: J. Pharm. Sc. 66, 1784 (1977)
85. Umagat, H., McGarry, P.F., Tscherne, R.J.: J. Pharm. Sc. 68, 922 (1979)
86. Rotsch, T.D., Sydor, R.J., Pietrzyk, D.J.: J. Chromatogr. Sc. 17, 339 (1979)
87. Singletary, R.O., Sancillio, F.D.: J. Pharm. Sc. 69, 144 (1980)
88. Hermansson, J.: Chromatographia 13, 741 (1980)
89. Wessely, K., Zech, K.: HPLC in pharmaceutical analysis Firmenschrift Hewlett Packard
90. Gilpin, R.K.: Anal. Chem. 51, 257R (1979)
91. Gilpin, R.K., Pachala, L.A., Ranweiler, J.S.: Anal. Chem. 53, 142R (1981)
92. Taylor, R.F., Gaudio, L.A.: J. Chromatogr. 187, 212 (1980)
93. Gupta, V. Das, Sachanandani, S.: J. Pharm. Sc. 66, 897 (1977)
94. Eiden, F., Greske, K.: Dtsch. Apoth. Ztg. 123, 958 (1983)
95. Greske, K.: Dissertation Uni München (1982)
96. Gupta, V. Das: J. Pharm. Sc. 66, 110 (1977)

 97. Bartlett, J.M., Segelman, A.B.: J. Chromatogr. 255, 239 (1983)
 98. Mannan, C.A., Krol, G.J., Kho, B.T.: J. Pharm. Sc. 66, 1618 (1977)
 99. Fu, C., Sibley, M.J.: J. Pharm. Sc. 66, 425 (1977)
100. Steward, J.T., Honigberg, I.L., Brant, J.P., Murray, W.A., Webb, J.L., Smith, J.B.: J. Pharm. Sc. 65, 1536 (1976)
101. Bailey, L.C., Abdou, H.: J. Pharm. Sc. 66, 564 (1977)
102. Butterfield, A.G., Lovering, E.G., Sears, R.W.: J. Pharm. Sc. 9, 222 (1980)
103. Fischer, L.J., Thies, R.L., Charkowski, D.: Anal. Chem. 50, 2143 (1978)
104. Leung, C.P.: J. Chromatogr. 178, 178 (1979)
105. Jansson, S.O., Andersson, J., Persson, B.E.: J. Chromatogr. 203, 93 (1981)
106. Pettersson, C., Schill, G.: J. Chromatogr. 204, 179 (1981)
107. Wildanger, W.: Dtsch. Lebensm. Rundschau 72, 160 (1976)
108. Firmenschrift Perkin-Elmer Nr. 102 (1979)
109. Frei, R.W., Santi, W., Thomas, M.: J. Chromatogr. 116, 365 (1976)
110. Nachtmann, F.: Z. Anal. Chem. 282, 209 (1976)
111. Nachtmann, F., Spitzy, H., Frei, R.W.: J. Chromatogr. 122, 293 (1976)
112. Gfeller, J.C., Frey, G., Frei, R.W.: J. Chromatogr. 142, 271 (1977)
113. Aigner, R., Spitzy, H., Frei, R.W.: J. Chromatogr. Sc. 14, 381 (1976)
114. Tscherne, R.J., Capitano, G.: J. Chromatogr. 136, 337 (1977)
115. Collet, G., Rocca, J.L., Sage, D., Beriticat, P.: J. Chromatogr. 121, 213 (1976)
116. Kessler, M.J.: J. Chromatogr. Sc. 20, 523 (1982)
117. Wal, Sj. Van der, Snyder, L.R.: J. Chromatogr. 255, 463 (1983)
118. Harzer, K., Barchet, R.: Dtsch. Apoth. Ztg. 116, 1229 (1976)
119. Halstead, G.W.: J. Pharm. Sc. 71, 1108 (1982)
120. Gupta, V. Das, Ghanekar, A.G.: J. Pharm. Sc. 66, 895 (1977)
121. Honigberg, J.L., Stewart, J.T., Smith, M.: J. Pharm. Sc. 67, 675 (1978)
122. Stewart, J.T., Honigberg, J.L., Coldren, J.W.: J. Pharm. Sc. 68, 32 (1979)
123. Gimet, R., Filloux, A.: J. Chromatogr. 177, 333 (1979)
124. Brown, N.B., Sleeman, H.K.: J. Chromatogr. 150, 225 (1978)
125. Stömbom, J., Bruhn, J.G.: J. Chromatogr. 147, 513 (1978)
126. Smith, R.N.: J. Chromatogr. 115, 101 (1975)
127. Smith, R.N., Vaughan, C.G.: J. Chromatogr. 129, 347 (1976)
128. Flora, K.P., Cradock, J.C., Davignon, J.P.: J. Chromatogr. 206, 117 (1981)
129. Görög, S., Herenyi, B., Javanovics, K.: J. Chromatogr. 139, 203 (1977)
130. Kohl, W., Witte, B., Höfle, G.: Planta med. 47, 177 (1983)
131. Szepesi, G., Gazdag, M.: J. Chromatogr. 205, 57 (1981)
132. Szepesi, G., Gazdag, M.: J. Chromatogr. 204, 341 (1981)
133. Maeden, F.P.B. van der, Rens, P.T. van, Buytenhuys, F.A., Buurmann, E.: J. Chromatogr. 142, 715 (1977)
134. Verzele, M., Taeye, L. de, Dyck, J. van, Decker, G. de: J. Chromatogr. 214, 95 (1981)
135. Forni, G., Massarani, G.: J. Chromatogr. 131, 444 (1977)
136. Klein, A.E., Davis, P.J.: J. Chromatogr. 207, 247 (1981)
137. Ahmed, M.S., Fong, H.H.S., Soejarto, D.D., Dobberstein, R.H., Waller, D.P., Moreno-Azorero, R.: J. Chromatogr. 213, 340 (1981)
138. Nobuhara, Y., Hirano, S., Namba, K., Hashimoto, M.: J. Chromatogr. 190, 251 (1980)
139. Wu, C.Y., Wittick, J.J.: Anal. Chem. 49, 359 (1977)
140. Olieman, C., Maat, L., Waliszewski, K., Beyerman, H.C.: J. Chromatogr. 133, 382 (1977)
141. Hansen, St.H.: J. Chromatogr. 212, 229 (1981)
142. Sasse, F., Hammer, J., Berlin, J.: J. Chromatogr. 194, 234 (1980)
143. Niwa, H., Ishiwata, H., Yamade, K.: J. Chromatogr. 257, 146 (1983)
144. Qualls, C.W. jr., Segall, H.J.: J. Chromatogr. 150, 202 (1978)
145. Ramsdell, H.S., Buhler, D.R.: J. Chromatogr. 210, 154 (1981)
146. Huizing, H.J., Boer, F. De, Malingré, Th.M.: J. Chromatogr. 214, 257 (1981)
147. Tittel, G., Hinz, H., Wagner, H.: Planta med. 37, 1 (1979)
148. Wagner, H., Neidhardt, U., Tittel, G.: Planta med. 41, 232 (1981)
149. Neidhardt, U.: Dissertation Uni München (1982)
150. Wurst, M., Flieger, M., Rekacek, Z.: J. Chromatogr. 150, 477 (1978)

151. Szepesy, L., Feher, J., Szepesi, G., Gazdag, M.: J. Chromatogr. 149, 271 (1978)
152. Yoshida, A., Yamazaki, S., Sakai, T.: J. Chromatogr. 170, 399 (1979)
153. Bethke, H., Delz, B., Stick, K.: J. Chromatogr. 123, 193 (1976)
154. Flieger, M., Wurst, M., Stuchlik, J., Rekacek, Z.: J. Chromatogr. 207, 139 (1981)
155. Szepesi, G., Gazdag, M., Terdy, L.: J. Chromatogr. 191, 101 (1980)
156. Dolinar, J.: Chromatographia 10, 364 (1977)
157. Wurst, M., Flieger, M., Rekacek, Z.: J. Chromatogr. 174, 401 (1979)
158. Verpoorte, R., Kodde, E.W., Bearheim, A. – Svendsen, Planta med. 34, 62 (1978)
159. Verpoorte, R., Bearheim-Svendsen, A.: J. Chromatogr. 109, 441 (1975)
160. Kingston, D.G.I.: Lloydia, 42, 237 (1979)
161. Verpoorte, R., Bearheim-Svendsen, A.: J. Chromatogr. 120, 203 (1976)
162. Urbanyi, Ti., Piedmont, A., Willis, E., Manning, G.: J. Pharm. Sc. 65, 257 (1976)
163. Weber, J.D.: J. AOAC, 59, 1409 (1976)
164. Noordam, A., Wasliewski, W., Olieman, C., Maat, L., Beyerman, H.C.: J. Chromatogr. 153, 271 (1978)
165. Kennedy, J.M., McNamara, P.E.: J. Chromatogr. 212, 331 (1981)
166. Sticher, O., Meier, B.: Pharm. Act. Helv. 53, 40 (1978)
167. Verzele, M., Potter, M. De: J. Chromatogr. 166, 320 (1978)
168. Johnson, E.L., Gloor, R., Majors, R.E.: J. Chromatogr. 149, 571 (1978)
169. Fisher, J.F.: J. Agric. Food. Chem. 26, 497 (1978)
170. Nestler, Th., Tittel, G., Wagner, H.: Planta med. 38, 204 (1980)
171. Evans, F.J.: J. Chromatogr. 88, 411 (1974)
172. Firmenschriften Merck (1976)
173. Castle, M.C.: J. Chromatogr. 115, 437 (1975)
174. Lindner, W., Frei, R.W.: J. Chromatogr. 117, 81 (1976)
175. Cobb, P.H.: Analyst, 101, 768 (1976)
176. Erni, F., Frei, R.W.: J. Chromatogr. 130, 169 (1977)
177. Rehm, K.D., Steinigen, M.: Pharm. Ztg. 127, 443 (1982)
178. Gfeller, J.C., Frei, G., Erni, R.W.: J. Chromatogr. 142, 271 (1977)
179. Fujii, Y., Fukuda, H., Saito, Y., Yamazaki, M.: J. Chromatogr. 202, 139 (1980)
180. Shimada et al.: J. Chromatogr. 124, 79 (1976)
181. Shimada et al.: Chem. Pharm. Bull. 24, 2995 (1976)
182. Tittel, G., Wagner, H.: Planta med. 39, 125 (1980)
183. Jurenitsch, J. et al.: Planta med. 39, 272 (1980)
184. Davydov, V.Ya., Gonzalez, E., Kiselev, A.V.: J. Chromatogr. 204, 293 (1981)
185. Tittel, G., Wagner, H.: Planta med. 43, 257 (1981)
186. Wichtl, M., Mangkudidjojo, M., Wichtl-Bleier, W.: J. Chromatogr. 234, 503 (1981)
187. Tittel, G., Habermeier, H., Wagner, H.: Planta med. 45, 207 (1982)
188. Wichtl, M., Wichtl-Bleier, W., Mangkudidjojo, M.: J. Chromatogr. 247, 359 (1982)
189. Desta, B., Kwang, E., McErlane, K.M.: J. Chromatogr. 240, 137 (1982)
190. Jurenitsch, J., Kapp, B., Bamberg-Kubelka, E., Kern, R., Kubelka, W.: J. Chromatogr. 240, 125 (1982)
191. Jurenitsch, J., Kapp, B., Bamberg-Kubelka, E., Kern, R., Kubelka, W.: J. Chromatogr. 240, 235 (1982)
192. Wu, F.F., Dobberstein, R.H.: J. Chromatogr. 140, 65 (1977)
193. Enson, J.M., Seiber, J.N.: J. Chromatogr. 148, 521 (1978)
194. Meier, B., Sticher, O.: J. Chromatogr. 138, 453 (1977)
195. Sticher, O., Meier, B.: Dtsch. Apoth. Ztg. 120, 1592 (1980)
196. Tittel, G., Wagner, H.: J. Chromatogr. 148, 459 (1978)
197. Tittel, G., Chari, V.M., Wagner, H.: Planta med. 34, 305 (1978)
198. Wulf, L.W., Nagel, C.W.: J. Chromatogr. 116, 271 (1976)
199. Adamovics, J., Stermitz, F.R.: J. Chromatogr. 129, 464 (1976)
200. Hostettmann, K., McNair, H.M.: J. Chromatogr. 116, 271 (1976)
201. Hostettmann, K., Jacot-Guillarmod, A.: J. Chromatogr. 124, 381 (1976)
202. Pettei, M.J., Hostettmann, K.: J. Chromatogr. 154, 106 (1978)
203. Carlson, R.E., Dolphin, D.: J. Chromatogr. 198, 193 (1980)
204. Lea, A.G.H.: J. Chromatogr. 194, 62 (1980)

205. Lea, A.G.H.: J. Chromatogr. 238, 253 (1982)
206. Okuda, T., Mori, K., Seno, K., Hatano, T.: J. Chromatogr. 240, 202 (1982)
207. Tittel, G., Wagner, H.: J. Chromatogr. 135, 499 (1977)
208. Tittel, G., Wagner, H.: J. Chromatogr. 153, 227 (1978)
209. Wagner, H., Tittel, G., Bladt, S.: Dtsch. Apoth. Ztg. 123, 515 (1983)
210. Tittel, G., Wagner, H.: Proceedings of the International Bioflavonoid Symposium, (D)-Munich, (1981) Elsevier Sc. Publ. Comp., Amsterdam-Oxford-New York (1982)
211. Daigle, D.J., Conkerton, E.J.: J. Chromatogr. 240, 202 (1982)
212. Castelee, K.V., Geiger, H., Sumere, Ch. Van: J. Chromatogr. 240, 81 (1982)
213. Rouseff, R.L., Ting, S.V.: J. Chromatogr. 176, 75 (1979)
214. Radaelli, C., Formentini, L., Santaniello, E.: Planta med. 42, 288 (1981)
215. Cairnes, D.A., Kingston, D.G.J., Rao, M.: Lloydia 44, 34 (1981)
216. Lim, C.K., Ayres, D.C.: J. Chromatogr. 255, 247 (1983)
217. Steinigen, M.: Pharm. Ztg. 128, 783 (1983)
218. Wagner, H., Heur, Y.H., Obermeier, A., Tittel, G., Bladt, S.: Planta med. 44, 193 (1982)
219. Komolafe, O.O.: J. Pharm. Sc. 70, 727 (1981)
220. Auterhoff, H., Graf, E., Eurisch, G., Alexa, M.: Arch. Pharm. 313, 113 (1980)
221. Graf, E., Alexa, M.: Planta med. 38, 121 (1980)
222. Bonati, A., Forni, G.: Fitoterapia 18, 159 (1977)
223. Forni, G.P.: Fitoterapia 53, 3 (1982)
224. Hayashi, S., Yoshido, A., Tanako, H., Mitani, Y., Yoshizawa, K.: Chem. Pharm. Bull. 28, 406 (1980)
225. Komolafe, O.O.: J. Pharm. Sc. 16, 496 (1978)
226. Görler, K., Mutter, S., Westphal, C.: Planta med. 37, 308 (1979)
227. Tittel, G., Tittel, Ch.: Bulletin de Liaison u° 11 du Groupe Polyphenols, F-Toulouse, 29.9.–1.10.1982
228. Fisher, J.F., Wheaton, T.A.: J. Agric. Food Chem. 24, 899 (1976)
229. Higgins, J.W.: J. Chromatogr. 121, 329 (1976)
230. Besso, H., Saruwatari, Y., Futamura, K., Kunihiro, K., Fuwa, T., Tanaka, O.: Planta med. 37, 226 (1979)
231. Sticher, O., Soldati, F.: Planta med. 36, 30 (1979)
232. Soldati, F., Sticher, O.: Planta med. 39, 348 (1980)
233. Schulten, H.R., Soldati, F.: J. Chromatogr. 212, 37 (1981)
234. Kimata, H., Hiyama, C., Yahara, S., Tanaka, O., Ishikawa, O., Aiura, M.: Chem. Pharm. Bull., 27, 1836 (1979)
235. Lin, J.-T., Heftmann, E.: J. Chromatogr. 207, 457 (1981)
236. Gracza, L., Ruff, P.: Dt. Apoth. Ztg. 121, 2817 (1981)
237. Herrmann, R.: Dt. Apoth. Ztg. 122, 1797 (1982)
238. Kraus, Lj., Stahl, E.: J. Chromatogr. 170, 269 (1979)
239. Sticher, O., Soldati, F., Lehmann, D.: Planta med. 35, 233 (1979)
240. Grevel, J.: Dissertation Uni München (1982)
241. Bauer, R., Wagner, H.: Dtsch. Apoth. Ztg. 123, 1313 (1983)
242. Nomeir, A.A., Abou-Donia, M.B.: J. AOCS 59, 546 (1982)
243. Becker, H., Bingler-Timeus, E.: Pharm. Ztg. 128, 1252 (1983)
244. Bauer, R., Tittel, G., Wagner, H.: Planta med. 48, 10 (1983)
245. Wurst, M., Pikryl, Z., Vankura, V.: J. Chromatogr. 191, 129 (1980)
246. Holland, J.A., McKerrell, E.H., Fuell, K.J., Burrows, W.J.: J. Chromatogr. 166, 545 (1978)
247. Eiden, F., Tittel, Ch.: Dtsch. Apoth. Ztg. 121, 1874 (1981)
248. Ahmed, M.S., Dobberstein, R.H., Farnsworth, N.R.: J. Chromatogr. 192, 387 (1980)
249. Ahmed, M.S., Dobberstein, R.H.: J. Chromatogr. 236, 523 (1982)
250. Ahmed, M.S., Dobberstein, R.H.: J. Chromatogr. 245, 373 (1982)
251. Schwanbeck, Kubeczka
252. Tittel, G., Bos, R., Wagner, H.: Vortrag: 1. Gesamtkongress der pharmazeutischen Wissenschaften, D-München, 17.–20.4.1983
253. Wal, Sj. Van der, Huber, J.F.K.: J. Chromatogr. 251, 289 (1982)
254. Verpoorte, R., Baerheim-Svendsen, A.: J. Chromatogr. 100, 227 (1974)

255. Verpoorte, R., Baerheim-Svendsen, A.: J. Chromatogr. 100, 231 (1974)
256. Gracza, L.: Dtsch. Apoth. Ztg. 120, 1859 (1980)
257. Ali, S.L.: Fresenius Z. Anal. Chem. 299, 124 (1979)
258. Verzele, M., Mussche, P., Qureshi, S.A.: J. Chromatogr. 172, 493 (1979)
259. Galetto, W.G., Walger, D.E., Levy, S.M.: J. AOAC, 59, 951 (1976)
260. Cleyn, R. De, Verzele, M.: Chromatographia 8, 342 (1975)
261. Sticher, O., Soldati, F., Joshi, R.K.: J. Chromatogr. 166, 221 (1978)
262. Henning, W.: Z. Lebensm. Unters. Forsch. 175, 345 (1982)
263. Freytag, W.E.: Pharm. Ztg. 128, 2869 (1983)
264. Bianchini, J.B., Gaydou, E.M.: J. Chromatogr. 190, 233 (1980)
265. Bauer, R.: Dissertation Uni München (1984)
266. Hänsel, R., Schulz, J.: Dtsch. Apoth. Ztg. 122, 215 (1982)
267. Tittel, G., Bos, R.: to be published
268. Hostettmann, K., Pettei, M.J., Kubo, J., Nakanishi, K.: Helv. Chim. Act. 60, 670 (1977)
269. Freytag, W.E.: Dissertation TU München (1983)
270. Vincent, P.G., Engelke, B.F.: J. Assoc. off. anal. Chem. 62, 310 (1979)
271. Doner, L.W., An-Fei Hsu: J. Chromatogr. (1982)
272. Tittel, G., Fiebig, M., Wagner, H.: to be published
273. Holland, J.A., McKerrell, E.H., Fuell, K.J., Burrows, W.J.: J. Chromatogr. 166, 545 (1978)
274. Kovar, K.A., Langlouis, H., Auterhoff, H.: Pharm. Weekblad Sc. Edit. 5, 134 (1983)

Appendix:
C = Column; F = Flow rate; T = Column temperature; Mph = Mobile phase

Application of HPLC for Analysis
of Psychotropic Drugs in Body Fluids

Hans-Joachim Kuss
Department of Neurochemistry Psychiatric Hospital (Director: Prof. H. Hippius)
University of Munich Nußbaumstraße 7 D-8000 München 2/FRG

1 Introduction

Psychopharmaca are a subgroup of medicaments which are characterized by their influence on the psyche. In order to develop their effect they must reach the central nervous system, and it is therefore necessary that they are rather lipophile substances. Due to their high enrichment in the tissue, plasma concentrations are relatively low and difficult to measure. Due to their physico-chemical properties it is impossible to separate psychotropic drugs from other medicaments. Rather the transitions are somewhat fluid as, for instance, to the beta blockers. In this article the following 3 groups are understood as psychotropic drugs: benzodiazepines, antidepressant and neuroleptic drugs.

The analysis of psychotropic drugs, including their chromatographic separation, is not principally different from that of other medicaments. Nevertheless, they are predominantly lipophilic substances with the particular qualities resulting therefrom. E.g., the renal excretion of most of the lipophilic substances is only inadequate. They undergo an intensive metabolism in the liver, and in urine they appear nearly exclusively in the form of their metabolites. Due to their chemical conversion in the liver and to their strong binding to liver cells they have mostly half life periods which are longer than the usual dosage intervals and thus accumulate in the body. Because of these properties, a particularly high interindividual variability of plasma concentrations of these substances is found at equal dosages. It is for this reason, that analysis of psychotropic drugs in patients is so important, even if – due to the complex mechanism of action – the relation to the therapeutic effect is not yet sufficiently clarified.

In patients the body fluids urine, plasma, cerebrospinal fluid, mother's milk and saliva are practically the only ones that can be used for chemical analysis. Just as for the clinical-chemical estimations, plasma is the medium usually taken, followed by urine. Since most of the psychotropic drugs are almost completely metabolized, a urinary assay allows no conclusion regarding the amount of the effective mother substance in the whole body. For this reason, estimations in urine are very rarely carried out in connection with psychotropic drugs, although for the qualitative evidence, e.g. for checking the compliance, a determination of the urinary concentration is sufficient. Furthermore, in basic pharmacokinetic investigations for the purpose of clarifying the fate of a substance, urinary assay will not be omitted. For correlation with the therapeutic outcome, however, only plasma concentrations are estimated.

Also, when measuring psychotropic drugs in human plasma their metabolites are particular complication. An extreme example is chlorpromazine, of which 150 metabolites are known and many of which are chemically most similar to the mother substance. Actually, a great many metabolites appear only in negligible concentrations. But fairly often a metabolite appears in plasma in a higher concentration than its mother substance. Many metabolites are not obtainable as a pure substance, so that it is impossible to determine their chromatographic separation. Therefore, one should always be prepared for interference from metabolites in the chromatogram.

There is a great deal of literature regarding analyses of psychotropic drugs in body fluids, which cannot be completely referred to here. In the following, the author will present a subjective selection of analytical methods, each of which he has verified.

2 Benzodiazepines

2.1 Extraction

The various benzodiazepines have weakly acidic up to markedly basic properties. In order to extract them with an organic solvent, it is not necessary to alter the pH-value of the plasma which is at about 8. Nevertheless, a concentrated buffer

Table 1. Determination of benzodiazepines in plasma

Publication	pH of Extrac- tion	Extraction with	Col- umn	Eluent	Detec- tion
Bugge (1976)	8; satu- ration with KCL	Benzene (back extraction)	Silica	heptane-isopropanol- methanol (40:10:1)	UV 232
Greizer- stein (1977)	9	1,5% isoamyl alco- hol in heptane	C18	1 mM $KH_2 PO_4$ (pH = 8) – methanol	UV 254
Harzer (1977)	9.5	Extrelut-diethyl- ether	C18	water – methanol	UV 254
Kabra (1978)	7	Diethyl ether	C18	0.01 M sodium acetate buffer ($P_4 = 4.6$)-aceto- mitrile (65:35)	UV 254
Mac Kichan (1979)	11	Chloroform	CN	0.5 mM KH_2PO_4 (pH = 4.5), acetonitrile (65:35)	UV 254

is generally added to the plasma before extraction. The extractability of the substances from plasma depends on several influences and is, in each case, empirically determined. In principal, all common organic solvents can be used for extraction of benzodiazepines. In Table 1 Benzol, Heptan with 1.5% Isoamylalcohol, Diethyl-ether and chloroform are mentioned. Often it is not the higher extraction outcome that is decisive for the choice of an organic solvent, but rather the possibility of extracting other substances which might disturb the analysis. The lower the concentrations to be determined are, the purer should be the organic solvent used. The extraction is nearly exclusively done with "end-over-end-shakers" in order to prevent emulsioning. To obtain a highly pure separation of phases, centrifugation must follow. Often the organic solvent is separated and evaporated after centrifugation. Usually the extract is then again solved in the eluens and injected into the HPLC instrument. The first extract often contains great amounts of lipids; they can be excluded by further prepurification steps, such as, for instance, reextraction into an acidic aqueous phase.

2.2 Chromatography

Except for the earliest study cited [7] (Table 1), separations were carried out with reversed phase columns. Buffer acetonitril or buffer methanol mixtures are used as eluent, a well functioning water-methanol mixture being the simplest. The aqueous phase may be buffered to pH 7 or slightly acidic. Benzodiazepines have a high absorption at 254 nm, only little advantage can be gained by using a different wavelength.

3 Antidepressants

3.1 Extraction

Antidepressants are always extracted under basic conditions. Sometimes the yield of extraction is mentioned (Table 2) to diminish using higher pH values. When a buffer was used, the pH value thereof was 10 or 10.5. Obviously, the pH value should be higher than the pKa value of the substances extracted, but should not exceed it by too much. In most of the cases extraction was carried out with hexane. Frequently hexane was mixed with a small amount of alcohol. Twice diethylether and once dichlormethane were used as organic solvents. Once a solid-liquid dispersion was employed.

3.2 Chromatography

According to Table 2 seven times a silica column and nine times a reversed phase column are used for separation. When using a silica column, mostly a small amount of an amine in an organic solvent is added, in order to prevent an irreversible absorption of the antidepressants. In one procedure a tertiary solution system and

Table 2. Determination of antidepressants

Publication	pH of Extraction	Extraction with	Column	Eluent	Detection
Watson (1975)	Basic	Light petroleum	Silica	1.5% ammonium in methanol	UV 240 UV 290
Watson (1977)	Strong basic	Dichloromethane, ethanol (protein precipitation)	Silica	Dichloromethane-propan-2-ol-conc.ammonia (100:2:0.25)	UV 240
Westenberg (1977)	10	Hexane	Silica	Hexane, dichloromethane, methanol (8:1:1)	UV 250

Table 2 (continued)

Publication	pH of Extraction	Extraction with	Column	Eluent	Detection
Van den Berg (1977)	Strong basic	Diethyl ether	Silica	0.2% methylamine in ethylacetate	UV 254
Brodie (1977)	Strong basic	Diethyl ether (back extraction)	C18	0.6% KH_2PO_4 (pH = 3) buffer 50%; 50% acetonitrile	UV 254
Kraak (1977)	Strong basic	Hexane	C8	1% propylamine in water-methanol-dichloromethane (13:8:3)	UV 254
Kuß (1978)	10.5	Hexane (back extraction)	Silica	0.2% propylamine in methanol	UV 254
Mellström (1978)	10.5	Hexane (back extraction)	Silica	0.1 M perchloric acid-methanol-dichloromethane-disisopropyl ether (0.9:5.2:10:30)	UV 254
Proelss (1978)	Strong basic	1% isoamyl alcohol in hexane	C18	Methanol/acetonitrile phosphate buffer (0.1 M; pH = 7.6) (41:15:44)	UV 254
Sutfin (1979)	Strong basic	20% n-butyl alcohol in hexane	Silica	0.4% ammonium hydroxyde in methanol-acetonitrile (1:5)	fluorescence; EX 240; EM 370
Preskorn (1980)	Strong basic	1% isoamyl alcohol in hexane (back extraction)	C18	56% perchlorate (pH = 2.5) 0.045 M; 44% acetonitril	UV 254
Jensen (1980)	Strong basic	Hexane (Dichlormethane)	C18	0.6% KH_2PO_4 (pH = 3) buffer + acetonitrile	UV 254
Breutzmann (1981)	Strong basic	2% isoamyl alcohol in hexane (back extraction)	C18	62.5% phosphate buffer 50 mM, pH = 5; 37.5% acetonitrile	UV 210
Kuß (1981)	10.5	Hexane (back extraction)	C18	65% 0.05 M phosphate buffer (pH = 2.7); 35% acetonitrile	UV 214
Bidlingmeyer (1982)	Basic	SEP-PAK C18 column	Silica	10 mM butylamine in methanol/water (90/10)	UV 254
Bock (1982)	Strong basic	2% isoamyl alcohol in hexane	C6	64% buffer: 50 mM KH_2PO_4, 36% acetonitrile	UV 205
Koteel (1982)	Basic	Hexane	CN	5 mM phosphate buffer (pH = 7), 25%; 15% methanol; 60% acetonitrile	UV 254

in another an ion pair mechanism are described. The hydrophobic columns operate mostly with a buffer-acetonitril mixture, generally with low pH values. Instead of the most often mentioned phosphate buffer, a perchloric acid is used in two studies. In another publication a water-methanol-dichlormethane mixture under addition of amines is used with a C8-column instead of a silica column.

3.3 Detection

With one exception, ultra violet absorption, mainly at 254 nm, was used for detection. Detection in distant UV is mentioned three times. Only in one study is fluorescence used for detection. It is only practicable in some antidepressants. The frequently applied antidepressant amitriptyline shows no fluorescence at all.

3.4 Analysis in Urine

In two studies analysis of antidepressants in urine is described (not included in Table 2). Since mainly the conjugates of hydroxylated metabolites are found in urine, both papers describe a conjugate separation. Biggs[2] uses acid deconjugation, i.e., urine is heated at 100° C for one hour after adding concentrated hydrochloric acid. In the case of amitriptyline this procedure has the disadvantage that aliphatic hydroxyl groups can be dehydrated, thus, on one hand resulting in the formation of a new compound and, on the other hand, allowing no further discrimination between the trans- and the cis-hydroxyl products. For this reason, the procedure of enzymatic separation with glusulase as proposed by Watson and Stewart[32] is superior.

4 Neuroleptics

4.1 Phenothiazines

Neuroleptic drugs are subdivided into the two groups of phenothiazines and butyrophenones. Phenothiazines are chemically quite similar to tricyclic antidepressants and can be extracted and chromatographically separated with analogous procedures. A review of the concentrations to be expected has been given by Muusze[22]. Many of the phenothiazines can be well identified with UV absorption. Particularly with regard to long-acting drugs (long chain esters of the phenothiazines), only very low concentrations are found in patient's plasma. In such cases UV absorption is not sensitive enough. Therefore, two papers using electrochemical detection for analysis are mentioned in Table 3a. For phenothiazines this detector has a higher sensitivity as compared to UV detectors; it is, however, much more susceptible to disturbances at the same time.

4.2 Butyrophenones

As regards butyrophenones, only for haloperidol, which is the most important of them, have two hplc-determination methods been described (Table 3b). Both the extraction procedures described and the chromatographic separation do not principally differ from the method applied for tricyclic antidepressants and for phenothia-

Table 3a. Determination of neuroleptics (phenothiazines)

Publication	pH of Extraction	Extraction with	Column	Eluent	Detection
Tjaden (1976)	Strong basic	Hexane	C2	0.05 M phosphate buffer – methanol	Amperometric
Muusze (1980)	Basic	Diethyl ether	Silica	Acetonitril – water – ammonia	UV 254
Wallace (1981)	Basic	0.2% methanol in hexane	CN	55% 0.02 M phosphate buffer; 45% acetonitril	Amperometric
Kilts (1982)	Strong basic	Diethyl ether – hexane (3:1) (back extraction)	Silica	0.036% methyl-amine in 2,2,4-trimethylpentane – methyline chloride – methanol (8:1:1)	UV 254
Taylor (1982)	Strong basic	1.5% isoamyl alcohol 10% dichloromethane in hexane (back extraction)	C2	0.05 M phosphate buffer (pH = 7.4) 30%; 70% methanol	UV 248

Table 3b. Determination of neuroleptics (butyrophenones)

Publication	pH of Extraction	Extraction with	Column	Eluent	Detection
Miyazaki (1981)	Strong basic (0.1 n HCL)	Chloroform (preextraction with diethyl ether)	C18	37% 0.2 M ammonium acetate buffer; 63% methanol	UV 250
Jatlow (1982)	Strong basic	2% isoamyl alcohol in hexane (back extraction)	C18	40% acetonitrile in phosphate buffer (pH = 3.8)	UV 250

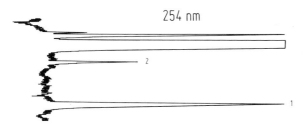

254 nm

Fig. 1. Chromatogram of patient's plasma after extraction

Nr.	Substance	Amount	Ret.time
1	cis-hydroxyamitriptyline (internal standard)	40 ng	7.85
2	Haloperidol	11 ng	3.79

Column: Nucleosil 50-5. Eluens: Methanol/ammonia 0.01%

zines. The concentrations to be measured are, however, lower. Therefore, when analyzing butyrophenones, special attention must be paid to using extremely pure extraction agents and to keeping the noise of the detector at a minimum (compare Fig. 1).

5 Discussion

In introducing a HPLC-method into a laboratory it is necessary to do some experiments beforehand. There are always three steps to follow in working out a method: 1. chromatography of an external standard, 2. extraction – experiments with a spiked sample of the biological fluid, and 3. test of the method with real samples from patients.

Using the method routineously the same scheme must be followed every day. To test the function of the whole apparatus an external standard must be injected first, similarly to the way it is done in clinical chemistry. In this way the actual retention times for this day are determined. After adjustment of the chromatography the extract of a control sample with a known amount of the substances to be analyzed is injected. This is a test for the extraction and gives the yield of extraction in this series of analyses. These controls are repeated after each tenth patient sample.

From the literature one can see that similar methods are always described. It is clear that hexane is very well suited for the extraction of the antidepressants. Often a reextraction in diluted acid is described. The by far most frequently used

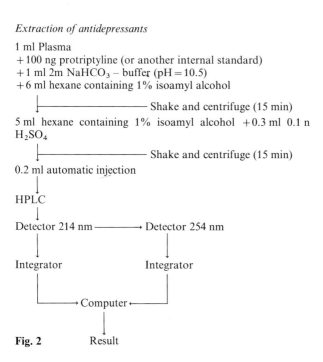

Extraction of antidepressants

1 ml Plasma
+ 100 ng protriptyline (or another internal standard)
+ 1 ml 2m $NaHCO_3$ – buffer (pH = 10.5)
+ 6 ml hexane containing 1% isoamyl alcohol

— Shake and centrifuge (15 min)

5 ml hexane containing 1% isoamyl alcohol + 0.3 ml 0.1 n H_2SO_4

— Shake and centrifuge (15 min)

0.2 ml automatic injection

HPLC

Detector 214 nm ⟶ Detector 254 nm

Integrator   Integrator

⟶ Computer ⟵

Fig. 2   Result

Table 4. Benzodiazepines

	Relative retention (time)	Absorption 214 nm/254 nm	Relative[a] absorption
Chlordiazepoxid	0.35	0.56	5.22
Fluordiazepam	0.49	1.70	1.19
Oxazepam	0.77	1.46	1.82
Lorazepam	0.87	2.33	1.23
Nitrazepam	0.98	1.40	1.80
Desmethyldiazepam	1.23	1.62	1.74

Column: 5 μ Nucleosil C18; 25 cm;
Eluens: 38% Acetonitrile, 62% 0.05 Phosphatebuffer, pH 2.8

[a] Absorption relative to amitriptyline at 254 nm

Table 5. Antidepressants

	Relative retention (time)	Absorption 214 nm/254 nm	Relative[a] absorption
Didesmethyldibenzepine	0.26	2.88	1.44
Desmethylzimelidine	0.28	0.75	1.68
Trans-OH-Nortriptyline	0.30	4.20	0.81
Desmethyldibenzepine	0.31	2.46	1.50
Trans-OH-Amitriptyline	0.32	4.10	1.13
Zimelidine	0.33	0.75	1.80
Dibenzepine	0.34	2.36	1.35
Cis-OH-Nortriptyline	0.35	4.13	1.21
Cis-OH-Amitriptyline	0.39	3.98	1.10
Desmethyloxaprotiline	0.45	41.13	0.095
Oxaprotiline	0.51	44.69	0.066
Mianserin	0.54	7.86	0.54
Doxepin	0.61	4.22	1.18
Desmethylnortriptyline	0.71	4.48	1.13
Desmethylmaprotiline	0.72	39.53	0.94
Desmethylimipramine	0.74	2.55	1.24
Protriptyline	0.76	7.93	0.52
Imipramine	0.86	2.55	1.06
Nortriptyline	0.86	4.28	1.12
Maprotiline	0.87	38.97	0.103
Desmethyltrimipramine	0.96	2.54	0.102
Amitriptyline	1.00	4.15	1.00
Trimipramine	1.10	2.61	0.93
Amitriptylin-N-Oxyd	1.11	4.19	0.92
Desmethylchlorimipramine	1.24	2.75	1.06
Chlorimipramine	1.46	2.75	1.04

Column: 5 μ Nucleosil C18; 25 cm;
Eluens: 38% Acetonitrile, 62% 0.05 M Phosphatebuffer pH 2.8

[a] Absorption relative to amitriptyline at 254 nm

Table 6. Neuroleptics

	Relative retention (time)	Absorption 214 nm/254 nm	Relative[a] absorption
Pipamperone	0.19	0.60	1.75
Sulpiride	0.20	6.29	1.21
Thioridazine	0.42	9.36	1.33
Benperidol	0.45	2.02	0.96
Perazine	0.49	0.42	2.14
Decentane	0.67	0.62	2.24
Neurocil	1.05	0.87	3.17
Haloperidol	0.81	0.96	1.74
Chlorpromazine	1.26	0.61	3.17
Fluspirilene	3.34	1.48	0.67

Column: 5 μ Nucleosil C18; 25 cm;
Eluens: 38% Acetonitrile, 62% 0.05 Phosphatebuffer pH 2.8

[a] Absorption relative to amitriptyline at 254 nm

chromatographic system is separation on a C-18 column and a buffer acetonitrile mixture as mobile phase. In Fig. 2, as well, a working extraction scheme for antidepressants is shown.

When using a method routineously, it is not enough to know the behaviour of the substance to be determined. It is important to have an overall view concerning the analysis of the three substance groups mentioned, because patients are mostly treated with more than one medicament. Sometimes two antidepressants are given together, but more frequent are combinations of antidepressants and neuroleptics. Also, it is possible that patients are treated with one neuroleptic from the butyrophenon group, together with one neuroleptic from the phenothiazine group. Psychiatric patients normally have sleep disturbances. In most cases, therefore, one can find, in addition to the medicament in question, benzodiazepines in plasma. However, one could try to exclude all these interferences by separating them chromatographically. But this is impossible due to the great number of psychotropic drugs on the market. Obviously, the retention time is not sufficient for the characterization of a substance. Using a second UV-detector, the specificity of the HPLC-analysis can improve. The second detector is simply connected to the outlet of the first detector. Using the information from both detectors, no problem identifying a medicament occurs in reality. It is then possible to use the quotient of the adsorption at both wavelengths. Considering the three Tables, 4, 5 and 6, one can see that neither the retention time nor the adsorption quotient are able to identify a substance exactly, but the combination of both is.

The Figures 3 and 4 show the chromatograms of some selected substances at two wavelengths before and after extraction, according to figure 2.

Generally, there is a need for an internal standard. With an internal standard one is sure that the extraction was not wrong in chromatograms with no measurable

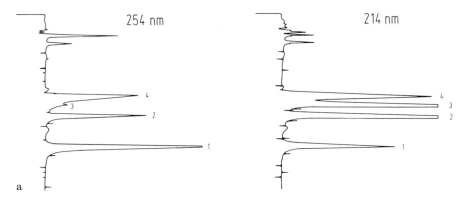

Fig. 3a. Chromatogram of external standards (detectors in series) 100 ng each

Nr.	Substance	Ret.time	Area 254 nm	Area 214 nm	214/254 nm
1	Chlorpromazine	17.46	15968	9924	0.62
2	Amitriptyline	13.37	7494	29207	3.90
3	Maprotiline	11.89	2084	27277	13.1
4	Oxazepam	10.79	11457	16273	1.42

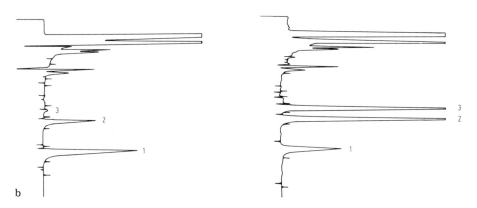

Fig. 3b. Chromatogram of spiked plasma after extraction of 100 ng

Nr.	Substance	Ret.time	Area 254 nm	Area 214 nm	214/254 nm
1	Chlorpromazine	17.48	8683	5178	0.60
2	Amitriptyline	13.41	3593	14818	4.13
3	Maprotiline	11.98	190	12017	60.3
4	Oxazepam				

Yield of extraction: Chlorpromazine 54%, Amitriptyline 48%, Maprotiline 45%, Oxazepam 0%

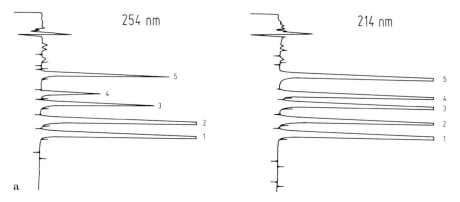

Fig. 4a. Chromatogram of external standards (detectors in series) 100 ng each

Nr.	Substance	Ret.time	Area 254 nm	Area 214 nm	214/254 nm
1	Desmethyldiaz.	15.81	19090	31179	1.63
2	Neurocil	13.91	21730	18952	0.87
3	Nortriptyline	11.77	7130	30358	4.26
4	Protriptyline	10.45	3400	27060	7.96
5	Doxepine	7.90	7256	30266	4.17

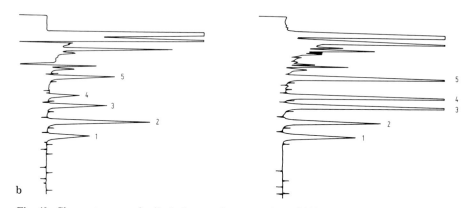

Fig. 4b. Chromatogram of spiked plasma after extraction of 100 ng

Nr.	Substance	Ret.time	Area 254 nm	Area 214 nm	214/254 nm
1	Desmethyldiaz.	15.82	3542	5787	1.63
2	Neurocil	13.91	7334	6471	0.88
3	Nortriptyline	11.81	3590	15405	4.29
4	Protriptyline	10.50	1703	13775	8.08
5	Doxepine	7.98	3569	14828	4.15

Yield of extraction: Desmethyldiazepam 19%, Neurocil 34%, Nortriptyline 50%, Protriptyline 50%, Doxepine 49%

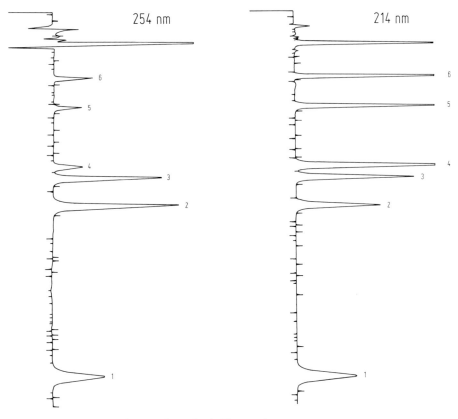

Fig. 5. Chromatogram of external standards 50 ng each

Nr.	Substance	Area 214/254 nm	Ret.time
1	Carbamazepine	4.51	28.86
2	10,11-dihydro-10-oxocarbamazepine	2.56	15.43
3	9-hydroxymethyl-10-carbamyl-acredine	4.12	13.20
4	10,11-epoxy-carbamazepine	20.9	12.30
5	10,11-dihydro-10-hydroxycarbamazepine	23.2	7.59
6	10,11-dihydro-10,11-trans-dihydroxy-carbamazepine	24.0	5.24

Column: 5 μ Nucleosil C18; Acetonitrile/Water 25:75

concentrations. More important, especially when using an automatic injector, is the information about the reproducibility concerning the extracted internal standard.

Concerning the yield of extraction in Figures 3 and 4, one has to take into account that the maximum yield is 56%, because there is a loss of extraction solvent in both extraction steps – 5 ml from 6 ml in step 1 and 200 μl from 300 μl in step 2. Obviously, the antidepressants amitriptyline, nortriptyline, protriptyline, doxepine and maprotiline, as well as the neuroleptics chlorpromazine and neurocil, are extracted nearly totally, whereas desmethyldiazepam is poorly extracted and oxazepam not at all.

Figure 5 shows a separation of carbamazepine and oxcarbazepine and their metabolites. These substances belong to the group of antiepileptic drugs, which is not referred to in this paper. A review concerning the determination of antiepileptic drugs in plasma can be found in Kabra et al. [12, 13], Riedmann et al. [26] and Rambeck et al. [25].

6 References

1. Berg, J.H. Van Den, Ruwe, J.J.M. De, Deelder, R.S., Plomp, Th.A.: Column liquid chromatography of tricyclic antidepressants. Journal of Chromatography, 138, 431–436 (1977)
2. Biggs, S.R., Chasseaud, L.F., Hawkins, D.R., Midgley, I.: Determination of amitriptyline and its major basic metabolites in human urine by high-performance liquid chromatography. Drug Metabolism and Disposition, Vol. 7, No. 4, 233–236 (1979)
3. Bidlingmeyer, B.A., Korpi, J., Little, J.N.: Determination of tricyclic antidepressants using silica gel with a reversed-phase eluent. Chromatographia, Vol. 15, No. 2, Febr. 1982, pp. 83–85
4. Bock, J.L., Giller, E., Gray, St., Jatolow, P.: Steady-state plasma concentrations of cis- and trans-10-OH amitriptyline metabolites. Clin. Pharmacol. Ther., Vol. 31, No. 5, May 1982, pp. 609–616
5. Breutzmann, D.A., Bowers, L.D.: Reversed-phase liquid chromatography and gas chromatography/Mass fragmentography compared for determination of tricyclic antidepressant drugs. Clinical Chemistry, Vol. 27, No. 11, 1907–1911 (1981)
6. Brodie, R.R., Chasseaud, L.F., Hawkins, D.R.: Separation and measurement of tricyclic antidepressant drugs in plasma by high-performance liquid chromatography. Journal of Chromatography, 143, 535–539 (1977)
7. Bugge, A.: Quantitative high-performance liquid chromatography of diazepam and N-desmethyldiazepam in blood. Journal of Chromatography, 128, 111–116 (1976)
8. Greizerstein, H.B., Wojtowicz, C.: Simultaneous determination of chlordiazepoxide and its N-demethyl metabolite in 50-μL blood samples by high pressure liquid chromatography. Analytical Chemistry, Vol. 49, No. 14, Dec. 1977, pp. 2235–2236
9. Harzer, K., Barchet, R.: Analyse von Benzodiazepinen und deren Hydrolysenprodukte, den Benzophenonen, durch Hochdruckflüssigkeitschromatographie in umgekehrter Phase und ihre Anwendung auf biologisches Material. Journal of Chromatography, 132, 83–90 (1977)
10. Jensen, K.M.: Determination of amitriptyline-N-oxide, amitriptyline and nortriptyline in serum and plasma by high-performance liquid chromatography. Journal of Chromatography, 183, 321–329 (1980)
11. Jatolow, P.I., Miller, R., Swigar, M.: Measurement of haloperidol in human plasma using reversed-phase high-performance liquid chromatography. Journal of Chromatography, 227, 233–238 (1982)
12. Kabra, P.M., Koo, H.Y., Marton, L.J.: Simultaneous liquid-chromatographic determination of 12 common sedatives and hypnotics in serum. Clinical Chemistry, Vol. 24, No. 4, 657–662 (1978)
13. Kabra, P.M., Stevens, G.L., Marton, L.J.: High-pressure liquid chromatographic analysis of diazepam, oxazepam and N-desmethyldiazepam in human blood. Journal of Chromatography, 150, 355–360 (1978)
14. Kraak, J.C., Bijster, P.: Determination of amitriptyline and some of its metabolites in blood by high-pressure liquid chromatography. Journal of Chromatography, 143, 499–512 (1977)
15. Koteel, P., Mullins, R.E., Gadsden, R.H.: Sample preparation and liquid-chromatographic analysis for tricyclic antidepressants in serum. Clinical Chemistry, Vol. 28, No. 3, 462–466 (1982)
16. MacKichan, J.J., Jusko, W.J., Duffner, P.K., Cohen, M.E.: Liquid-chromatographic assay of diazepam and its major metabolites in plasma. Clinical Chemistry, Vol. 25, No. 6, 856–859 (1979)

17. Kilts, C.D., Patrick, K.S., Breese, G.R., Mailman, R.B.: Simultaneous Determination of thioridazine and its S-oxidized and N-demethylated metabolites using high-performance liquid chromatography on radially compressed silica. Journal of Chromatography, 231, 377–391 (1982)
18. Kuss, H.-J., Nathmann, M.: Quantitative Bestimmung trizyklischer Antidepressiva mit der Hochdruckflüssigkeitschromatographie. Arzneimittelforschung 28, 1301 (1978)
19. Kuss, H.-J., Feistenauer, E.: Quantitative high performance liquid chromatographic assay for the determination of maprotiline and oxaprotiline in human plasma. Journal of Chromatography 204, 349–353 (1981)
20. Mellström, B., Braithwaite, R.: Ion-pair liquid chromatography of amitriptyline and metabolites in plasma. Journal of Chromatography, 157, 379–385 (1978)
21. Miyazaki, K., Arita, T., Oka, I., Koyoma, T., Yamashita, I.: High performance liquid chromatographic determination of haloperidol in plasma. Journal of Chromatography, 223, 449–453 (1981)
22. Muusze, R.: Determination of serum-levels of psychotropic drugs by liquid-solid column chromatography. In: Usdin, Eckert, Forrest (eds): Phenothiazines and structurally related drugs: basic and clinical studies. Elsevier North Holland, Inc., Amsterdam 1980, pp. 125–128
23. Preskorn, S.H., Leonard, K., Hignite, Ch.: Liquid chromatography of amitriptyline and related tricyclic compounds. Journal of Chromatography, 197, 246–250 (1980)
24. Proelss, H.F., Lohmann, H.J., Miles, D.G.: High-performance liquid-chromatographic simultaneous determination of commonly used tricyclic antidepressants. Clinical Chemistry, Vol. 24, No. 11, 1948–1953 (1978)
25. Rambeck, B., Riedmann, M., Meijer, J.W.A.: Systematic Method of Development in liquid chromatography applied to the determination of antiepileptic drugs. Therapeutic Drug Monitoring, Vol. 3, No. 4, 377–395 (1981)
26. Riedmann, M., Rambeck, B., Meijer, J.W.A.: Quantitative simultaneous determination of eight common antiepileptic drugs and metabolites by liquid chromatography. Therapeutic Drug Monitoring, Vol. 3, No. 4, 397–413 (1981)
27. Roth, W., Beschke, K., Jauch, R., Zimmer, A., Koss, F.W.: Fully automated high-performance liquid chromatography. A new chromatograph for pharmacokinetic drug monotoring by direct injection of body fluids. Journal of Chromatography, 222, 13–22 (1981)
28. Sutfin, T.A., Jusko, W.J.: High-performance liquid chromatographic assay for imipramine, desipramine, and their 2-hydroxylated metabolites. Journal of Pharmaceutical Sciences, Vol. 68, No. 6, June 1979, pp. 703–705
29. Wallace, J.E. et al.: Determination of promethazine and other phenothiazine compounds by liquid chromatography with electrochemical detection. Analytical Chemistry, Vol. 53, No. 7, June 1981, pp. 960–962
30. Watson, I.D., Stewart, M.J.: Separation and quantitative determination of tricyclic antidepressants by high-performance liquid chromatography. Journal of Chromatography, 110, 389–392 (1975)
31. Watson, I.D., Stewart, M.J.: Quantitative determination of amitriptyline and nortriptyline in plasma by high-performance liquid chromatography. Journal of Chromatography, 132, 155–159 (1977)
32. Watson, I.D., Stewart, M.J.: Assay of tricyclic structured drugs and their metabolites in urine by high-performance liquid chromatography. Journal of Chromatography, 134, 182–186 (1977)
33. Westenberg, H.G.M., Drenth, B.F.H., Zeeuw, R.A. De, Cuyper, H., Praag, H.M. Van, Korf, J.: Determination of Clomipramine and desmethylclomipramine in plasma by means of liquid chromatography. Journal of Chromatography, 142, 725–733 (1977)
34. Taylor, G., Houston, J.B.: Simultaneous determination of promethazine and two of its circulating metabolites by high-performance liquid chromatography. Journal of Chromatography, 230, 194–198 (1982)
35. Tjaden, U.R., Lankelma, J., Poppe, H.: Anodic Coulometric Detection with a glassy carbon electrode in combination with reversed phase high-performance liquid chromatography. Journal of Chromatography 125, 275–286 (1976)

HPLC of Amino Acids and Proteins

H. Engelhardt
Angewandte Physikalische Chemie Universität des Saarlandes
D-6600 Saarbrücken/FRG

1 Chromatography of Amino Acids

The separation of amino acids was the first completely automated chromatographic process with post-column derivatization. The separation was carried out on a cation exchanger and the detection in the eluate after separation was performed by means of ninhydrin[1]. This method has had its place for more than 30 years in every protein laboratory. Amino acid analysis has taken a further step forward with the appearance of HPLC and the requirement for more rapid and sensitive separation and detection techniques. At first the classical method was merely adapted to the conditions of HPLC by modifying the separation resins and changing and improving the detection methods, more recently the whole area of amino acid analysis has been optimized with regard to the separation system, derivatization before or after separation, improvement of the sensitivity of detection and including automation of the individual steps.

Amino acid analysis is a good example of the fact that there is seldom one optimal separation system for analytical problems, but rather that the choice of a suitable separation system depends on the analytical problem, e.g. on the matrix, in which the amino acids are to be determined, on the desired and necessary sensitivity and on the necessary (possible) speed of analysis. Then the question also arises as to whether the amino acid derivatives are to be prepared first and then separated or whether the separated amino acids are to be converted afterwards (on-line) into absorbing or fluorescent derivatives (e.g. in a chemical reaction detector, post-column derivatization).

Refractometric detection or UV detection at 205 nm would be possible in principle. The sensitivity and selectivity of detection are insufficient in both cases. The necessity for gradient elution because the amino acids possess a wide range of polarities means that refractometry is not suitable for detection.

1.1 Separation Systems for the Analysis of Amino Acids

The good water solubility of amino acids and their derivatives means that two separation systems come into question, separation on ion exchangers and on reversed phases.

Moore and Stein employed ion-exchange chromatography in the first fully automated LC system which they developed over 30 years ago incorporating derivatization after separation; this is employed today in automatic amino acid analysers[1,2]. It can also be transfered to HPLC systems. This method is employed exclusively for the separation of amino acids and small peptides. Derivatization is performed in the column eluate after separation on the column. At least one further reagent pump is necessary for this. The mixture of eluate and reagent has to be stored until the reaction is complete. When this is being done the separation achieved on the column must not be destroyed by additional band broadening in the reaction system. Very fast reactions such as derivatization with fluorescamin[4,5] and o-phthalaldehyde (OPA)[6,7] are being employed increasingly alongside the classical ninhydrin method[3].

Chromatography on reversed phases is more suitable for the separation of amino acid derivatives than for free amino acids. Because of the charge possessed by free amino acids and their derivatives it is necessary to improve the efficiency and selectivity of the chromatographic milieu by means of the eluent. The lowering of the pH or the addition of suitable counterions (ion-pair chromatography) are simple methods of optimization [61, 62].

The reproducibility of the retention data and their comparability from column to column are not different for amino acid analysis than for other chromatographic separation systems. For this reason no precise details will be given concerning the separation systems, i.e. columns and precise eluent composition, neither will chromatograms be reproduced, since most publications describe model separations of standard mixtures and the properties of the stationary phases are often not precisely defined. As is well known the reproducibility of the stationary phases also leaves something to be desired.

1.2 Derivatization of Amino Acids

The problem of amino acid derivatization is concentrated on the selective and sensitive detection of the amino acids. The amounts of sample available are very small, whether they be proteins whose structures are to be determined or samples from investigations in the field of nutritional science, so that the derivatization reaction should allow detection in the pmol range. The concentration of amino acids in the body fluids lies in and below the ppb range.

As already mentioned, the derivatization can be carried out before the separation. Derivatization should occur quantitatively to yield single products for each amino acid, the reagent should not interfere. It is likely that derivatization will yield products that are more like each other, however, the high separational selectivity of the chromatographic system means that there are seldom problems. Pre-column derivatization can often be carried out automatically immediately before separation.

Derivatization after separation requires, in general, a higher instrumental complexity. Commercial instruments often need to be optimized in order to achieve higher detection sensitivity. The advantages and disadvantages of the various methods of derivatization are listed in Table 1 and will be discussed in what follows in connection with the various separation systems.

1.2.1 Precolumn Derivatization

1.2.1.1 Separation of Amino Acid Derivatives

1. After Degradation Reactions

During the Edman degradation the phenylthiohydantoins (PTH) of the amino acids are formed by reaction with phenyl isothiocyanate; these exhibit adequate absorption in the UV. Even though the absorption maximum lies at 269 nm the sensitivity of detection obtained at 254 nm remains adequate. The detection limit seems to lie in the 5 pmol range. The starting quantities employed in the analysis of a protein

Table 1. Pre- and post-column derivatization

Pre-column derivatization	Post-column derivatization
Advantages	
Simple method	Continuous on-line method
Chromatographic system unchanged	Easily automated
Reaction time unimportant	Separation of the unchanged substance
Free choice of reaction conditions	Separation not affected by excess reagent
	No time-consuming and loss-causing
	Sample preparation
	Reaction does not have to be complete
	Separation not interfered with by artefacts
	Better reproducibility
	Detectors in series are possible
Disadvantages	
Sample substances generally become more alike	Instrumental complexity
	Additional band broadening in reactor
Reaction must be quantitative	No free choice of reaction conditions (eluent)
Interference can occur to separation	Reagent and reaction products should not
– reagent	interfer at detection
– reaction medium	
– formation of artefacts	
– several derivatives from one component	

s sequence should be between 1–10 nmol[9]. Microcolumn techniques allow an improvement of a factor of 10^{10}[10].

The separation itself is the greater problem here. Although in theory, only one amino acid derivative is produced at each step of the degradation (if degradation is quantitative) the separation system still ought to possess the capacity to completely separate and identify all of the 20 PTH derivatives of interest. The separation can be performed on RP columns (e.g. octadecyl phases) with aqueous acetonitrile mixtures as eluents. Gradient elution is indispensable for the complete separation of all 18–22 PTH derivatives. But also isocratic separation in 15 minutes seems possible[60]. The basic PTH derivatives of arginine and histidine in particular, cause problems their retention being greatly influenced by the presence of residual silanol groups of the stationary phase. Being very polar both PTH derivatives ought to be eluted at the start of the chromatograms. However the elution is very dependent on the properties of the stationary phase and the working age of the column. The addition of ion-pair formers or acids to inhibit the dissociation can prove helpful here[11].

It seems feasible to couple an automatic sequencer with HPLC[12].

Other isothiocyanates (e.g. 4-dimethylaminoazobenzene 4-isothiocyanate which yields DABTH derivatives with absorption maximum at 436 nm) have been employed. Separation is possible on reversed phases employing gradients[13].

2. Reaction with Phenyl Isothiocyanate

When free amino acids are reacted with the Edman reagent phenyl isothiocyanate, then the phenyl thiocarbamyl derivatives of the amino acid are formed[14], which can be separated on reversed phases and detected with UV detectors (254 nm) or by an electrochemical detector[10]. The derivatization can be automated and integrated with a commercial HPLC system[15]. Derivatization requires 10–20 min, while the separation (gradient) is complete in ca. 12 min. both primary and secondary amino acids can be reliably identified and quantified down to the 1 pmol level.

3. Dansyl Derivatives

The derivatization destroys the zwitterionic character of the amino acids, which reduces the problems during the chromatographic separation. Astonishingly the selectivity is sufficient to ensure the separation of the compounds which are very similar after derivatization. The derivatization with dimethylaminonaphthalene sulphonyl chloride occurs at nitrogen, the dissociation of the free carboxyl group is easily prevented, so that the efficient chromatography of the dansyl amino acids on reversed phases does not cause any problems. However gradient elution is essential.

Dansyl amino acids can be detected by both UV and fluorescence detectors (excitation at 360 nm, emission at 480 nm). Detection limits in the fmol region have been described for fluorescence[16]. Analogous derivatives contain the azobenzene chromophore (DABS = 4-dimethylaminoazobenzene-4′-sulphonyl chloride) and are detected in the visible (436 nm).

The problem with dansylation lies in the fact that it is difficult to obtain reproducible unequivocal reaction products of dansyl chloride with amino acids; this is particularly the case for lysine and arginine. Additionally the reagent which is present in excess, reacts to form products which can interfere with separation and identification.

The dansyl derivatives of the amino acids can be separated into their enantiomers on suitable chiral phases[17] or by the addition of chiral reagents to the eluents[18]. The dangers of racemization of the amino acids during their reaction with dansyl chloride is small, so that this system can be employed to check the optical purity of amino acids. The selectivity of the enantiomer separation is large. Relative retentions greater than 2 are not uncommon. However the enantiomer separation phases are not suitable for all dansyl amino acids. A two-dimensional method of operation is necessary, e.g. first separation of the dansyl amino acid on reversed phases and then separation of the enantiomers of the individual dansyl amino acids on chiral phases.

4. Derivatives with o-phthalaldehyde (OPA)

The reaction of o-phthalaldehyde with primary amines in the alkaline medium and in the presence of mercapto derivatives leads to the formation of intensely fluorescent isoindole derivatives (OPA derivatives). The reaction is rapid and unequivocal at room temperature. The separation on reversed phases followed by fluorescence detection constitutes a rapid sensitive and selective detection method for all amino acids with primary amino groups. This derivatization method was first employed by Roth[6, 7], mainly for post-column derivatization, but it has also been recommended for pre-column derivatization[19, 20]. The problems involved in pre-column derivatization will now be discussed. Along with the ethyl mercaptan originally employed, mercaptoethanol, ethanedithiol, propyl mercaptan and 3-mercaptopropionic acid have also been employed. The mercaptan (whose function is also that of a reducing agent) is incorporated into the isoindole structure, so that the various OPA derivatives differ in their retention properties and their stabilities.

The reaction only occurs with *primary* amino acids, so that it is not possible to detect proline and hydroxyproline. The reaction products of cysteine and cystine only exhibit a weak fluorescence. The situation can be improved in the latter case by alkylation of the cysteine with iodoacetic acid before the OPA reaction[21] or by oxidation to cystine sulphonic acid[22]. Oxidation with chloramine T (sodium hypochlorite is effective) is employed for the post-column derivatization of secondary amino acids. A modification of this method has been described for pre-column derivatization[14] whereby an additional reduction step with sodium borohydride is included[23].

The amino acids (10–400 nmol/ml) are derivatized by the addition of a 1 molar borate buffer (pH 9.5–10) after the addition of a methanolic solution of the mercaptan and of OPA. Since OPA is not stable in alkali and mercaptans easily oxidize in this medium it is necessary to make up the solutions (in methanol) separately and add them to the sample (10–400 nmol/ml) in the order buffer (0.5 ml), mercaptan (20–25 µl pure substance possibly in methanolic solution) and OPA (10–20 mg/ml methanol). It is important that the thiol be added before or with the OPA, whereby the molar concentration of thiol should be twice that of the OPA solution. In our experience it is possible to store the mixture of buffer. OPA and mercaptan for at least 24 h without affecting the quantitative result. Because of the oxidation of aldehyde and mercaptan the solution should be stored in the absence of air. It has been demonstrated[6] that the fluorescence yield is ca. 10 to 30 times higher at pH 9.5 than at pH 6.0. This is probably a result of the increased rate of reaction in strongly alkaline medium (investigation employing post-column derivatization). The chromatographic separations are always carried out at a pH between 6 and 7.5, certainly in order to avoid shortening the working life of the column unduly. There is no indication in the literature as to whether increasing the pH afterwards (post-column addition) makes it possible to increase the fluorescence markedly.

In contrast to the situation with PTH and dansyl derivatives the lifetime of OPA derivatives is limited. Even during formation the derivatives begin to decompose. The short half-lives of the OPA derivatives can cause some problems during quantitative determination. Half-lives of between 6 and 60 min been determined at pH 4–6[24]. Automation may help to reduce these causes of error. The rate of decomposition is dependent, amongst other things, on the excess of OPA and on the identity of the mercaptan. It has been demonstrated that nonpolar mercaptans (ethyl mercaptan as compared with mercaptoethanol) increase the lifetime[25]. The lifetime is reported as being a few hours for tert-butyl mercaptan[26]. This, however, does not seem to be completely clarified, since very long lifetimes were also obtained with 3-mercaptopropanol[27] and 3-mercaptopropionic acid[28, 29]. the latter compounds possess the advantage of a low vapour pressure and hence cause less malodorous inconvenience.

Since the reaction with OPA proceeds very rapidly at room temperature while, on the other hand, the derivatives are very unstable, the method seems tailor-made for on-line derivatization immediately before injection onto the separation column.

Reaction times of between 90 and 120 seconds are sufficient in all cases. Reagent and sample are taken up one after the other by means of an automatic sample uptake system, mixed in a mixing chamber and applied to the column after the reaction is complete. Various companies now offer additional accessories and programs for the automation of these pre-column derivatizations. The reproducibility of the automatic pre-column derivatization ought to be better than manual sample preparation. Standard deviations of 0.8% for retention times and of 1.9% for the quantitative determination of the areas have been reported [15].

The separation is carried out by gradient elution on RP columns. The polarity of the derivative is a function of the mercaptan employed, so it is necessary to employ various different gradient systems, e.g. 20% CH_3OH to 80% CH_3OH when mercaptoethanol is employed. A gradient of up to 50% acetonitrile is adequate for 3-mercaptopropionic acid.

The derivatives can be detected by UV (absorption maximum 320–340 nm) or preferably and more sensitively by their fluorescence. The excitation can be performed at 340 nm but appreciably higher fluorescence is obtained on excitation at 220 nm. Emission occurs at 450 nm. A few nmol can be detected by means of UV absorption. While the detection limit is at 0.5 pmol (500 fmol) for fluorescence excitation at 340 nm the limit can be reduced 10-fold (40 fmol) by excitation at 220 nm [30]. Similarly low detection limits have been reported for the electrochemical detection of isoindole derivatives [26].

Currently pre-column derivatization with OPA seems to be the most sensitive and rapid method for the analysis of amino acids, however the instability of the derivatives of some amino acids and the impossibility of the direct derivatization of the secondary amino acids are disadvantages.

5. Other Derivatizing Reagents (NBD, FMOC)

Other reagents are and will, with certainty, be employed and described (detailed reviews are to be found in [59] and elsewhere). However it is questionable whether these will be so economically priced and simple to employ as to displace the OPA method. Thus 7-chloro and 7-fluoro-4-ni-trobenzo-2-oxa-1,3-diazole (NBD) [31, 32] are reagents that can compete in sensitivity of detection with OPA and whose amino acid derivatives are simple to separate on RP. One problem which is common to all derivatizing reagents is that they must not produce additional peaks in the area of the chromatogram that is of interest, which are attributable to the reagent or its decomposition products. Only those reagents where these products can be simply separated from the amino acid derivatives are of interest.

During the reaction with FMOC (9-fluorenylmethyl chloroformate) [33, 34] which reacts with both primary and secondary amino acids to yield fluorescent derivatives, which can also be separated on reversed phases, reaction products of the reagent are produced that mask the amino acid FMOC derivatives that are to be separated. It is reported that these interfering products can be almost quantitatively removed by extraction with pentane [35]. The necessity for this extraction makes the employment of this derivatization method in automated pre-column derivatization procedures more difficult. However the combination of OPA and FMOC derivatization is reported to make possible the sensitive automated determination of both primary and secondary amino acids [36].

1.2.2 Post-column Derivatization

1.2.2.1 Separation of the Amino Acids

The specially developed automatic amino acid analysers employ ion-exchange chromatography and post-separation reaction with ninhydrin for detection[1]. Cation exchangers based on polystyrene are employed. Because of their compressibility and the change in swelling behaviour (the height of the packing in the column changes) the separation and, in particular, the regeneration have to be carried out at slow flow rates. An analysis cycle with separation and regeneration takes up a great deal of time, so that fully automatic instruments with automatic sample application were developed at an early stage. The newer, more heavily crosslinked ion exchangers[37] which are also manufactured with a small particle size (10–30 µm) allowed the shortening of the cycle time (1–2 hours analysis time). Elevated temperatures are often employed to increase the efficiency of the separation since they shorten the diffusion times of the substances within the crosslinked resins. Cycle times of about 60 min ought to be achievable by employing cation exchangers based on silica gel (5–10 µm particle diameters)[38] since here it is possible to shorten the regeneration time appreciably because of the better pressure stability[39].

The components for normal HPLC can be employed without difficulty to assemble an automatic amino acid analyser. At least one additional dosing pump is required for the dection system (see below) alongside an automatic sample applier, a gradient system and possibly thermostatting facilities. Since citrate buffers which also contain halogen ions are employed almost exclusively it is necessary to take account of the corrosion resistance of the stainless steel components to citric acid and halides in weakly acid medium. Lithium citrate buffer solutions yield better resolution than sodium-containing buffers.

The separation of free amino acids is also possible with reversed phase systems. To improve separation efficiency it is recommended to work either at low pH (all amino acids are positively charged) or to add ion pairing agents (mostly alkyl sulfonic acids have been employed)[61,62]. Of course, in both cases post-column derivatization is not affected by the eluent additives.

1.2.2.2 Reaction Detector

Extra experimental complication is necessary for the post-column derivatization system, which depends on the detection reaction and reaction conditions to be employed.

The reaction system consists of the following components in addition to the chromatographic separation system: – Reagent pumps, mixing chamber, reaction system with the possibility of operating at elevated temperature and detector. The construction of the reaction detector should be modular since, depending on the reaction employed, up to two reaction systems must be coupled together. Since many reactions are performed at elevated temperature (increase of the reaction rate) a heat exchanger is also necessary, in order to adjust the temperature of the eluent stream to suit the detector. With efficient reaction detectors band broadening should be minimal to reduce peak dilution during reaction.

Three different reactions are employed for the detection of amino acids: – Only one pump is required for reagent supply in the case of the classical ninhydrin reaction. The reaction takes place relatively slowly but can be accelerated by increasing the temperature (sometimes to over 100° C); the formation of disruptive gas bubbles can be avoided by means of a capillary after the detector cell. The detection of the reaction products is performed photometrically at 570 nm. Ninhydrin not only reacts with primary amines but also with secondary amines, whereby however a yellow dye is formed which is detected at 440 nm. Therefore, either a detector is required that can simultaneously detect at two wavelengths or a programmable detector, that can be switched to 440 nm for the detection of proline and hydroxyproline.

If detection is restricted to primary amino acids then a simple one-reagent pump system is also suitable for the OPA reaction. The reaction proceeds rapidly at room temperature so that thermostatting is unnecessary here. If it is also desired to detect secondary amino acids then a double reaction system is necessary: – The first dosing with chloramine T to oxidize proline etc. in a separate reaction system, then the addition of the OPA reagent in a second mixing chamber. The oxidation reduces the limits of detection of the primary amino acids so that it seems preferable only to dose with chloramine T during the time that proline or hydroxyproline is being eluted[40]. As previously mentioned a fluorescence detector is to be preferred for sensitivity of detection.

Reaction with fluorescamine[3–5] (Fluram) requires a system with two pumps, one of which supplies the reagent and the second a strongly alkaline buffer that increases the yield of fluorescent derivatives. The reaction occurs rapidly so that it is sometimes even possible to dispense with a reaction chamber. The reaction products themselves decompose in the medium so that it is necessary to keep the reaction time very constant. The reagent is also not suitable for the detection of secondary amino acids. The problems of optimization and the very high price have certainly hindered the broad acceptance of this reagent. The sensitivity is a factor of 10–100 higher than that of the ninhydrin reaction and thus is of the same order of magnitude as that for the OPA reaction.

1.2.2.2.1 Reaction Detection System

The individual components of a reaction detection system should fulfil the following requirements: –

1. Reagent Pump

A continual pulse-free reagent stream is necessary. Pulse damping can be achieved with classical RC components (Bourdon tubes – restriction capillaries) because a small dead volume is not necessary. The system should be resistant to the often very corrosive reagents. The flow constancy of the reagent pump directly affects the accuracy of the determination when quantitating with concentration-sensitive detectors via the peak areas. In addition the relationship between reagent and eluent changes, thus altering the degree of reaction. Short-term flow constancy of the pump which is worse than 1% turns out to be unacceptable. For this reason we employ low cost HPLC pumps as reagent pumps.

2. Mixing

The column eluate and reagent are usually mixed by means of a T piece or Y piece, whereby in the first instance the two flows meet each other at an angle of 180°. There are various opinions concerning the mixing angle, however, it does not seem to be of great importance[41]. The problem is that in such T pieces, as with open tubes, there is scarcely any radial mixing, rather the flows of reagent and eluate run parallel to each other for a long time so that the commencement of the reaction is delayed. In this type of mixing it is necessary that the reagent and eluate be mixed in approximately equal volumes. This leads to a 1:1 dilution and hence to a reduction in sensitivity. The large dead volume is a further disadvantage. Mixers of the cyclone type are appreciably more effective[42] and can be constructed with relatively small dead volumes (10 nl). Another advantage of this mixer is that good mixing still occurs when the reagent flow is only 1/25 that of the eluent flow[43]. The dilution of the substance zones can be appreciably reduced in this manner.

3. Reactor

The eluate and reagent have to be held in the reactor during the course of the reaction without losing the separation that has already been achieved on the chromatographic column because of axial mixing in the reactor. Three different types of reactor are employed: – a) packed columns, b) open tubes with segmentation and c) open tubes without segmentation, but with forced radial mixing, e.g. by simple coiling (coiled open tubes) or knitting[42]. The advantages and disadvantages of individual types of reactor are presented in Table 2.

The packed reactors are chromatographic columns with 3–6 mm diameters in order to achieve sufficient volume with relatively short lengths (additional pressure drop!). They are packed with inert material usually glass spheres. This packing is only partially suitable for amino acid analyses employing strongly alkaline reagens (pH 9–10). Tantalum spheres could be an alternative here. The problems with this type of reactor lie in the dangers of blockage with micro-particles, which are often formed during the reaction and in the high additional pressure drop, since the contribution of the reactor to band broadening is the more negligible the smaller the particle diameter.

Open tubes with segmentation of the flow by air bubbles to reduce axial mixing (which reduces the separation already achieved) find widespread application in auto analyser systems. This type of reactor is only of limited utility for coupling with HPLC, since an intolerable further band broadening would be brought about during the removal of the air bubbles. This type of reactor is excellently suited for very slow reactions (reaction time longer than 5 min). However, the derivatization methods employed in amino acid analysis are all appreciably more rapid (reaction time ca. 1 min) so that it does not seem suitable in this case.

It is possible to construct simple reaction detection systems for amino acid analysis with the appreciably easier to construct coiled or knitted capillaries. If open tubes are employed without segmentation then axial mixing leads to destruction of the separation produced in the column. Suitable bending can be employed to stimulate a secondary flow in the capillaries which leads to radial mixing and appre-

Table 2. Reactor types

Packed	Air-segmented	Open tubes wound or knitted
Advantages		
Instrumentally simple T > 100° C possible	Low pressure drop long reaction times possible	Instrumentally simple T > 100° C possible (steel tube)
Packing can participate in reaction (bound enzymes, catalysts)	Stopped flow possible multi-stage reaction detector cheaper	Band broadening inde- pendent of flow-rate for knitted
Disadvantages		
Higher pressure drop danger of blockage	Band broadening when removing bubbles	Pressure drop a factor of 2–3 higher then in straight tubes
Thermostatting problems	Instrumentally complex	
Problems with aggressive reagents	Vapour pressure of eluent interferes: reaction temperature = b.p. − 30°	
Limited packing life	Limitations to choice of material (wetting)	
	Constancy of reaction time problematical be- cause of gas segmentation	
	Electronic elimination of gas bubbles complex and expensive	
Optimal particle size 20–30 μm	Optimal diameter 1–2 mm	Optimal 0.25–0.35 mm i.d.

ciably reduces the band broadening. Simply coiling the tube (coiled open tubes) leads to an appreciable reduction in band broadening. Knitted[42] and stitched[43] capillaries are more efficient bent tubes, where the secondary flow is completely developed at low flow velocities. This means that the band broadening is independent of the flow rate. These knitted capillaries are also efficient heat exchangers because of the good radial mixing.

The optimal capillary diameter is 0.25–0.35 mm i.d. (commercial Teflon capillaries of 0.3 mm nominal i.d. are in fact 0.33–0.38 mm i.d.). A standard length of 10 m is sufficient for a reaction time of up to 80 sec at the normal chromatographic flow rates of 0.8–1.0 ml/min. Such reactors are compatible with the separation efficiencies obtained nowadays with good 5 or 7 μm separation columns. Short capillaries (60 cm) are adequate additional heat exchangers for bringing the eluents back to the detector temperature (lowering noise, improvement of fluorescence yield). Such reactors can be constructed for use with 3 μm columns or micro-columns of 1–2 mm diameter by employing 0.1 mm i.d. stainless steel tubing.

It is difficult to discuss the advantages and disadvantages of the various types of reactor system. No experimental results are available for such a comparison, rather theoretical observations[44] which however only allow a limited comparison of reactors under their individual optimal conditions. A comprehensive monograph on post-column derivatization is in preparation[45]. Nevertheless, open tubes (coiled or knitted) are probably optimal for amino acid derivatization. They can be employed for both the ninhydrin reaction and for derivatization with OPA. A comparison of the ninhydrin and the OPA post-column reaction has been made with a commercial system[40]. The additional oxidation of secondary amines was also investigated in the case of the OPA reaction. The separation systems were identical (polystyrene cation exchanger; dp 9 μm, lithium citrate buffer) so that a real comparison is possible of the post-column derivatization. Detection limits (signal-to-noise ratio = 2) below 100 pmol were obtained with ninhydrin. Detection limits of 225–250 pmol were reported for proline and tryptophan. Under the conditions employed the OPA reaction is a factor of 20 better. Detection limits of 5 pmol were obtained, the detection limit for cysteine was 30 pmol. The addition of chloramine T and reaction (oxidation) at 40° C allowed the detection of 100 pmol of proline. The authors recommended the OPA reaction when it is not necessary to detect proline. The reproducibility of the retention times corresponded to a standard deviation of 1–4%, that for the peak areas was 3%. With an optimized reaction detection system (cyclone mixer, efficiently deformed "knitted" open tubes) it is possible to reduce the detection limit for the OPA reaction (excitation at 220 nm) to 20 pg amino acid[43].

1.2.3 Summary

The most important derivatization reactions and the reported detection limits are listed in Table 3 for comparison.

The instrumental complexity for pre-column derivatization is insignificantly higher than for a normal HPLC gradient system. In the case of the on-line derivatization

Table 3. Pre-column derivatization

		OPA	Dansyl	Dabsyl	PTH
Detection of sec. amino acids		No	Yes	Yes	Yes
Simple to automate		Yes	No	No	Yes
Stable products		No	Yes	Yes	Yes
Interference from side reactions		No	Yes	Yes	No
Detector		UV	UV	VIS	UV
		Fluoresc.	Fluoresc.		
Detection limits (pmol)	UV	1000	2	2	5
	F	0.5	0.5	–	–

	OPA	OPA/chloramine	Ninhydrin	Fluram
Post-column derivatization				
Detection of sec. amines	No	Yes	Yes	No
Stable products	Yes	No	Yes	No
Detector	Fluoresc.	Fluoresc.	VIS	Fluoresc.
Detection limits (pmol)	5	10–100	100	10

with OPA it is necessary to arrange for the sequential uptake and mixing of the reagent and the sample from different vessels into the storage loop, and application of the mixture to the column after a standardized reaction time. Only fluorescence can be employed as a sensitive detector. An instrument with the possibility of excitation at 220 nm with a deuterium lamp is to be desired.

The expense of post-column derivatization is much higher. Here at least one additional HPLC pump is required as well as the mixer and reactor. The separation and detection reaction are unambiguous, but the time required for the separation can be larger than in RP separation after pre-column derivatization.

Because of the different problems encountered in amino acid analysis there can be no one answer to the question of which process is universally applicable.

In the case of the determination of the structures of proteins the Edman degradation itself produces PTH derivates which are separable without any difficulty on RP columns.

If the total amino acid composition of proteins and peptides is to be investigated and if a peptide synthesis is to be monitored analytically then the separation of OPA derivatives after pre-column derivatization is a suitable method, however all the secondary amino acids are undetectable by this method. This method provides rapid analyses while employing standard chromatographic columns and instruments. Because of the decomposition of the OPA derivatives the quantitative reproducibility is only adequate if the derivatization step is automated and on-line and takes place immediately before sample application and the separation is performed under standardized conditions.

Automatic amino acid analysis with post-column derivatization is still the method of choice for the determination of total content of amino acids in biological fluids (plasma, urine, etc.). The components of the HPLC apparatus can be employed here without any modification, or put otherwise, a gradient HPLC apparatus is able to substitute for an automatic amino acid analyser. The analysis times are a factor of 3–6 longer than is the case for the RP separation of OPA derivatives. There is no difficulty in detecting secondary amino acids when ninhydrin is employed. It would be possible to improve on the analysis times and the detection sensitivity here by optimization of the ion exchangers (silica gel-based, reduction of the particle size), of the visible light detector and the reaction conditions (optimization of the mixing chamber, reactor and reaction conditions) and by using microcolumns.

2 Chromatography of Proteins

Proteins differ from each other in their molecular size, shape, charge and in their hydrophobic properties. These differences are exploited in the classical protein isolation and purification methods in order to isolate and separate proteins by employing sedimentation (ultracentrifuge), ion exchange and salting out. Chromatographic separation techniques are available which separate according to size, charge and hydrophobicity and are employed with success for the separation and purification of proteins. The choice of the analytical method depends on the aims of the analysis.

2.1 Separation on Reversed Phases

If, for example, only the degradation products of large protein molecules are to be analysed, where there is no question of needing to retain biological activity, then chromatography is possible on reversed phases[46, 47]. The best separation results are obtained when working at about pH 2, since all proteins then possess the same charge and interferences (reversed elution, asymmetrical peaks etc.) caused by difference in charge are avoided. The elution is performed by employing a gradient of increasing acetonitrile concentration. C8 or C18 reversed phases preferably with a mean pore size of ca. 50 nm are suitable stationary phases. The coating of the surface silanol groups should be as complete as possible[48]. The influence of unoccupied silanol groups makes itself evident at high acetonitrile concentrations, particularly above 50%. Here the retention begins to increase with increasing acetonitrile content – in contrast to the usual behaviour[47, 49]. An optimal reversed phase should possess as few accessible silanol groups as possible (good end-capping necessary), the whole surface should be available to the proteins without inhibition (large mean pore size, 30–60 nm).

Because of the low specific surface area and hence the carbon content per unit volume of column that this involves the proteins begin to be eluted by even small proportions of acetonitrile in aqueous buffer solution.

A few percent change in the composition of the eluent leads to a complete switch from total retention to inert elution. Very high demands are placed on the reproducibility of the gradient elution. It is almost impossible to perform an isocratic determination of the k' values or retention times. The characterization of the retention is best performed by means of the eluent composition at which the individual proteins are eluted within the gradient. Increasing the temperature leads to an improvement of the peak shape (reduction in eluent viscosity, increase in diffusion coefficients), but scarcely affects the retention behaviour (a shortening of the retention time would have been expected). Denaturation of the proteins (precipitation) is always to be feared at temperatures above 50° C.

Since these separations succeed best at strongly acid pH and because the interaction between the sample molecules and the stationary phase is so strong, the separating capacity, on the one hand, is very good, on the other hand, there is a great danger of changing the structure of the proteins. Hence, it is to be feared that the proteins will lose their biological activity on reversed phase separation. Often the elution is not quantitative, which can lead to a ghost chromatogram with a blank gradient without sample application. Absorbed proteins naturally alter the selectivity of the column.

2.2 Separation on Ion Exchangers

Synthetic resin-based ion exchangers for HPLC are usually highly cross-linked in order to improve the pressure stability. This means that the pore diameter of the resins is very low, so that these resins are of only limited application in the separation of proteins. Furthermore in aqueous media the separation by ion exchange is often superimposed or overlayed by hydrophobic interactions, this, however, can be reduced by the addition of organic eluents. Nevertheless, ion exchangers based on hydrophilic resins (polyethers) with large diameter pores have been commercially available for several years as anion and cation exchangers[55]. As with the classical soft gel ion exchangers the separations can be performed by buffers alone or in combination with organic eluents. Chromatofocusing is also possible with ampholytes as buffer[56]. Large pore silica gel, coated with polyethylene imine, can also be employed as an ion exchanger for the separation of peptides and proteins[57].

The combination of HPLC column technology with affinity chromatography is likely to extend the range of application of both methods[58], whereby the complete coverage of the surface of the stationary phase in order to avoid loss of substance is likely to prove to be the most critical problem.

2.3 Separation by Exclusion Chromatography

Exclusion chromatography is still the most elegant method of protein separation, whose low pressure version, gel filtration, has been and is still very successfully employed for the purification and isolation of proteins. By definition no interaction should occur in this separation method between the surface of the stationary phase and the sample molecules. The separation takes place exclusively according to molecular size because of the difference in accessibility of the pore volume to the protein molecules. Unfortunately this method only possesses a small separating capacity, which is limited to the pore volume of the stationary phase in the separating column. Chemically modified silica gels are available for the separation of proteins under HPLC conditions[47, 50, 51]. Ionogenic interactions between the non-reactable silanol groups and the protein molecules, which could lead to ion exchange or ion exclusion, can be reduced or avoided by the addition of neutral salts (0.3–0.5 M) to the eluent. Aqueous buffer solutions can be employed as eluents in exclusion chromatography, i.e. it is possible to work under physiological conditions. The recovery rates are high and the biological activity of the proteins is completely retained during this method[51].

Since the separating capacity is very limited and, on the other hand, the end of the separation can be predicted (the exclusion chromatography is completed with the elution of the smallest molecules, this corresponds to the inert peak in absorption chromatography) other optimization strategies must be employed than is the case for retention chromatography. The separations have to be carried out at appreciably lower eluent velocities because of the lower diffusion coefficients. If the optimal eluent velocity for samples of molecular weight up to 500 daltons is 1 mm/sec (corresponding to a flow rate of 1 ml/min in a 4 mm diameter column), then the optimal flow rate for samples of molecular weight 40000 to 100000 daltons is a factor of 100 less. The increased analysis times this involves can be compensated for by employing shorter and more efficient separating columns packed, for example, with small particles (dp < 5 μm)[52].

An increase in the pore volume of the silica carrier (increasing the ratio of the pore volume to the interparticle volume in the separation column) also leads to an improvement of the separation efficiency of exclusion chromatography and allows a further reduction in the length of the column. Unfortunately this direction of increasing the pore volume has natural limits, since it occurs after a certain level at the expense of the silica structure and, hence, of the mechanical stability of the carrier.

2.4 Separation by Salting-out Chromatography

Salting-out chromatography, also known as hydrophobic interaction chromatography, is a method that combines the almost physiological conditions of exclusion chromatography with the separating capacity of chromatography on reversed phases. Here silica gel phases are employed which are not modified by non-polar alkyl groups but by polar groups, such as N-acetylpropylamide[50, 51, 53] or glycol ethers[51]. The hydrophobic properties are small, no sorption of the proteins occurs in pure buffers (I <0.3). Under these conditions such phases are also suitable for exclusion chromatography. In salting-out chromatography the samples are applied in a buffer that also contains a high concentration of neutral salt, e.g. ammonium sulphate (2–2.5 M). The ammonium sulphate concentration is then reduced steadily in a gradient where solvent B contains less salt or is pure buffer and the proteins are eluted according to their hydrophobicity. The pH of the solvent can also be employed for optimization. The sorption of a protein is always greatest in the region of its isoelectric point. The recovery rate of the proteins and the retention of their biological activities is similar to that in exclusion chromatography, however the separational efficiency of salting-out chromatography is superior to that of the latter.

It is to be expected that in the near future HPLC technology will simplify the preparative isolation of proteins from complex matrices and that the rapid and efficient analytical possibilities of HPLC will allow the monitoring of biotechnological processes.

3 References

1. Speckman, D.H., Stein, W.H., Moore, S.: Anal. Chem. 30, 1190 (1958)
2. Zimmermann, C.L., Pisano, J.J.: in Methods in Enzymology, Vol. 47, C.W.H. Hiss, S.N. Timascheff, Eds., Academic Press, New York 1947, p. 45ff.
3. Samejima, K., Dairman, W., Udenfried, S.: Anal. Biochem. 42, 222 (1971)
4. Udenfried, S. et al.: Scienc 178, 871 (1972)
5. Stein, S., Bohlen, P., Stone, J., Dairman, W., Udenfried, S.: Anal. Biochem. Biophys. 155, 202 (1973)
6. Roth, M.: Anal. Chem. 43, 880 (1971)
7. Roth, M., Hampai, A.: J. Chromatogr. 83, 353 (1973)
8. Engelhardt, H., Müller, H., Schön, U.: "The Column" in A. Henschen et al. eds., HPLC in Biochemistry, VCH, Weinheim (1985)
8a. Seiler, N., Knodgen, B.: J. Chromatogr. 341, 11 (1985)
9. Margolies, M.N., Brauer, H.: J. Chromatogr. 148, 429 (1978)
10. Granberg, R.R.: LC Magazine 2, 276 (1984)
11. Engelhardt, H., Dreyer, B., Schmidt, H.: Chromatographia 16, 11 (1982)
12. Kohr, W.J., Rodriguez, H., Harkins, R.N.: Abstract 502, Baltimore 1982
13. Chang, J.Y., Lehmann, A., Wittmann-Leibold, B.: Anal. Biochem. 102, 380 (1980)
14. Koop, D.R., Morgan, E.T., Tarr, G.E., Coon, M.J.: J. Biol. Chem. 257, 8472 (1982)

15. Bidlingmeyer, B.A., Cohen, S.A., Tarvin, T.L.: J. Chromatogr. 336, 93 (1984)
16. Bayer, E., Grom, E., Kaltenegger, B., Uhman, R.: Anal. Chem. 48, 1106 (1976)
17. Kromidas, S.: Dissertation, Saarbrücken 1983
18. Page, J.N. Le, Lindner, W., Davies, G., Seitz, D.E., Karger, B.L.: Anal. Chem. 51, 433 (1979)
19. Hill, D.W., Walters, F.M., Wilson, T.D., Stuart, J.D.: Anal. Chem. 51, 1338 (1979)
20. Hodgin, J.C.: J. Liquid Chromatogr. 2, 1047 (1979)
21. Cooper, J.D.H., Turnell, D.C.: J. Chromatogr. 227, 158 (1982)
22. Lee, K.S., Drescher, D.G.: J. Biol. Chem. 82, 250 (1977)
23. Cooper, J.D.H., Lewis, M.T., Turnell, D.C.: J. Chromatogr. 285, 484 und 490 (1984)
24. Dong, M.W., DiCesare, J.I., Steinwand, M.: Perkin Elmer Series, Angew. Chromatographie, Heft 42, 1985
25. Pfeifer, R.F., Hill, D.W.: in Advances in Chromatogr., J.C. Giddings et al. Eds. Vol. 22, p. 61, Marcel Dekker, New York 1983
26. Allison, L.A., Mayer, G.S., Shoup, R.E.: Anal. Chem. 56, 1089 (1984)
27. Stobaugh, J.F. et al.: Anal. Biochem. 135, 495 (1983)
28. Földi, Dr., LKB München: Private communication
29. Godel, H., Graser, T., Földi, P., Pfaender, P., Fürst, P.: J. Chromatogr. 297, 49 (1984)
30. Turnell, D.C., Cooper, J.D.H.: Clin. Chem. 28, 527 (1982)
31. Imain, K., Watanabe, Y., Toyooks, T.: Chromatographia 16, 214 (1982)
32. Linblad, W.J., Diegelman, R.F.: Anal. Biochem. 138, 390 (1984)
33. Carpino, L.A., Han, G.Y.: J. Org. Chem. 37, 3404 (1972)
34. Moye, H., Boning, A.J.: Anal. Lett. 12, 25 (1979)
35. Einarsson, S., Josefsson, B., Lagerkvist, S.: J. Chromatogr. 282, 609 (1983)
36. Lecture A. Apffel: Symposium über Kopplungsverfahren Saarbrücken 1985
37. Hamilton, P.B.: Anal. Chem. 35, 2055 (1963)
38. Dissertation P. Orth: Saarbrücken 1979
39. Dissertation B. Lillig: Saarbrücken 1984
40. Cunico, R.L., Schlabach, T.: J. Chromatogr. 266, 461 (1983)
41. Scholten, A.H.M.T., Brinkman, U.A.Th., Frei, R.W.: J. Chromatogr. 218, 3 (1981)
42. Engelhardt, H., Neue, U.: Chromatographia 15, 403 (1982)
43. Engelhardt, H., Lillig, B.: High Res. Chrom. & Chrom. Comm. 8, 531 (1985)
44. Deelder, R.S., Kuipers, A.T.J.M., Van den Berg, J.H.M.: J. Chromatogr. 255, 545 (1983)
45. I. Krull Editor, Post column derivatization in HPLC, M. Dekker, New York to be published 1986
46. Hancock, W.S., Sparrow, J.T.: in "High Performance Liquid Chromatography Advances and Perspectives", Vol. 3, p. 49–85, Cs. Horvath ed., Academic Press, New York 1983
47. Hearn, M.T.W.: in Advances in Chromatography, Vol. 20, J.C. Giddings et al. eds., Dekker, New York 1982
48. Engelhardt, H., Müller, H.: Chromatographia 19, 77 (1984)
49. Hearn, M.T.W., Grego, B.: J. Chromatogr. 218, 497 (1981)
50. Engelhardt, H., Mathes, D.: J. Chromatogr. 185, 305 (1979)
51. Engelhardt, H., Mathes, D.: Chromatographia 14, 325 (1981)
52. Engelhardt, H., Ahr, G.: J. Chromatogr. 282, 385 (1983)
53. Schön, U.: Dissertation, Saarbrücken (1984)
54. Miller, N.T., Feibush, B., Karger, B.L.: J. Chromatogr. 316, 519 (1985)
55. Product information FPLC, Pharmacia Fine Chemicals, Uppsala, Schweden
56. Proceedings of the 30th Symposium, Protides of Biological Fluids 1982, Pergamon Press, Oxford
57. Alpert, A.J., Regnier, F.E.: J. Chromatogr. 185, 375 (1979)
58. Larsson, P.-O., Glad, M., Hansson, L., Mansson, M.-O., Ohlson, S., Mosbach, K.: in Advances in Chromatography, Vol. 21, J.C. Giddings et al. eds., Dekker, New York 1983
59. Lottspeich, F., Henschen, A., in Henschen A. et al. eds.: HPLC in Biochemistry; VCH, Weinheim, 1985
60. Lottspeich, F.: J. Chromatogr. 326, 321 (1985)
61. Molnar, J., Horvath, Cs.: J. Chromatogr. 142, 623 (1977)
62. Radjai, M.K., Hatch, R.T.: J. Chromatogr. 196, 319 (1980)

Chromatographic Methods in the Separation
of Coal and Oil Products

Th. Crispin
Bayer AG, PH-EP-AQ-QL, 5090 Leverkusen, Bayerwerk/FRG

I. Halász
Universität des Saarlandes, Physikalische Chemie, 6600 Saarbrücken/FRG

1 Introduction

A further efficient employment of the enormous reserves of fossil fuels will only be made possible by detailed investigation of the components and their properties and behaviour during different applications. Chromatographic methods have played a dominating role here since the 1960s.

Because of the multiplicity of stationary and mobile phases within the various chromatographic techniques (LC, GC, thin-layer, supercritical-fluid chromatography) and because of the large numbers of specific detection possibilities it is possible to obtain comprehensive and reproducible evidence concerning the composition of crude oils and coal products in a very short time.

The following chapter is intended to provide the reader, who is not very familiar with this area of application, with a synopsis of the most important routinely applied, mainly chromatographic, methods for the separation and characterization of petroleum and coal products, with particular emphasis on new trends and developments.

2 Distillates

2.1 Gas Chromatography

2.1.1 Light Products (Boiling Point $\leq 230\,°C$)

Alongside gas analysis[1] which is employed, amongst other things to determine the calorific value and to check the highest permissible emission limits, GC is principally employed for the separation and characterization of gasoline and naptha (raw gasoline) distillates, which are almost exclusively employed as automotive fuels and as feed stocks for reformers. The types and amounts of the constituents are decisive for product quality (calorific value, volatility, octane number, storage properties).

A high resolution chromatographic separation column is necessary for complete analysis. Parallel to this the detection methods should, to a first approximation, be substance-independent. GC fulfils both requirements to a large degree. Variously coated and thereby selective glass or quartz capillaries[2-4] with up to 3000 plates per meter solve most separations satisfactorily up to a molecular weight of 400 (for paraffins to 750). A "universal" mass detector is available in the guise of the flame ionization detector. Other specific detectors (S, N, O) are the state of the art[5,6].

Additionally coupling is possible to a mass spectrometer, which enables direct identification of peaks. This "on-line" mode and the processing of the data which are generated by means of a computer-controlled GC-MS system considerably eases the task of the analyst.

Whittemore[7] provides a typical example for the efficiency of the GC-MS analysis of a gasoline. Within 4 hours a total of 377 components were separated, often excellently, and over 90% were positively identified by mass spectrometry.

Packed columns of lower efficiency are completely adequate for the solution of certain specific problems such as, for example, the determination of the benzene and toluene contents[8]. In fact a high resolution is not desired for the GC simulation of a distillation[9]. The boiling ranges can very quickly be determined from such GC data and other important bulk physical properties can be predicted, which is of importance, for instance, in process control.

2.1.2 Middle Distillates, Vacuum Gas Oil, Shale Oil, Coal Tar, SRC Products

The great complexity in this boiling point range allows of two possibilities for the meaningful application of GC.

a) The production of GC fingerprints, particularly in paraffin wax analysis[10, 11].

Schomburg[12, 13] has described high efficiency glass capillary columns for the separation of coal tar, which have a working temperature up to 390 °C. Figure 1 illustrates a typical GC pattern for multinuclear aromatics from coal tar. More than 140 components with boiling points up to 550 °C (molecular weights up to 302) can be separated and identified or characterized by GC-MS coupling.

b) Previous enrichment of interesting components by suitable methods (precipitation, extraction, chromatography). This second possibility will be looked at more closely in the following section:

Marquardt[14] separated paraffins from the matrix (heavy gas oil) as their urea adducts and then determined them gas chromatographically. A "pure" paraffin chromatogram with excellent resolution can be produced from $n\text{-}C_{12}-n\text{-}C_{35}$ within 30 minutes. This technique can also be employed for other samples (Fig. 2).

Liquid chromatographic "group" separations on silica gel (aluminium oxide) stationary phases offer an excellent opportunity for enriching certain classes of substance prior to the actual gas chromatography. Here the elution capacity of the eluent is increased from nonpolar to polar in a step-wise manner. The individual classes of substance can then be analysed gas chromatographically.

Di Sanzo[15] has investigated high boiling (boiling point > 170 °C) Fischer-Tropsch products from various process conditions. Paraffins, olefins, alcohols, esters, ketones and acids are obtained in the liquid chromatographic preseparation and individually characterized by high resolution GC (the stationary phase is OV 101). Important evidence can be obtained concerning the course of the process by the employment of secondary analytical methods (IR, NMR).

Figure 3 illustrates the separation scheme employed. This process has also been employed for the shale oils which are of very complex composition; here slight variations in the scheme are necessary because of the nature of the samples[16]. Asphaltenes[39] and other very polar materials must be removed before commencement.

Alexander[17] has enriched the aliphatic and aromatic hydrocarbons of various types of coal by means of an extraction step followed by LC fractionation. It can be demonstrated by means of capillary GC (glass capillary with OV1 stationary phase), that the distribution of n-paraffins differs in samples from differing types of sources (Fig. 4). The GC fingerprints of the aromatic fractions are also characteristic of individual types of coal.

A similar separation method has been described by Romanowski et al.[18]. Substances up to a molecular weight of 456 were identified by GC-MS coupling as the last stage in separation.

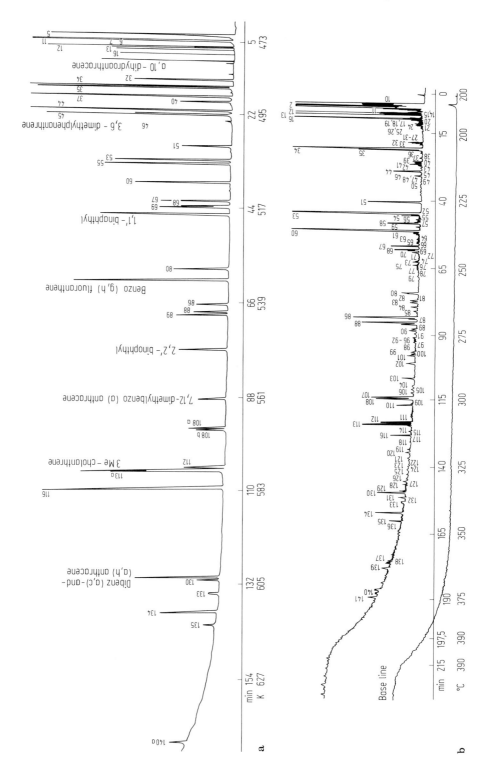

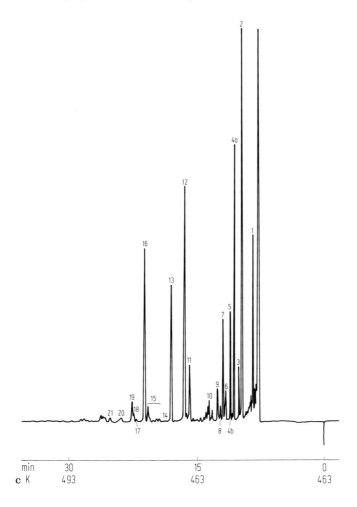

Fig. 1a–c. Gas chromatography of PNAs and coal tar on Poly S 179. **a** PNA test mixture. Column, 92-m Poly S 179, I.D. 0.27 mm alkali-glass, HCl-etched. Column temperature, programmed from 473 to 663 °K at 1 °K/min. Carrier gas, hydrogen, 0.35 m/sec at 663 °K. 34 = Phenanthrene; 35 = anthracene; 53 = fluoranthene; 60 = pyrene; 86 = benzo[a] anthracene; 88 = chrysene; 112 = benzo[e]pyrene; 113a = benzo[a]pyrene; 116 = perylene; 130 = indeno[1,2,3-cd]pyrene; 134 = benzo[ghi]perylene; 135 = anthanthrene; 140a = coronene. **b** Coal tar. Column, 92-m Poly S 179, I.D. 0.27 mm. Column temperature, programmed from 473 to 663 °K at 1 °K/min. Carrier gas, hydrogen, 0.35 m/sec at 663 °K. Peaks listed as in (a). **c** Coal tar (low-boiling fraction). Column, 92-m Poly S 179, I.D. 0.27 mm. Column temperature, programmed from 463 to 495 °K at 1 °K/min. Carrier gas, helium. 0.20 m/sec at 463 °K. 1 = Indene; 2 = naphthalene; 3 = benzo[b]thiophene; 4a = 2-methylnaphthalene; 4b = azulene; 5 = 1-methylnaphthalene; 7 = biphenyl; 11 = acenaphthene; 12 = acenaphthylene; 13 = dibenzofuran; 16 = fluorene. (d) Coal tar. Instrument, Finnigan 4000. Column, 92-m Poly S 179, I.D. 0.27 mm. Column temperature, 600–610 °K. Interface, Pt capillary, I.D. 0.1 mm. Interface temperature, 540 °K. Ion source temperature, 573 °K. Benzpyrene fraction of the computer-reconstructed total ion chromatogram. 107 = Benzo[b]fluoranthene; 108a = benzo[j]fluoranthene; 108b = benzo[k]fluoranthene; 112 = benzo[e]pyrene; 113a = benzo[a]pyrene; 116 = perylene. (Reprinted with permission from [13])

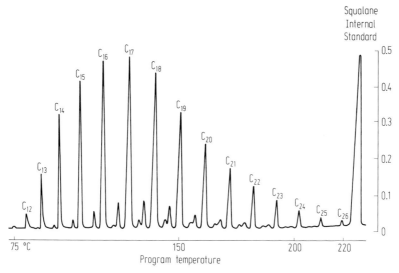

Fig. 2. GLC spectrum from HGO-A (HGO = heavy gas oil). (Reprinted with permission from [14])

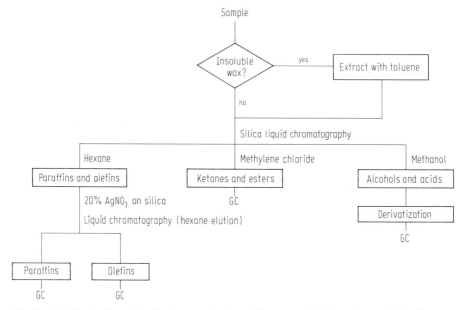

Fig. 3. Analytical scheme for the characterization of heavy-end Fischer-Tropsch Liquids. (Reprinted with permission from [15])

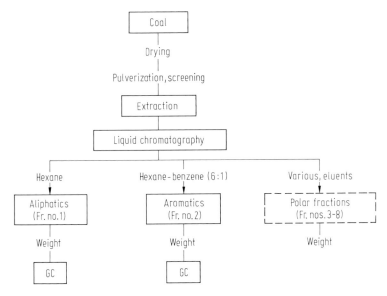

Fig. 4. Analytical scheme for eluents used with polar fractions. (Reprinted with permission from [17])

2.2 HPLC Methods

The time-consuming classical LC methods[19-21] are being replaced more and more by modern HPLC. Liquid chromatographic methods possess the advantage over GC that the sample is not destroyed (→ secondary analysis). The basic ideas behind the application of liquid chromatography (adsorption chromatography) in the fields of petroleum and coal originate from L.R. Snyder[21,22].

2.2.1 Silica Gel (Aluminium Oxide) as Stationary Phases

The investigations of Suatoni and Swab have pointed the way[23-25]; their rapid and reproducible HPLC technique for the analytical group separation of petroleum and coal products is applicable within a wide boiling point range.

Gasoline fractions[23] are separated on a 10 μm silica gel column with a very apolar perfluorinated eluent (FC-78) into paraffins, olefins and aromatics. The latter are recovered as a sharp peak by back flushing the column. This offers an alternative to the standard FIA analysis[26], whose application is subject to certain restrictions. The total analysis time can be reduced by a factor of 20–30 while retaining the same high degree of accuracy and reproducibility. If paraffins and olefins occur in one fraction[24], this mixed fraction can be separated on an activated silica gel column with i-octane/FC-78 (70:30) v/v as mobile phase whereby the olefins are eluted by back flushing. n-Hexane is employed as apolar mobile phase instead of FC-78 for fractions with a boiling point at 270 °C. Aromatics are again obtained by back flushing the column, while polar compounds are displaced from the column by a mixture of $CH_2Cl_2|(CH_3)_2CO$ (1:1)$_{v|v}$. A typical apparatus for such separa-

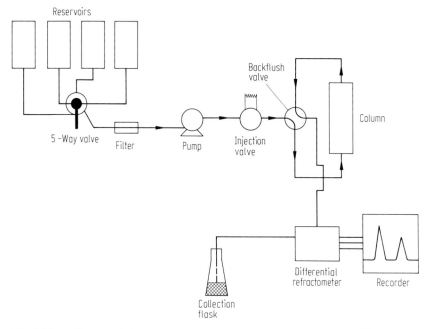

Fig. 5. Flow diagram of chromatograph. (Reprinted with permission from [25])

tions is illustrated in Fig. 5. The overlap of fractions is minimal in all applications, which is of particular importance when working preparatively[25], (column 100×1.6 cm) (0.25–5 g sample) with secondary analysis of the fractions.

Vogh and Thomson[27] have developed efficient reusable preparative columns (SiO_2, Al_2O_3) for the separation of vacuum gas oils (370–535 °C) and similar products. Depending on its composition between 1 and 6 g of sample can be separated by gradient elution (diethyl ether in hexane) and back flushing in slightly less than 2 hours. The maximum loading and cutting times (time of operating the valve) for the individual fractions are determined in an analytical system with suitable model substances and employed in the preparative separation. The following fractions are obtained: – paraffins and monoaromatics, diaromatics, polyaromatics, polar compounds. The first fraction (paraffins + monoaromatics) is separated into paraffins and monoaromatics on a silica gel column. The columns can be reused without loss of sharpness of separation. In all work in this molecular weight range the unequivocal assignment to chemically defined classes (paraffins, aromatics, etc.) is difficult, since, of necessity, the systems are calibrated with low molecular weight compounds. The properties of these substances then have to be extrapolated to the high molecular weight range, because here there is only a very small number of suitable model substances available. However this is of secondary importance for relative conclusions or the comparison of various products.

2.2.2 Specialized Stationary Phases

A short review of the possible stationary phases and eluents for employment in HPLC is contained in[28]. It is emphasized here that the employment of SiO_2 and

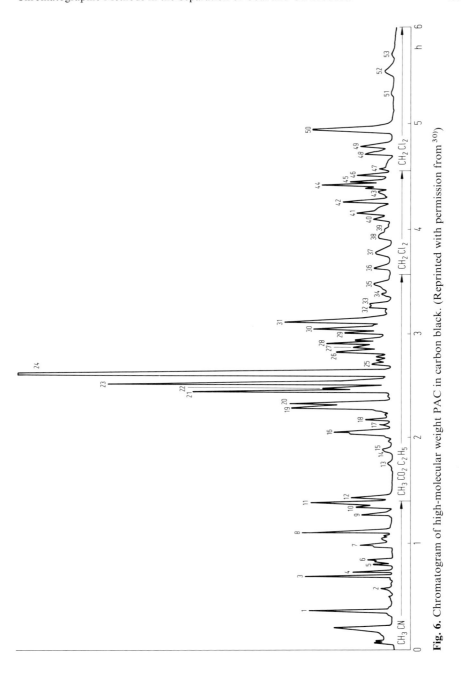

Fig. 6. Chromatogram of high-molecular weight PAC in carbon black. (Reprinted with permission from [30])

Al_2O_3 in HPLC applications is only of utility in group analysis where the separation of individual substances does not play a decisive role. The separation of individual, defined fractions (paraffins, aromatics, etc.) requires phases with specialized properties. The main field of interest is in the further chromatographic separation of aromatic fractions, since aromatics have a decisive effect on product quality and on the behaviour of a product in a conversion process [29]. Certain aromatics are powerful carcinogens and hence of considerable importance as far as safety at the place of work is concerned.

Lee[30] extracted coal with CH_2Cl_2. The high molecular weight aromatic extract was applied to an analytical C-18 reversed-phase column (250×4.1 mm) and separated by gradient elution. The chromatogram (Fig. 6) shows an excellent separation. It was possible to identify a total of 53 individual components positively on the basis of their fluorescence spectra and mass spectroscopy. Furthermore, the author provides an extensive literature survey of the separation of polycyclic aromatics.

Mourrey et al.[31] have described an aromatic-specific phase. This phase can be employed with both apolar ("normal mode") as well as with typical RP eluents ("reversed-phase mode"). Aromatics are obtained by the preparative separation of n-hexane or toluene extracts of shale oil on silica gel (Al_2O_3). The following separation on pyrrolidone phase occurs via a charge transfer mechanism strictly according to the number of aromatic rings. Thiophenes and heterocyclic nitrogen compounds scarcely interact with the phase. The reversed-phase mode (eluent: CH_3OH/H_2O (60:40) v/v) is best suited for molecules containing up to six rings, while the normal mode (eluent: n-heptane) has proved of value for samples with more than three aromatic rings. The retention range is determined with the aid of model substances, since the chromatographic separation of the complex shale oil extracts is only slight.

Holstein[32] describes the synthesis of a tetrachlorophthalimido phase (TCl phase), that is suitable for the fractionation of coal-liquification products and other technical aromatic mixtures by HPLC. It was possible to show, on the basis of investigations on more than 40 model substances and N-heteroaromatics, that con-

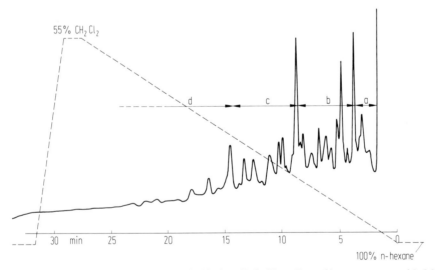

Fig. 7. Chromatographic separation of a lignite oil distillate (free of heterocompounds). Mobile phase: a) n-hexane; b) dichloromethane. Gradient profile: linear from 100% a) to 55% b) in 27.5 min; Detector: UV 280 nm 0.4 A.U.F.S.; Flow rate: 3.0 cm³/min; Column: TCl. a) monoaromatic range, b) diaromatic range, c) triaromatic range, d) polyaromatic range. (Reprinted with permission from [32])

densed aromatics are eluted strictly according to the numbers of their aromatic rings independent of any alkyl substitution. The principle of the separation mechanism is donor-acceptor (DA) interaction between the phase and the individual sample molecules (donor-acceptor-complex chromatography).

One chromatographic application with a coal derived product is illustrated in Fig. 7. It was possible to identify 21 individual compounds by employing mass spectrometry [33]. In a recent investigation [34] there is a discussion of some theoretical aspects of DA complex chromatography together with a review of various DA phases and possibilities for their application.

2.3 New Methods (Column Switching)

When a multicomponent mixture or a particular substance class from a matrix is fractionated on several columns, possibly with different packings, this is referred to as multidimensional chromatography.

It is possible to obtain extremely high plate numbers (e.g. by recycling) for the portion of the chromatogram that is of interest and to remove the uninteresting sample components rapidly by the employment of appropriate column switching. Additional band broadening is avoided by the employment of switching valves, so that the sample is no longer passed through the pump. Too high a pressure drop is also avoided. An excellent review article concerning possible column switching techniques, including multidimensional, recycling, back flush, on-column extraction, is to be found in [35].

Ogan and Katz [36] have coupled RP chromatography with exclusion chromatography for the purpose of separating polycyclic aromatics in coal-liquification products. The method includes three chromatographic steps, whereby the first two divide the sample into appropriate fractions, so that the final analytical, high-resolution step is "simplified". In the first step of the separation the sample is prefractionated on a low resolution reversed phase (C-18) in order to obtain the interesting retention range. This column is online coupled to an exclusion chromatographic system which separates the fraction further from the C-18 pre-column according to molecular weight. The exclusion chromatography fractions are collected and then applied individually to a high resolution analytical C-18 column, where the components are separated. "Interfering" compounds do not affect the separation, since they are removed during prefractionation. Figure 8 illustrates the apparatus employed. The calibration of the system, that is the choice of eluents and valve circuits, is made with the aid of standard substances. Emphasis is placed on the fact that the system is of general application to complex mixtures. Analytical fractions can be obtained in less than 20 minutes from such samples. the retention times and peak areas obtained in the last separation step are highly reproducible.

The application of column switching methods to the isocratic analysis of gasoline, light and heavy gas oils and SRC products (solvent refined coal) has been discussed by Alfredson [37], primarily from the viewpoint of the rapid evaluation of product quality by means of the HPLC group separation of paraffins, olefins, aromatics and polar compounds. Figure 9 illustrates the apparatus employed and the possible connection paths.

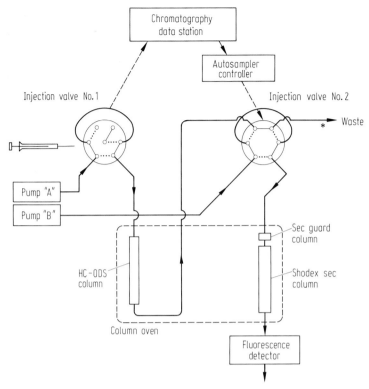

Fig. 8. Block diagram of the arrangement of the pumps, columns and valves used for the coupled-column chromatography system (Reprinted with permission from [36])

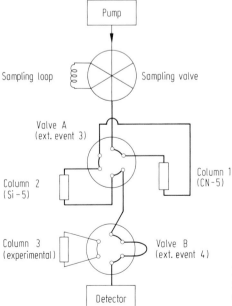

Fig. 9. Column switching flow diagram: normal flow path Note that solid black line indicates path of flow. Det = Detector; Ext = external. (Reprinted with permission from [37])

In all cases the eluent was anhydrous n-hexane. The optimal column combinations are investigated for the desired separation into saturated Hydrocarbons (paraffins + naphthenes), olefins, aromatics and polar compounds. General polar compounds can be separated out by the pre-column packed with chemically bound nitrile phase and eluted later by back flushing. The actual separation into saturated Hydrocarbons (paraffins + naphthenes), olefins and aromatics takes place on the second silica gel column, which is connected in series with the first. The system is again calibrated with the aid of suitable model substances (valve switching times). In the case of light products, such as gasoline, the saturated components are separated by means of a third column via a second switching valve. This column contains a specific packing for paraffin/naphthene separations. Fig. 10 illustrates the separation of a light and a heavy gas oil by the method described above.

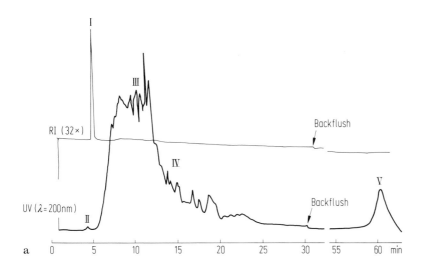

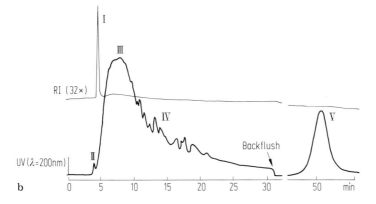

Fig. 10a and b. Hydrocarbon group type separation of light (**a**) and heavy gas oil (**b**). Columns: MicroPak CN-5 plus Si-5 in series. Mobile phase: dry n-hexane at 0.8 ml/min. Detectors: UV at 200 nm (1.0 a.u.f.s.) and RI (32×). I = Saturates; II = Olefins; III = aromatics (alkyl benzenes); IV = PAHs; V = Polars. (Reprinted with permission from [37])

As can be seen from the two investigations described, this new chromatographic method can evaluate the most various products on the basis of their group composition, without great instrumental expense, in a very short time (cf. FIA analysis[26]), while obtaining high selectivity and a large number of theoretical plates for a relatively small pressure drop.

3 Heavy Products (Vacuum Residues, etc.)

3.1 Introduction

The increasing scarcity of world-wide oil supplies and the parallel price explosion in the 1970s made it necessary to develop and exploit the heavy and heaviest products (heavy oils, tar sands, distillation residues) as valuable sources of raw materials. The total global exploitable reserves of such products is estimated to be 150×10^9 t[38]. The problems encountered in processing them are mainly caused by their content of metals (Ni, V) and heteroatom-rich asphaltenes[39] (n-heptane-insoluble components). These exert a de-activating effect upon the cracking catalysts employed. Furthermore, such compounds have an effect on the production of coke. A complex processing technology is necessary because of these factors[40]. Since the properties and compositions of heavy oil and coal products are closely related to their behaviour during conversion processes, an efficient analysis of their components is necessary for an economic technology, i.e. in order to guarantee the "quantitative" production of a particular product range. Heavy oils and related products are characterized by the following features: low H/C ratio, high mean molecular weight ($M_n \sim 900$), increased metal, nitrogen and sulphur content, the presence of asphaltenes[39], high boiling point ($> 400\ °C$). Furthermore, it is established that the components cover a wide polarity scale (paraffins, naphthenes, aromatics, polar heterocomponents, etc.). The exact characterization provides difficulties since the number of individual components in such products is very large because of the high mean molecular weight, i.e. the variety and complexity of the chemical composition makes unequivocal characterization (separation into defined groups) difficult.

The following section reviews the methods available at present for the fractionation and characterization of heavy petroleum and coal products.

3.2 Classical Characterization

Key Fractions

Two precisely defined fractions (key fractions) are taken under standardized conditions and classified as paraffin-based, naphthene-based or mixed-based on the basis of their densities[29]. The solidifying point[29] of a key fraction is employed as a criterion for its classification as either "paraffin-containing" or "paraffin-free". Most naphthene or mixed-based oils (high naphthene number) do not contain "paraffins" according to this definition. Such investigations are important for the processing of heavy oils to lubricating oils, since the paraffin must be removed.

Conradson Coking Test[29]

This test serves for making predictions concerning the behaviour, at high temperatures, of the product employed, which is of great importance in the majority of conversion processes. The sample is heated to a high temperature with the exclusion of air and the coked residue reported in terms of percentage by weight. This can amount to up to 25% for distillation residues.

Elementary Analysis

The H/C atomic ratio and the heteroatom (O, N, S) content and the trace metals (V, Ni) can also allow conclusions to be made concerning the behaviour of the sample in a conversion process (coke formation, catalyst poisoning).

3.3 Precipitation Processes

Determination of Asphaltene According to DIN[39]

High molecular weight substances are concentrated in petroleum fractions with boiling points higher than 300 °C, especially in distillation residues; these make themselves evident as detrimental (coke formation on thermal stress, poisoning of catalysts by nitrogen compounds and metals) when heavy products are employed as lubricating oils or as feed stocks for catalytic crackers and, hence, they have to be removed. These so-called *asphaltenes* are precipitated by n-heptane under standardized condition and are, therefore, defined in terms of the precipitation process. The heteroatoms and metals are heavily enriched in the precipitate. n-Propane (liquified under pressure at ambient temperature) has proved the most suitable for technical scale crude oil processing, since it produces the largest yields of "asphaltene-free" oil.

Colloid precipitation

Neumann[41] has described vacuum residues as colloid disperse solutions of "asphaltenes" and "petroleum oil resins" in a dispersion phase, i.e. a homogeneous, oily phase mainly consisting of carbohydrates. The colloid particles can be removed completely by ultrafiltration and ultracentrifugation[42]. The colloids cannot be solubilized in polar liquids and, hence, are precipitated.

The following separation scheme was recommended:

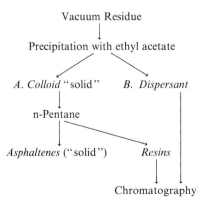

De-waxing

Paraffin-based residue oils are preferred as lubricants. They have, however, to be freed from high molecular weight waxes. This is basically performed by the employment of selective solvents[29], in which the undesirable components are insoluble (methyl ethyl ketone, propane for de-asphalted oils).

3.4 Chromatographic Processes

Heavy petroleum and coal products can only be incompletely analysed by GC at the present state of the art, since the vapour pressure of the substances, which are to be separated, is too low and very high separation temperatures would, therefore, be necessary. The lifetime of a commercially available column is limited at temperatures above 300 °C (column bleeding). Further, there is a danger of catalytic decomposition of the sample e.g. in the injection block. LC is independent of molecular weight within wide limits. The high molecular residual oils cannot, however, be "completely" analysed without suitable preliminary separations. A multitude of stationary phases and eluents are available, with which it is possible to carry out meaningful liquid chromatographic characterizations of heavy products. The UV, fluorescence or RI signal are employed for detection in the overwhelming number of cases. A mass sensitive detection method is at the investigational stage (LC-MS[43]). The available analytical and preparative LC methods for the group separation of heavy petroleum and coal products will now be reviewed.

3.4.1 Classical Low Pressure Methods

Almost all methods have in common the separation of the sample into "paraffins, aromatics and polar compounds" (acids, bases, etc.) on silica gel, aluminium hydroxide or ion exchangers with the aid of stepwise elution.

DIN 51384[44]. the sample is dispersed in n-pentane and separated on activated silica gel. n-Pentane elutes the nonaromatic components (paraffins, naphthenes). The aromatic components are estimated by calculation (amount of sample is 3.6 g).

ASTM D 2549[45]. A glass column filled with bauxite and silica gel is loaded with a 2–10 g sample. The saturated compounds are eluted by means of n-pentane. The aromatic and the polar compounds are recovered by diethyl ether, chloroform and ethanol.

Fehr[46] separated vacuum distillates and atmospheric residues (3 g) on a combined silica gel-alumina-packed glass column (200 × 2 cm) into 6 fractions (separation time 8 h). The polar fraction was then separated again into two subfractions on silica gel. The determination of the sulphur, nitrogen and metal contents of the seven fractions revealed differences in the composition, which was employed to characterize the residue. The behaviour of the residue in a hydrogenating residue desulphurization can be evaluated on the basis of this classification.

A similar separation has been suggested by Sawatzky[47]. The analysis time was drastically shortened by the employment of a low pressure pump. Polar compounds such as acids and bases were concentrated in quite specific fractions. The recovery rate amounted to between 84% and 97%.

The comprehensive book "Chromatography in Petroleum Analysis" by Altgelt and Gouw provides an excellent review of the various separation processes for heavy petroleum products on SiO_2 and Al_2O_3.

API methods. The investigations of the USBM-API (US Bureau of Mines, American Petroleum Institute) are today's basis for most of the separation schemes in the field of heavy petroleum products. The method separates the whole sample into chemically significant fractions, which are suitable for a further characterization of the particular class of compound. Most of the publications are by Latham and Jewell[48]. The most recent investigation[49-51] describes the fractionation of extremely heavy residues (boiling point > 675 °C) into "acids", "bases", "neutral nitrogen compounds", "aromatic hydrocarbons" and "saturated hydrocarbons". A recycling elution apparatus was employed, which makes possible a low solvent consumption and a continual elution of sample components. The heart of the apparatus consists of an API glass column filled with the appropriate stationary phase. Each separation step must be made separately. Acids and bases are removed with the aid of macroporous polystyrene-based anion or cation exchangers. Neutral nitrogen compounds are isolated by complexation with Fe-III-loaded clay. The remaining aromatics and paraffins are separated by normal column chromatography on SiO_2. Figure 11 illustrates the fractionation scheme employed.

The separation of 4 different residues according to the API scheme yielded useful information concerning their composition (including the distribution of heteroatoms). The polar fractions (acids, bases, neutral N compounds) were fractionated further by exclusion chromatography and adsorption chromatography on basic Al_2O_3 and characterized by means of IR spectroscopy, elemental analysis and various methods for molecular weight determination. Carboxylic acids, phenols, pyrroles, amides and pyridines were identified and partially quantified as the major classes of compound.

It is necessary to make the following reservations: – The acids and bases, the neutral nitrogen compounds, the aromatics and paraffins are themselves defined

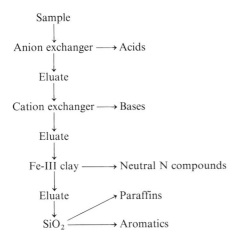

Fig. 11. API fractionation scheme

in petroleum analysis by the process itself. Investigations with "suitable" model compounds and comprehensive IR and NMR measurements and potentiometric titrations demonstrate that these classical chemical groups are enriched by the methods applied. However, it is in no way guaranteed that, for instance, the acid fraction consists exclusively of acids or the base fraction represents exclusively bases in the classical sense.

Sara[52]. This consists of a modification of the API method with the aim of economy in time and standardization. Both ion exchanger resins and the Fe III-loaded clay are packed into one column. The decisive improvement of this method lies in the fact that the asphaltenes[39] are obtained in one fraction. The name Sara is derived from the major fractions which are obtained in the separation (*Saturates, Aromatics, Resins, Asphaltenes*). A rapid HPLC application (20 minutes) is described in[53].

3.4.2 HPLC Applications

The complexity of distillation residues and similar products is so large that it seems impossible to expect the resolution of individual substances even by the employment of high efficiency HPLC techniques. Neither are meaningful fingerprints to be expected without special preliminary group separations to enrich particular classes

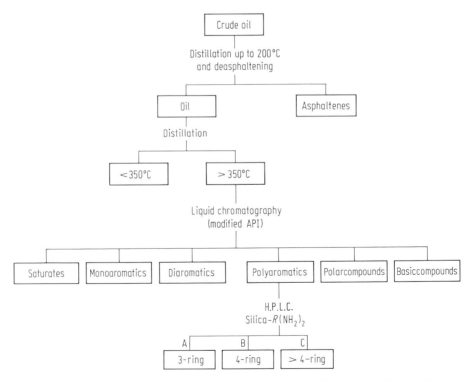

Fig. 12. Scheme of the separation and characterization. – Spectrofluorescence characterization of A, B and C – GC-ms identification of A, B. (Reprinted with permission from[54])

of substance (e.g. aromatics). As a result of this, detailed publications devoted to this field have yet to appear.

Chmielowiec et al.[54] separated subfractions of heavy petroleum products according to the number of condensed rings by means of a complex fractionation scheme (Fig. 12) and demonstrated the differences between different starting materials. Seven fractions with differing polyaromatic ring systems were obtained and analysed by HPLC separation on a diamine phase (column: 250×4.6 mm, eluent: n-hexane/CH_2Cl_2 (96:4)). Secondary analytical methods such as GC-MS, fluorescence spectroscopy and ^{13}C-NMR support the conclusions made. Fig. 13 illustrates

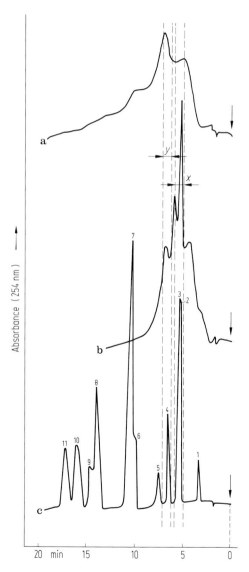

Fig. 13. Chromatograms obtained on the silica-diamine column with mobile phase of methylene chloride-hexane (4:96). (A) Lloydminster polynuclear aromatic hydrocarbon fraction; (B) Medicine River polynuclear aromatic hydrocarbon fraction; (C) Mixture of PAH standards: 1, naphthalene; 2, anthracene; 3, phenanthrene; 4, pyrene; 5, benzo[b]-fluorene; 6, benzo[a]anthracene; 7, chrysene; 8, benzo[a]pyrene; 9, perylene; 10, anthanthrene; 11, benzo[ghi]perylene. (Reprinted with permission from [54])

the HPLC fractionation of two samples including a model mixture. C-18 reversed phases seem less suitable because of fraction overlap.

Boduszynski[55, 56)] separated group analytically prefractionated material from heavy coal liquification products on an amine phase according to the number of double bonds in the ring system (9 fractions). The fractions so obtained were characterized by field ionization mass spectroscopy (FIMS). The HPLC system consisted of an analytical amine phase (column: 300×3.9 mm) with n-heptane as eluent. The cutting points between the "double bond fractions" were determined with the aid of 50 model compounds. An excellent correlation was found between log k' and the number of double bonds in the ring system. A heavy distillate (boiling point 260–427 °C) and a residue (boiling point > 427 °C) were separated under these conditions and investigated mass spectrometrically. The FIMS data confirmed the separation according to the number of double bonds. Furthermore, a molecular weight distribution was obtained for each individual fraction, since the method involves molecular ions exclusively. the paper concludes with a comprehensive literature survey of similar problems.

Bollet et al.[57)] have described the rapid quantitative HPLC method (30 minutes) for the separation of petroleum vacuum distillation residues into saturated, aromatic and polar compounds. There is a discussion of the choice of a suitable stationary phase (SiO_2, Al_2O_3, alkylamine, alkylnitrile), the detection methods (RI, UV) and the advantages of the back-flush technique. SiO_2 and Al_2O_3 are less suitable, since the H_2O content of the eluents has a decisive effect on the separation. Emphasis is laid on the necessity for an accurate chromatographic analysis, since the various classes of compound (type and quantity) react differently in conversion processes (optimal process control). Calibration curves were constructed for the "quantitative" determination of apolar and "polar" components (aromatic and polar).

The next step for the meaningful application of HPLC is likely to be the employment of the specific carriers described in 2.2.2 to the heaviest products, after these have been prefractionated by suitable methods.

3.4.3 Exclusion Chromatography EC

EC offers the advantage, that the samples can, in principle, be separated not according to their chemical functionality but according to their molecular size. The first aim of the development of an EC system must be to exclude the sorption effects by means of a suitable choice of support and eluent. Since substances classes with large differences in polarity are represented in heavy petroleum products it is scarcely possible to achieve a pure exclusion mechanism. EC is usually employed to determine molecular weight distributions of heavy products in the range 400 to 4000 Daltons.

Albaugh et al.[58)] have determined the S, N, Ni and V content as a function of the molecular size. Characteristic weight distribution curves were constructed for various residues. the Ni and V distributions exhibited maxima at 500 and 1900 Daltons.

Done et al.[59)] have developed a rapid EC method for the characterization of various petroleum products. both UV and RI detection are employed. The aromaticity and paraffin distribution can be estimated from 50 fingerprint chromatograms. Impurities in technical products were rapidly recognized and characterized.

Hall and Heron[60] investigated the molecular mass and molecular size distribution of Ni, V, and S-containing asphaltene and maltene fractions (heptane-soluble components). The comparison of the molecular size distribution with the pore size distribution curve of the cracking catalysts employed for conversion, revealed that the metal and sulphur-containing molecules are too large to enter the pores of the catalyst.

Steuer[61] has suggested a preparative EC scheme, with which vacuum and hydrogenation distillation residues can be fractionated. Sorption effects can usually be ignored. Table 1 lists the significant system parameters.

As well as fingerprint chromatograms the distribution of the heteroatoms (N, S) in the individual (molecular mass) fractions was determined. Up to 1 g sample per hour could be separated in a separation step. Furthermore it was possible to construct a universal, product-independent calibration curve (log M versus Elution volume)[68]. This generally valid calibration curve made it possible to carry out an exclusion chromatographic molecular mass determination of chemically differing samples. Combination with the extrographic fractionation method[62-64] – cf. also 3.4.4 – enabled comprehensive conclusions to be made concerning the composition of heavy petroleum products.

Table 1. EC system from[61]

Eluent: THF/CH$_3$OH (9:1) v/v	
Stationary phase	Styragel – column combination 1 × 7 cm (1000 Å): pre-column 2 × 25 cm (1000 Å) 3 × 25 cm (500 Å) 1 × 25 cm (100 Å)
Total length: 1.57 m (i.d. = 16 mm)	
Flow rate: 5 ml/min	
Amount of sample applied: 0.8 g	

3.4.4 Chromatographic Methods with Specialized Sample Preparation

In the overwhelming proportion of LC processes the apolar sample components (paraffins, naphthenes) are obtained first with n-pentane, n-hexane or n-heptane. Heavy petroleum and coal products, however, contain considerable quantities of asphaltenes (cf. 3.1), which are insoluble in these apolar eluents. A portion of the matrix is, therefore, excluded from exact chromatographic processes, since the sample must be dissolved in the first solvent of the stepwise elution series. For these reasons most samples are de-asphalted before the first actual chromatographic separation[39,41] or subjected to similar processes. By this means "superimposed" peaks are avoided in the chromatogram because the sample does not have to be dissolved any longer in a "strong" solvent (e.g. CH$_2$Cl$_2$) before starting chromatography. If it is wished to avoid this complicated procedure an alternative chromatographic method has proven itself, which circumvents the problem described and possesses a range of other advantages for group analysis.

Halász et al.[62-64] have suggested extrography as a method of preparative group separation according to polarity criteria; this is based on the experiences of classical sorption chromatography and selective solvent extraction. The name is derived from the components *extraction* and *chromatography,* since both extractive and chromatographic processes play their part in the separation.

The sample (e.g. vacuum residue) is dissolved in CH_2Cl_2 and mixed with a defined amount of a porous polar support (silica gel Si-100, filled with 50–200 µm particles). After evaporation of the solvent a thin layer of substance remains on pourable silica gel. Then, when dry, this material is packed into a steel column (sample application column, *SAC*) whose internal diameter is more or less a matter of choice (11–34 mm). In order to improve the separational performance the SAC is directly coupled to a second column of the same internal diameter (filter column, *FC*) containing an uncoated active sorbent (Al_2O_3 neutral, filled with 50–200 µm particles). Eluents of increasing polarity are pumped through this column system and, after removal of the solvent, characteristic fractions are obtained, which vary very greatly in their polarity. The apparatus employed is constructed exclusively of simple components (Fig. 14).

The considerable polarity differences within the sample together with the polar surface of the sorbent mean that a form of preclassification (preseparation) can take place within the thin substance layer during the evaporation of the sample solvent. Later on the specific properties of the individual eluents (solvent capacity, elution power) mean, that a particular amount of sample will be displaced from the SAC by "its" specific eluent. This "extract" is fractionated further on the FC. Table 2 reproduces the order of the solvents as they are employed.

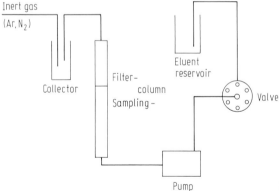

Fig. 14. Apparatus for extrography

Table 2. Order of elution (order of fractions)

1. n-Heptane (anhydrous)
2. n-Heptane/carbon tetrachloride (2:1) (v/v)
3. Carbon tetrachloride
4. Methylene chloride
5. Toluene/propan-2-ol/water (500:100:5) (v/v/v)
6. Toluene/triethylamine (85:15) (v/v)

The process of analysis for the characterization of heavy products should be substance-independent, in order to be able to correlate the available analysis data with each other. That is the separation into the various substance classes should be independent of the distribution (concentration) of the various components within the product to be investigated. If this is the case than it is possible to make correlations with other methods or with technical scale processes. This means for the extrographic separation process: –

1. Optimal loading, i.e. how many g sample can be tolerated per g silica gel of the SAC without reduction in sharpness of separation. The mass yield per fraction must be *independent* of the loading of the system.
2. An optimal ratio between the lengths of the SAC and the FC.
3. Optimal duration of elution for each eluent.

   Comprehensive extrographic measurement data reveal two possibilities in view of the essential nature of the above requirements.

a) Loading: 60 mg sample/g silica gel
   Length ratio SAC/FC = 3:2 → analytical system
b) Loading: 0.3–0.5 g sample/g silica gel
   Length ratio SAC/FC = 1:3 → preparative system

   The results (mass yield, elementary analysis) obtained for a particular sample are practically identical for each method (Tables 3 and 4).

   Independent of the starting material a particular eluent produces chemically similar components (the same H/C ratio, N-content of the particular fractions, appearance, consistency), however, different quantities are produced from different products. Decreasing the column loading in the analytical system does not alter

Table 3. Gravimetric yield of various extrographic fractions obtained with an analytical (a) and a preparative (p) system from different products. Sequence of fractions (1–6) according to Table 2

Sample	System	Fraction					
		1	2	3	4	5	6
Kirkuk VR	a	8.40	7.00	9.10	35.2	27.8	5.5
	p	8.30	6.90	10.60	35.8	28.2	3.8
Ceuta VR	a	8.80	8.80	8.00	27.2	27.1	7.4
	p	8.80	8.40	7.60	28.2	27.1	6.7
B 200 VR	a	9.30	9.00	12.40	33.2	23.1	6.2
	p	8.90	9.00	13.00	34.1	23.0	5.4
Athabaska	a	13.00	7.70	8.90	25.0	26.0	4.1
Top	p	12.60	7.70	8.30	27.0	26.0	3.6
Ceuta HR*	a	15.80	8.40	8.30	32.0	21.3	1.7
	p	15.50	8.60	7.90	34.4	22.9	1.4
CPAL*	a	29.20	8.90	10.90	29.4	17.4	0.6
	p	30.20	9.80	11.60	26.4	16.8	0.4

VR   = Vacuum residue
*HR   = Hydrogenation residue
*CPAL = Cracking Product of Arabian Light vacuum residue

Table 4. H/C* ratios (data for nitrogen (weight %) in brackets) from both systems. See Table 3 for further explanation

Sample	System	Fraction					
		1	2	3	4	5	6
Kirkuk VR	a	1.95	1.89	1.64	1.34	1.28	1.12
		(−)	(−)	(−)	(0.33)	(1.16)	(0.88)
	p	1.96	1.87	1.66	1.35	1.28	1.16
		(−)	(−)	(−)	(0.40)	(1.02)	(1.01)
Ceuta VR	a	1.95	1.81	1.61	1.38	1.33	1.10
		(−)	(−)	(−)	(0.35)	(1.04)	(1.32)
	p	1.98	1.87	1.64	1.43	1.37	1.16
		(−)	(−)	(−)	(0.33)	(1.18)	(1.48)
B200 VR	a	1.97	1.86	1.64	1.45	1.35	1.12
		(−)	(−)	(−)	(0.23)	(0.78)	(1.00)
	p	1.94	1.83	1.64	1.38	1.29	1.10
		(−)	(−)	(−)	(0.15)	(0.92)	(0.91)
Ceuta HR*	a	2.04	1.85	1.53	1.12	1.03	0.91
		(−)	(−)	(−)	(0.30)	(2.00)	(1.74)
	p	1.98	1.85	1.57	1.10	1.02	1.02
		(−)	(−)	(−)	(0.50)	(1.75)	(1.37)
CPAL*	a	1.95	1.73	1.38	0.95	0.94	
		(−)	(−)	(−)	(0.40)	(0.70)	
	p	1.94	1.63	1.32	0.95	0.92	
		(−)	(−)	(−)	(0.61)	(0.70)	

*H = Hydrogen content (mole %);
 C = Carbon content (mole %)

the analyses any more, which allows the conclusion that the system parameters are optimal. With suitable column dimensions the preparative system can achieve a mass throughput of 10 g/hour. Fig. 15 illustrates the extrographic separation of Kirkuk vacuum residue under optimal analysis conditions. As can be seen a pronounced fractionation takes place. The elementary analyses (Table 4) underline the separation according to polarity.

The combination of extrography with the API separation scheme[64], i.e. each extrographic fraction was subjected to the API separation scheme, was employed to characterize the individual fractions according to their chemical composition. the following can be said:

– the first fraction is purely paraffinic.
– The aromatics are concentrated with increasing numbers of rings in fractions 2–4.
– Polar compounds (acids, bases, neutral N compounds) are present in fractions 5 and 6.

It was possible to confirm these results independently of the API method on the basis of IR, NMR and GC investigations[64, 69].

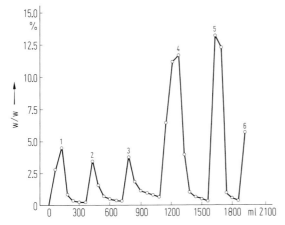

Fig. 15. Vacuum residue fractions from an optimized analytical extrographic system. See Table 2 for sequence of eluents (1–6). Sample: Kirkuk Vacuum Residue (KVR); Sampling column: 51 cm × 2.2 cm; Filter column: 34 cm × 2.2 cm; Loading: 0.06 g of KVR/g stationary phase (SiO$_2$); Linear velocity: 0.7 mm/sec

The advantages of extrography are briefly as follows:
– Pronounced fractionation according to polarity → combination with EC
– High mass throughput per unit time without loss of separational sharpness
– Asphaltene precipitation unnecessary
– The sample does not have to be soluble in the first eluent of the stepwise series.
– Substance-independent method
– Simple components
– Low costs
Since the introduction of extrography[62] several publications have appeared where the sample is brought onto an inert or polar support before the commencement of the actual separation.

Boduszynski[65, 66] has described a method for fractionating SRC (solvent refined coal) products. SRC is defined as the pyridine-soluble vacuum residue from the distillation of coal liquification products (boiling point ≥ 427 °C). The aim of the investigation was to exploit the solubility and adsorption properties (chemical properties) of the product in one step for its separation into characteristic fractions.

The sample (0.3 g) was applied to an *inert support* (Fluoropack Teflon powder: 22 g) and packed into a glass column (500 × 9 mm). The column was eluted stepwise with n-hexane, toluene and pyridine. Fractions were obtained which were separated on the basis of the difference in their solubility in the various eluents. Interactions with the solid support are negligible. *Note* this has a disadvantageous effect because the limit of loading is very low (14 mg/g). This is understandable when it is remembered that Teflon powder possesses scarcely any pore volume.

An investigation was made of the effect of flow rate on the sharpness of separation. Up to 8 ml/min (u = 0.7 mm/s) can be tolerated (analysis time including pre and post-treatment: 5 h). Six SRC residues were compared with each other. The further separation of the solubility fractions was performed by coupling the first named process with adsorption chromatography on basic aluminium oxide. Two variants are possible:

1. Separation of the solubility fractions by the column switching technique (Fluoropak column with $3 Al_2O_3$ columns connected in parallel)
2. Direct coupling of the Fluoropak column to an Al_2O_3 column.

The column system was calibrated by means of 50 model substances. The subfractions were characterized by means of elemental analysis and IR spectroscopy. The
advantage of this method lies in the fact that sample components, which are strongly
or irreversibly retarded on polar phases, are partially removed during the extraction
step.

Lee et al.[67] have developed a two-stage method for the fractionation of heavy
coal liquification products. A sample of 0.1–0.3 g is coated onto 3 g Al_2O_3 (neutral)
and packed into a steel column, whose lower portion is packed with 6 g uncoated
Al_2O_3. Four fractions are collected by stepwise elution. They consist of aliphatic
hydrocarbons, neutral polycyclic aromatics, nitrogen-containing polycyclic aromatics, hydroxypolycyclic aromatics. The nitrogen fraction is separated further on SiO_2
as support according to the same principles. The fractions were characterized by
means of a high performance GC-MS system, whereby it was possible to identify
many individual components.

4 Concluding Remarks

Today it is possible to characterize heavy petroleum and coal products quite rapidly
by the rational combination of suitable chromatographic techniques, that is to separate the most various starting materials into individual substance classes (paraffins,
aromatics, etc.). The following general scheme seems to be applicable in the majority
of cases.

    I. Preparative group separation according to polarity
       a) Classical column chromatography on SiO_2, Al_2O_3 or ion exchangers
       b) Extrography
    II. Separation of the individual groups according to molecular size
           Exclusion chromatography
   III. Separation of the EC-fractions by HPLC on group-specific phases

The coupling of modern spectroscopic methods (IR, NMR, MS) leads to the acquisition of informative data, with whose aid it should be possible to produce correlations
with technological parameters. The detection of individual substances will continue
to provide problems for the foreseeable future, because the numbers of compounds
present in this boiling point range (molecular weight range) is so enormous.

5 References

1. Jeffery, P.G., Kipping, P.J.: "Gas Analysis by Gas Chromatography", 2nd ed., Pergamon,
   New York, 1972
2. Schomburg, G., Husman, H.: Chromatographia 8, 517–530 (1975)
3. Schomburg, G., Husman, H., Weeke, F.: J. Chromatogr. 99, 63–79 (1974)
4. Schomburg, G., Husman, H., Weeke, F., Borwitzky, H., Behlam, H.: J. High Resol.
   Chromatogr. and Chromatogr. Comm. 2, 461–474 (1979)

 5. David, D.I.: "Gas Chromatographic Detectors", John Wiley and Sons, New York, London, Sydney, Toronto (1974)
 6. Schneider, W., Frohne, J.Ch., Bruderreck, H.: J. Chromatogr. 245, 71–83 (1982)
 7. Whittemore, M.: In "Chromatography in Petroleum Analysis", K.H. Altgelt, T.H. Gouer, Chromatographic Science Series, Vol. 11, Marcel Dekker, New York, Basel (1979)
 8. ASTM D 3606
 9. Eggerston, F.T., Groennings, S., Holst, I.I.: Anal. Chem. 32, 904 (1960)
10. Stuckey, C.L.: Anal. Chem. Acta 60, 47 (1972)
11. Nakajawa, H., Tsuje, S., Itho, T., Komoto, M.: J. Chromatogr. 260, 391 (1983)
12. Schomburg, G., Dielmann, R., Borwitzky, H., Husman, H.: J. Chromatogr. 167, 337 (1978)
13. Borwitzky, H., Schomburg, G.: J. Chromatogr. 170, 99 (1979)
14. Marquart, I.R., Dellow, G.B., Freitas, E.R.: Anal. Chem. 40, 1633 (1968)
15. Sanzo, F.P. Di: Anal. Chem. 53, 1911 (1981)
16. Sanzo, F.P. Di, Uden, P.C., Siggia, S.: Anal. Chem. 52, 906 (1980)
17. Alexander, G., Hazai, I.: J. Chromatogr. 217, 19 (1981)
18. Romanowski, T., Funcke, W., Grossmann, I., König, I., Balfaus, E.: Anal. Chem. 55, 1030 (1983)
19. Snyder, L.R.: Anal. Chem. 36, 128 (1964)
20. Snyder, L.R.: Anal. Chem. 36, 779 (1964)
21. Snyder, L.R.: Anal. Chem. 37, 713 (1965)
22. Snyder, L.R.: "Principles of Adsorption Chromatography", Marcel Dekker, New York, 1968
23. Suatoni, I.C., Garber, H.R., Davis, B.R.: J. Chromatogr. Sci. 13, 367 (1975)
24. Suatoni, I.C., Swab, R.E.: J. Chromatog. Sci. 18, 375 (1980)
25. Suatoni, I.C., Swab, R.E.: J. Chromatog. Sci. 14, 535 (1976)
26. ASTM D 1319, 17 (1973) 476, Hydrocarbon Types in Liquid Petroleum Products by Fluorescent Indicator Adsorption
27. Vogh, I.M., Thomson, I.S.: Anal. Chem. 53, 1345 (1981)
28. Holstein, W., Severin, D.: Anal. Chem. 53, 2356 (1981)
29. Riediger, B.: "Die Verarbeitung des Erdöls" Springer Verlag, Berlin 1971
30. Lee, M.L., Peaden, P., Hirata, Y., Novotny, M.: Anal. Chem. 52, 2268 (1980)
31. Mourey, T.H., Siggia, S., Uden, P.C., Crowley, R.I.: Anal. Chem. 52, 885 (1980)
32. Holstein, W.: Chromatographia 14, 468 (1981)
33. Holstein, W., Severin, D.: Chromatographia 15, 231 (1982)
34. Holstein, W., Hemetsberger, H.: Chromatographia 15, 186 (1982)
35. Miller, R.L., Poile, A.F.: International Laboratory 9, 60 (1981)
36. Ogan, K., Katz, E.: Anal. Chem. 54, 169 (1982)
37. Alfredson, T.v.: J. Chromatogr. 218, 715 (1981)
38. Meyer, R.F., Fulton, P.A.: "Toward on estimate of world heavy crude oil and tar sands resources", Discussionscontribution 2. Unitar Conference (Caracas) 1981
39. DIN 51 595 (1978)
40. Weitkamp, J.: Erdöl u. Kohle, Erdgas, Petrochemie 35, 460 (1982)
41. Neumann, H.I.: Erdöl u. Kohle, Erdgas, Petrochemie 34, 336 (1981)
42. Whiterspoon, P.A.: "Die kolloidale Natur des Erdöls", Vorträge der wissenschaftlichen Tagung für Erdöl", Budapest 1962
43. Games, D.E.: "Advances in Chromatography" 21 (1983)
44. DIN 51 384 (1978)
45. ASTM – D 2549
46. Fehr, E.: Compendium zur DGMK-Fachgruppentagung, Hanover 1975
47. Sawatzky, H., George, A.: Fuel 55, 16 (1976)
48. Jewell, D.M., Latham, D.R.: Anal. Chem. 44, 1391 (1972)
49. Mc Kay, J.F. et al.: Fuel 60, 14 (1981)
50. Mc Kay, J.F. et al.: Fuel 60, 17 (1981)
51. Mc Kay, J.F. et al.: Fuel 60, 27 (1981)
52. Jewell, D.M.: I + EC Fundamentals 13, 278 (1974)
53. Galye, L.G., Suatoni, I.C.: J. Liquid Chromatogr. 3 (2), 229 (1980)

54. Chimielowiec, I., Beshai, I.E., George, A.E.: Fuel 59, 838 (1980)
55. Boduszynski, M.M., Hurtubise, R.I., Allen, T.W., Silver, H.F.: Anal. Chem. 55, 225 (1983)
56. Boduszynski, M.M., Hurtubise, R.I., Allen, T.W., Silver, H.F.: Anal. Chem. 55, 232 (1983)
57. Bollet, C., Escalier, I.C., Sonteyrand, C., Caude, M., Rosset, R.: J. Chromatogr. 206, 289 (1981)
58. Albaugh, E.W., Talarico, P.C., Davis, B.E.: In "Gel Permeation Chromatography", K.H. Altgel, Segal, Marcel Dekker, New York, Basle 1971
59. Done, I.N., Reid, W.K.: In "Gel Permeation Chromatography" K.H. Altgelt, Segal, Marcel Dekker, New York, Basle 1971
60. Hall, G., Herron, S.P.: In "Advances in Chemistry Series" Vol 195
61. Steuer, W.: Dissertation, Saarbrücken 1983
62. Halász, I.: DGMK-Forschungsbericht zum Projekt 81-09 (1978)
63. Halász, I., Bogdoll, B.: Erdöl + Kohle, Erdgas, Petrochemie 34, 549 (1981)
64. Crispin, Th.: Dissertation, Saarbrücken 1983
65. Boduszynski, M.M., Hurtubise, R.I., Silver, H.F.: Anal. chem. 54, 372 (1982)
66. Boduszynski, M.M., Hurtubise, R.I., Silver, H.F.: Anal. Chem. 54, 375 (1982)
67. Later, D.M., Lee, M.L., Bartle, K.D., Kong, R.C., Vassilaros, D.L.: Anal. Chem. 53, 1612 (1981)
68. Steuer, W., Jost, K., Halász, I.: Chromatographia 20(1) S. 13–19 (1985)
69. Maldener, G., Crispin, Th., Halász, J.: to be published

Subject Index

Solute Index

H. Engelhardt

High Performance Liquid Chromatography

Chemical Laboratory Practice

Translated from the German by G. Gutnikov

1979. 73 figures, 13 tables. III, 248 pages
ISBN 3-540-09005-3

Contents: Chromatographic Processes. – Fundamentals of Chromatography. – Equipment for HPLC. – Detectors. – Stationary Phases. – Adsorption Chromatography. – Partition Chromatography. – Ion-Exchange Chromatography. – Exclusion Chromatography. Gel Permeation Chromatography. – Selection of the Separation System. – Special Techniques. – Purification of Solvents. – Subject Index.

This simple and non-mathematical introduction to high-performance-liquid chromatography (HPLC) emphasizes the practical aspects of achieving a successful separation. This method usually permits analyses to be carried out more rapidly than by gas chromatography and is, moreover, eminently suited for the separation of heatlabile, high-boiling, or nonvolatile substances, without lengthy or tedious derivatization. In principle, all substances that are stable in solution are amenable to separation by HPLC. HPLC equipment is described in terms of the individual components, their expected performance capabilities and suitability for certain applications.
The areas of applications of the various separation techniques (adsorption, partition, ion-exchange, exclusion) are pointed out in order to facilitate selection of the most appropriate technique by the worker for his particular problem. Considerable discussion is devoted to the parameters that are important in optimizing or improving a given separation.
The application of HPLC to actual problems in organic chemistry, pharmacological research, medicine, biochemistry and petrochemistry are illustrated by numerous relevant examples. This book is a translation of the well-known and very successful German edition.

Springer-Verlag
Berlin Heidelberg
New York Tokyo

Tables of Spectral Data for Structure Determination of Organic Compounds

By **E. Pretsch, T. Clerc, J. Seibl, W. Simon**

Translated from the German by K. Biemann
Chemical Laboratory Practice

1983. IX, 316 pages. ISBN 3-540-12406-3

Contents: Introduction. – Abbreviations and Symbols. – Summary Tables. – Combination Tables. – ^{13}C-Nuclear Magnetic Resonance Spectroscopy. – Proton Resonance Spectroscopy. – Infrared Spectroscopy. – Mass Spectrometry. – UV/VIS (Spectroscopy in the Ultraviolet or Visible Region of the Spectrum). – Subject Index.

Prompted by the suggestions of many readers, the publisher has undertake the production of an English translation of the 2nd German edition of this succesful work. The first edition has already appeared in Japanese, Spanish and Croatic. Prof. Klaus Biemann, Massachusetts Institute of Technology, Cambridge, USA kindly translated the textual parts of this work into English.
This book represents a compilation of spectroscopic data – in the form of tables and charts – and their correlation to molecular structure. It is intended to aid in the interpretation of ^{13}C-and ^1H-nuclear magnetic resonance, infrared, ultravioloet-visible and mass spectra. The material was compiled for courses and exercises which the authors offered over a ten year period to their students at the Federal Institute of Technology (ETH) in Zurich.
The book offers a compact summary of reference data in a form that can be grasped easily. It is to be viewed as a supplement to texts and reference books dealing with these spectroscopic techniques and is designed for those who are routinely faced with the task of interpreting such spectral information.
The use of this book requires only the knowledge of the basic principles of spectroscopic techniques, but its content is structured in such a way that will also serve as a reference work for specialist in the field.

Springer-Verlag
Berlin Heidelberg
New York Tokyo

Springer